Handbook of Experimental Pharmacology

Volume 150

Springer

Berlin
Heidelberg
New York
Barcelona
Hong Kong
London
Milan
Paris
Singapore
Tokyo

Pharmacology of GABA and Glycine Neurotransmission

Contributors

B.E. Alger, J. Auta, E.A. Barnard, D. Belelli, B. Bettler
H. Betz, J. Bormann, N.G. Bowery, N.J. Brandon, J.E. Clark
W.A. Clark, E. Costa, S.J. Enna, A. Feigenspan, A. Guidotti
S.C. Harney, R. Adron Harris, N.L. Harrison, R.J. Harvey
B.K. Kanner, K. Kaupmann, H.Y. Kim, M.D. Krasowski
J.J. Lambert, F.E.N. Le Beau, B.S. Meldrum, H. Möhler
S.J. Moss, R.W. Olsen, J.A. Peters, P. Schloss, T.G. Smart
P. Thomas, P. Whiting

Editor
Hanns Möhler

Professor
Dr. Hanns Möhler
ETH and University of Zürich
Institute of Pharmacology and Toxicology
Winterthurerstr. 190
CH-8057 Zürich
e-mail: mohler@pharma.unizh.ch

With 48 Figures and 11 Tables

ISBN 3-540-67616-3 Springer-Verlag Berlin Heidelberg New York

Library of Congress Cataloging-in-Publication Data
Pharmacology of GABA and Glycine Neurotransmission / Hanns Möhler, editor; contributors, B.E. Alger . . . [et al.].
p. cm. – (Handbook of experimental pharmacology; v. 150)
Includes bibliographical references and index.
ISBN 3540676163 (alk. paper)
1. GABA – Physiological effect. 2. Glycine – Physiological effect. 3. Amino acid neurotransmitters. 4. GABA – Receptors. I. Möhler, Hanns. II. Alger, B.E. III. Series.
QP905.H3 vol. 150
[QP563.G32]
615′.1 s – dc21
[615′.7] 00-063583

Springer-Verlag Berlin Heidelberg New York
a member of BertelsmannSpringer Science+Business Media GmbH

Cover design: design & production GmbH, Heidelberg
Typesetting: Best-set Typesetter Ltd., Hong Kong

SPIN: 10699411 27/3020-5 4 3 2 1 0 – printed on acid-free paper

Preface

Neuronal transmission by GABA and glycine has become one of the most fascinating areas in neuroscience in recent years. New roles for synaptic inhibition have been identified and are increasingly being incorporated into functional models of neuronal networks in order to understand complex brain functions. Similarly, basic issues of the molecular architecture of inhibitory synapses have been addressed by resolving the multiplicity and mode of operation of receptors, transporters, signal transduction mechanisms and synapse formation. These developments are helping to shape the strategies for the investigation of disease states and for the pharmacological and therapeutic intervention in inhibitory processes.

In the present volume, recognized experts in the field of neuronal transmission by GABA and glycine give an appraisal of these recent developments. Specific topics include:

1. The physiology of the GABA and glycine systems.
2. The structure, pathophysiology and regulation of $GABA_A$ receptors.
3. The pharmacological modulation of $GABA_A$ receptors by benzodiazepines, steroids, general anaesthetics, alcohols and anticonvulsants.
4. The structure, signal transduction and pharmacology of $GABA_B$ receptors.
5. The role of $GABA_C$ receptors.
6. The function of GABA transporters.
7. The structure, diversity and pharmacology of glycine receptors and glycine transporters.
8. The heightened therapeutic potential arising from the new evidence on the regulation of inhibitory signal transduction at the molecular, cellular and systems level.

This volume is intended for neuroscientists as well as for pharmacologists, psychiatrists, neurologists and medicinal chemists. It aims to serve as a state of the art reference on the role of neuronal inhibition in brain function and on the therapeutic strategies available for CNS disorders. It may also provide an incentive for further research, in particular on the integration of the structural and functional aspects of inhibitory transmission.

Hanns Möhler

List of Contributors

ALGER, B.E., Department of Physiology and Program in Neuroscience, University of Maryland School of Medicine, 655 West Baltimore Street, Baltimore, MD 21201, USA
e-mail: balger@umaryland.edu

AUTA J., Psychiatric Institute, Department of Psychiatry, University of Illinois at Chicago, College of Medicine, 1601 West Taylor Street, Chicago, IL 60612, SA

BARNARD, E.A., Department of Pharmacology, University of Cambridge, Tennis Court Road, Cambridge CB2 1QJ, UK
e-mail: eb247@cam.ac.uk

BELELLI D., Department of Pharmacology and Neuroscience, Ninewells Hospital and Medical School, The University of Dundee, Dundee DD1 9SY, Scotland, UK

BETTLER B., Novartis Pharma AG, Therapeutic Area, Nervous System Research, K-125.6.08, CH 4002 Basel, Switzerland
bernhard.bettler@pharma.novartis.com

BETZ H., Max-Planck-Institut für Hirnforschung, Abteilung Neurochemie, Deutschordenstrasse 46, D-60528 Frankfurt am Main, Germany
e-mail: Betz@mpih-frankfurt.mpg.de

BORMANN, J., Ruhr-Universität Bochum, Lehrstuhl für Zellphysiologie, ND4/132, Universitätsstr. 150, D-44780 Bochum, Germany
e-mail: Joachim.Bormann@ruhr-uni-bochum.de

BOWERY, N.G., Department of Pharmacology, Division of Neuroscience, The Medical School, The University of Birmingham, Edgbaston, Birmingham B15 2TT, UK
e-mail: n.g.bowery@bham.ac.uk

Brandon, N.J., MRC Laboratory of Molecular Cell Biology and Dept. of Pharmacology, University College London, Gordon Street, London WC1E 6BT, UK

Clark, J.E., Endocrinology and Chemical Biology, Merck & Co., Inc., P.O. Box 2000, RY80T-126, Rahway, NJ 07065, USA
e-mail: janet_clark@merck.com

Clark, W.A., Laboratory of Cellular Biology, National Institute of Deafness and Communication Disorders, 2A-03, 5 Research Court, Rockville, MD 20850, USA

Costa, E., Psychiatric Institute, Department of Psychiatry (MC-912), University of Illinois at Chicago, College of Medicine, 1601 West Taylor Street, Chicago, IL 60612, USA
e-mail: Costa@psych.uic.edu

Enna, S.J., Department of Pharmacology, Toxicology and Therapeutics, University of Kansas School of Medicine, 3901 Rainbow Boulevard, Kansas City, KS 66160 USA
e-mail: senna@kumc.edu

Feigenspan A., Carl-von-Ossietzky-Universität Oldenburg, Department of Neurobiology, D-26111 Oldenburg, Germany

Guidotti, A., Psychiatric Institute, Department of Psychiatry, University of Illinois at Chicago, College of Medicine, 1601 West Taylor Street, Chicago, IL 60612, USA
e-mail: AGuidotti@psych.uic.edu

Harris, R.A., Institute for Cellular and Molecular Biology, The University of Texas, Austin, TX, USA

Harney, S.C. Department of Pharmacology and Neuroscience, Ninewells Hospital and Medical School, The University of Dundee, Dundee DD1 9SY, Scotland, UK

Harrison, N.L., C.V. Starr Laboratory for Molecular Neuropharmacology, Dept. of Anesthesiology, A-1050, Weill Medical College of Cornell University, 525 East 68th Street, New York N.Y. 10021, USA
e-mail: neh2001@med.cornell.edu

Harvey, R.J., Department of Pharmacology, The School of Pharmacy, 29–39 Brunswick Square, London WC1N 1AX, UK
e-mail: Harvey@cob.ulsop.ac.uk

Kanner, B.I., Department of Biochemistry, Hadassah Medical School, The Hebrew University, P.O. Box 12272, Jerusalem, Israel 91120
e-mail: kannerb@cc.huji.ac.il

Kaupmann K., Novartis Pharma AG, Nervous System Research, CH-4002 Basel, Switzerland

Kim, H.Y., Department of Molecular and Medical Pharmacology, University of California, Los Angeles, CA 90095-1535, USA

Krasowski, M.D., Department of Neurobiology, The University of Chicago, Chicago, IL 60637, USA

Lambert, J.J., Department of Pharmacology and Neuroscience, Ninewells Hospital and Medical School, The University of Dundee, Dundee DD1 9SY, Scotland, UK
e-mail: j.j.lambert@dundee.ac.uk

Le Beau, F.E.N., School of Biomedical Sciences, University of Leeds, Leeds, LS2 9NQ, UK
e-mail: F.E.N.LeBeau@leeds.ac.uk

Meldrum, B.S., Dept of Clinical Neurosciences, Institute of Psychiatry, DeCrespigny Park, London SE5 8AF, UK
e-mail: spgtbsm@iop.kcl.ac.uk

Möhler, H., Institute of Pharmacology and Toxicology, University of Zurich and Swiss Federal Institute of Technology (ETH) of Zurich, CH-8057 Zurich, Switzerland
e-mail: mohler@pharma.unizh.ch

Moss, S.J., MRC Laboratory of Molecular Cell Biology and Dept of Pharmacology, University College London, Gordon Street, London WC1 E 6BT, UK

Olsen, R.W., Department of Molecular and Medical Pharmacology, University of California School of Medicine, Center for the Health Sciences, 23-120 CHS, P.O. Box 951735, Los Angeles, CA 90095-1735, USA
e-mail: rolsen@mednet.ucla.edu

Peters, J.A., Department of Pharmacology and Neuroscience, Ninewells Hospital and Medical School, The University of Dundee, Dundee DD1 9SY, Scotland, UK

Schloss, P., Zentralinstitut für Seelische Gesundheit, Abteilung Pyschopharmakologie, D-68159 Mannheim, Germany

Smart, T.G., The School of Pharmacy, Department of Pharmacology, University of London, 29–39 Brunswick Square, London WC1 N 1AX, UK
e-mail: tsmart@ulsop.ac.uk

Thomas, P., The School of Pharmacy, Department of Pharmacology, University of London, 29–39 Brunswick Square, London WC1N 1AX, UK

Whiting, P., Biochemistry & Molecular Biology, Neuroscience Research Centre, Merck Sharp & Dohme Research Laboratories, Eastwick Road, Harlow, Essex, CM20 2QR, UK
e-mail: paul_whiting@merck.com

Contents

Section I: Physiology of the Neurotransmitters GABA and Glycine

CHAPTER 1

Physiology of the GABA and Glycine Systems

Section II: Pharmacology of the GABA System
GABA$_A$ Receptors

CHAPTER 2

The Molecular Architecture of GABA$_A$ Receptors

CHAPTER 3

Functions of GABA$_A$-Receptors: Pharmacology and Pathophysiology

CHAPTER 4

Steroid Modulation of $GABA_A$ Receptors

J.J. LAMBERT, J.A. PETERS, S.C. HARNEY, and D. BELELLI.

CHAPTER 8

Tolerance and Dependence to Ligands of the Benzodiazepine Recognition Sites Expressed by $GABA_A$ Receptors

CHAPTER 9

$GABA_A$ Receptors and Disease

CHAPTER 10

$GABA_C$ Receptors: Structure, Function and Pharmacology

$GABA_B$ Receptors

CHAPTER 11

Structure of $GABA_B$ Receptors

CHAPTER 12

Pharmacology of $GABA_B$ Receptors

CHAPTER 13

$GABA_B$ Receptor Signaling Pathways

Section III: Pharmacology of the Glycine System

CHAPTER 16

Structures, Diversity and Pharmacology of Glycine Receptors and Transporters

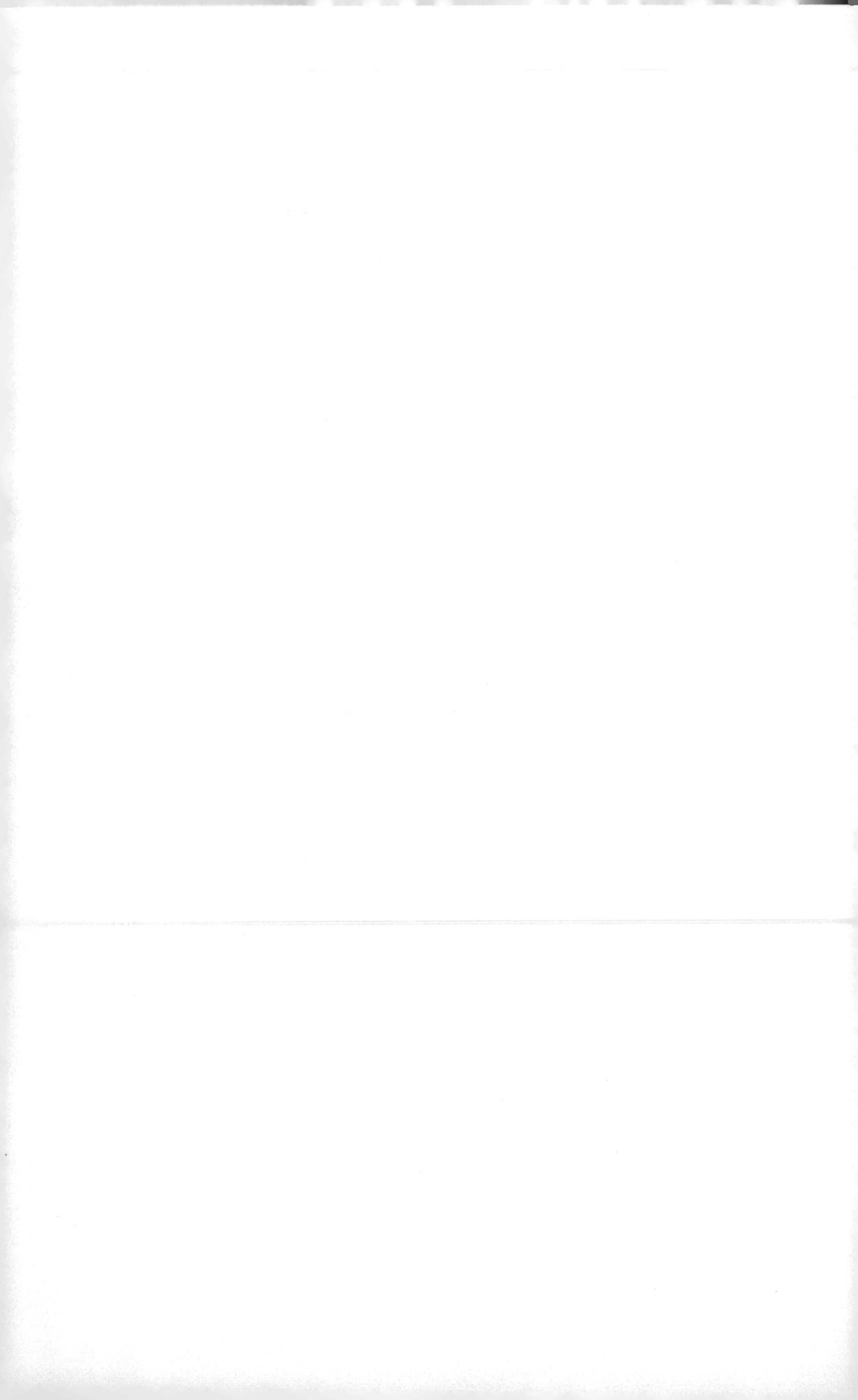

Section I
Physiology of the Neurotransmitters GABA and Glycine

CHAPTER 1
Physiology of the GABA and Glycine Systems

B.E. ALGER and F.E.N. LE BEAU

A. Introduction

An explosion of information about the roles of inhibition mediated by GABA and glycine has made this area one of the richest and most fascinating in neurophysiology. This chapter will survey many themes in synaptic inhibition in the vertebrate central nervous system, but will concentrate on developments since 1995. Several excellent reviews can be consulted for details of earlier work, e.g., see KAILA 1994; MACDONALD and OLSEN 1994; MODY et al. 1994; THOMPSON 1994. We focused on neurophysiological effects and did not attempt to cover the literature on exogenously applied modulators or biochemical modifiers of synaptic inhibition except occasionally to help illuminate another point. We discuss $GABA_A$ and $GABA_B$ responses mainly, leaving $GABA_C$ to be discussed in another chapter in this volume. Glycinergic systems are considered throughout, but the bias is still towards GABA. In vitro cellular studies have led to major advances and, because the hippocampus and cerebellum are most immediately adaptable to in vitro slice preparations, a disproportionate percentage of the work has been done on these structures. This is rapidly changing, and investigation of other brain areas is widening and deepening our knowledge of inhibition in information processing there as well.

Early work on inhibitory synapses dealt with transmitter identification and ionic mechanisms. Beginning about ten years ago, the application of gigaohm-seal recording techniques to slices (BLANTON et al. 1989; EDWARDS et al. 1989) stimulated high-resolution studies of ionic currents and the microphysiology of synapses. We will discuss a wide, though not exhaustive, range of phenomena to demonstrate the variety of functions carried out by GABA and glycine. Inhibitory neurotransmitters affect neuronal activity mainly by gating ion channels either directly or indirectly through second-messenger systems, although "direct" effects on transmitter release processes may also occur (see Sect. G.II.2). Much work in the past decade has filled in details of these factors. However, some of the most significant advances in the neurophysiology of inhibitory systems have come about because of increased understanding of the cellular and circuit-level actions of the neurotransmitters on target cells, and of regulation of inhibitory systems. Even as the basic issues of receptor subtypes, ionic mechanisms, and second messenger systems are becoming resolved, for every new type of excitatory process, function, or interconnection discovered, a new role for synaptic inhibition seems to arise. A great deal remains to be done before details of inhibitory processes will be fully incorporated into functional models of large neuronal ensembles, but much progress has been made, and information now being developed should help direct strategies useful for pharmacological and therapeutic investigations.

B. Subtypes of Interneurons

Although inhibitory principal neurons such as cerebellar Purkinje cells exist, inhibitory responses are generally produced by inhibitory interneurons, and an accelerating research effort has been directed towards cataloguing interneurons and their properties.

The great majority of interneurons in the brain use GABA as their neurotransmitter, whereas in the spinal cord and brainstem glycine is the major interneuron neurotransmitter. Simple generalities beyond this are difficult to make, however. Interneuronal somata tend to be scattered rather than clustered. Microelectrode studies conducted by patient experimenters blindly moving electrodes through brain slices were informative, but the random distribution and low packing density of many interneuron systems impeded rapid progress. Development of optical techniques that permit visualization of cells (Dodt and Zieglgansberger 1990) and the application of patch-clamp technology to brain slices (Blanton et al. 1989; Edwards et al. 1989) have accelerated the pace, and kinds, of discoveries in the central nervous system but investigations of interneurons in particular have benefited. A growing number of interneuron classes has been identified based on one or a few criteria. A thorough recent compendium is the review of hippocampal interneurons by Freund and Buzsaki (1996), and reviews of work in cortex (Kawaguchi 1995; Kawaguchi and Kubota 1996; Azouz et al. 1997; Gonchar and Burkhalter 1997; Kawaguchi and Kubota 1997), cerebellum (Voogd and Glickstein 1998), and olfactory bulb (Devries and Baylor 1993; Shepherd 1994) are available.

Interneurons in hippocampus and cortex are typically non-pyramidal in shape and assume a wide variety of morphological forms, with differently shaped somata, dendritic branching patterns, spine investment, and axonal arborizations. They are distinguished by their voltage- and ligand-gated channels and by their complement of co-localized neuropeptides, calcium-binding proteins, afferent input, and target cell populations. Useful schemes for classification of interneurons are often based on localization of their somata, dendrites, and axonal arborizations. The orientation and distribution of their dendrites and axons help define the specificity of afferent input and target populations and function. Interneurons selectively innervate certain target cells at well-defined cellular regions, the chandelier cell of neocortex and hippocampus (which innervates initial axon segments very specifically) being a good example. In many cortical regions, basket cells form dense networks of terminals on somata of principal neurons. The specificity of the chandelier and basket cell output suggests that control of somatic integration and axonal action potential initiation are their major functions.

At the other extreme are groups of interneurons that terminate at such a distance on the distal dendrites of principal cells that a direct influence on action potential threshold is virtually precluded. Their main role may be in local dendritic integration. Interneurons contact other interneurons specifically in, e.g., hippocampus (FREUND and BUZSAKI 1996; GULYAS et al. 1996), neocortex (FREUND and MESKENAITE 1992; TAMAS et al. 1998) and cerebellum (VOOGD and GLICKSTEIN 1998). Because of their often broad axonal distributions, interneurons that control other interneurons may affect large populations of principal cells. Networks of interneurons play major roles in the generation of rhythms in circuit activity (see Sect. I). However, the details of the functional roles of the interneurons cannot be readily inferred even in cases in which generalizations such as these are possible. Even in the hippocampus, apart from the classification of isolated interneuronal properties and a developing nomenclature for different types of interneurons, precisely defined, non-overlapping classes of interneurons have not yet been identified. A vivid illustration of the difficulties in classification comes from a study of interneurons in the hippocampal CA1 region (PARRA et al. 1998). Sixteen morphological and 28 physiological and pharmacological phenotypes were distinguished. However, clustering of morphological and physiological properties did not occur. If an interneuron "class" was defined narrowly as consisting of only those cells in which all properties were held in common, the 26 cells completely characterized by all criteria implied the existence of at least 26 classes. Twenty-six additional, incompletely characterized cells suggested the existence of a total of 52 classes, with the number probably increasing as more properties were examined. The authors concluded that each hippocampal interneuron might be unique, i.e., classification is not possible. Apparently the idea that there are rigidly definable classes of interneurons subserving specified functions must be abandoned, at least for the cortex and hippocampus. This is not to say that the concept is useless, however, as classes of interneurons may still be defined dynamically according to their participation in various states of brain activity, and each cell may be part of many classes. The groupings could change as a result of physiological and morphological plasticity.

I. Electrophysiological Properties of Interneurons

The physiological response properties that distinguish different subtypes of interneurons are determined in large part by the different complements of ion channels that they possess. There are differences in ligand-gated channels (McBAIN and DINGLEDINE 1993; TOTH and McBAIN 1998; KATONA et al. 1999; SVOBODA et al. 1999) and non-ligand-gated channels (ZHANG and McBAIN 1995a,b; MACCAFERRI and McBAIN 1996; MARTINA et al. 1998).

1. Voltage-Dependent Channels

A distinctive electrophysiological signature of many interneurons is a "fast-spiking" firing pattern (Schwartzkroin and Mathers 1978; McCormick et al. 1985; Connors and Gutnick 1990; Buhl et al. 1994; Buhl et al. 1996; Morin et al. 1996; Ali et al. 1998; Ali and Thomson 1998). The action potential of these cells is less than one-half the duration at half-maximal amplitude of that of principal cells, 0.6 ms vs 1.5–2 ms, respectively, and is followed by a large, sharp, and relatively brief afterhyperpolarization (AHP). See Fig. 1 for examples. High input resistance and an absence of a slow, Ca^{2+}-dependent K^+ conductance permit these interneurons to fire at high spontaneous firing rates and to discharge repetitively, without accommodating, when depolarized. Cells with these properties can be identified with near certainty as interneurons. However, because many interneurons do not show these properties, a converse argument cannot be made. Another common firing pattern found in interneurons of stratum lacunosum moleculare in the hippocampus (Ali et al. 1998) and neocortex (Kawaguchi and Kubota 1997; Xiang et al. 1998) is the burst-firing mode indicative of action potential initiation by a low-threshold Ca^{2+} spike.

Differences in Na^+ channel properties contribute to the distinctions between principal cell and fast-spiking interneuron action potentials (Martina and Jonas 1997). Na^+ currents recorded in nucleated patches from identified hippocampal-slice interneurons have faster deactivation kinetics and differences in voltage dependence of inactivation when compared to those in pyramidal cells. The molecular bases of the differences in current were not clear, although there is a precedent for differences in kinetic properties among Na^+ channels with different subunit composition in other systems, among other possibilities. Nevertheless, different Na^+ channel gating could contribute significantly to the fast-spiking pattern of interneuronal firing. Interneurons possess an array of high-voltage-activated Ca^{2+} currents, which resemble those of pyramidal cells (Lambert and Wilson 1996).

Strides towards the identification of the ion channel complement of interneurons are being made with the combination of single-cell RT-PCR techniques and electrophysiological analysis. Application of these methods has revealed that the fast-spiking properties of basket cells in the dentate gyrus are probably explained in large part by enhanced expression of mRNA for the Kv3.1/Kv3.2, vs Kv4.2/Kv4.3, subunits (Weiser et al. 1995; Du et al. 1996). The former are expressed in almost all interneurons, yet in only a small fraction of the regularly spiking CA1 pyramidal cells (Martina et al. 1998). Conversely, whereas the great majority of pyramidal cells express Kv4.2 and Kv4.3, only about half of the basket cells do. The Kv3 channels are activated at very depolarized membrane potentials and hence do not affect action potential initiation, although they do affect spike firing. By rapidly repolarizing the membrane, these channels contribute to very brief action potentials and large, fast AHPs. The A-type K channels constituted from Kv4 subunits regulate

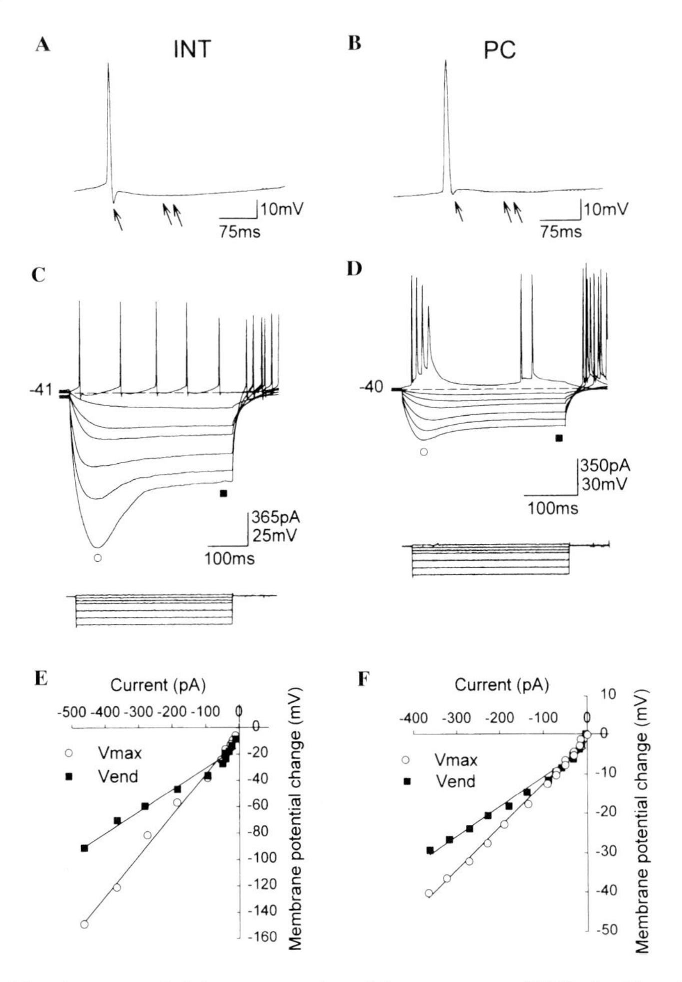

Fig. 1A–F. Membrane and firing properties of interneurons (INT: **A**, **C** and **E**) and pyramidal cells (PC: **B**, **D**, and **F**). **A,B** Single action potentials elicited in an interneuron in L-M (**A**) and a pyramidal cell (**B**). The action potential duration (measured at the base) was shorter in the interneuron. Both fast-duration afterhyperpolarizations (fAHPs) (↑) and medium-duration afterhyperpolarizations (mAHPs) (↑↑) were larger in amplitude in interneurons. The membrane potential was –52 and –45 mV in **A** and **B**, respectively. **C,D** Responses to current injection. A depolarizing current pulse elicited a regular train of action potentials in interneurons (**C**). Burst firing followed by a period of accommodation was evoked in the pyramidal cell (**D**). In both cell types, during large-amplitude hyperpolarizing pulses membrane potential reached an initial peak value (O), which was followed by a sag to a steady level (■). Resting membrane potential was more depolarized than usual in these 2 cells. **E,F** Graph of membrane potential changes, at the peak (O; V_{max}) and at the end of the pulse (■; V_{end}), vs current injected, with their respective linear regression, for the cells shown in **C** and **D**. In all cell types, with large current injection membrane responses were clearly smaller at the end of the pulse than at the peak. Cell input resistance was obtained from the slope of the regression lines, at the peak, and at the end of the pulse. In both interneurons and pyramidal cells there was a significant reduction in input resistance at the end of the pulse. (Reprinted from Morin et al. 1996, with permission)

spike firing in pyramidal cells at a slower frequency by delaying spike onset. Thus, the higher firing frequencies common in interneurons would be favored by the absence of Kv4.

The Kv3 subunits do not appear to be expressed in the oriens-alveus (O-A) interneurons, which, nevertheless, also have faster spiking properties than pyramidal cells. In the O-A interneurons of CA1, Ca^{2+}-dependent K^+ conductances, consisting of both iberiotoxin- and apamin-sensitive components, underlie the fast and slow AHP components (ZHANG and MCBAIN 1995a). By activating these channels Ca^{2+} influx, which increases during periods of high activity, may limit the output of the interneurons and, by doing so, increase the excitation of the pyramidal cells. In the O-A cells the prominent Ca^{2+}-dependent conductances repolarize the interneuronal action potentials and regulate the inter-spike interval. These contrasting results emphasize that similar physiological characteristics can result from different underlying ionic mechanisms.

The hyperpolarization-activated, anomalous rectifier current, I_h, confers pacemaker properties on the O-A interneurons of CA1 MACCAFERRI and MCBAIN (1996) found that the specific I_h antagonist, ZD7288, attenuated the spontaneous firing frequency by increasing the intraspike voltage trajectory, while having minimal effects on the interneuronal action potential properties. Because the action potential waveform was not altered, a decrease in I_h would not lead to changes in the amount of GABA release per action potential, but rather to changes in the neuronal firing frequency. I_h could be increased by norepinephrine, which then increased the interneuronal firing frequency.

2. Ligand-Gated Channels

Differences between interneurons produced by differences in ligand-gated channels are exemplified by the glutamate receptors. The excitatory postsynaptic potentials (EPSPs) in interneurons have a markedly faster time course than the EPSPs in principal cells in hippocampus (MILES 1990a) or neocortex (THOMSON et al. 1993). Absence of NMDA receptors (at some interneuron synapses), precise timing of glutamate release, and rapid deactivation kinetics of the interneuronal AMPA receptors contribute to the brevity of the interneuron EPSPs (HESTRIN 1993; GEIGER et al. 1997). Because the duration of a synaptic potential determines the time course of temporal summation possible for that potential, it appears that interneurons may act as coincidence detectors, requiring a number of precisely timed excitatory inputs for their activation. However, this interpretation may require modification for some interneurons having kainate-receptor-dependent EPSPs (COSSART et al. 1998; FRERKING et al. 1998). An unusual feature of the kainate EPSP is its very slow time course, lasting over 100 ms. Temporal summation of kainate-mediated EPSPs is quite marked. It is possible that a given cell could act as a coincidence detector when fast, AMPA-only synapses are activated and as integrators when kainate receptors are also activated. In any case, the presence of slow kainate responses suggests that precise coincidence of multiple EPSPs is not always a requirement for activation of interneurons. Furthermore, in some cases both NMDA and non-NMDA components of the EPSPs are present on interneurons (MORIN et al. 1996), providing additional scope for regulation.

A given pyramidal cell often makes only a single synapse with an interneuron (GULYAS et al. 1993). The distribution of single-pyramidal-cell-to-interneuron EPSPs tends to be broad in hippocampus (MILES 1990a; ALI et al. 1998; ALI and THOMSON 1998) and neocortex (THOMSON et al. 1993), showing frequent failures of transmission, and yet very large individual EPSPs as well. The EPSP distribution in pyramidal cells tends to be more uniform, with both fewer failures and larger events. Pronounced paired-pulse facilitation occurs at the excitatory synapses onto some interneurons, larger than that seen at pyramidal-cell–pyramidal-cell contacts. The reasons for these differences were not clear, but a model involving a greater probability of branch-point failure in the axonal projections to the interneurons could explain the data (ALI et al. 1998). In other cases (basket and bistratified cells) the pyramidal-cell–interneuron EPSPs became depressed with repetitive stimulation. A presynaptic locus for EPSP plasticity was identified in all cases.

In contrast to this picture, paired recordings of synaptically coupled principal cells and GABAergic interneurons reveal that the interneuron-to-pyramidal-cell transmission proceeds with few failures (MILES and WONG 1984; MILES 1990b), probably because the interneurons tend to make multiple synapses on a principal cell (MILES and WONG 1984; BUHL et al. 1994; TAMAS et al. 1997b) and because the probability that a given interneuron terminal will trigger transmitter release is relatively high (MILES 1990b; TAMAS et al. 1997b).

Interneurons receive inputs from a variety of pathways; however, receptors with very different properties can be selectively targeted by a single postsynaptic cell to synapses made by some pathways and not those made by others (TOTH and MCBAIN 1998). Interneurons express a different set of AMPA-type glutamate receptors than do principal cells (RACCA et al. 1996). Synaptic innervation of certain interneurons is effected by Ca^{2+}-permeable AMPA receptors (cf. references in TOTH and MCBAIN 1998). In some CA1 interneurons, kainate causes an essentially linear conductance increase and, in others, an inwardly rectifying conductance (MCBAIN and DINGLEDINE 1993). Inward rectification suggests that the glutamate receptors on these cells lack the GluR2(R) subunit. Unlike most AMPA receptors, those lacking a GluR2 subunit are highly Ca^{2+} permeable, and show a strong inward rectification. The inward rectification is conferred by the susceptibility of the channels to voltage-dependent block by intracellular polyamines. Principally found in stratum lucidum, these calretinin-containing interneurons receive input from the mossy fibers. Using the selective polyamine neurotoxin, philanthotoxin-433, TOTH and MCBAIN (1998) showed that certain s. lucidum interneurons expressed inwardly rectifying Ca^{2+}-permeable glutamate receptors at about half of all mossy fiber synapses, but only Ca^{2+}-impermeable, largely non-rectifying, receptors at recurrent collateral synapses. Analogously, Purkinje cells express the $\delta 2$ glutamate receptor subunit at parallel fiber, but not at climbing fiber, synapses. As discussed below (see Sect. H.IV), the presence of Ca^{2+}-permeable AMPA receptors at synapses of amygdalar interneurons enables them to express an NMDA independent form of LTP. High Ca^{2+} permeability, however, is also associated with the great vulnerability of cells possessing these receptors to cell death following ischemia or seizures (references in TOTH and MCBAIN 1998). Selective receptor expression at certain synapses is a developing theme that is likely to enrich further the computational complexity of interneuronal networks.

C. Physiological Responses Mediated by Inhibitory Neurotransmitters

As the number and sophistication of studies of inhibitory systems have increased, so has appreciation of the subtlety and complexity of the roles of inhibitory neurotransmitters. Earlier views of the function of inhibition in neuronal integration emphasized: (1) its ability to sculpt the constant barrage of amorphous excitatory input, and thus give form to the state of excitability of the cell, (2) the importance of disinhibition as a regulatory principle that could confer great flexibility on the actual contribution of the normally inhibitory inputs to the firing pattern of cells, (3) the ability of inhibition to "gate" the throughput of excitatory influences in a spatially and temporally specific way (see ALGER 1991, for review). In the following sections we review some of the major aspects of inhibitory transmission that are undergoing advances and stimulating revisions of traditional views of inhibition. Table 1 gives a brief list of neuronal functions thought to be subserved by the GABA and glycine systems.

I. Membrane Effects of GABA and Glycine

GABA and glycine receptors are both members of the same ligand-gated channel superfamily (JO and SCHLICHTER 1999) and therefore share many similarities. Study of large hyperpolarizing inhibitory postsynaptic potentials (IPSPs) evoked by afferent stimulation of $GABA_A$ or glycine revealed that prevention of action potential firing was an important role of inhibition (ALLEN et al. 1977). Prominent fast inhibitory postsynaptic currents (IPSCs) are the result of opening channels permeable mainly to Cl^- ions (although, as discussed below, also to HCO_3^-). The predominance of Cl^- conductance, together with the concentration gradient for Cl^-, which in adult cells is directed from the outside to the inside of the cell, means that usually $GABA_A$ergic and glycinergic transmission increases the postsynaptic membrane conductance and hyperpolarizes the cell. These two factors constitute two different forms of inhibitory influences: the former by moving the membrane potential away from the range of activation of voltage-dependent currents, e.g., Na^+ or NMDA currents, the latter by decreasing the input resistance of the cell and thereby

Table 1. Some neuronal functions of GABA and glycine. Even cursory consideration of the issue leads to a list of functions for GABA and glycine such as shown. Divided somewhat arbitrarily into groups of various neurotransmitter actions, this table, by no means complete, nevertheless suggests the broad diversity of sometimes contradictory functions served by GABA and glycine. Considerations such as cellular location, developmental state, frequency of and history of use and placement in a given neuronal circuit all influence their roles

Membrane effects

a. Ionotropic ($GABA_A$, $GABA_C$ and glycine) Increase Cl^- conductance (depolarizing in juveniles, hyperpolarizing, or depolarizing in adults)
b. Ionotropic ($GABA_A$ and glycine) Increase HCO_3^- conductance (depolarizing)
c. Metabotropic (G-protein dependent; $GABA_B$) Inhibit voltage dependent Ca^{2+} conductance, enhance K^+ conductance, directly inhibit release mechanism

Effects on cellular excitability

a. Inhibit activation of voltage-dependent Na^+ and Ca^{2+} conductance in soma and dendrites (prevent action potential initiation, alter spontaneous firing pattern)
b. Inhibit NMDA responses (reduce NMDAR-dependent Ca^{2+} influx and downstream sequelae)
c. Deinactivate K^+ currents (inhibit or delay subsequent action potential firing)
d. Regulate synaptic integration by altering passive membrane properties (reduce summation)
e. Increase excitability through membrane depolarization (Cl^--dependent response in juveniles; HCO_3^- response in adults)

Effects on signaling

a. Inhibit neurotransmitter release by blocking action potential conduction in preterminal axon
b. Inhibit neurotransmitter release through metabotropic receptors on terminal ($GABA_B$)
c. Preserve relative strength of release during a train of stimuli by reducing probability of release
d. Retard development of long-term changes in synaptic strength (e.g., LTP, LTD)
e. Promote development of LTD

Circuit level effects

a. Promote synchronous firing by removing inward current inactivation (rebound firing)
b. Reduce afferent stimulation through feedforward inhibition
c. Promote rhythmic firing through depolarizing membrane effects
d. Regulate network switching by differential control of afferent inputs
e. Reorganize sensory and motor systems
f. Excite targets through disinhibition (inhibition of inhibition)
g. Disrupt synchrony through inhibition of recurrent excitatory circuits

decreasing the voltage response caused by other currents. This shunting inhibition is always effective, but is dominant when the transmitter equilibrium potential is close to the resting potential of the cell and hence the transmitter cannot affect the membrane potential much. Contrary to initial impressions based on observations of large hyperpolarizing IPSPs, which emphasized the membrane potential change, recognition of the importance of the membrane potential shunt has increased, particularly because of persistent uncertainty about whether or not the IPSP in the unimpaled cell actually alters the membrane potential at rest (see Sect. C.III). The increase in Cl^- conductance caused by activation of $GABA_A$ergic and glycinergic synapses located on the somata of principal cells is ideally suited to control the membrane potential at the axonal trigger zone.

A very slow $GABA_A$ IPSC has been detected in pyramidal cells in the hippocampus (PEARCE 1993; PEARCE et al. 1995) and piriform cortex (KAPUR et al. 1997b). By virtue of its long time course and dendritic site of generation, this IPSC may be especially important in regulating the slow EPSP mediated by activation of the NMDA receptor, and phenomena, e.g., LTP, controlled by this receptor (KAPUR et al. 1997b). The slow IPSC is subject to regulation by $GABA_B$ autoinhibition (see Sect. G.II), whereas the fast IPSC is not (PEARCE et al. 1995; KAPUR et al. 1997b). The slow IPSC is often difficult to detect in somatic IPSC recordings in hippocampal CA1 pyramidal cells, although its time course reflects a slow conductance change and not simply cable filtering. The slow IPSC could be mediated by a $GABA_A$ receptor with a subunit composition different from that which mediates somatic IPSCs. In support of this, the fast component of the IPSC is blocked by furosemide, whereas the slow component is not.

GABA, but not glycine, also activates a K^+ conductance by acting on a G-protein-coupled $GABA_B$ receptor, as discussed below (see Sect. G.II).

II. Depolarizing GABA and Glycine Responses

While GABA and glycine are the main inhibitory neurotransmitters in the adult mammalian central nervous system, activation of their receptors does not always lead to a membrane hyperpolarization and neuronal inhibition. In the early development of the brain, GABA acts as the main excitatory transmitter (GAIARSA et al. 1995; BEN-ARI et al. 1997), and in adult neurons activation of $GABA_A$ receptors can depolarize as well as hyperpolarize cells (e.g., ANDERSEN et al. 1978; ALGER and NICOLL 1979, 1982a,b; WONG and WATKINS 1982; PERREAULT and AVOLI 1988). Glycine receptors are also transiently expressed in higher brain regions, including the hippocampus, during the first two weeks of postnatal life, and activation of glycine receptors induces a depolarizing chloride-dependent response (ITO and CHERUBINI 1991).

1. Depolarizing GABA and Glycine Responses in Young Tissue

During the first postnatal week, spontaneous, bicuculline-sensitive, giant depolarizing potentials (GDPs) that trigger action potentials predominate in hippocampal CA3 cells (GAIARSA et al. 1995). This excitatory effect of GABA is a general feature of developing CNS neurons. In immature neurons the Cl^- gradient is outward, rather than inward, as it is in mature neurons. When the Cl^- channels open, Cl^- ions leave the cell, thus depolarizing it. In slices, GDPs elicit synchronous neuronal activity that is, via activation of voltage-gated Ca^{2+} channels, associated with synchronous Ca^{2+} oscillations (LEINEKUGEL et al.

1995). In the developing CNS these Ca^{2+} oscillations can trigger intracellular signaling cascades and appear to be important for the growth of pyramidal cells and the formation of synaptic connections. Depolarizing $GABA_A$ responses can even initiate NMDA-dependent LTP (BEN-ARI et al. 1997). Similarly, in developing brain stem and spinal cord, glycine acts as an excitatory transmitter (KIRSCH and BETZ 1998).

The switch from a depolarizing to a hyperpolarizing $GABA_A$ response in the developing rat hippocampus is correlated with the induction of the expression of a specific K^+/Cl^- co-transporter, KCC2 (RIVERA et al. 1999). These results support the prevailing view that fast hyperpolarizing GABA inhibition is dependent on an efficient mechanism for the extrusion of Cl^- (ZHANG et al. 1991; THOMPSON 1994).

2. Depolarizing $GABA_A$ Responses in Adult Tissue

In adult neurons exogenous GABA can elicit a depolarization when applied to dendrites. However, synaptically released GABA can also depolarize adult cells under certain conditions: for instance, when pentobarbital (which prolongs GABA responses) (ALGER and NICOLL 1979, 1982a), the K channel blocker 4-aminopyridine (4-AP) (PERREAULT and AVOLI 1988), or zinc (XIE and SMART 1991) is present. Depolarizing $GABA_A$ responses are also caused by tetanic stimulation (WONG and WATKINS 1982; PERREAULT and AVOLI 1988) or brief high-frequency stimulation (GROVER et al. 1993) or by single stimuli following block of $GABA_B$ receptors (THALMANN 1988).

The early observations showed that even dendritic depolarizing GABA responses (ALGER and NICOLL 1979, 1982a; WONG and WATKINS 1982; STALEY and MODY 1992) had an inhibitory function in adults. Single depolarizing responses were not large enough to reach threshold for action potential generation and, on the contrary, prevented antidromic action potential invasion of the soma. The associated conductance increase shunts more intense depolarizations and prevents firing (STALEY and MODY 1992). However, with intense repetitive stimulation $GABA_A$ depolarizations can be very large and are capable of activating NMDA responses and eliciting action potentials (STALEY et al. 1995).

Demonstrations of $GABA_A$-initiated hyperpolarizations superimposed on $GABA_A$ depolarizations in adult neurons showed that a reversed Cl^- gradient cannot account for the depolarizing $GABA_A$ response (ALGER and NICOLL 1979, 1982a,b). Noise analysis argued that conductance increases to two ionic species were necessary to account for the $GABA_A$ response (DJORUP et al. 1981). These observations implied the participation of some other ion besides Cl^- in the depolarizing response in adult cells. Bicarbonate ions (HCO_3^-) permeate the $GABA_A$ and glycine channels about one-fifth as efficiently as do Cl^- ions (BORMANN et al. 1987). The HCO_3^- concentration is a function of pH. At normal pH inside and outside the cell there is a strong outward driving force on HCO_3^-, and an inward driving force on Cl^-. The reversal potential for

$GABA_A$ current is at a balance point between these opposing forces on Cl^- and HCO_3^-, and is normally tilted towards the Cl^- equilibrium potential because the Cl^- permeability of the channel is so much higher than is the HCO_3^- permeability. Several hypotheses that incorporate a role for HCO_3^- ions in the $GABA_A$ response have been put forward (KAILA and VOIPIO 1987):

1. STALEY et al. (1995) propose that both Cl^- and HCO_3^- permeate the same $GABA_A$ channel. Following intense activation of the $GABA_A$ channel the chloride gradient is less effectively maintained than the HCO_3^- gradient, which is preserved by the action of carbonic anhydrase. Accumulation of Cl^- within the cell thus leads to a depolarizing shift in the $GABA_A$ reversal potential towards the HCO_3^- equilibrium potential. The $GABA_A$-mediated response becomes depolarizing.
2. PERKINS and WONG (1996) propose that the depolarizing GABA-mediated IPSCs induced in hippocampal CA3 pyramidal neurons by 4-AP might be mediated by a subtype of $GABA_A$ receptor that is preferentially selective for HCO_3^-. They detected shifts in the $GABA_A$ reversal potential under conditions in which Cl^- gradient collapse could not occur.
3. KAILA et al. (1997) and SMIRNOV et al. (1999) propose a two-stage process. Following a high-frequency train, an activity-induced increase in external K^+ results in an inhibition or reversal of Cl^- extrusion from the cell, via the K-Cl^- co-transporter (THOMPSON et al. 1988) and again a positive shift in the $GABA_A$ reversal potential (KAILA 1994). The increased extracellular K^+ concentration also has a direct depolarizing effect. SMIRNOV et al. (1999) have recently found that the depolarizing and hyperpolarizing phase of the high-frequency-stimulated biphasic $GABA_A$ response can be pharmacologically distinguished. Intracellular QX-314 abolishes the depolarization without affecting the hyperpolarization, while intracellular F-, and omission of added intracellular ATP, has the converse effect. Together with the data of PERKINS and WONG (1996), the results of SMIRNOV et al. (1999) argue that the simple form of the Cl^- accumulation cannot account for the biphasic, high-frequency-activated $GABA_A$ response. Whether or not separate $GABA_A$ receptors or some other factors are involved remains to be seen. Regardless of the ionic mechanism, it is clear that depolarizing $GABA_A$ responses are not curiosities, but have physiological effects and perhaps different pharmacological properties that will surely be important to understand.

III. Membrane Potential Changes Caused by GABA in Unimpaled Cells

An interesting challenge to conventional interpretations of the polarity of $GABA_A$ responses has recently arisen. Non-invasive techniques have been used to infer the membrane potential changes caused by $GABA_A$ in unimpaled cells by using cell-attached patch recordings. When the pipette K^+ concentration is roughly equal to the intracellular K^+ concentration, the current through single K^+ channels will reverse when the transpatch potential is ~0 mV. Recordings of single-channel currents at various transpatch potentials in the presence and absence of GABA agonists implied that indeed GABA does depolarize the membranes of pituitary nerve terminal (ZHANG and JACKSON 1995) and dentate hilar neurons (SOLTESZ and MODY 1994).

Reasoning that disturbances caused by invasive electrode techniques might so distort cellular properties as to render measurements of the neuronal membrane potential, and therefore the direction and degree of membrane potential change caused by neurotransmitters, incorrect, VERHEUGEN et al. (1999) extended this method to estimate membrane potentials in unimpaled cells. Because the membrane potential of the cell is not affected by the cell-attached patch, the transpatch potential will be 0 mV when the command potential of the patch pipette is equal to the membrane potential of the cell. Brief voltage ramps delivered to the pipette elicited voltage-dependent K current through the K channels in the membrane patch. The reversal potential of the K current through the patch was then equal (with only slight error) to the membrane potential of the cell. Subsequent break-in to the whole-cell mode permitted a comparison between results obtained with the two techniques. The direct result was the finding of a systematic error in membrane potentials measured with whole-cell, as against cell-attached, methods, with the whole-cell values being about 15 mV more depolarized than the cell-attached values. With this non-invasive technique VERHEUGEN et al. (1999) found that activation of $GABA_A$ receptors by muscimol produced an equivalent depolarization in younger and older cells, suggesting the Cl^- gradient might be the same at both ages, contrary to the usual interpretation based on intracellular experiments. While the possibility of a systematic error in measurement of membrane potential is worrisome, it is not yet clear if errors of this magnitude are generally a problem. The non-invasive studies were performed at cooler temperatures that tend to enhance the magnitude of depolarizing $GABA_A$ responses, which could have contributed to the absence of a clear $GABA_A$-induced hyperpolarization. Moreover, the temporal resolution of the non-invasive method (~1 s) would be insufficient to detect an initial transient $GABA_A$-induced hyperpolarization. Intracellular measurements provide estimates of resting potentials that are not substantially different from those often obtained with the non-invasive technique. Hence, although the issue cannot be regarded as resolved, it is important, and the non-invasive technique will be a useful addition to the electrophysiologist's arsenal.

D. Miniature Inhibitory Postsynaptic Currents

I. Saturation of Receptor Patches by Quantal Release

Evoked IPSCs, or action-potential-dependent spontaneous IPSPs, generally represent the synchronous occurrence of many quantal events, and accordingly are influenced by factors that affect interneuronal action potential firing. However, effects of drugs, or of biochemical and molecular processes, are often exerted at the quantal level. A neurophysiological question with important pharmacological implications is whether or not quantal release of neurotransmitter is sufficient to saturate the receptors at synaptic receptor patches. If receptors in a patch are not saturated by a quantum of transmitter, then changes in the amount of neurotransmitter released by a presynaptic action potential could be functionally important. If the receptor patches are saturated, then multiquantal release (see below) or drugs, for example, benzodiazepines that affect receptor binding affinity, may be limited to influencing duration, but not peak amplitude, of synaptic responses (MODY et al. 1994). Recent work suggests that this issue may not have a simple resolution: it may be necessary to determine for individual classes of inhibitory synapses whether or not the receptor patches are saturated by quantal amounts of transmitter.

The amplitude distribution of quantal release at central synapses rarely has the Gaussian form that the distribution of MEPP amplitudes at the neuromuscular junction has. The distribution of miniature IPSCs (mIPSCs) is typically skewed positively towards large quantal amplitudes (COLLINGRIDGE et al. 1984; EDWARDS et al. 1990; ROPERT et al. 1990; OTIS et al. 1991, 1994; DE KONINCK and MODY 1994; PITLER and ALGER 1994a; THOMPSON et al. 1997). For example, the mean $GABA_A$ mIPSC in CA1 cells is 20–40 pA, but mIPSCs as large as 100 pA occur. In cerebellar Purkinje cells the skew is more pronounced, and TTX-insensitive mIPSCs several hundred pA in amplitude are common (LLANO et al. 1991; LLANO and GERSCHENFELD 1993; AUGER and MARTY 1997; NUSSER et al. 1997). The potential neurophysiological importance of spontaneous quantal release necessitates understanding the determinants of quantal size.

At the neuromuscular junction, a quantum of ACh falls on a broad field of ACh receptors. As the vesicular ACh content is relatively constant, the quantal size is determined by the number of receptors activated, usually ~2500 for a MEPP. At $GABA_A$ synapses on CA1 (EDWARDS et al. 1990; ROPERT et al. 1990) and neocortical pyramidal cells (GALARRETA and HESTRIN 1997), estimates based on the mean conductances of mIPSCs and single $GABA_A$ channels are that opening of 10–30 channels produces the mIPSC. Variability in quantal size is small, suggesting that the receptors in the postsynaptic receptor patch are saturated by a quantum of GABA (EDWARDS et al. 1990). Benzodiazepines that enhance GABA binding to $GABA_A$ receptors should increase the numbers of channels that are opened by subsaturating levels of GABA, but benzodiazepines did not increase mIPSC amplitudes in CA1 (DE KONINCK and MODY 1994), which also argued that synaptic receptor patches were saturated by a quantum of GABA.

Given that small numbers of $GABA_A$ channels can account for an mIPSC and that the receptors in a patch are saturated by the contents of a single vesicle, then the observed variability in quantal size would probably not be due to variable amounts of GABA packaged in the vesicles, but to variability in the numbers of receptors in a patch. This issue was addressed in cerebellar stellate cells in a technical tour-de-force that combined patch-clamp measurements of mIPSCs with quantitative immunogold localization of $GABA_A$ receptors to receptor patches identified with electron microscopy (NUSSER et al. 1997). It was found that variation in quantal size was mirrored by variation in numbers of synaptic receptors per patch. Receptor density across patches was uniform and receptor subtype homogeneous. Larger patches were associated with more GABA receptors and larger mIPSCs. The receptors in smaller patches, i.e., those with <80 channels, were evidently saturated by the GABA contained in a single vesicle, because a benzodiazepine did not affect the amplitudes of small responses. It did, however, enhance the amplitudes of large mIPSCs, suggesting that in the larger patches the receptors are not saturated. One limitation of this study was its inability to associate individual mIPSCs with the synapses from which they originated; comparisons of mIPSC amplitudes and receptor patch size had to be made statistically.

A similar conclusion follows from studies on a glycinergic synapse in rat brainstem (LIM et al. 1999). Measurements of a glycine synaptic patch area with immunolabeling of gephyrin, a protein required for clustering of glycine receptors (FROEHNER 1998), revealed a large variability in patch area. Both patch and mIPSC sizes varied across cells and there was a good correlation between them, suggesting that much of the variability in glycinergic mIPSC sizes can be accounted for postsynaptically. Glycinergic mIPSC amplitudes increased from neonatal to juvenile ages, yet changes in single-channel properties appeared to play no role in the increase (SINGER and BERGER 1999). Again increases in number of receptors in a patch appeared to be responsible.

A different approach to the question used α-latrotoxin, a spider toxin, which induces bursts of mIPSCs (AUGER and MARTY 1997). Essentially all of the mIPSCs in a single α-latrotoxin-induced burst of mIPSCs at cerebellar synapses originate from a single release site, permitting calculations to be made of mean receptor occupancy, numbers of receptors in a patch, and single-channel conductances. At single sites the mIPSC distributions were more symmetrical, and narrower, than typical distributions of mEPSCs recorded from cell somata, which represent activity from diverse synapses distributed across the cell (but see TANG et al. 1994). In partial agreement with the conclusions of the Nusser study, α-latrotoxin-induced bursts revealed a range in numbers of receptors, and a correspondingly wide range in mIPSC sizes, across patches. There was also a three- to fourfold range in single-$GABA_A$-channel conductance; hence, significant between-site variability in single-channel properties may contribute to broad mIPSC distributions. In general, quantal variation in cerebellum seemed to be satisfactorily accounted for by the variation in the properties of receptor patches. Thus, in both hippocampus and cerebellum postsynaptic factors at inhibitory synapses determined quantal size.

In cultures of hippocampal neurons (Vautrin et al. 1994) and of retinal amacrine cells (Frerking et al. 1995), a single presynaptic terminal can make two synapses, one on a postsynaptic cell and an "autapse" on the cell from which the terminal originated, the entire arrangement being called a "dinapse". Simultaneous recordings from two cells involved in a dinapse revealed simultaneous mIPSCs in both cells, implying that a single quantum of neurotransmitter released from the terminal affected both patches. In a study on amacrine cells the amplitudes of the simultaneous mIPSCs were highly correlated, and a benzodiazepine enhanced their amplitudes, implying the receptors in the two patches were not saturated. Thus, a presynaptic factor, variation in the amount of GABA released from different vesicles (reflecting variation in vesicle size, as the vesicular GABA concentration is thought to be constant) accounted for the variance in mIPSC amplitudes in amacrine cells.

In recording from dissociated tissue-cultured cells, a single nerve terminal can be trapped beneath the tip of a patch pipette sealed onto the cell membrane (Lewis and Faber 1996a; Forti et al. 1997). The spontaneous synaptic currents detected in the cell-attached position must originate from the trapped terminal. In rat spinal cord and medullary neurons such recordings reveal that the same variability and skewness that characterize mIPSCs in the whole-cell recording mode are also properties of the single-terminal mIPSCs (Lewis and Faber 1996b). Clearly this within-site variability cannot be explained by the postsynaptic, between-site factors revealed by Nusser et al. (1997), and variations in amount of transmitter released, or in the state of postsynaptic receptors, must be responsible. The same conclusions have been reached in the study of glutamatergic transmission for synapses in tissue-cultured hippocampal neurons (Bekkers et al. 1990; Forti et al. 1997; Liu et al. 1999). Moreover, an important caveat to the use of the benzodiazepine, zolpidem, in these studies has arisen recently. Perrais and Ropert (1999) found that the mean amplitude of mIPSCs recorded in layer V cells of rat visual cortex in vitro is increased when the experiments are done at room temperature. This effect, which was attributed to the activation of more synaptic receptors because of the increase in the GABA binding affinity caused by zolpidem, implied that $GABA_A$ receptors were not saturated by single GABA quanta. However, when the experiment was performed at a warmer temperature (35 °C) zolpidem did not increase the mIPSC amplitude, suggesting that this drug does not accurately reveal the degree of receptor occupancy at the warmer temperatures sometimes used in these studies.

Therefore, the question of whether or not receptor patches are saturated by a quantum of transmitter seems to have no simple answer; high-resolution, well-controlled studies arrive at opposing conclusions. The actual significance of these differences is not yet understood; however, when a drug acting at $GABA_A$ receptors is globally applied to a heterogeneous array of synapses, its effects on inhibitory synaptic responses could vary across different synapses even if the receptor subtype is exactly the same at each synapse.

II. Co-Release of GABA and Other Transmitters

It is now well established that in different sets of interneurons, GABA colocalizes with a variety of neuropeptides (Freund and Buzsaki 1996), including somatostatin, neuropeptide Y, cholecystokinin, and vasoactive intestinal peptide. Some of these peptides do affect GABA responses and modulate GABA actions physiologically, but there is much work to do before the details of the GABA-neuropeptide interactions become clear.

1. GABA and Glycine

An intriguing aspect of the study of quantal responses at spinal cord and medullary neurons (Lewis and Faber 1996a,b) was that, even at single synapses, the mIPSCs were sensitive to both strychnine and bicuculline at low concentrations, implying that glycine and GABA could be released by a given terminal and that receptors for both were present in the same patch. $GABA_A$ergic and glycinergic receptors are colocalized at synaptic contacts in the spinal cord, and GABA and glycine can be taken up into the same synaptic vesicle (cf. references in Lewis and Faber 1996a,b). Definitive evidence that both GABA and glycine can be co-released from the same synaptic vesicle has now come from a careful pharmacological analysis of mIPSCs in the spinal cord (Jonas et al. 1998). Individual quantal responses had, in variable proportions, properties of both glycine- and $GABA_A$-mediated responses; fast mIPSC rises were blocked by strychnine, slow decays by bicuculline. At the low concentrations used, strychnine and bicuculline were confirmed to be selective antagonists at glycine and GABA receptors, leading to the conclusion that both neurotransmitters can be released from the same synaptic vesicle. The studies were done on cells isolated from young animals, so it is not yet clear if co-transmission of glycine and GABA represents a developmental stage or whether it is also a property of the mature nervous system. An open question is also what determines the proportion and variability of the glycine vs the GABA components, and both pre- and postsynaptic mechanisms are possible. It will be most interesting to learn if these findings represent a rare exception to "Dale's Principle" or whether similar co-release of different amino acids also occurs at other synapses, and what the physiological significance of this mode of transmission is. Are the effects of the two transmitters simply additive, or do they interact in some way? If co-release of GABA and glycine does occur in adults, then drugs acting at one or the other receptor can potentially shift the synaptic influence towards one or the other, so answers to these questions will be important.

2. Co-Release of GABA and ATP

Recordings from synaptically coupled pairs of cultured spinal cord cells revealed that about 50% of the presynaptic cells released ATP, and yet all of these cells released GABA (Jo and Schlichter 1999), leading to the conclusion that GABA and ATP are released from the same cells. Various fast neurotransmitters are colocalized with neuropeptides, and both peptide and neurotransmitter can be released, although often with different stimulation regimes. The demonstration of co-release of GABA and glycine from single vesicles discussed above showed that simultaneous secretion of two fast neurotransmitters could occur. Unlike GABA and glycine, which had similar effects on membrane potential, GABA and ATP have opposing effects: ATP acts as an excitatory neurotransmitter, while GABA is inhibitory. If GABA and ATP are released from the same terminal (not determined), then their physiological effects could offset each other's unless special conditions of receptor placement, receptor responsiveness, etc., are met. In principle, though, it would be possible to change the sign of such a synapse, i.e., from inhibition to excitation, by altering the release conditions.

III. Multiquantal Release

At many central synapses one active zone is typically found at each presynaptic nerve terminal, and one quantum of neurotransmitter is generally thought to be released by an action potential. Multiquantal release would cause variability in mIPSC size, and could prolong mIPSCs by delaying transmitter clearance, if postsynaptic receptor patches are not saturated by one quantum of transmitter. Multiquantal release (perhaps from several sites at a single synapse) has been detected as a change in unitary EPSC size under varying conditions of release probability (Silver 1998), with higher probability of release favoring multiquantal release. From estimates of open-channel probability, receptor occupancy has been calculated at 0.45–0.6 for the AMPA responses at uniquantal cerebellar mossy-fiber synapses. Thus, at these synapses the postsynaptic receptor patches are not saturated.

If the receptors in the postsynaptic receptor patch are saturated by the contents of one vesicle, then it seems difficult to know if one or more than one vesicle is released, as mIPSC size cannot fluctuate; nevertheless, even in this case, multiquantal release can be detected. When multiple quanta are released into a single synaptic cleft, the concentration of neurotransmitter in the cleft will be higher than when a single quantum is released, and therefore a low-affinity competitive antagonist will have a diminished effect on the postsynaptic responses in instances of multiquantal release. Tong and Jahr (1994) demonstrated multiquantal release at glutamatergic synapses with this approach. Similarly, multiquantal release has been inferred by the dependence of the potency of low-affinity receptor antagonists in blocking EPSCs on the probability of transmitter release (Silver 1998). If one quantum is released per terminal, the potency of the antagonist should be independent of probability of release. Thus far multiquantal release at GABAergic or glycinergic synapses does not appear to have been detected using this method.

When Sr^{2+} is substituted for extracellular Ca^{2+}, the synchronous quantal release of transmitter (Miledi 1966; Goda and Stevens 1994), including

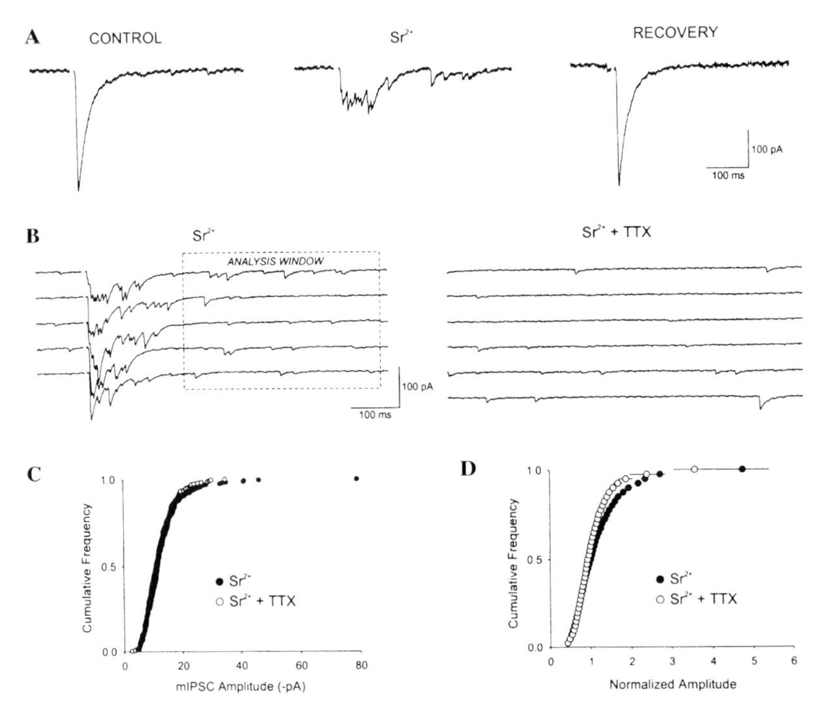

Fig. 2A–D. Sr^{2+} induces asynchronous quantal release of GABA. **A** *Traces* from left to right show the effects of replacing Ca^{2+} with Sr^{2+} on an evoked monosynaptic IPSC recorded from a hippocampal CA1 pyramidal cell. All experiments were done in APV and CNQX to block ionotropic glutamate responses. Notice that, following 10 min of application of Sr^{2+}, the IPSC is mainly composed of asynchronous events and is reduced in amplitude. Full recovery was observed 15 min after switching back to Ca^{2+}-containing saline solution. **B** *Left panel* shows five consecutive *traces* of evoked IPSCs recorded in the presence of Sr^{2+}. Spontaneous asynchronous events were measured within a 400-ms-wide analysis window. *Right panel* shows six consecutive *traces* of spontaneous mIPSCs from the same cell in Sr^{2+} with 0.5 μmol/l TTX. Stimulus artifacts in **A** and **B** are blanked out for clarity. **C** Comparison of the amplitude distributions of spontaneous IPSCs recorded from a pyramidal neuron in the presence of Sr^{2+} (232 events; mean ± S.E.M., –12.7 ± 0.5 pA) with the mIPSCs observed over a period of 1 min in Sr^{2+} and TTX (233 events; mean ± S.E.M., –12.7 ± 0.5 pA). **D** Summary of the average amplitude distributions obtained from five cells in Sr^{2+} and in Sr^{2+} with TTX. The distributions of mIPSCs in Sr^{2+} with and without TTX in C and D are not statistically different from each other (p>0.3 and 0.05, respectively) as determined by the Kolmogorov-Smirnov (K-S) test. (Reproduced from Morishita and Alger 1997, with permission)

GABA (Morishita and Alger 1997; Behrends and ten Bruggencate 1998), is disrupted, e.g., Fig. 2. Action potentials still induce release, but quantal release is asynchronous and the miniature events occur spread out in time over intervals of ~ 1 s after the action potential. This dispersion makes it possible to measure quantal parameters, amplitude, total number, and frequency directly. At $GABA_A$ synapses between cultured striatal neurons (Behrends and TEN BRUGGENCATE 1998), the amplitudes of asynchronous evoked mIPSCs in Sr^{2+} change as a function of conditions that alter the probability of release. This effect is easily explained by assuming that presynaptic factors determine quantal size, and is not obviously consistent with an exclusively postsynaptic mechanism. Multiquantal release (onto unsaturated receptor patches) and graded (rather than all-or-none) release of GABA from synaptic vesicles were suggested as possible mechanisms.

Single-site synaptic connections on cerebellar interneurons (AUGER et al. 1998) were identified in paired-cell recordings under conditions of reduced transmitter release (see Fig. 3). Multiquantal release was detected in the excess of mIPSCs occurring in doublets (i.e., within 1–5 ms of each other) and in the non-linear summation of the doublet mIPSCs. Non-linear summation indicates the mIPSCs are not independent (cf. TANG et al. 1994). The calculated mean

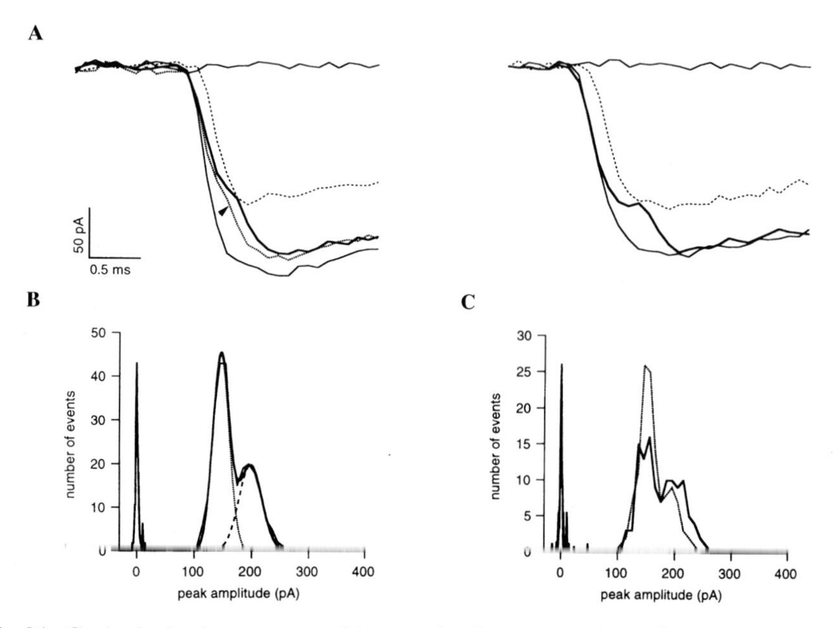

Fig. 3A–C. A single-site synapse with two closely separated amplitude components. **A** In this single-site recording, two distinct amplitude levels were observed. In several traces, double events were seen to jump from one level to the other (*thick line* responses), or to display an inflection point near the lower amplitude level (*arrowhead*). **B** Overall amplitude histogram from this experiment (480 trials), showing two distinct peaks. In dual component traces only the peak amplitude of the second event was entered. The histogram was fitted to the sum of two Gaussian curves (*thick line*; *dotted lines* indicate each curve separately) with mean amplitudes and SD values of 147 ± 14 pA and 198 ± 20 pA, respectively. The scaled noise histogram is also shown (failure rate was 0.50). **C** Histograms for first (*thick line*) and second (*dotted line*) halves of the data. Although the proportion of events in the higher amplitude peak decreased from the first to the second data range, the two peaks appear in both cases. (Reproduced from AUGER et al. 1998, with permission)

receptor occupancy at the dual release patches was 0.7, indicating that these patches were not saturated. The mIPSC doublets seem to represent cases of slight disparity in the timing of release of multiple vesicles. Truly synchronous, single-site multivesicular release would have been revealed as larger mIPSCs (provided, again, receptor occupancy is less than 1.0). KIRISCHUK et al. (1999) studied $GABA_A$ergic transmission at single boutons in cultured superior collicular neurons, correlating electrophysiological measurements with simultaneous measurements of presynaptic Ca^{2+} concentration. Under conditions of action-potential block, the bouton could be directly depolarized by current passed through a closely apposed glass pipette. The amplitudes of the resulting single-bouton IPSCs varied greatly and were correlated, although imperfectly, with the magnitudes of the presynaptic Ca^{2+} transients. The results demonstrated not only lack of receptor patch saturation by a vesicle, but that variation in vesicle release contributed to evoked IPSC variability. Hence, variation in synaptic transmitter release appears to contribute to the substantial variability of mIPSCs at various CNS synapses. There appears to be no simple pattern followed in the CNS, and both pre- and postsynaptic factors have to be considered in issues relevant to quantal transmission.

IV. Tonic Inhibition

GABA is continually released spontaneously at synapses, to some extent because of TTX-insensitive quantal release (EDWARDS et al. 1990), but often to a greater degree because of interneuronal action potential activity (ALGER and NICOLL 1980; OTIS et al. 1991). In some cells, e.g., hippocampal CA1, spontaneous quantal release occurs at a low rate (~1 Hz); in other cells it occurs at a considerably higher rate. A steady-state "tonic" form of inhibition, mainly a shunting inhibition, is caused by the summation of conductances resulting from the temporal overlap of spontaneous events (OTIS et al. 1991; SALIN and PRINCE 1996). When this summated background conductance is blocked by $GABA_A$ antagonists, an increase in cell excitability results.

Through the use of whole-cell recordings, lesions that truncated the dendritic tree, and computational modeling, SOLTESZ et al. (1995) showed that the tonic barrage of spontaneous IPSPs originated mainly from somatic synapses in dentate granule cells. BANKS et al. (1998) arrived at the same conclusion using selective application of $GABA_A$ antagonists to different parts of the dendritic tree of CA1 pyramidal cells. The lack of evidence for spontaneous release from dendritic synapses could not be attributed to cable filtering and is not understood, although differences in the properties of dendritic versus somatic terminals have been observed (MILES 1996).

The effects of tonic inhibition may be more subtle than simply preventing action potential firing. In the cerebellum, action-potential-dependent tonic inhibition determines the irregular firing pattern observed in Purkinje cells and interneurons in the molecular layers (HAUSSER and CLARK 1997). With glutamate receptors blocked, both Purkinje cells and interneurons are spontaneously active, and $GABA_A$ antagonists cause an increase in firing rates, as well as a dramatic reduction in the variability of the interspike intervals. Paired recordings showed that an action potential in the interneuron delays the occurrence of an action potential in the target cells, with the magnitude of the effect being directly proportional to the magnitude of the variable IPSP. The constant barrage of IPSPs thus introduces irregularity into the spontaneous action potential discharge of the postsynaptic cells. Simultaneous dendritic and somatic Purkinje cell recordings also showed that, by altering the passive cell properties, tonic inhibition increases the electrotonic length of the cell such that dendritic EPSPs have a lesser effect in the soma when inhibition is intact than when it is blocked. Tonic inhibition limits the interval over which temporal summation with a given EPSP is possible.

Tonic inhibition can also result from TTX-insensitive events. Especially in spatially restricted regions, such as cerebellar glomeruli, small amounts of GABA that spill over from synapses to surrounding extrasynaptic receptors can accumulate and influence cellular firing (Brickley et al. 1996). Originally liberated as a result of synaptic activity, GABA nevertheless achieves a level steady-state concentration without obvious fluctuations caused by individual events. The result is a shunting inhibition equivalent to the persistent activation of only a few $GABA_A$ receptors. Despite its small magnitude, this form of tonic inhibition produces clear effects that become more significant throughout development.

E. Dendritic Inhibition

I. Control of Dendritic Electroresponsiveness

Synaptic interconnections among interneurons and principal cells determine the kinds of roles that inhibition can play. Feedback or recurrent circuitry was first emphasized in the hippocampus and other structures, although possibilities for afferent collateral inhibition (a form of "feedforward inhibition") had clearly been recognized (Eccles 1964). The dense GABAergic innervation of the somatic regions of pyramidal cells, coupled with the dipole theory interpretation of somatic positive extracellular field potentials (Andersen et al. 1964a), led to the conclusion that the primary form of interneuron activation was through feedback from pyramidal cell firing and was directed principally at cell somata. This emphasized the role of basket cells (Andersen et al. 1964b), which were well known to make dense networks of somatic terminations. Evidence that GABAergic inhibition was activated by feedforward as well as feedback pathways and innervated dendrites as well as somata (e.g., Wong and Prince 1979; Alger and Nicoll 1979, 1982a; Buzsaki 1984; Miles et al. 1996) led to numerous additional possibilities for neuronal integration.

Dendrites have become increasingly recognized as active participants in neuronal integration, and dendritic inhibition has correspondingly risen in importance. Studies of olfactory (Jahr and Nicoll 1980) hippocampal (Wong and Prince 1979), cerebellar (Llinas and Sugimori 1980) and neocortical (Johnston et al. 1996, for review) dendrites using sharp microelectrodes revealed not only that dendrites possessed active properties, and did far more than receive and passively propagate excitatory signals, but that these properties were under the control of synaptic GABAergic inhibition. Complex burst potentials, the result of voltage-dependent Na^+ and Ca^{2+} currents, are prominent features of principal cell dendrites in the CNS and are often under the control of synaptic $GABA_A$ inhibition (Wong and Prince 1979; Miles et al. 1996; Miura et al. 1997).

Intradendritic recordings from alligator cerebellar cells (Llinas 1988) revealed a complex burst potential with small regenerating potentials evidently originating at branch points in the dendritic tree. It was suggested that, by acting at branch points, dendritic inhibition could "functionally amputate" portions of the dendritic tree, and hence isolate the soma from certain afferent inputs. Wong et al. (1979) showed that a correctly timed IPSP could completely abort a burst potential in a pyramidal cell dendrite, emphasizing the role of dendritic inhibition in the all-or-none regulation of the burst response, which was known to be a basic property of hippocampal pyramidal cell dendrites. A combined morphological and physiological study in hippocampal pyramidal cells showed directly that inhibitory cells making perisomatic contacts suppressed repetitive Na^+-dependent action potential firing, whereas dendritically terminating inhibitory cells controlled dendritic electrogenesis directly and initiation of axonal action potentials indirectly (Miles et al. 1996).

II. Dendrodendritic Inhibition

Anatomical evidence had suggested that mitral cells in the olfactory bulb could participate in an unusual form of dendritic interaction, dendrodendritic inhibition, with the granule cells. Upon mitral cell depolarization, excitatory transmitter (ultimately shown to be glutamate) would be released from synaptic specializations on the mitral cell dendrites, where it would act on the opposing granule cell gemmule (spine). The granule cells contain GABA, and activation of the granule-to-mitral-cell synapse on the gemmule would release GABA and inhibit the mitral cell. Confirmation of this hypothetical scheme was provided by Jahr and Nicoll (1980, 1982) and others (Nowycky et al. 1981) in intracellular studies in the turtle in vitro olfactory bulb preparation.

This finding has been extended to mammals, with the development of the rat olfactory bulb slice technique (Isaacson and Strowbridge 1998; Schoppa et al. 1998). Dendrodendritic inhibition has been found in other areas of the nervous system, as well. Unlike many glutamatergic synapses in which AMPA receptors play the predominant role in mediating fast synaptic transmission, glutamate released from the mitral cells can cause GABA release from the granule cells by activating NMDA, as well as non-NMDA, receptors (Isaacson and Strowbridge 1998; Schoppa et al. 1998). The prominent role of

NMDA receptors notwithstanding, GABA release from the granule cell is triggered by Ca^{2+} influx through P/Q- and N-type Ca channels and not through NMDA channels. Lateral inhibition in the olfactory bulb can also be mediated by the dendrodendritic circuit. In simultaneous recordings from pairs of mitral cells, activation of dendrodendritic inhibition of one sets up an IPSP in a neighboring cell, even in the presence of TTX (Isaacson and Strowbridge 1998).

The dendrodendritic circuit thus gives rise to a highly localized reciprocal inhibition of the mitral cells. More recently it has been proposed that the olfactory circuit provides the basis of lateral inhibition and odor discrimination (Yokoi et al. 1995; Brennan and Keverne 1997). The lateral inhibition that is produced suppresses a weak excitatory response in neighboring mitral cells, thus sharpening the tuning specificity for odorants and enhancing the resolution of the olfactory system (Yokoi et al. 1995).

Autoreception also occurs in excitatory neurons when glutamate or an analog released from a cell acts on extrasynaptic receptors on that same cell. The release of glutamate from mitral cells in the olfactory bulb can cause a long-lasting self-excitatory response (Nicoll and Jahr 1982) that is mediated by NMDA receptors. Autoexcitation is under the control of the recurrent dendrodendritic IPSP, however, and is not obvious unless $GABA_A$ receptors are blocked.

III. Back-Propagating Action Potentials

Simultaneous whole-cell recordings from the somata and dendrites of neocortical (Stuart et al. 1997) and hippocampal pyramidal cells Tsubokawa and Ross 1996, as well as from mitral cells of the olfactory bulb (Chen et al. 1997), have revealed new features of dendritic processing. Dendritic potentials that summed to action potential threshold at the initial segment triggered axonal spikes, as was expected, but the action potential propagated backwards into the dendrites as well as forwards down the axon. The extent of back-propagation is controlled by synaptic inhibition: the action potentials increase in size and reach farther into the dendrites when $GABA_A$ IPSCs are blocked (Chen et al. 1997).

With strong synaptic stimulation, excitatory dendritic synaptic inputs can sometimes elicit dendritic Na^+ spikes prior to triggering axonal Na^+ spikes (Golding and Spruston 1998). Usually the dendritic spikes triggered axonal spikes, but occasionally they did not, suggesting they did not infallibly propagate to the soma. The occurrence of primary dendritic spiking was regulated by GABA-mediated inhibition and NMDA-dependent synaptic potentials. When inhibition was blocked, spike initiation shifted to the dendritic locus; when the NMDA receptors were subsequently blocked, spike initiation shifted back to the axon. Somewhat surprisingly (in view of the relatively short time from EPSP onset to spike initiation, ≤ 5 ms), both $GABA_A$ and $GABA_B$ receptors exerted similar control.

A recent description of a new function for back-propagating action potentials revealed another role for dendritic inhibition (LARKUM et al. 1999); see Fig. 4. If an axonally initiated back-propagating action potential was followed within a few milliseconds by initiation of an EPSP in the apical dendrites, a

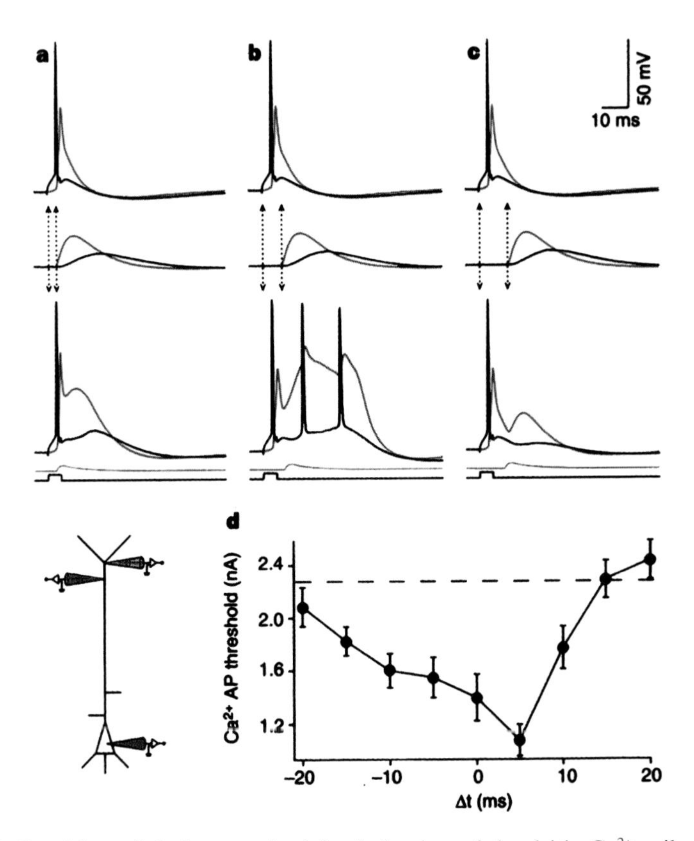

Fig. 4a–d. Precision of timing required for induction of dendritic Ca^{2+} spike. Experimental configuration shown diagrammatically (*lower left*). Recordings were made from the dendrite (*red*; 600 μm from the soma) and the soma (*black*) of an L5 pyramidal neuron. A third dendritic electrode (*pink*; 700 μm from the soma) was used for injecting current (electrode colors correspond to recording traces). Time intervals: **a** 3 ms; **b** 7 ms; **c** 11 ms elicited a burst of APs only in **b** at threshold. Δt was taken as the time between the start of the somatic current injection and that of the dendritic current injection. Note, however, that the AP due to the somatic current injection followed the onset by ~3 ms in this case. **d** A burst of APs could be generated by the combination of dendritic current injection and a back-propagating AP at other times, but the threshold for this was least at $\Delta t = 5$ ms. *Each point* is the average of eight neurons (*error bars*, S.E.M.) and represents the threshold for current injection needed to elicit a dendritic Ca^{2+} AP. *Dashed line* represents the Ca^{2+}-AP threshold without a back-propagating AP (2.28 ± 0.14 nA). For ~100 ms after $\Delta t = 10$ ms, the threshold was even slightly higher than without the back-propagating AP. (Reproduced from LARKUM et al. 1999, with permission)

large Ca^{2+} spike was generated in the dendrites. The Ca^{2+} spike could, in turn, trigger additional axonal action potentials. Evidently the summation of the two sources of depolarization lowered the threshold for Ca^{2+} spike initiation. Generation of the Ca^{2+} spike was facilitated by blocking either $GABA_A$ or $GABA_B$ receptor. Triple recordings (dendritic and somatic electrodes on one cell, the third electrode on a synaptically coupled interneuron) showed that, conversely, the Ca^{2+} spike could be abolished by a correctly timed IPSP, even if the IPSP did not affect the back-propagating action potential itself. Generation of the Ca^{2+} spike by the summation of the EPSP and back-propagating action potential could provide a critical mechanism for detecting and reporting synchronous activity in two distinct cortical regions. Prevention of the Ca^{2+} spike initiation by the IPSP in these cells decouples the two distinct spike initiation zones and disrupts this coincidence detection mechanism. The many roles played by Ca^{2+} in cellular processes highlight the significance of dendritic inhibition.

IV. Control of Persistent Cation Currents

Inhibition serves many functions by regulating voltage-dependent currents. Conversely, the occurrence of non-inactivating conductances can also amplify inhibitory potentials. In the thalamus (WILLIAMS et al. 1997) and neocortex (STUART 1999), the turning off of persistent Ca^{2+} and Na^+ conductances by a hyperpolarizing IPSP, respectively, magnifies the apparent amplitude and duration of the IPSP if the IPSP is evoked when the persistent cation currents are activated. The underlying $GABA_A$ currents are not themselves altered; rather the IPSP hyperpolarization closes some of the open cation-permeable channels. Truncation of the standing inward current is equivalent to an outward current that sums with the outward $GABA_A$-induced current. The enhancement can be prevented by blocking the voltage-dependent cation channels. The kinetics of the enhanced IPSP thus reflect the kinetics of inactivation and reactivation of the persistent cation-dependent current as well as those of the $GABA_A$ current. IPSP amplification caused by the persistent non-inactivating Na^+ current can help synchronize action potential firing at membrane potentials near rest.

F. Somatic-Axonal Inhibition

I. Conduction Block Along the Preterminal Axon

Conduction block refers to the interruption of action potential propagation along an axon. It was first postulated as a factor for modulation of neuronal signaling many years ago, but was often attributed to features of axonal geometry or changes in extracellular milieu.

In the spinal cord, conduction block was identified in recordings at points rostral and caudal to the point of entry of dorsal root fibers into the cord

(Wall 1995). Single afferent fibers make a T junction and project a rostral and a caudal branch. Wall (1995) observed that, whereas the action potential propagated rostrally without failure, the action potential typically failed to propagate along the caudal portion of the bifurcating axons. When bicuculline or picrotoxin was applied, however, caudal propagation also occurred, thus implying a conduction block caused by $GABA_A$ receptor activation, probably resulting from spontaneous activation of GABA interneurons in the cord. Strychnine was not effective, ruling out a role for glycinergic inhibition.

Most vertebrate axons are too small for impalement by electrodes and hence for direct study, so the mechanism by which IPSPs block conduction in the cord is not clear. Zhang and Jackson (1993, 1995) show that presynaptic depolarization of pituitary nerve terminals caused by activating $GABA_A$ channels could reduce and block the preterminal action potential. The depolarization (caused by a reversed Cl^- gradient in the terminal) inactivated voltage-dependent Na channels and prevented action potential conduction further along the terminal arborizations. Conduction block in the axon is potentially very powerful, as all synapses downstream from the point of block would be effectively inactivated. It is not known if conduction block by $GABA_A$ receptors takes place at synapses, or whether extrasynaptic receptors play a role. Interestingly, different types of $GABA_A$ receptors may be targeted to synaptic vs extrasynaptic regions (Brickley et al. 1999).

Debanne et al. (1997) provided evidence for another type of mechanism for conduction block caused by inhibitory transmitters in cultured hippocampal slices. IPSP hyperpolarizations removed the inactivation of the transient, voltage-dependent A-type potassium current in axons, and termination of the IPSP was followed by the activation of I_A, a transient outward current. When I_A was activated depolarizations normally sufficient to induce action potentials could not do so, and hence axonal conduction was prevented. The I_A antagonist 4-AP blocked the effect. Clustering of A channels near axonal branch points may enhance the potency of this mechanism (Kopysova and Debanne 1998). This work not only illustrates that the preterminal axon can be a target of $GABA_A$-mediated inhibition, but also that the $GABA_A$ conductance can act in concert with other factors to produce its effects.

Variable conduction block may also occur in the complex axonal arborizations of the inhibitory axons themselves. Simultaneous recordings from two cerebellar Purkinje cells showed that many spontaneous IPSCs occurred synchronously in both cells, suggesting that they from a single interneuron originated (Vincent and Marty 1996). If the interneuronal action potential propagated faithfully to both cells, then there would have been a reasonably constant ratio of the synchronous IPSC amplitudes. Instead, when the synchronous IPSC amplitudes in cell 1 were plotted against those in cell 2, there appeared to be no relationship between them, a result that could be explained by variable success in propagation of action potentials along the axonal branches to cell 1 or cell 2, although other interpretations are possible. Variability in IPSC amplitudes in the postsynaptic cell was very much reduced when, in paired recordings, the presynaptic interneuron was filled with the K^+ channel blocker Cs^+, suggesting a role for K^+ channel activation in modulating variable release. Mutant mice lacking the Kv1.1 type of K^+ channel have an increased frequency of sIPSC firing, a phenomenon conceivably caused by a decrease in axonal conduction block, because the firing frequency of the interneurons was not changed (Zhang et al. 1999).

Examination of the extremely large, complex interneuron axonal arbors, and consideration of the myriad factors (ionic concentrations, pH, osmolarity, neurotransmitters, and modulators) that can affect CNS axons lead to an alternative interpretation, namely that action potentials could arise at numerous "ectopic" sites along the axon, and not only at the initial-segment region. Different axonal segments could then act independently of each other and the cell body, and the observed differences between IPSCs caused in different target cells by a common interneuron would be caused not only by variability in conduction but also by variability in the sites of action potential initiation. Other explanations are of course also possible. Nevertheless, these studies emphasize the axon as a site of regulation in neuronal interactions, and provide interesting counterpoint to the new focus on the role of action potential propagation in dendrites in neuronal integration.

II. Depolarization-Induced Suppression of Inhibition (DSI)

Depolarization of a hippocampal pyramidal cell (Pitler and Alger 1992b, 1994a; Alger et al. 1996; Lenz et al. 1998; Ohno-Shosaku et al. 1998) or a cerebellar Purkinje cell (Llano et al. 1991; Vincent et al. 1992; Vincent and Marty 1993; Glitsch et al. 1996) causes a transient suppression of monosynaptic $GABA_A$ergic IPSCs recorded in that cell. The process, called DSI, is initiated by voltage-dependent Ca^{2+} influx into the postsynaptic cells (Llano et al. 1991; Pitler and Alger 1992b; Lenz et al. 1998; Ohno-Shosaku et al. 1998); however, it is not prevented by NMDA antagonists, as is the dephosphorylation-dependent $GABA_A$ receptor down-regulation that has been described (Stelzer and Shi 1994; Chen and Wong 1995; Wang and Stelzer 1996). DSI of evoked IPSCs appears as an increase in number of failures of quantal release, suggesting a presynaptic mechanism (see Fig. 5). In fact, a substantial body of evidence shows that there is no change in postsynaptic $GABA_A$ receptor responsiveness during DSI, whether this is assessed by iontophoretic GABA application or various forms of quantal analysis, including coefficient of variation, quantal content (Alger et al. 1996), or direct counting of asynchronous mIPSCs induced in Sr^{2+}-containing extracellular solutions (Morishita and Alger 1997). On the contrary, all of these measurements lead to the conclusion that the mechanism of DSI expression is a reduction in release of GABA from presynaptic nerve terminals, i.e., that a retrograde signal must pass between the postsynaptic cell and the interneuron to cause the interneuron to reduce its release of GABA for a brief time. Vincent and Marty (1993) provided compelling evidence for a messenger by showing that,

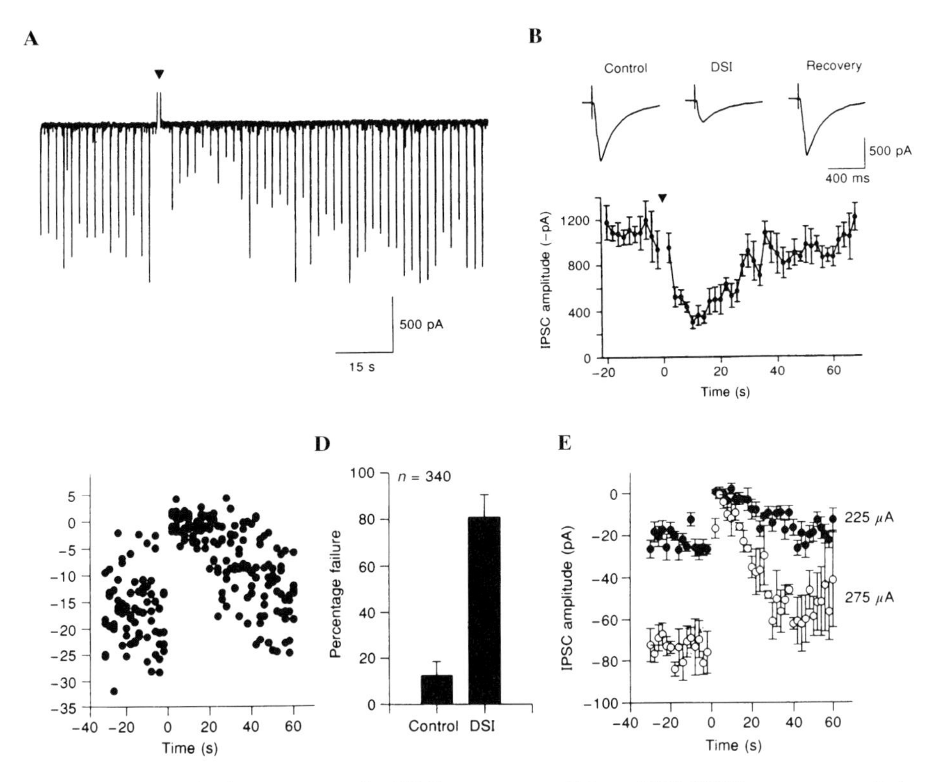

Fig. 5. A,B Evoked monosynaptic IPSCs are susceptible to DSI. IPSCs were recorded under whole-cell voltage clamp in the presence of 10 μmol/l CNQX and 50 μmol/l APV (carbachol was not present) with CsCl-containing pipettes. IPSCs were elicited continuously at 0.5 Hz with extracellular electrical stimulation in the vicinity of the recorded cell. At 90-s intervals a 1-s 70-mV depolarizing voltage step from –60 mV was delivered (*arrowhead at time zero*) to the pyramidal cell. **A** A typical complete DSI trial on an evoked IPSC (*downward strokes*, note time scale). **B** Combined data from same cell as in **A**. *Traces* at the top are averages of five responses each in control conditions (prior to the DSI pulse), during the DSI period and following recovery from DSI. The *graph* shows the entire time course from this experiment; *each point* represents the mean ± S.E.M. of five responses. **C-E** Failures of quantal IPSCs evoked with minimal stimulation to stratum radiatum increase during DSI. **C** Graph shows IPSC amplitudes of five DSI episodes with a 90-mV depolarizing step occurring at time 0 from one cell. Minimal IPSCs, recorded with KCl-filled electrodes, were evoked at 0.5 Hz. Peak amplitude measurements were made in the window from 0 ms to 14 ms following the extracellular stimulus. For the ten sweeps immediately prior to the voltage step, the stimulus was ineffective in eliciting an IPSC in only 6 of 50 trials, but failed to elicit a response in 50 of 60 trials during the DSI period. **D** Histogram illustrates percentage failure of transmission during the control and DSI period for three cells as in **C**, comparing the ten traces immediately prior to and following the depolarizing voltage step, for a total of 340 trials. **E** Failures of transmissions during the DSI period were evident when quantal-sized IPSCs were evoked with a stimulus intensity of 225 μA (●), as in **C** (minimal stimulation data from this cell included in **D**). Failures were still observed when larger, multicomponent IPSCs were evoked with a stimulus intensity of 275 μA (○). Data from multicomponent IPSCs not included in histograms in **D**. (Reproduced from Alger et al., 1996, with permission.)

if simultaneous recordings were made from two Purkinje cells, then IPSCs that were synchronous in both, i.e., that therefore probably were produced by a common interneuron, were suppressed in both cells if the DSI-inducing voltage step protocol was delivered to only one. Clearly, effects of the DSI in this case were not confined to the single postsynaptic cell but spread by some means to neighboring cell(s).

Recent evidence suggests the retrograde messenger in DSI may be glutamate, or a glutamate analog, and may produce effects by acting on a presynaptic metabotropic glutamate receptor (LLANO and MARTY 1995). DSI can be mimicked and occluded by agonists of metabotropic glutamate receptors (mGluRs), group II mGluRs being implicated in cerebellum (LLANO and MARTY 1995; GLITSCH et al. 1996) and group I in the hippocampal CA1 region (MORISHITA et al. 1998). That DSI involves the presynaptic activation of a G protein (PITLER and ALGER 1994a; MORISHITA et al. 1997) is consistent with a role for mGluRs. In CA1, DSI can be blocked by bath-applying 50 μmol/l 4-AP, or 250 nmol/l veratridine, agents that block certain types of K channels and Na channels, respectively (ALGER et al. 1996). Because, in these experiments, postsynaptic K and Na channels are blocked by Cs^+, TEA and QX-314, 4-AP and veratridine must act at a presynaptic site. One possibility is that DSI induces a type of axonal conduction block in the interneuronal axonal plexus and thereby prevents GABA release. This hypothesis is compatible with an intriguing observation about DSI, namely that, unlike many forms of presynaptic inhibition, DSI is not associated with a change in the probability of release at individual GABA-releasing nerve terminals. This conclusion is supported by repeated observations that DSI is not accompanied by a change in the paired-pulse depression (PPD) ratio (ALGER et al. 1996). Usually, when two IPSCs are evoked in quick succession, the amplitude of the second is reduced by ~50% when compared with the first. A process that changes the probability of release at a nerve terminal will typically alter the PPD ratio. One mechanism that would depress release without altering the PPD would be conduction block, but there are others. Interestingly, in lamprey axons, a group I mGluR increases the activation of a voltage-dependent 4-AP-sensitive K^+ current (COCHILLA and ALFORD 1998), and, as noted above, 4-AP-sensitive currents regulate axonal conduction (DEBANNE et al. 1997). Such effects would be compatible with a conduction-block for model hippocampal CA1 DSI.

DSI has also been observed to occur in dissociated tissue-cultured hippocampal neurons (OHNO-SHOSAKU et al. 1998), and, although the process seems generally similar, DSI in culture shows some differences from DSI in acute slices. For example, in culture DSI is associated with a change in the paired-pulse ratio, suggesting the possibility of a different expression mechanism. GABAergic inhibitory interneurons in culture also express DSI, as do Purkinje cells and other GABAergic cells in cerebellum. Hippocampal and cerebellar DSI are not identical (ALGER and PITLER 1995), and it appears that there will be several different manifestations of this regulatory process.

Although a functional role for DSI has not yet been demonstrated, it is clear that DSI can cause an increase in EPSCs (WAGNER and ALGER 1996b) and can be induced by low-Mg^{2+}-induced burst potentials (LE BEAU and ALGER 1998). It is likely that the coupling between a principal cell and its inhibitory inputs allows for selective feedback regulation of individual cells in a population.

III. Autoreception and Inhibition

Several forms of inhibition may be considered together under the concept of autoreception, i.e., when signals released from a cell act on the cell's own receptors. Autoreception may occur via:

1. "Autapses," fully developed synapses made by axonal terminals on the somato-dendritic regions of the cell originating the axon (VAN DER LOOS and GLASER 1972).
2. Transmitter released from nonsynaptic regions that acts on extrasynaptic receptors.
3. Transmitter released from synaptic terminals that acts on extrasynaptic receptors on or near that terminal.

In GABA-releasing cells, the first two cases involve activation of $GABA_A$ receptors. The third, the activation of presynaptic $GABA_B$ receptors, is discussed later. All cases of inhibitory autoreception have in common the functional effect of decreasing the inhibition exerted on postsynaptic cells.

1. Autaptic Transmission

Autapses form readily on dissociated tissue-cultured neurons (BEKKERS and STEVENS 1991; VAUTRIN et al. 1994); only recently, however, have suspected autapses been confirmed, with electron microscopy, to exist on GABAergic interneurons in fully differentiated tissue from adult animals in both hippocampus (COBB et al. 1997) and neocortex (TAMAS et al. 1997a). In a large study, TAMAS et al. (1997a) found that, whereas basket- and dendrite-targeting cells were very likely to form autapses, double bouquet cells were less likely, and autapses were rare or nonexistent on pyramidal and stellate cells. When they were found, autapses were made on the same cellular regions (dendrite, soma) as those on which the cell made synapses on other cells. Selective expression of this type of synapse by certain neurons, and the precise cellular localization of the autapses, suggested these are not random phenomena, but are part of a specific regulatory system. Autaptic autoinhibition seems poised to inhibit firing of the interneuron and thereby perhaps to contribute to phasic output from the cell.

2. Preterminal Extrasynaptic Receptors

The second type of autoreception involves release of GABA from presumed synaptic sites that acts on extrasynaptic sites along the preterminal axon. In cerebellar stellate and basket cells, an action potential initiated at either somatic or axonal sites is immediately followed by a slow Ca^{2+}-dependent Cl^- conductance mediated by $GABA_A$ autoreceptors (POUZAT and MARTY 1999). All-or-none linkage of this conductance with the action potential indicated it was triggered in the recorded cell, and modeling studies, coupled with morphological observations, pointed to the axon as the site of initiation; yet true autaptic transmission could be excluded by the absence of the morphological specializations of autapses. The main functional distinction between this conductance and the $GABA_B$-mediated autoreceptor action at the nerve terminal, besides the receptor subtype, is its distributed nature, which would permit this effect to regulate subsequent action potential conduction along the axons. Thus, whereas the $GABA_B$ autoreceptor action regulates release from the releasing terminal (and a few other terminals in the neighborhood), the distributed axonal conductance could, by preventing propagation of the action potential, prevent release from all downstream synapses. The suggested effect is similar to the preterminal axonal conduction block in the spinal cord (Sect. F.I). Although this mechanism of presynaptic autoregulation has not been shown to occur naturally, it appears capable of making a significant contribution to control of inhibition.

G. $GABA_B$ Responses

I. Postsynaptic Inhibition

$GABA_B$ receptors are found at pre- and postsynaptic sites. There is general agreement that postsynaptic $GABA_B$ receptors activate a pertussis-toxin-sensitive G-protein coupled to an inwardly rectifying K channel (GIRK) (NEWBERRY and NICOLL 1984a,b; GAHWILER and BROWN 1985; ANDRADE et al. 1986; MISGELD et al. 1995). When the channel is opened, the membrane is hyperpolarized and the cell is inhibited. The GIRK, which can be blocked by extracellular Ba^{2+} ions, can be activated by other G-protein-coupled receptors, as well as $GABA_B$, including adenosine and 5-HT_{1a} (ANDRADE et al. 1986). Extracellular stimulation leads to a sequential $GABA_A$-$GABA_B$-mediated response.

An important issue is whether or not $GABA_A$ and $GABA_B$ responses can be produced by the same interneuron. The hippocampal CA1 inhibitory circuits involved in producing feedforward compound $GABA_A$ – $GABA_B$ IPSPs are clearly distinct from those producing solely recurrent $GABA_A$ IPSPs (ALGER and NICOLL 1982b; ALGER 1984; NEWBERRY and NICOLL 1984a). Basket cells had long been thought to mediate recurrent inhibition (ANDERSEN et al. 1964b), although other interneurons are now known to fulfill this role as well (FREUND and BUZSAKI 1996). A simple hypothesis is that at least two groups of interneurons are involved, one of which is incapable of producing $GABA_B$ IPSPs. Nevertheless, paired recordings from interneurons and pyramidal cells have not yet unambiguously identified any interneuron that produces $GABA_B$

IPSPs in pyramidal cells, even though gross applications of excitants such as glutamate or 4-AP to specific areas in CA1 do elicit $GABA_B$ responses (NURSE and LACAILLE 1997). In neocortex, micro-application of glutamate also produced fast ($GABA_A$) and slow ($GABA_B$) IPSPs, but these always appeared in isolation; mixed fast-slow IPSPs were not seen, and it was also suggested that the two responses were produced by separate classes of interneurons (BENARDO 1994). An alternative model is that GABA spillover from the synapse, or GABA release from several interneurons, would be necessary to induce $GABA_B$ responses via extrasynaptic receptors. It is not clear whether or not $GABA_B$ receptors are clustered in postsynaptic receptor patches, or are distributed more broadly in extrasynaptic regions. $GABA_B$ mIPSCs have not been reported (OTIS and MODY 1992), although spontaneously released GABA does affect $GABA_B$ receptors, as can be inferred from effects of $GABA_B$ antagonists on spontaneous neuronal activity (MCLEAN et al. 1996b; OUARDOUZ and LACAILLE 1997). Recent reports of cloning (KAUPMANN et al. 1997) and expression (JONES et al. 1998; KAUPMANN et al. 1998; WHITE et al. 1998) of $GABA_B$ receptors will play a major role in addressing these important issues.

The magnitude of the $GABA_B$ response is sometimes very small (e.g., PITLER and ALGER 1994b), and this may be the result of using Cl^- salts in the recording electrode. Whole-cell pipettes containing salts of methylsulfonate or gluconate permitted full-sized baclofen- or serotonin-mediated responses, whereas when Cl^- was the predominant anion, these responses, and the synaptically evoked $GABA_B$ IPSC, were very significantly reduced (LENZ et al. 1997). The Cl^- effect had a hyperbolic dose-response curve with an EC_{50} of about 40 mmol/l Cl^-. The effect was exerted on the K channel or perhaps the G protein, as membrane responses to intracellular GTPγS, which typically produce a hyperpolarized membrane potential and decreased neuronal input resistance (ANDRADE et al. 1986), were also significantly reduced by high internal $[Cl^-]$.

Although the biophysical basis of the inhibitory action of Cl^- is not known, the finding may be of physiological relevance, as during spreading depression (LUX et al. 1986) $[Cl^-]_i$ concentrations rise greatly. Moreover, during development $[Cl^-]$ in many neurons is elevated because of the different expression of the K^+/Cl^- transporter in young tissue (ZHANG et al. 1991; RIVERA et al. 1999). The lack of $GABA_B$-mediated responses early in development has been noted (LUHMANN and PRINCE 1991; GAIARSA et al. 1995), although often attributed to the lack of $GABA_B$ receptors. The absence of PPD (i.e., presynaptic $GABA_B$ function) in neonatal hippocampal slices may be the result of too little of the released GABA accessing the $GABA_B$ receptors (CAILLARD et al. 1998). Perhaps the higher $[Cl^-]_i$ in young neurons plays a role in the apparent absence of postsynaptic $GABA_B$ responses as well. Most intriguingly, LOPANTSEV and SCHWARTZKROIN (1999) have recently found that the synaptically activated, evoked $GABA_B$ IPSP is modulated by the preceding $GABA_A$ IPSP. Evidently the increase in intracellular Cl^- concentration induced by the $GABA_A$ IPSP is sufficient to affect the $GABA_B$ response. The close temporal coupling between these conductances suggests interesting possibilities for postsynaptic interactions.

II. Presynaptic Inhibition

Presynaptic inhibition in the spinal cord, first correctly identified by FRANK and FUORTES in 1957 (see NICOLL and ALGER 1979 for review), was associated with primary afferent depolarization (PAD) produced in one dorsal root by prior stimulation of other nearby dorsal roots. Although the mechanism of PAD is complex, GABA is involved. The high internal [Cl^-] in these fibers causes activation of the $GABA_A$ receptors to depolarize the terminals and reduce release. Although axonal conduction block (see Sect. F.I) can be seen as a kind of presynaptic inhibition, and, in the cases discussed, is mediated by $GABA_A$ receptors, most GABA-mediated presynaptic inhibition occurs through activation of $GABA_B$ receptors. This topic has been reviewed (THOMPSON et al. 1993; THOMPSON 1994; WU and SAGGAU 1997), and many of the major principles are well established. Interestingly, glycine seems to act only at postsynaptic sites and not to mediate presynaptic inhibition.

1. $GABA_B$ Autoreceptor Activation

The role of $GABA_B$ autoreceptors on GABAergic nerve terminals in controlling $GABA_A$ IPSCs is well established (THOMPSON and GAHWILER 1989; DAVIES et al. 1990; MOTT and LEWIS 1994; THOMPSON 1994; MISGELD et al. 1995). GABA released by an action potential activates presynaptic autoreceptors and reduces release caused by subsequent action potentials. The suppression of IPSCs mediated by this form of autoreception can be critically important for the induction of LTP (DAVIES et al. 1991; MOTT and LEWIS 1991); the NMDA component of the EPSP that is normally suppressed by the $GABA_A$ responses is disinhibited by the $GABA_B$ action (MOTT and LEWIS 1991; DAVIES and COLLINGRIDGE 1996).

Although presynaptic $GABA_B$ receptors exist on both inhibitory and excitatory nerve terminals in hippocampus, defined axo-axonic synapses have not been described in the brain, despite being prevalent in brain stem and spinal cord. In the brain, autoreceptors on the inhibitory terminals are activated by GABA, which is released synaptically at the nerve terminal (DAVIES et al. 1990) (see THOMPSON 1994 for review). Paired-pulse stimulation of monosynaptic $GABA_A$ IPSCs (i.e., IPSCs evoked in the presence of blockers of fast ionotropic glutamate receptors, CNQX and APV) reveals a significant depression of the amplitude of the second pulse when compared to the first. Most PPD of $GABA_A$ IPSCs is blocked by $GABA_B$ receptor antagonists, confirming the major prediction of the autoreceptor model. (Not all GABAergic terminals have $GABA_B$ receptors (LAMBERT and WILSON 1993b; PEARCE et al. 1995), and paired-pulse depletion of neurotransmitters evidently accounts for residual PPD when $GABA_B$ receptors are blocked.) The $GABA_B$ receptors on excitatory terminals ("heteroreceptors") are activated by synaptic GABA spillover from the synaptic cleft to nearby glutamatergic axons (ISAACSON et al. 1993). Similar heteroreceptors are present on nerve terminals from which other neurotransmitters are released, e.g., dopaminergic, noradrenergic, serotonergic, etc. (VIZI and KISS 1998).

2. Mechanism of Presynaptic $GABA_B$ Inhibition

In contrast to the numerous unequivocal demonstrations that $GABA_B$ receptor activation does have presynaptic inhibitory effects, it has been difficult to elucidate the actual mechanism of presynaptic inhibition. An obvious possibility is that the GIRK channels, if coupled to the presynaptic receptors, could shunt action potentials and prevent propagation to the terminals. Activation of $GABA_B$ receptors can also inhibit voltage-dependent Ca^{2+} currents in a variety of neurons (DOZE et al. 1995; WU and SAGGAU 1997), and so could inhibit release by preventing Ca^{2+} influx into the terminal. Attempts have been made to distinguish pre- from postsynaptic mechanisms. Initial reports of differences in pertussis toxin sensitivity and antagonist blockade between pre- and postsynaptic inhibition, which supported a distinction between the pre- and postsynaptic receptor types, were questioned because of the possibility that receptor-effector coupling, or "receptor reserves," might account for the differences (DUTAR and NICOLL 1988; YOON and ROTHMAN 1991). Barium ions, which block GIRK channels and postsynaptic $GABA_B$ effects (NEWBERRY and NICOLL 1985), significantly reduced presynaptic baclofen actions on IPSPs in the CA3 region of organotypic hippocampal slices (THOMPSON and GAHWILER 1992). However, in the CA1 region of acute slices Ba^{2+}, which dramatically reduced postsynaptic $GABA_B$ responses, had only slight effects on presynaptic baclofen effects on inhibitory nerve terminals or on PPD (LAMBERT et al. 1991; PITLER and ALGER 1994b; ROHRBACHER et al. 1997). Some of the disparate data could be explained by the use of bath-applied baclofen to activate presynaptic $GABA_B$ receptors, because this does not distinguish between presynaptic $GABA_B$ receptors directly involved in regulating release and $GABA_B$ receptors located at other presynaptic sites. Paired-pulse depression mediated by $GABA_B$ autoreceptors is the ideal assay for the physiologically relevant receptors controlling release, and neither Ba^{2+} nor phorbol ester (PITLER and ALGER 1994b) nor tetrahydroaminoacridine (THA) (LAMBERT and WILSON 1993a) had any effect on PPD. All three agents affect mIPSC release and postsynaptic channels, and hence differences in access to receptors on postsynaptic, vs presynaptic, sites cannot explain the data. The most likely conclusion is that the presynaptic and postsynaptic $GABA_B$ effects are mediated by different effector mechanisms. Indeed, postsynaptic outward currents mediated by transmitters that activate G-protein-coupled receptors, including baclofen, are absent in transgenic mice lacking GIRK2 (LUSCHER et al. 1997), whereas presynaptic inhibition by the bath-applied transmitters is unaffected in these mutants. Thus, it is clear that the GIRK2 channel is coupled only to the postsynaptic $GABA_B$ receptor. The same receptor may mediate different cellular actions based on effector coupling and subcellular localization. It remains possible that pre- and postsynaptic $GABA_B$ receptors represent different subtypes.

$GABA_B$ receptors inhibit voltage-dependent Ca^{2+} currents in a variety of cell types, including hippocampus, and, hence, probably induce presynaptic inhibition at least partly in this way (DOZE et al. 1995; WU and SAGGAU 1995, 1997). However, block of Ca^{2+} influx clearly cannot fully account for the presynaptic effects of baclofen as spontaneous miniature excitatory postsynaptic potentials (mEPSCs), which are insensitive to block of voltage-sensitive Ca^{2+} channels by Cd^{2+}, are nevertheless inhibited by baclofen (SCANZIANI et al. 1992). In rat midbrain culture, $GABA_B$ receptors inhibit TTX-, Ba^{2+}-, and Cd^{2+}-sensitive mIPSC release (ROHRBACHER et al. 1997). In view of possible differences in spontaneous and evoked release, it is important that baclofen also blocks release evoked by the secretagogues gadolinium, ionomycin, and α-latrotoxin that is independent of Ca^{2+} influx through voltage-gated Ca^{2+} channels (CAPOGNA et al. 1996). The effects on α-latrotoxin-induced release were especially important as α-latrotoxin may act downstream of all Ca^{2+}-requiring steps. Recent evidence of a direct interference in the exocytotic process by baclofen may lead to understanding the mechanism of Ca^{2+}-independent presynaptic inhibition (ISAACSON and HILLE 1997).

III. $GABA_B$ Enhancement of Synaptic Activity

Paradoxically, presynaptic $GABA_B$ receptors can also enhance the efficacy of synaptic transmission (BRENOWITZ et al. 1998). Cells in the nucleus magnocellularis (nMAG) are activated by glutamatergic synapses from the auditory nerve and receive a GABAergic projection from the superior olive. The nMAG cells receive EPSCs that occur at frequencies up to several hundred Hz. Typically at these frequencies the EPSCs undergo marked depression, evidently because of synaptic depletion or receptor desensitization. Baclofen reduces excitatory transmission by acting on the presynaptic receptors.

BRENOWITZ et al. (1998) found that the reduction in transmission caused by baclofen depended on the frequency of occurrence of EPSCs; when they were evoked at frequencies <100 Hz, all of the EPSCs (in a train of 10) were reduced. When the EPSC frequency was >200 Hz in the presence of baclofen, the EPSCs in the train, after the first 2–3, were actually larger than corresponding EPSCs in the absence of baclofen. By reducing excitatory transmitter release, and thus frequency-dependent synaptic depression, baclofen caused a relative enhancement of transmission during the train. There was little difference in the depression during a train in baclofen whether the train was elicited at 20 Hz or 500 Hz, whereas without baclofen much greater depression occurred during the high-frequency train. A decrease in the probability of release appeared to be responsible, because the results were mimicked by reducing initial transmission with a low-Ca^{2+}/high-Mg^{2+} solution to a degree similar to that caused by baclofen. Baclofen also prevented the delay in population spike peaks that occurred during a train of stimuli and thereby also maintained the timing of action potentials during the train. In this case, presynaptic activation of the $GABA_B$ receptor has the apparently paradoxical effect of preserving a state of excitation. These experiments reinforce the concept that it may be difficult to assign fixed labels such as inhibitory or excitatory to particular neurotransmitters. Their actions within a circuit are dependent on the context in which they act.

H. Response Plasticity and IPSPs

At least two major issues must be considered:

1. The role of inhibition in regulating plasticity of other synapses
2. The plasticity of the inhibitory synapses themselves

As will be evident, an important emerging issue is whether the GABAergic cell under study is a principal neuron, e.g., the cerebellar Purkinje cell, or an interneuron. In many ways GABAergic principal cells resemble excitatory principal neurons more than they do GABAergic inhibitory interneurons in their capacity for undergoing response plasticity.

Long-term induction of response plasticities is usually dependent on a rise in internal Ca^{2+} in the postsynaptic cell. Except in the case of depolarizing GABA responses (McLean et al. 1996a), activation of a GABAergic or glycinergic synapse would not be expected to increase postsynaptic Ca^{2+}, so an important question in understanding plasticity at inhibitory synapses is what is the origin of the necessary Ca^{2+}. Multiple answers to this question are possible. Co-activation of NMDA receptors, high- and low-voltage-activated Ca^{2+} channels, as well as IP_3-dependent release from intracellular stores, all appear to play a role in different cases.

IPSP plasticity may be involved in “homeostatic plasticity” (Turrigiano 1999), i.e., those non-Hebbian changes in synaptic strength that occur within a network that tend to maintain cell firing rates within a given range, while preserving disparities in individual synaptic weights. The neurotrophic factor, BDNF (brain-derived neurotrophic factor), represents an example of a possible homeostatic regulator. BDNF release itself is activity dependent, and, once released, BDNF reduces excitatory synaptic strengths while increasing inhibitory synaptic strengths, thus reducing the heightened excitability and its own release.

I. Short-Term Plasticity of Interneuron Output

Output of GABA interneurons is typically reduced for a short period after repetitive stimulation. Numerous mechanisms (reviewed in Alger 1991;

Stelzer 1992; Thompson 1994) of short-term IPSP plasticity have been discovered. They include shifts in E_{IPSP}, presynaptic inhibition via $GABA_B$ autoreceptor activation, receptor desensitization, NMDA-dependent $GABA_A$ receptor down-regulation, and transmitter depletion. Usually these factors reduce the strength of inhibition and render the population of affected principal cells transiently more excitable. Often use dependent, these factors cause graded decreases in inhibition and thereby "gate" various forms of excitatory processes (Alger 1991; Thompson 1994; Bear and Abraham 1996). Conversely, short-term potentiation of applied glycine responses in rat sacral dorsal commissural nucleus neurons, which is mediated by a Ca^{2+}-permeable type of AMPA receptor (Xu et al. 1999), transiently enhances inhibition.

Paired-pulse stimulation of monosynaptic IPSPs typically causes a marked depression of the second response when the interstimulus interval is 20–2000 ms (Davies et al. 1990). Under conditions of low release, e.g., when, due to stochastic processes, the first response of the pair happens to be small, PPD is reduced and may turn into PPF. Similarly, when release is reduced by substitution of extracellular Ca^{2+} by Sr^{2+} (Morishita and Alger 1997), paired-pulse stimulation elicits PPF instead of PPD. Nevertheless, in most of these studies depression of inhibition was seen in response to extracellular stimulation which activates surrounding tissues as well as the interneuronal axon. Stimu-

Fig. 6a–e. Frequency-dependent depression of unitary excitatory and inhibitory synaptic connections. **a** Depression of PSCs in response to sustained activation at 20 Hz (1000 action potentials) in three types of unitary synaptic connections: pyramidal neuron to pyramidal neuron (P→P, $n = 7$), pyramidal neuron to fast-spiking neuron (P→FS, $n = 11$), and fast-spiking neuron to pyramidal neuron (FS→P, $n = 7$). Results are presented as percentage of the PSC amplitude during the baseline period (0.25 Hz). Each symbol represents the average of 20 consecutive PSCs. Note similar depression during the transient period in the three types of connection, but smaller depression of inhibitory PSCs during the steady-state period. *Symbol code in this panel* applies to entire figure. **b** Time course of the recovery of the PSC amplitude after switching back to baseline frequency (0.25 Hz) following 1000 action potentials at 20 Hz (see **a**). *Lines* represent fits with single-exponential functions to average values of individual PSCs from 6 P→P (τ = 12.1 s), 8 P→FS (τ = 12.7 s), and 7 FS→P (τ = 4.3 s) synaptic connections. **c** The initial decline in the PSC when the frequency of synaptic stimulation increased to 20 Hz was studied with brief trains of 20 action potentials. Baseline was obtained at 0.25 Hz. Data from the three types of unitary synaptic connections, P→P ($n = 3$), P→FS ($n = 3$), and FS→P ($n = 3$), are superimposed. Symbols represent the average response of individual PSCs after 15 to 25 repetitions of the same protocol. **d** Experiments similar to those described in **a** were done over a range of presynaptic action potential frequencies (5–40 Hz). Transient PSCs, defined as the average amplitude of the first 50 unitary PSCs (see corresponding *line* in **a**), were not significantly different at any frequency among the three types of unitary synaptic connections. **e** Steady-state PSCs, defined as the average amplitude of the 800th to the 1000th responses (see *line* in **a**), showed statistically significant differences between inhibitory and excitatory synaptic connections at 10 Hz and 20 Hz (same symbol code as in **a**). Data in **d** and **e** were obtained from a total of 12 P→FS, 8 P→P, and 7 FS→P synaptic connections. Each symbol represents the mean of 3–11 experiments. (Reproduced from Galarreta and Hestrin 1998, with permission)

lation of individual, visually identified interneurons in hippocampal CA1 induces unitary IPSCs in synaptically coupled pyramidal cells (CARMANT et al. 1997). The IPSCs show little or no PPD, suggesting that some of this plasticity may be a function of coactivation of other cells in the preparation, which could cause greater liberation of GABA and hence greater activation of $GABA_B$ autoreceptors.

II. Balance Between Excitation and Inhibition

In general, the balance of excitation and inhibition is a critical parameter for normal system function. As noted above, many factors decrease the strength of inhibition. However, as too great a decrease in inhibition leads to pathological hyperexcitability (see MELDRUM and WHITING, chap. 6, this volume), the question arises how appropriate balance between the two is maintained in

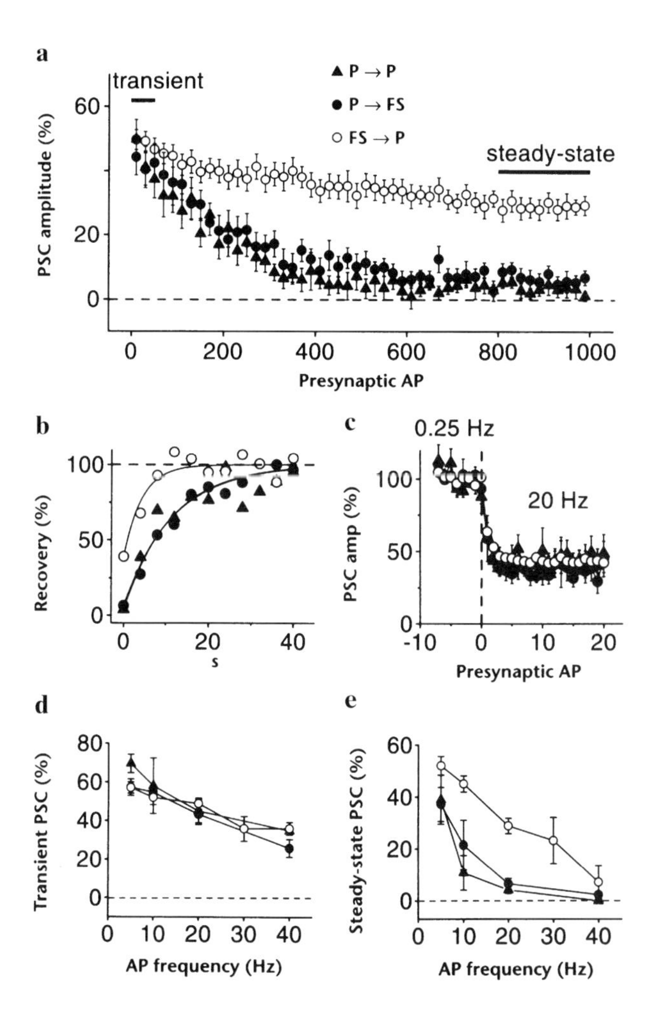

the face of sustained neuronal activation. One answer, in the neocortex, is that repetitive activation causes a greater depression of excitatory, than of inhibitory, synaptic currents, and prevents imbalance towards excitation (Galarreta and Hestrin 1998; Varela et al. 1999). Recordings of synaptically coupled pairs of pyramidal cells and of coupled pyramidal-cell-interneuron pairs in layer I as well as layer II/III showed that the monosynaptically recorded unitary EPSCs and IPSCs induced by prolonged intracellular stimulation of the presynaptic cell differed notably in their susceptibility to depression. Not only did EPSCs depress to a greater extent than IPSCs, but IPSCs recovered from depression much faster (see Fig. 6). Differences in depression between EPSCs and IPSCs have also been detected in the rapid depression that occurs with brief stimulus trains (Varela et al. 1999). Thus, electrophysiological stability can be maintained because of the different properties of inhibitory and excitatory synapses.

This relationship could differ from place to place in the brain, however, as the innervation of interneurons varies. Whereas the amplitude of evoked $GABA_A$ IPSCs in neocortex quickly increases to a maximum with increases in stimulus strength, amplitudes of evoked EPSCs do not (Ling and Benardo 1995). In the neocortex, IPSCs were activated exclusively via non-NMDA receptor activation, whereas in the hippocampus some IPSCs can be evoked by both NMDA- and non-NMDA-dependent mechanisms (Freund and Buzsaki 1996). Again, there may be regional variability in seemingly basic properties.

III. The Roles of IPSPs in Regulating Plasticity at Excitatory Synapses

1. LTD of $GABA_A$ergic IPSPs in Hippocampus

There is a long and controversial history of the role of inhibition in LTP, the lasting change in excitability thought to underlie learning and memory. In principle, a persistent reduction in IPSPs, in effect a long-term depression, LTD, of IPSPs, could be involved in LTP of excitatory systems. Various conditions cause long-lasting depression of $GABA_A$ IPSPs, including tetanic stimulation (in young guinea pig CA3 cells (Stelzer 1992)) and activation of mGluR following long-duration bath application of t-ACPD (Liu et al. 1993). In some studies the somatically recorded IPSPs did not change, or even increased, as a result of the LTP-inducing stimulation. When lasting plasticity of IPSPs in principal cells occurs, the first question is: where did the change occur? There are at least three classes of synapses to consider: the interneuron-principal-cell synapse, the excitatory synapses onto the interneurons, and other synapses in polysynaptic networks that innervate the interneurons.

In one study (Stelzer et al. 1994) IPSPs recorded in the presence of CNQX from CA1 pyramidal cell dendrites were persistently depressed by repetitive stimulation, while somatically recorded IPSPs showed no consistent change. The mechanism of the IPSP depression in this case was postsynaptic, i.e., involving a decrease in $GABA_A$ receptor responsiveness following NMDA receptor activation, because responses to iontophoretically applied GABA were also reduced. The challenge in this instance is to identify the factors underlying the selective sensitivity of dendritic $GABA_A$ receptors to down-regulation.

In other cases the actual site of the long-lasting modification was not the $GABA_A$ synapse. In CA3, repetitive bouts of low-frequency stimulation produced a lasting suppression of IPSP in CA3 pyramidal cells (MILES and WONG 1987), evidently because of an mGluR-mediated action on the interneurons (MILES and PONCER 1993). A stimulus train delivered to s. radiatum produced LTD of the s.-radiatum-evoked EPSCs in the interneurons. Even when the EPSC in the interneuron is suppressed, however, the actual synaptic locus of the LTD mechanism is not clear; it could either be at the EPSC synapse onto the interneuron or, as argued by MACCAFERRI and MCBAIN (1995), the effect could be "passively propagated" via the pyramidal cell to the interneuron. That is, the LTD could actually be expressed at the input to the pyramidal cells, which, in turn, activate the interneurons.

While passive propagation can readily account for feedback or recurrent inhibition, it cannot explain LTD of feedforward activation of interneuronal IPSPs. In some CA3 interneurons induction of LTD of one specific excitatory input could be established (MCMAHON and KAUER 1997). Interestingly, this LTD generalized to other non-stimulated excitatory inputs on the same cells, a finding that could be explained by a postsynaptic model whereby the LTD induction process induced at one set of synapses on the interneuron caused a widespread depression of excitatory synapses on the cell. This in turn led to a depressed output from the cell. The mechanism by which this novel form of LTD (iLTD) occurs is not clear, but is unlike those producing LTD of pyramidal cell inputs.

In the absence of evidence (see Sect. H.IV) that the same IPSPs can undergo persistent enhancement, it may be difficult to integrate persistent IPSP depression into network models, because the inhibitory synapses would tend to accumulate in the depressed state, leading to unbalanced excitation. In general, there does not seem to be widespread support for the proposition that persistent IPSP suppression, specific to interneuron inputs or outputs, accompanies LTP expression in hippocampus. As most LTP studies are done in the presence of $GABA_A$ antagonists, it is clear that the glutamatergic synapse is the primary site of LTP expression in CA1. Nevertheless, there are exceptions to this rule, and it may prove necessary to investigate each system of interest.

2. LTD of $GABA_A$ergic IPSPs in Cerebellum

Cerebellar Purkinje cells make monosynaptic inhibitory contacts with, among others, cells in the deep cerebellar nuclei (DCN). MORISHITA and SASTRY (1996) showed that tetanic stimulation of the Purkinje cell axons produced a long-

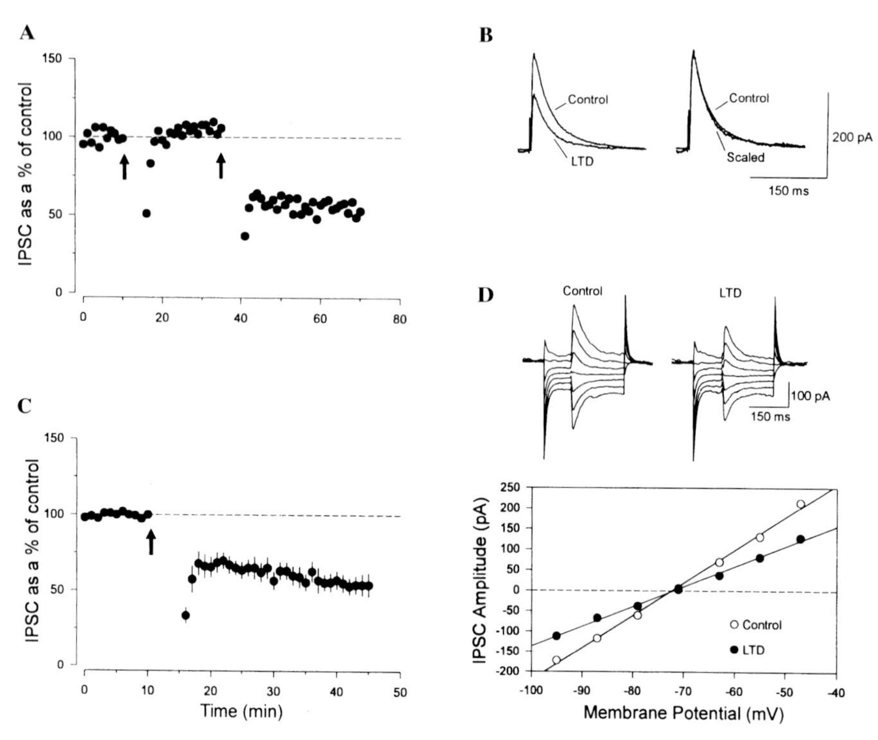

Fig. 7A–D. A 10-Hz stimulation delivered in current-clamp mode induces long-term depression (LTD) of deep nuclear inhibitory postsynaptic currents (IPSCs). **A** Graph shows that a 10-Hz, 5-min train (*arrow*) does not induce LTD when delivered in voltage-clamp mode; however, if the same stimulation is given in current-clamp mode, LTD occurs. **B** Average of 3 IPSCs recorded before (*control*) and 30 min after the 10-Hz stimulation (*LTD*) are superimposed to illustrate the magnitude of the sustained depression. Note when the amplitude of the depressed IPSC is scaled to match the amplitude of the control IPSC, there is no appreciable difference in their shape, indicating that there is little change in the kinetics of the IPSC during LTD. **C** Summary of 14 experiments illustrating the time course of the depression after the 10-Hz stimulation (*arrow*) in current-clamp mode. **D** Consecutive IPSCs evoked at various holding potentials before (*control*) and 30 min after the 10-Hz stimulation (*LTD*). The corresponding current-voltage *plot* is shown below the *traces*. The calculated reversal potential for the IPSC in control is 72.3 mV, whereas during LTD it is 72.9 mV. IPSCs were recorded with a nystatin-containing pipette solution. IPSCs in **B** were voltage clamped at –54 mV. (Reproduced from Morishita and Sastry 1996, with permission)

lasting depression of the IPSPs recorded in the DCN cells (see Fig. 7). The locus of the LTD_{IPSP} expression appeared to be postsynaptic as it was heterosynaptic and responses to iontophoretically applied GABA were also persistently depressed. The mechanism of LTD_{IPSP} resembled that of LTD_{EPSP} in its dependence on intracellular Ca^{2+} and Ca^{2+}-dependent phosphatase activity. Recent evidence (Aizenman et al. 1998) has extended the findings in the DCN cells by showing that Ca^{2+} entering these cells as a result of rebound depolarizations from preceding brief hyperpolarizations is responsible for its initiation. Rebound depolarizations are reliably triggered by a high-frequency train of IPSPs. Manipulations producing modest rebound firing produced LTD_{EPSP}, while more vigorous stimulation produced LTP_{IPSP}.

IV. Long-Lasting Enhancement of $GABA_A$ IPSPs

1. LTP of $GABA_A$ergic IPSPs

In the adult hippocampus, with some exceptions, there is little evidence for plasticity of $GABA_A$ synapses under normal conditions. Experiments to determine if excitatory synapses onto interneurons in hippocampus were potentiated produced equivocal results, perhaps because distinctions were not made between the input to the interneuron and other polysynaptic factors (McBain and Maccaferri 1997). In an excellent recent review McBain et al. (1999) discuss the morphological and neurochemical differences between the excitatory synapses on pyramidal cells and those on interneurons that mitigate against the LTP-expressing capability of the latter. Major factors include lack of spines (the small space promotes Ca^{2+} sequestration), spine apparatus, and differences between the glutamate receptors of the interneurons and those of pyramidal cells. The absence of the Ca^{2+}-dependent phosphatase calcineurin in interneurons may be partly responsible for a general lack of interneuron LTD, as calcineurin is an important mediator of pyramidal cell LTD (Mulkey et al. 1994). In hippocampus, excitatory synapses onto GABAergic cells differ from those onto pyramidal cells. Citron, a protein effector of the G-protein Rho, is found exclusively in GABAergic interneurons, where it binds the NMDA receptors in the postsynaptic density (Zhang et al. 1999). Another protein, p135 SynGAP, performs this function in pyramidal cells. CaMKII is present in pyramidal cells, but not in GABAergic interneurons (Sik et al. 1998). Indeed, direct measurements show that these synapses on interneurons do not undergo LTP. When LTP of IPSPs does occur it appears largely to be passively propagated from upstream sites onto the interneurons. That is, enhanced activation of cells that make excitatory synapses on the interneurons cause the evoked IPSP to become larger. The "pairing protocol" for LTP induction (Gustafsson et al. 1987), in which tetanic stimulation is not used and LTP induction is confined to the single postsynaptic cell being studied, is useful for distinguishing between an NMDA-dependent effect on the synapses on the interneuron and others elsewhere in the circuit. McBain et al. (1999) offer the interpretation that, inasmuch as interneurons often pace various rhythmic firing behaviors (see Sect. I, below), having reliable, relatively unmodifiable interconnections to the interneurons may serve this clocklike function best.

Nevertheless, there are exceptions to the rule that IPSPs provide a stable regulatory signal. Some excitatory synapses onto inhibitory interneurons do exhibit LTP. An NMDA-receptor-independent, postsynaptically induced form of LTP can be induced at synapses containing Ca^{2+}-permeable AMPA receptors on amygdalar interneurons (MAHANTY and SAH 1998). LTP of the inputs to the interneurons resulted in an enhanced disynaptic IPSC recorded from the amygdalar pyramidal cells, but the $GABA_A$ synapse itself did not change. The LTP was blocked by high intracellular concentrations of the Ca^{2+} chelator BAPTA in the interneuron, implying the induction process has a postsynaptic Ca^{2+}-dependent component, and yet Ca^{2+} influx through voltage-gated Ca^{2+} channels was insufficient to produce LTP. Interestingly, mossy fiber synapses onto s. lucidum interneurons in the hippocampal CA3 region do not express LTP (MACCAFERRI et al. 1998) despite the fact that these same afferents express a presynaptic form of LTP on the CA3 pyramidal cells. Thus, although the induction and expression of LTP appear to be presynaptic at mossy fiber synapses, the postsynaptic target nevertheless has some influence on the process. The assumption here is that induction and expression really are solely presynaptic. It is not known if the postsynaptic form of LTP observed in the amygdala (MAHANTY and SAH 1998) can be induced at the Ca^{2+}-permeable synapses on s. lucidum interneurons.

Bidirectional plasticity of GABAergic IPSPs occurs in neonatal rat hippocampus (MCLEAN et al. 1996a), with $LTD_{GABA\text{-}A}$ being NMDA-receptor dependent and $LTP_{GABA\text{-}A}$ NMDA-receptor independent. These results show that in the developing hippocampus, $GABA_A$-mediated responses are subject to long-lasting plasticity. However, inasmuch as at these early stages $GABA_A$ responses are actually excitatory, the results do not address the issue of plasticity of inhibition.

Actual LTP of IPSPs in developing tissue is seen in other parts of the brain. LTP of monosynaptic $GABA_A$ergic IPSPs in slices from layer V of young rat visual cortex appears to have a presynaptic origin and is not affected by changes in postsynaptic membrane potential or activation of NMDA receptors, although it has many phenomenological similarities to LTP of EPSPs (KOMATSU 1994, 1996). LTP_{IPSP} is dependent on postsynaptic Ca^{2+} and G proteins and is blocked by $GABA_B$ antagonists. LTP_{IPSP} is induced by coactivation of $GABA_B$ and either α-adrenoreceptor or 5-HT2 receptors, which causes Ca^{2+} release from intracellular stores and an as-yet-unknown biochemical process.

2. LTP of Glycinergic IPSPs

The goldfish Mauthner cell receives inputs from glycinergic inhibitory interneurons; however, in paired recordings from an interneuron and a Mauthner cell many anatomically well-defined synaptic contacts are found to be physiologically silent, i.e., small or no responses result from activating these synapses (CHARPIER et al. 1995). Tetanic stimulation of afferents to the silent inhibitory cells produces a dramatic and lasting appearance of robust IPSPs at the previously ineffectual synapses. The mechanism of this strengthening was presumed to be LTP-like and presynaptic in locus, because it could be mimicked by manipulation of intracellular processes that affect transmitter release in the interneuron. LTP of normal glycinergic synapses on the Mauthner cell was directly demonstrated in paired interneuron-Mauthner-cell recordings (ODA et al. 1995). LTP at these synapses may involve a retrograde messenger as it is blocked by postsynaptic Ca^{2+} chelation, but expressed as an increase in quantal release.

3. Long-Lasting Enhancement of IPSPs – Not LTP

Other types of lasting enhancements of IPSPs, probably not caused by traditional LTP mechanisms, have also been described. Both pre- and postsynaptic mechanisms are implicated. $GABA_A$ergic IPSPs recorded from the dorsomedial nucleus of the solitary tract in transverse medullary slices showed a sustained "tetanus-induced potentiation" (TIP) (GLAUM and BROOKS 1996) that resembled somewhat the LTP_{IPSP} in visual cortex: it is independent of NMDA receptor activation and dependent on the activation of $GABA_B$ receptors. TIP is a long-lasting, but not permanent, state of potentiation with a duration of ~45 min, i.e., resembling "early" rather than "late" LTP. Activation of the $GABA_A$ receptors during the tetanus was inessential – TIP was evident following bicuculline washout. A role for a presynaptic site of modification perhaps involving P/Q-type Ca^{2+} channels was suggested, but the issue of a role for postsynaptic Ca^{2+} was not addressed.

A long-lasting rebound potentiation of IPSPs is produced in cerebellar Purkinje cells following a brief tetanic stimulation to the climbing fiber axons, or a train of voltage pulses given in the cell soma (LLANO et al. 1991; KANO et al. 1992). The potentiation, which decayed with a slow time course, represented a Ca^{2+}-dependent up-regulation of $GABA_A$ receptors (as iontophoretic GABA responses were also increased) and was associated with a measured rise in intracellular Ca^{2+} concentration. Intracellular application of BAPTA prevented the response. Activation of CaMKII is thought to be responsible for the $GABA_A$ receptor up-regulation (KANO 1996).

Kindling is a lasting change in excitability produced by repeated, daily bouts of an initially subliminal stimulation that eventually causes full-blown seizures. Kindling is used as a model of an epileptic state (MCNAMARA et al. 1984; MCINTYRE and RACINE 1986). Because decreases in $GABA_A$ inhibition often cause epileptifom discharges, it is somewhat surprising that kindling in the dentate gyrus caused a potentiation of inhibitory responses (SHIN et al. 1985; OTIS et al. 1994). Quantitative immunogold receptor labeling revealed that both receptor density and total synaptic junction area increased, so the number of receptors activated by a quantum of $GABA_A$ increased (NUSSER et al. 1998).

V. Target-Cell Specificity of Action

The balance between excitation and inhibition may be modified by target-specific plasticity determined in part by the identity of the postsynaptic cells.

Interneurons differ markedly in the degree to which excitatory synapses on them facilitate in response to repetitive stimulation (ALI and THOMSON 1998). Although synaptic facilitation and depression are largely functions of the presynaptic excitatory terminals in this case, whether a given synapse facilitates or depresses seems to be under the control of the postsynaptic (GABAergic) interneurons. Evidence for this was provided by simultaneous triple recordings from a neocortical pyramidal cell and two different classes of interneurons (REYES et al. 1998). The synapses onto bitufted cells facilitated, while the synapses onto multipolar interneurons depressed. Because the presynaptic cell provided both types of nerve terminals, their physiological difference appeared to be determined by a retrograde signal from the interneuron.

Presynaptic, long-term plasticities can also be controlled by the postsynaptic interneurons. Mossy fibers in CA3 contact both pyramidal cells and interneurons. However, whereas repetitive stimulation induced LTP at the pyramidal cell synapses, no change, or long-term depression (LTD), was simultaneously induced at the interneuron synapses (TOTH and MCBAIN 1998). Interestingly, although the postsynaptic receptors on these two cell types differ (see Sect. B.I.2), the induction of LTP rather than LTD was a function of the different properties of the presynaptic terminals. Glutamate release from terminals on pyramidal cells was influenced by cAMP-dependent processes and was enhanced by forskolin, whereas forskolin had no effect on the EPSCs produced in the interneurons. Thus, basic properties of presynaptic terminals of a given input pathway are coordinated with the nature of the postsynaptic cell. In this case, the excitatory connection from dentate gyrus to CA3 will be enhanced, and the inhibitory connection weakened by repetitive activation of granule cells. The computational properties of the system will be correspondingly altered by appropriate afferent input.

VI. Facilitation of LTD Induction at Other Synapses by IPSP Depression

It is clear that $GABA_A$ IPSPs can regulate the expression of NMDA-dependent plasticities in the hippocampus. $GABA_A$ antagonists facilitate LTP and LTD induction by disinhibiting NMDA responses (ABRAHAM and WICKENS 1991; TOMASULO et al. 1993; ZHANG and LEVY 1993; BEAR and ABRAHAM 1996). However, for IPSP depression to have this effect, lasting suppression of IPSPs is not required. Rather the IPSPs need to be suppressed only long enough to permit Ca^{2+} influx into the cells through the NMDA channels. Typically the induction period is brief. Thus the short forms of IPSP depression that have been described are especially important in regulating long-term response plasticity.

Some forms of LTD, like some forms of LTP, involve NMDA receptor stimulation. In CA1, LTD produced by 1-Hz stimulation given for 15 min is much more prominent in young hippocampal tissue (10–21 days) than it is in adult, ≥35-day hippocampus, where it is either much reduced (DUDEK and

BEAR 1992; DUDEK and FRIEDLANDER 1996) or absent (WAGNER and ALGER 1995). WAGNER and ALGER (1995) showed that pharmacological antagonism of IPSPs facilitates the induction of NMDA-dependent LTD of excitatory transmission in adult, although not juvenile, animals (but, cf. THIELS et al. 1994). The difference in susceptibility to LTD induction appeared to be related to a developmental difference in the maturation of inhibition. Evidently a more potent inhibitory influence is maintained during the stimuli in adult slices than is maintained in younger tissue, and, by weakening $GABA_A$ergic inhibition, bicuculline rendered the adult tissue capable of evincing LTD. These results do not depend on the resolution of the issue of LTD of IPSPs, but rather on the strength of inhibition during the LTD-inducing stimulus train. Interestingly, adult and juvenile slices alike were susceptible to "depotentiation," an LTD-like effect that removes a previously established LTP (STAUBLI and LYNCH 1990). This finding would be compatible with the concept that the LTP-inducing stimulation somehow weakens inhibition in a way that is not always detectable with a somatic electrode (WAGNER and ALGER 1996a).

I. Synaptic Inhibition and the Generation of Rhythmic Firing Patterns in Populations of Cells

Long suspected on the basis of morphological and immunohistochemical data, paired electrophysiological recordings have confirmed that interneurons synapse onto other interneurons. While this clearly paves the way for disinhibitory effects on principal cells, as envisioned by ROBERTS (1991) and KRNJEVIC (1981), further consideration of the interconnectivity among groups of interneurons has deepened the complications. This topic is well discussed in FREUND and BUZSAKI (1996), and it suffices to mention here that, depending on the complexity of these interconnections, straightforward principal-cell disinhibition may be only one of a set of possible outcomes.

The role of synaptic inhibition in generating the rhythmic waves recorded in the thalamocortical system and hippocampus was recognized by the late 1950s and early 1960s (ECCLES 1964). Feedback inhibition via the recurrent inhibitory circuits that had been discovered in the spinal cord and various brain regions appeared to provide an ideal substrate for rhythm generation. Excitation of the principal cells would be cut off by the recurrent IPSP which would itself cease as the principal cell firing stopped, permitting excitation to rise again.

Hyperpolarizing inhibition can synchronize principal cell firing by imposing periodic membrane potential fluctuations, which control the timing of action potential generation, on cells. Different patterns of rhythmic activity, including theta (4–12 Hz), gamma (30–100 Hz) and fast (>200 Hz) oscillations, involving the synchronous firing of principal neurons and interneurons, subserve many functions in the developing and adult CNS (for reviews see CHERUBINI et al. 1991; SINGER and GRAY 1995). Cortical interneuron networks may generate both slow and fast cortical oscillatory activity (e.g., Whittington et al. 1995, 1997; Buhl et al. 1998; Fisahn et al. 1998; Penttonen 1998; Rinzel et al. 1998; Zhang et al. 1998). Similarly, inhibitory neurons of the thalamic reticular and perigeniculate nuclei generate the synchronized activity of thalamocortical networks (McCormick and Bal 1997). Gamma oscillations (30–100Hz) occur in various brain structures and several different species (Singer and Gray 1995; Laurent 1996). Synchronous cortical gamma oscillations can occur over large distances and could, therefore, provide a substrate for "binding" together spatially separated areas of cortex, a hypothetical process whereby disparate aspects of a complex object, for example, are combined to form a unitary perception of it (Traub et al. 1996).

I. Gamma Oscillations

Gamma activity is especially evident in the hippocampus and entorhinal cortex, and gamma oscillations recorded in vivo occur synchronously in each subdivision of the hippocampus (Bragin et al. 1995). In vitro models of gamma activity in the hippocampus and somatosensory cortex exist (Whittington et al. 1995; Buhl et al. 1998; Fisahn et al. 1998). Inhibitory interneurons appear to play a critical role in all cases. Gamma oscillations in CA1 pyramidal cells depend on metabotropic glutamate receptor activation and can occur in the absence of fast excitatory transmission (Whittington et al. 1995). The oscillations can be blocked by bicuculline, suggesting that they are produced within an interneuron network and then entrain pyramidal cell firing. Although some gamma oscillations persist in the presence of ionotropic glutamate receptor blockers, these oscillations are spatially restricted, with a maximum range of 1.2mm (Whittington et al. 1995). Longer-range synchrony arises when, as would be expected to occur under more physiological conditions, pyramidal cells participate in the gamma oscillations (Traub et al. 1996). An important problem in understanding long-range synchrony is how coherence is established over distances sufficient to involve significant delays caused by axonal conduction time. The model of Traub et al. (1996) proposes that, when interneurons fire doublets, rather than single, spikes, coherent long-range synchrony is established over many millimeters. Experimental observations of interneuron firing patterns support the model. When higher-intensity stimulation is used, a switch from gamma to beta (10–25Hz) rhythms occurs, and this is associated with a decrease in gamma frequencies.

On the other hand, in the hippocampal CA3 region muscarinic cholinergic activation causes gamma oscillatory activity which is completely blocked by bicuculline (Fisahn et al. 1998), as well as by the non-NMDA receptor antagonist NBQX, thus implicating CA3 recurrent excitatory connections in this case. Although the mechanisms for their generation may vary, the frequency of the gamma oscillations is dependent on the magnitude of the unitary inhibitory postsynaptic conductance and its time course. Barbiturates, which prolong the decay of the IPSP (Nicoll et al. 1975), decrease the frequency of the oscillations (WHITTINGTON et al. 1995; FISHAHN et al. 1998; BUHL et al. 1998).

II. Theta Rhythms

Theta oscillations (4–7 Hz) are prominent in the rat hippocampus and are thought to be important in integrative and memory function (BLAND and COLOM 1993). During theta activity, rhythmically firing interneurons produce GABA-mediated fluctuations of the membrane potential of CA1 pyramidal cells (LEUNG and YIM 1986; FOX 1989). Rhythmic chloride-mediated conductances originate close to the cell body (FOX 1989; SOLTESZ et al. 1993). Intracellular recordings of hippocampal pyramidal cells and interneurons show that theta frequency is voltage independent, but that theta amplitude and phase are voltage dependent (YLINEN et al. 1995b). Complete phase reversal occurs at the Cl^- equilibrium potential, supporting the conclusion that rhythmic IPSPs contribute markedly to the generation of theta. COBB et al (1995), using paired intracellular recordings, showed that rhythmic activation of a presynaptic basket or axo-axonic interneuron at theta frequency instantly phase locked the spontaneous firing of the pyramidal cells in CA1 (see Fig. 8). Because GABAergic interneurons have extensive axonal arborizations, this synchronized inhibition could then be imposed onto a large population of principal neurons (DEKKER and PARKER 1994; COBB et al. 1995). In some cases theta can be produced by blocking $GABA_A$ and $GABA_B$ receptors (KONOPACKI et al. 1997), so other factors are also important.

III. Single-Unit Studies In Vivo

Isolation of single-unit firing from the hippocampi of behaving rats, using simultaneous recordings from multiple electrode arrays, has been reported (CSICSVARI et al. 1998). Interneuron action potentials could be distinguished from pyramidal cell action potentials. During rhythmic firing behaviors (sharp waves, or theta activity), synchronous firing of both pyramidal cells and interneurons occurred. Cross-correlational analysis revealed single pyramidal firing was directly coupled to interneuron firing, and complex spikes were more effective in driving the interneurons than were single spikes. The efficiency with which interneurons were driven varied as a function of the neuronal population activity.

Which interneurons are capable of synchronizing principal cell activity, and whether different interneuronal subpopulations are responsible for the generation of the different frequency patterns of activity, remain to be determined. Intracellular recordings from hippocampal basket cells in vivo showed that these cells, which innervate the perisomatic region of pyramidal cells, are capable of firing action potentials at gamma frequency in vivo (PENTTONEN 1998). In the dentate gyrus the firing of morphologically identified interneuronal types was phase-locked to gamma activity (SIK 1997).

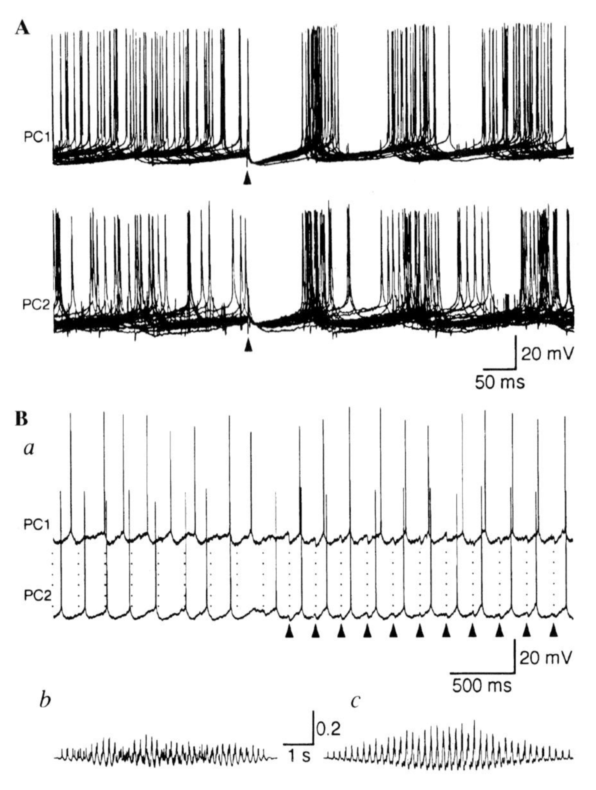

Fig. 8A,B. Synchronization of pyramidal cell (PC) firing in the presence of ionotropic glutamate-receptor antagonists. **A** Two simultaneously recorded pyramidal neurons were depolarized to elicit action potentials during which single IPSPs (*triangles*), evoked at 0.2–0.5 Hz by minimal stimulation, reset the regular firing of both cells (30 consecutive sweeps; $n = 5$). The stimulation strength was adjusted to evoke an IPSP of amplitude equivalent (<3 mV) to that produced by an individual, intracellularly recorded interneuron. In addition, rebound depolarization as in Fig. 2A,B could be evoked in both pyramidal cells (data not shown). **B** Rhythmic IPSPs (*a*) evoked by minimal stimulation at 5 Hz (*triangles*) synchronize the firing of two simultaneously recorded pyramidal neurons. *Dotted lines* indicate intervals of 0.2 s. Cross-correlogram (*b*) for the two neurons in a 5-s period before rhythmic minimal stimulation. Corresponding cross-correlogram (*c*) for the 5-s period following the start of rhythmic minimal stimulation. Note more pronounced cross-correlation during entrainment. (Reproduced from COBB et al. 1995, with permission)

IV. Thalamic Rhythms

In the thalamus, inhibition is involved in generating rhythmic oscillations that occur in non-rapid-eye-movement sleep (non-REM) (MCCORMICK and BAL 1997). Hyperpolarizing GABAergic IPSPs activate a depolarizing, mixed Na^+- and K^+-dependent current that turns on at negative membrane potentials (I_h) (see PAPE 1996 for review). The activation of I_h slowly depolarizes the neuron to threshold for activation of a low-threshold Ca^{2+} current (I_T) that causes a Ca^{2+} spike and a high-frequency burst of action potentials. The falling phase of the low-threshold Ca^{2+} spike helps activate I_h. Spindle waves in the thalamus are generated by an interaction between the GABAergic neurons of the thalamic reticular nucleus and the excitatory thalamic relay cells (see McCormick and Bal 1997 for review). Reticular cells evoke a barrage of IPSPs that activates I_h in relay neurons. When elicited at depolarized membrane potentials, the offset of the hyperpolarizing IPSP in thalamocortical cells is followed by a rebound low-threshold Ca^{2+} spike. When elicited at the hyperpolarized membrane potentials associated in these neurons with sleep, the IPSP is reversed to a relative depolarization and directly triggers the low-threshold spike and causes burst firing in the reticular nucleus cells (Bazhenov et al. 1999). Activation of I_h also leads to the initiation of low-threshold Ca^{2+} spikes and action potentials in the thalamic relay cells. These action potentials re-excite the reticular cells to which they are reciprocally connected, and so a rhythmic pattern of activity is established. Many cortical and hippocampal cells, both pyramidal cells and interneurons, exhibit I_h (Maccaferri and McBain 1996; Pape 1996), but whether this current is responsible for the generation of oscillatory activity in these structures is unclear.

Certain neurotransmitters simultaneously suppress evoked GABA release (Pitler and Alger 1992a; Behrends and ten Bruggencate 1993), while enhancing spontaneous action-potential-dependent release. While this apparent paradox is not yet resolved, it does suggest that these transmitters might, by shifting the mode of GABA release from pulsatile to tonic, also shift the function of GABA with in neuronal circuits. The population rhythmicity fostered by pulsatile release (Cobb et al. 1995) might be switched to desynchronized, irregular firing that is induced by tonic GABA release (Hausser and Clark 1997).

V. Depolarizing $GABA_A$ Responses and Rhythmic Firing

Depolarizing GABA responses have important physiological roles in immature and adult CNS; e.g., they can help relieve the voltage-dependent block of the NMDA channel by Mg^{2+} (Staley et al. 1995). Depolarizing GABA responses can give rise to synchronous excitatory activity. 4-AP induces slow potentials in the hippocampus and entorhinal cortex that are mediated by $GABA_A$ receptors (Perreault and Avoli 1989, 1992), and that persist in the absence of excitatory transmission (Perreault and Avoli 1989; Michelson and Wong 1991, 1994). These $GABA_A$-mediated potentials could be important in certain forms of epilepsy, as they are able to facilitate the onset of ictal discharges in the entorhinal cortex (Avoli et al. 1996). In thalamic reticular nucleus, depolarizing $GABA_A$ responses are capable of triggering low-threshold spikes which propagate and initiate sleep spindle oscillations (7–14 Hz) throughout the thalamocortical network (Bazhenov et al. 1999). Both cholinergic and GABAergic projections to the hippocampus originate in the medial septum. By recording from pairs of theta-related septohippocampal cells, Bland et al. (1999) studied cellular activity during transition into, and out of, hippocampal theta activity. Inhibition of key hippocampal cell types was critical in both transitions. Depolarizing $GABA_A$ responses activate other interneurons, giving rise to inhibitory effects on principal cells. In the hippocampus, depolarizing $GABA_A$-mediated events synchronously activate a population of interneurons, which in turn causes large-amplitude IPSPs in the pyramidal cells (Michelson and Wong 1991, 1994).

VI. Hypersynchrony and Pathology

While synchronized cortical network rhythms are thought to subserve normal physiological functions, including sensory processing (Gray and McCormick 1996), consciousness (Llinas and Ribrary 1993) and memory storage (Lisman and Idiart 1995), decreases in inhibition resulting in hypersynchronized activity occur in the pathological condition of epilepsy. In the glycinergic system, mutations that affect the glycine receptor are associated with inherited "startle" syndromes (Rajendra and Schofield 1995). These hyperexcitability reactions to sensory stimuli are thought to occur because of impaired glycine-mediated inhibition.

During epileptic activity a high degree of synchronized firing of populations of cortical principal cells leads to sharp waves or spikes in the EEG. Blockade of inhibitory function is a common approach for inducing epileptiform activity (see Meldrum and Whiting, chap. 6, this volume). Such large-scale reductions in inhibition are, however, unlikely to occur physiologically, and more subtle variations in inhibitory and excitatory strength are to be expected. Studies in the hippocampal slice preparation have shown that a reduction of inhibition allows latent recurrent excitatory connections in CA3 to become functional (Miles and Wong 1987). Under conditions of reduced excitation, activation of a few pyramidal cells can therefore entrain additional pyramidal cells within the hippocampus and subsequently drive neurons in other limbic structures from which epileptic activity can become widespread.

As well as providing an inhibitory input to the thalamic relay cells, thalamic reticular cells also provide inhibition within the reticular nucleus via axon collaterals. Recurrent inhibition may prevent hypersynchrony and generalized epilepsy (Huguenard and Prince 1994; Kim et al. 1997). Indeed, in mice devoid of the $\beta 3$ $GABA_A$ receptor subunit, $GABA_A$ responses within the reticular nucleus are dramatically reduced and, concomitantly, thalamic synchrony is greatly increased (Huntsman et al. 1999). Thus, reciprocal inhibitory connections can desynchronize as well as synchronize activity in neuronal populations. Muscarinic, cholinergic induction of synchronous epileptiform activity in principal cells of the entorhinal cortex involves rhythmic firing of principal cells and GABA interneurons (Dickson and Alonso 1997). Synchronous firing in interneuron networks was not abolished when fast excitatory transmission was pharmacologically blocked. Evidently, IPSPs can pace epileptic activity as it can other rhythmic firing. Through activation of nicotinic receptors, acetylcholine excites neocortical interneurons that fire low-threshold spikes and target pyramidal cell dendrites (Xiang et al. 1998). Acetylcholine inhibits fast-spiking interneurons that target pyramidal cell somata. Conceivably, this transmitter can direct the flow of information through cortical circuits by switching on and off interneuronal networks. A detailed understanding of the role of interneurons in the generation and maintenance of epileptiform activity is, however, lacking.

VII. Control of Rhythmic Firing Through Inhibition of Gap Junctional Connections

GABA regulates the pattern-generation properties in the olivocerebellar system (Lang et al. 1996). The cerebellar nuclei provide a major source of GABAergic input to the inferior olive. Disruption of the integrity of this transmission either by picrotoxin injection into the olive or chemical lesioning of the nuclei altered the rate, synchrony, and rhythmicity of complex spikes induced in Purkinje cells by the climbing fibers that originate in the inferior olive. The basis of rhythmic complex spiking was the gap-junction-mediated electrotonic coupling among olivary cells. Blockade of $GABA_A$ inhibition, by increasing the input resistance of the coupled cells, would increase the effective electrotonic coupling between cells, thus synchronizing their activity to a greater degree. Whereas simple inhibition would decrease firing rates without necessarily altering the degree of synchronous activity, an effect on synchrony, mediated via alterations in electrotonic coupling, would not necessarily be accompanied by changes in firing rates.

Electrical coupling among interneurons is also implicated in the generation of oscillatory patterns of activity. Fast (>200 Hz) oscillations in CA1 pyramidal cells depend on synaptic inhibition (Ylinen et al. 1995a). In the hippocampal slice preparation, however, fast oscillatory activity persists in the presence of antagonists of both excitatory and inhibitory transmission, but is abolished by gap junction blockers (Draguhn et al. 1998). Gap junctions and reciprocal connections among interneurons have also been proposed to be responsible for the slow (<1 Hz) activity recorded in CA1 pyramidal neurons (Zhang et al. 1998). In the molecular layer of the cerebellar cortex, synchronized activity between adjacent interneurons is mediated by electrotonic junctions; chemical transmission plays no role (Mann-Metzer and Yarom 1999). Electrical coupling currents initiated the synchronous action potential firing in coupled cells. However, prolonged voltage-dependent intrinsic currents, triggered by the action currents, widened the temporal window in which synchronized firing could occur. The coupling ratio between the cells varied with the input resistance of the postsynaptic cell and not with the coupling resistance, which was constant. The functional organization of the compact networks of interneurons that were revealed with dye injections would be subject to ready modulation by a variety of influences on the cells, and this in turn would alter the dynamics of the rhythmic firing patterns generated by the networks. In neocortex simultaneous recordings from pairs of interneurons revealed separate electrically coupled networks of fast-spiking (GALARETTA and HESTRIN 1999; GIBSON et al. 1999) or low-threshold-spiking (GIBSON et al. 1999) interneurons; pyramidal cells were not electrically coupled to these cells or to each other. The coupling coefficient, although modest (~0.1 for low sinusoidal current injection frequencies and less for higher frequencies), was nevertheless sufficient to promote synchronous firing in connected cells. The two networks of interneurons received separate synaptic inputs, which should foster their participation in distinct rhythmic activities. Interestingly, the incidence of "dye coupling" (the passage of dye molecules through gap junctional channels, a commonly used test for the presence of electrical connections) was rare, in contrast to the frequent occurrence (70%–80%) of electrical coupling detected in paired recordings (GIBSON et al. 1999). This suggested that novel, dye-impermeable gap junctions may be involved, and showed that the absence of dye coupling is not definitive evidence against electrical coupling.

J. The Role of Inhibition in Sensory Processing

A central role for inhibition in sensory processing has long been appreciated. In addition to reducing or blocking neuronal responses, inhibition shapes neuronal responses to specific stimuli (DYKES et al. 1984; SILLITO 1984; CROOK and EYSEL 1992) and governs the temporal response properties of sensory neurons (BUONOMANO and MERZENICH 1995, 1998). In the spinal cord and brainstem both GABAergic and glycinergic inhibition affect the response properties of sensory neurons, whereas, in the cerebral cortex, GABA determines the stimulus-specific responses and the receptive field properties of sensory neurons. In the cortex, GABA inhibition is also involved in the plasticity of receptive field properties and cortical topography which occur in somatic sensory, auditory, visual cortex and motor cortex (for reviews see KAAS 1991; SCHEICH 1991; SCHREINER 1992; GILBERT 1993; JONES 1993; WEINBERGER 1995; BUONOMANO and MERZENICH 1998).

I. Receptive Field Shape

A detailed discussion of inhibition and the shaping of stimulus-specific responses within the sensory pathways is beyond the scope of this review. Suffice it to say that inhibition shapes receptive fields in all sensory modalities, including orientation selectivity in the visual system (SILLITO 1984) and frequency tuning in the auditory system (SUGA et al. 1997). Extracellular application of bicuculline increases the size of receptive fields and reduces the sharpness of tuning (e.g., DYKES et al. 1984; SILLITO 1984). However, the precise role of inhibition in the generation of sensory neuronal responses remains controversial. NELSON et al. (1994) found that blocking inhibition in a single neuron in the visual cortex (with an intracellular perfusion of cesium fluoride, SITS, and picrotoxin) had only a very minor effect on that cell's orientation selectivity. Both simple and complex cells retained much of their orientation selectivity, suggesting that inhibitory synaptic inputs were not essential for this response characteristic. DOUGLAS et al. (1995) proposed a model of recurrent excitation within the cortex that could generate the receptive field properties of visual neurons. Blocking GABA currents within a single cell would have a limited effect as the majority of the excitatory inputs arise through cortical connections from other cells, whose orientation selectivity had not been modified. In this model inhibition served only to prevent "runaway excitation." Even when specific inhibitory inputs are important, they need not depend on a direct inhibitory action upon the principal cells, as disinhibition has also been proposed to play an important part in the integration of afferent inputs. In the somatosensory cortex, for example, GABAergic axons arising from the basal forebrain preferentially make synaptic contacts with GABAergic neurons which would, therefore, be expected to result in a powerful disinhibition of pyramidal cells (DYKES et al. 1984). Disinhibition is probably important in all sensory cortical areas as inhibitory inputs from the basal forebrain terminate on GABAergic cells throughout the neocortex (FREUND and MESKENAITE 1992). However, the final functional consequences of what would appear to be straightforward disinhibition depend on the complete details of the neural circuits involved.

An unresolved issue concerns the mechanism by which sensory neurons integrate their mixed excitatory and inhibitory inputs. Several studies addressed the issue of whether or not linear or nonlinear synaptic mechanisms are involved in sensory computations. Inhibition can suppress excitation as a result of the linear summation of excitatory and inhibitory currents onto the cell. Recordings from simple cells in the cat visual cortex (JAGADEESH et al. 1993) showed that the responses to moving stimuli in these cells could be predicted by the linear summation of their responses to stationary stimuli. Alternatively, other response properties may be generated by non-linear inhibitory mechanisms, such as shunting inhibition, which cause an increase in membrane conductance and reduce the amplitudes of the excitatory responses. Non-linear inhibitory mechanisms allow more complex sensory computations to occur. BORG-GRAHAM et al. (1998) have proposed that shunting inhibition is important for the generation of on/off opponency in visual cortical neurons.

II. Dynamic Modulation of Receptive Fields

1. Deafferentation Plasticity

Receptive field properties of neurons in adult mammals are not fixed and can be dynamically modulated following injury and also with learning, experience and stimulus conditions. It has been suggested that GABA is also involved in the plastic changes that underlie the physiologically observed reorganization of cortical topography that occurs in adults. Major reorganizations of cortical maps in the adult brain were initially demonstrated following either peripheral nerve damage, or amputation, in which areas of cortex lose their normal sensory inputs ("deafferentation plasticity"). More subtle reorganizations are related to changes in neuronal activity (e.g., SCHEICH 1991; BUONOMANO and MERZENICH 1998). These modifications, which result in a change in the size or shape of receptive fields, can occur immediately, within minutes, or arise on a much longer time scale, with changes developing over days, weeks, or months. Long-term changes in cortical representations may involve the growth of new connections (DARIAN-SMITH 1994), but rapid changes in cortical topography could arise through changes in inhibition, particularly through the unmasking of existing excitatory connections (JACOBS and DONOGHUE 1991) and changes in synaptic efficacy.

The down-regulation of GABA inhibition is also thought to be important in deafferentation plasticity (JONES 1993), which could reflect decreases in glutamic acid decarboxylase (GAD) or in GABA receptor number (for reviews see KAAS 1991; GARRAGHTY and KAAS 1992; JONES 1993). That these changes are related to changes in activity, and not to loss of GABA neurons, is demonstrated by the fact that GABA and GAD levels can recover if normal inputs are restored (JONES 1993). Reductions in GABAergic inhibition also mediate plastic changes occurring after amputation in humans (e.g., CHEN et al. 1998), with the rapid removal of inhibition being essential for deafferentation plasticity to occur in the human cortex (ZIEMANN et al. 1998). Cortical $GABA_A$ and $GABA_B$ may suppress expression of reordered cortical somatic maps induced by deafferentation plasticity (LANE et al. 1997). A better understanding of the mechanisms involved in cortical reorganization could help in rehabilitation programs. For example, loss of cortical representations following cochlea damage may be reversible or preventable with the appropriately timed implantation of cochlea implants (KLINKE et al. 1999).

2. Activity-Dependent Receptive Field Modifications

Modifications in cortical representations can also occur as a result of altered patterns of sensory afferent activity. For example, in the auditory system the best frequency of a cortical neuron can be shifted towards a conditioned stimulus frequency (WEINBERGER 1995). An increase in activity in a subset of inputs can also result in representational expansions of the cortical maps in adults (WEINBERGER 1995; BUONOMANO and MERZENICH 1998). Use-dependent changes in cortical representations can occur within minutes and thus facilitate rapid adaptations to changes in the sensory input and can subsequently increase responses to, and representations of, behaviorally significant inputs. As for the large-scale changes in cortical maps that can occur following injury, adaptation of inhibitory inputs is one mechanism proposed to play a role in use-dependent plasticity (GILBERT 1993). Within the cortex, intrinsic horizontal afferent pathways connect different representational areas. HIRSCH and GILBERT (1993) found that these connections evinced use-dependent changes in synaptic strength that could contribute to cortical reorganizations. Although most studies on maps and plasticity have been concerned with changes evoked by spatially or spectrally specific stimuli, there is a growing body of evidence showing that the temporal responses of cortical neurons can also be altered by experience (BUONOMANO and MERZENICH 1995; BUONOMANO et al. 1997). Despite the wealth of information on synaptic plasticity, and cortical map plasticity, it remains to be determined if LTP of excitatory and/or inhibitory connections is essential for cortical reorganizations (BUONOMANO and MERZENICH 1998). It is clear that in subcortical systems increased inhibition can alter topographic sensory maps.

Long-term reorganization of topographic sensory maps also involves $GABA_A$ergic inhibition. Auditory space is mapped in the external nucleus of the inferior colliculus of the barn owl through a topographic organization of neurons with sharply tuned responsiveness to interaural time differences (ZHENG and KNUDSEN 1999). Connections between the inferior colliculus and the optic tectum direct the animal's gaze towards important sounds. Alteration of the normal relationship between auditory space and correct gaze direction during development produces an abnormal representation of auditory space, which is, however, appropriate for the altered auditory-visual relationship. Normal auditory responsiveness to interaural time differences is not permanently lost, but is suppressed by enhanced $GABA_A$ergic inhibition, and reappears when bicuculline is applied to the inferior colliculus.

Unlike reorganizations of maps, the initial establishment of topographic sensory maps may not be dependent on inhibition. Olfactory neurons expressing a given odorant receptor project invariantly to one of only two glomeruli in the bulb, thus establishing a spatial mapping of olfactory qualities on the olfactory epithelium. Mutant mice in which the homeobox genes, *DIx-1* or *DIx-2*, have been knocked out lack the GABAergic interneurons of the bulb, yet the topographical maps form normally (BULFONE et al. 1998), so inhibitory responses early in development are not an absolute prerequisite for correct mapping.

3. Glycine and Motor Reorganization

Glycinergic transmission plays an analogous role in the reorganization of locomotor activity in the prenatal rat (KUDO and NISHIMARU 1998). Coordinated motor activity recorded in the ventral roots of the isolated spinal-cord–hindlimb preparation was unaffected by glutamate receptor antagonists, but abolished by strychnine and mimicked by glycine application. Like GABA, glycine is thought to act as an excitatory neurotransmitter in developing nervous systems, and the rhythmic activities triggered by glycine were lost as its inhibitory functions emerged.

K. Conclusions

Neurophysiological actions of GABA and glycine clearly encompass much more than simple inhibition of neuronal action potential firing, and include a wide variety of very subtle effects. Indeed, in numerous instances it is a misnomer to consider them "inhibitory" neurotransmitters in view of the direct excitatory effects they can have. Diverse and extensive regulatory effects on rhythmic firing patterns abound. If the past is a guide, continued study of GABAergic and glycinergic systems will yield more surprises, with the functional diversity rivaling the morphological and neurochemical diversity of these systems that has long been recognized. The main details of the microphysiology of inhibitory synapses will be understood before too long, but new complexities will arise as different receptor subunit combinations are localized to specific synaptic locations and found to have distinctive functional properties. It is likely that novel aspects of the various response plasticities will be discovered. Most significantly, as more neuronal networks are investigated, and those under investigation become larger and more intricate, the scope for neurophysiological influences mediated by GABA and glycine will undoubtedly grow. Broad generalizations about the functions of these systems continue to be hard to come by. For better or worse, detailed cellular investigation of specific systems of interest will be required for the foreseeable future.

Acknowledgments. Work in B.E.A.'s laboratory is supported by NIH grants NS30219 and NS36612. We thank Dr. Scott Thompson for his comments on a draft of this manuscript. We thank E. Elizabeth for expert word processing and editorial assistance.

References

Abraham WC, Wickens JR (1991) Heterosynaptic long-term depression is facilitated by blockade of inhibition in area CA1 of the hippocampus. Brain Res 546:336–340
Aizenman CD, Manis PB, Linden DJ (1998) Polarity of long-term synaptic gain change is related to postsynaptic spike firing at a cerebellar inhibitory synapse. Neuron 21:827–835
Alger BE (1984) Characteristics of a slow hyperpolarizing synaptic potential in rat hippocampal pyramidal cells in vitro. J Neurophysiol 52:892–910
Alger BE (1991) Gating of GABAergic inhibition in hippocampal pyramidal cells. Ann N Y Acad Sci 627:249–263
Alger BE, Pitler TA, Wagner JJ, Martin LA, Morishita W, Kirov SA, Lenz RA (1996) Retrograde signalling in depolarization-induced suppression of inhibition in rat hippocampal CA1 cells. J Physiol (Lond) 496:197–209
Alger BE, Nicoll RA (1979) GABA-mediated biphasic inhibitory responses in hippocampus. Nature 281:315–317
Alger BE, Nicoll RA (1980) Spontaneous inhibitory post-synaptic potentials in hippocampus: mechanism for tonic inhibition. Brain Res 200:195–200
Alger BE, Nicoll RA (1982a) Pharmacological evidence for two kinds of GABA receptor on rat hippocampal pyramidal cells studied in vitro. J Physiol (Lond) 328:125–141
Alger BE, Nicoll RA (1982b) Feed-forward dendritic inhibition in rat hippocampal pyramidal cells studied in vitro. J Physiol (Lond) 328:105–123

Alger BE, Pitler TA (1995) Retrograde signaling at GABAA-receptor synapses in the mammalian CNS. Trends Neurosci 18:333–340
Ali AB, Deuchars J, Pawelzik H, Thomson AM (1998) CA1 pyramidal to basket and bistratified cell EPSPs: dual intracellular recordings in rat hippocampal slices. J Physiol (Lond) 507:201–217
Ali AB, Thomson AM (1998) Facilitating pyramid to horizontal oriens-alveus interneurone inputs: dual intracellular recordings in slices of rat hippocampus. J Physiol (Lond) 507:185–199
Allen GI, Eccles J, Nicoll RA, Oshima T, Rubia FJ (1977) The ionic mechanisms concerned in generating the i.p.s.ps of hippocampal pyramidal cells. Proc R Soc Lond B 198:363–384
Andersen P, Eccles JC, Loyning Y (1964a) Location of postsynaptic inhibitory synapses on hippocampal pyramids. J Neurophysiol 27:592–607
Andersen P, Eccles JC, Loyning Y (1964b) Pathway of postsynaptic inhibition in the hippocampus. J Neurophysiol 27:608–619
Andersen P, Bie B, Ganes T, Laursen AM (1978) Two mechanisms for effects of GABA on hippocampal pyramidal cells. In Ryall and Kelly (eds), Iontophoresis and transmitter mechanisms in the mammalian central nervous system. Elsevier/North-Holland Biomedical Press, pp 179–181
Andrade R, Malenka RC, Nicoll RA (1986) A G protein couples serotonin and GABAB receptors to the same channels in hippocampus. Science 234:1261–1265
Auger C, Kondo S, Marty A (1998) Multivesicular release at single functional synaptic sites in cerebellar stellate and basket cells. J Neurosci 18:4532–4547
Auger C, Marty A (1997) Heterogeneity of functional synaptic parameters among single release sites. Neuron 19:139–150
Avoli M, Barbarosie M, Lucke A, Nagao T, Lopantsev V, Kohling R (1996) Synchronous GABA-mediated potentials and epileptiform discharges in the rat limbic system in vitro. J Neurosci 16:3912–3924
Azouz R, Gray CM, Nowak LG, McCormick DA (1997) Physiological properties of inhibitory interneurons in cat striate cortex. Cereb Cortex 7:534–545
Banks MI, Li T-B, Pearce RA (1998) The synaptic basis of $GABA_{A,slow}$. J Neurosci 18:1305–1317
Bazhenov M, Timofeev I, Steriade M, Sejnowski TJ (1999) Self-sustained rhythmic activity in the thalamic reticular nucleus mediated by depolarizing $GABA_A$ receptor potentials. Nature Neurosci 2:168–174
Bear MF, Abraham WC (1996) Long-term depression in hippocampus. Ann Rev Neurosci 19:437–462
Behrends JC, ten Bruggencate G (1993) Cholinergic modulation of synaptic inhibition in the guinea pig hippocampus in vitro: excitation of GABAergic interneurons and inhibition of GABA-release. J Neurophysiol 69:626–629
Behrends JC, ten Bruggencate G (1998) Changes in quantal size distributions upon experimental variations in the probability of release at striatal inhibitory synapses. J Neurophysiol 79:2999–3011
Bekkers JM, Richerson GB, Stevens CF (1990) Origin of variability in quantal size in cultured hippocampal neurons and hippocampal slices. Proc Natl Acad Sci USA 87:5359–5362
Bekkers JM, Stevens CF (1991) Excitatory and inhibitory autaptic currents in isolated hippocampal neurons maintained in cell culture. Proc Natl Acad Sci USA 88:7834–7838
Ben-Ari Y, Khazipov R, Leinekugel X, Caillard O, Gaiarsa J-L (1997) $GABA_A$, NMDA and AMPA receptors: a developmentally regulated 'menage a trois'. Trends Neurosci 20:523–529
Benardo LS (1994) Separate activation of fast and slow inhibitory postsynaptic potentials in rat neocortex in vitro. J Physiol (Lond) 476:203–215

Bland BH, Oddie SD, Colom LV (1999) Mechanisms of neural synchrony in the septohippocampal pathways underlying hippocampal theta generation. J Neurosci 19:3223–3237
Bland BH, Colom LV (1993) Extrinsic and intrinsic properties underlying oscillation and synchrony in limbic cortex. Prog Neurobiol 41:157–208
Blanton MG, Lo Turco JJ, Kriegstein AR (1989) Whole cell recording from neurons in slices of reptilian and mammalian cerebral cortex. J Neurosci Meth 30:203–210
Borg-Graham LJ, Monier C, Fregnac Y (1998) Visual input evokes transient and strong shunting inhibition in visual cortical neurons. Nature 393:369–373
Bormann J, Hamill OP, Sakmann B (1987) Mechanism of anion permeation through channels gated by glycine and gamma-aminobutyric acid in mouse cultured spinal neurones. J Physiol (Lond) 385:243–286
Bragin A, Jando G, Nadasdy Z, Hetke J, Wise K, Buzsaki G (1995) Gamma (40–100 Hz) oscillation in the hippocampus of the behaving rat. J Neurosci 15:47–60
Brennan PA, Keverne EB (1997) Neural mechanisms of mammalian olfactory learning. Prog Neurobiol 51:457–481
Brenowitz S, David J, Trussell L (1998) Enhancement of synaptic efficacy by presynaptic $GABA_B$ receptors. Neuron 20:135–141
Brickley SG, Cull-Candy SG, Farrant M (1996) Development of a tonic form of synaptic inhibition in rat cerebellar granule cells resulting from persistent activation of $GABA_A$ receptors. J Physiol (Lond) 497:753–759
Brickley SG, Cull-Candy SG, Farrant M (1999) Single-channel properties of synaptic and extrasynaptic $GABA_A$ receptors suggest differential targeting of receptor subtypes. J Neurosci 19:2960–2973
Buhl EH, Halasy K, Somogyi P (1994) Diverse sources of hippocampal unitary inhibitory postsynaptic potentials and the number of synaptic release sites. Nature 368:823–828
Buhl EH, Han Z-S, Lorinczi Z, Stezhka VV, Karnup SV, Somogyi P (1994) Physiological properties of anatomically identified axo-axonic cells in the rat hippocampus. J Neurophysiol 71:1289–1307
Buhl EH, Szilagyi T, Halasy K, Somogyi P (1996) Physiological properties of anatomically identified basket and bistratified cells in the CA1 area of the rat hippocampus in vitro. Hippocampus 6:294–305
Buhl EH, Tamas G, Fisahn A (1998) Cholinergic activation and tonic excitation induce persistent gamma oscillations in mouse somatosensory cortex in vitro. J Physiol (Lond) 513:117–126
Bulfone A, Wang F, Hevner R, Anderson S, Cutforth T, Chen S, Meneses J, Pedersen R, Axel R, Rubenstein JLR (1998) An olfactory sensory map develops in the absence of normal projection neurons or GABAergic interneurons. Neuron 21:1273–1282
Buonomano DV, Hickmott PW, Merzenich MM (1997) Context sensitive synaptic plasticity and temporal-to-spatial transformations in hippocampal slices. Proc Natl Acad Sci USA 94:10,403–10,408
Buonomano DV, Merzenich MM (1995) Temporal transformation transformed into a spatial code by a neural network with realistic properties. Science 267:1028–1030
Buonomano DV, Merzenich MM (1998) Cortical plasticity: from synapses to maps. Ann Rev Neurosci 21:149–186
Buzsaki G (1984) Feed-forward inhibition in the hippocampal formation. Prog Neurobiol 22:131–153
Caillard O, McLean HA, Ben-Ari Y, Gaiarsa J-L (1998) Ontogenesis of presynaptic $GABA_B$ receptor-mediated inhibition in the CA3 region of the rat hippocampus. J Neurophysiol 79:1341–1348
Capogna M, Gahwiler BH, Thompson SM (1996) Presynaptic inhibition of calcium-dependent and -independent release elicited with ionomycin, gadolinium, and a-latrotoxin in the hippocampus. J Neurophysiol 75:2017–2028
Carmant L, Woodhall G, Ouardouz M, Robitaille R, Lacaille J-C (1997) Interneuron-specific Ca^{2+} responses linked to metabotropic and ionotropic glutamate receptors in rat hippocampal slices. Eur J Neurosci 9:1625–1635

Charpier S, Behrends JC, Triller A, Faber DS, Korn H (1995) "Latent" inhibitory connections become functional during activity-dependent plasticity. Proc Natl Acad Sci USA 92:117–120
Chen QX, Wong RKS (1995) Suppression of GABAA receptor responses by NMDA application in hippocampal neurones acutely isolated from the adult guinea-pig. J Physiol (Lond) 482:353–362
Chen R, Corwell B, Yaseen Z, Hallett M, Cohen LG (1998) Mechanisms of cortical reorganization in lower-limb amputees. J Neurosci 18:3443–3450
Chen WR, Midtgaard J, Shepherd GM (1997) Forward and backward propagation of dendritic impulses and their synaptic control in mitral cells. Science 278:463–467
Cherubini E, Gaiarsa JL, Ben-Ari Y (1991) GABA: an excitatory transmitter in early postnatal life. Trends Neurosci 14:515–519
Cobb SR, Buhl EH, Halasy K, Paulsen O, Somogyi P (1995) Synchronization of neuronal activity in hippocampus by individual GABAergic interneurons. Nature 378:75–78
Cobb SR, Halasy K, Vida I, Nyiri G, Tamas G, Buhl EH, Somogyi P (1997) Synaptic effects of identified interneurons innervating both interneurons and pyramidal cells in the rat hippocampus. Neuroscience 79:629–648
Cochilla AJ, Alford S (1998) Metabotropic glutamate receptor-mediated control of neurotransmitter release. Neuron 20:1007–1016
Collingridge GL, Gage PW, Robertson B (1984) Inhibitory post-synaptic currents in rat hippocampal CA1 neurones. J Physiol (Lond) 356:551–564
Connors BW, Gutnick MJ (1990) Intrinsic firing patterns of diverse neocortical neurons. Trends Neurosci 13:99–104
Cossart R, Esclapez M, Hirsch JC, Bernard C, Ben-Ari Y (1998) GluR5 kainate receptor activation in interneurons increases tonic inhibition of pyramidal cells. Nature Neurosci 1:470–478
Crook JM, Eysel VT (1992) GABA-induced inactivation of functionally characterized sites in cat visual cortex (area 18) effects on orientation tuning. J Neurosci 12:1816–1825
Csicsvari J, Hirase H, Czurko A, Buzsaki G (1998) Reliability and state dependence of pyramidal cell-interneuron synapses in the hippocampus: an ensemble approach in the behaving rat. Neuron 21:179–189
Darian-Smith C (1994) Axonal sprouting accompanies functional reorganization in adult cat striate cortex. Nature 368:737–740
Davies CH, Davies SN, Collingridge GL (1990) Paired-pulse depression of monosynaptic GABA-mediated inhibitory postsynaptic responses in rat hippocampus. J Physiol (Lond) 424:513–531
Davies CH, Starkey SJ, Pozza MF, Collingridge GL (1991) GABAB autoreceptors regulate the induction of LTP. Nature 349:609–611
Davies CH, Collingridge GL (1996) Regulation of EPSPs by the synaptic activation of $GABA_B$ autoreceptors in rat hippocampus. J Physiol (Lond) 496:451–470
De Koninck Y, Mody I (1994) Noise analysis of miniature IPSCs in adult rat brain slices: properties and modulation of synaptic GABAA receptor channels. J Neurophysiol 71:1318–1335
Debanne D, Guerineau NC, Gahwiler BH, Thompson SM (1997) Action-potential propagation gated by an axonal I_A-like K^+ conductance in hippocampus. Nature 389:286–289
Dekker LV, Parker PJ (1994) Protein kinase C – a question of specificity. Trends Biochem Sci 19:73–77
DeVries SH, Baylor DA (1993) Synaptic circuitry of the retina and olfactory bulb. Neuron 10 [Suppl]:139–149
Dickson CT, Alonso A (1997) Muscarinic induction of synchronous population activity in the entorhinal cortex. J Neurosci 17:6729–6744
Djorup A, Jahnsen H, Laursen AM (1981) The dendritic response to GABA in CA1 of the hippocampal slice. Brain Res 219:196–201
Dodt H-U, Zieglgansberger W (1990) Visualizing unstained neurons in living brain slices by infrared DIC-videomicroscopy. Brain Res 537:333–336

Douglas RJ, Koch C, Mahowald M, Martin KAC, Suarez HH (1995) Recurrent excitation in neocortical circuits. Science 269:981–985
Doze VA, Cohen GA, Madison DV (1995) Calcium channel involvement in GABAB receptor-mediated inhibition of GABA release in area CA1 of the rat hippocampus. J Neurophysiol 74:43–53
Draguhn A, Traub RD, Schmitz D, Jefferys JGR (1998) Electrical coupling underlies high-frequency oscillations in the hippocampus in vitro. Nature 394:189–192
Du J, Zhang L, Weiser M, Rudy B, McBain CJ (1996) Developmental expression and functional characterization of the potassium-channel subunit Kv3.1b in parvalbumin-containing interneurons of the rat hippocampus. J Neurosci 16:506–518
Dudek SM, Bear MF (1992) Homosynaptic long-term depression in area CA1 of hippocampus and effects of *N*-methyl-D-aspartate receptor blockade. Proc Natl Acad Sci USA 89:4363–4367
Dudek SM, Friedlander MJ (1996) Developmental down-regulation of LTD in cortical layer IV and its independence of modulation by inhibition. Neuron 16:1097–1106
Dutar P, Nicoll RA (1988) Pre- and postsynaptic GABAB receptors in the hippocampus have different pharmacological properties. Neuron 1:585–591
Dykes RW, Landry P, Metherate R, Hicks TP (1984) Functional role of GABA in cat primary somatosensory cortex: shaping receptive fields of cortical neurons. J Neurophysiol 52:1066–1093
Eccles JC (1964) The physiology of synapses. Springer Verlag, New York
Edwards FA, Konnerth A, Sakmann B, Takahashi T (1989) A thin slice preparation for patch clamp recordings from neurones of the mammalian central nervous system. Pflugers Arch 414:600–612
Edwards FA, Konnerth A, Sakmann B (1990) Quantal analysis of inhibitory synaptic transmission in the dentate gyrus of rat hippocampal slices: a patch-clamp study. J Physiol (Lond) 430:213–249
Fisahn A, Pike FG, Buhl EH, Paulsen O (1998) Cholinergic induction of network oscillations at 40 Hz in the hippocampus in vitro. Nature 394:186–189
Forti L, Bossi M, Bergamaschi A, Villa A, Malgaroli A (1997) Loose-patch recordings of single quanta at individual hippocampal synapses. Nature 388:874–878
Fox SE (1989) Membrane potential and impedance changes in hippocampal pyramidal cells during theta rhythm. Exp Brain Res 77:283–294
Frank K, Fuortes MGF (1957) Presynaptic and post-synaptic inhibition of monosynaptic reflex. Fed Proc 16:39–40
Frerking M, Borges S, Wilson M (1995) Variation in GABA mini amplitude is the consequence of variation in transmitter concentration. Neuron 15:885–895
Frerking M, Malenka RC, Nicoll RA (1998) Synaptic activation of kainate receptors on hippocampal interneurons. Nature Neurosci 1:479–486
Freund TF, Buzsaki G (1996) Interneurons of the hippocampus. Hippocampus 6:347–470
Freund TF, Meskenaite V (1992) γ-Aminobutyric acid-containing basal forebrain neurons innervate inhibitory interneurons in the neocortex. Proc Natl Acad Sci USA 89:738–742
Froehner SC (1998) Gathering glycine receptors at synapses. Science 282:1277–1279
Gahwiler BH, Brown DA (1985) GABAB-receptor-activated K^+ current in voltage-clamped CA_3 pyramidal cells in hippocampal cultures. Proc Natl Acad Sci USA 82:1558–1562
Gaiarsa J-L, McLean H, Congar P, Leinekugel X, Khazipov R, Tseeb V, Ben-Ari Y (1995) Postnatal maturation of gamma-aminobutyric acidA and B-mediated inhibition in the CA3 hippocampal region of the rat. J Neurobiol 26:339–349
Galarreta M, Hestrin S (1997) Properties of $GABA_A$ receptors underlying inhibitory synaptic currents in neocortical pyramidal neurons. J Neurosci 17:7220–7227
Galarreta M, Hestrin S (1998) Frequency-dependent synaptic depression and the balance of excitation and inhibition in the neocortex. Nature Neurosci 1:587–594
Galarreta M, Hestrin S (1999) A network of fast-spiking cells in the neocortex connected by electrical synapses. Nature 402:72–75

Garraghty PE, Kaas JH (1992) Dynamic features of sensory and motor maps. Curr Opin Neurobiol 2:522–527
Geiger JRP, Lubke J, Roth A, Frotscher M, Jonas P (1997) Submillisecond AMPA receptor-mediated signaling at a principal neuron-interneuron synapse. Neuron 18:1009–1023
Gibson JR, Beierlein M, Connors BW (1999) Two networks of electrically coupled inhibitory neurons in neocortex. Nature 402:75–79
Gilbert CD (1993) Rapid dynamic changes in adult cerebral cortex. Curr Opin Neurobiol 3:100–103
Glaum SR, Brooks PA (1996) Tetanus-induced sustained potentiation of monosynaptic inhibitory transmission in the rat medulla: evidence for a presynaptic locus. J Neurophysiol 76:30–38
Glitsch M, Llano I, Marty A (1996) Glutamate as a candidate retrograde messenger at interneurone-Purkinje cell synapses of rat cerebellum. J Physiol (Lond) 497:531–537
Goda Y, Stevens CF (1994) Two components of transmitter release at a central synapse. Proc Natl Acad Sci USA 91:12,942–12,946
Golding NL, Spruston N (1998) Dendritic sodium spikes are variable triggers of axonal action potentials in hippocampal CA1 pyramidal neurons. Neuron 21:1189–1200
Gonchar Y, Burkhalter A (1997) Three distinct families of GABAergic neurons in rat visual cortex. Cereb Cortex 7:347–358
Gray CM, McCormick DA (1996) Chattering cells: superficial pyramidal neurons contributing to the generation of synchronous oscillations in the visual cortex. Science 274:109–113
Grover LM, Lambert NA, Schwartzkroin PA, Teyler TJ (1993) Role of HCO^-_3 ions in depolarizing GABAA receptor-mediated responses in pyramidal cells of rat hippocampus. J Neurophysiol 69:1541–1555
Gulyas AI, Miles R, Sik A, Toth K, Tamamaki N, Freund TF (1993) Hippocampal pyramidal cells excite inhibitory neurons through a single release site. Nature 366:683–687
Gulyas AI, Hajos N, Freund TF (1996) Interneurons containing calretinin are specialized to control other interneurons in the rat hippocampus. J Neurosci 16:3397–3411
Gustafsson B, Wigstrom H, Abraham WC, Huang Y-Y (1987) Long-term potentiation in the hippocampus using depolarizing current pulses as the conditioning stimulus to single volley synaptic potentials. J Neurosci 7:774–780
Hausser M, Clark BA (1997) Tonic synaptic inhibition modulates neuronal output pattern and spatiotemporal synaptic integration. Neuron 19:665–678
Hayashi Y, Momiyama A, Takahashi T, Ohishi H, Ogawa-Meguro R, Shigemoto R, Mizuno N, Nakanishi S (1993) Role of a metabotropic glutamate receptor in synaptic modulation in the accessory olfactory bulb. Nature 366:687–690
Hestrin S (1993) Different glutamate receptor channels mediate fast excitatory synaptic currents in inhibitory and excitatory cortical neurons. Neuron 11:1083–1091
Hirsch JA, Gilbert CD (1993) Long-term changes in synaptic strength along specific intrinsic pathways in the cat visual cortex. J Physiol (Lond) 461:247–262
Huguenard JR, Prince DA (1994) Clonazepam suppresses $GABA_B$-mediated inhibition in thalamic relay neurons through effects in nucleus reticularis. J Neurophysiol 71:2576–2581
Huntsman MM, Porcello DM, Homanics GE, DeLorey TM, Huguenard JR (1999) Reciprocal inhibitory connections and network synchrony in the mammalian thalamus. Science 283:541–543
Isaacson JS, Solis JM, Nicoll RA (1993) Local and diffuse synaptic actions of GABA in the hippocampus. Neuron 10:165–175
Isaacson JS, Hille B (1997) $GABA_B$-mediated presynaptic inhibition of excitatory transmission and synaptic vesicle dynamics in cultured hippocampal neurons. Neuron 18:143–152
Isaacson JS, Strowbridge BW (1998) Olfactory reciprocal synapses: dendritic signaling in the CNS. Neuron 20:749–761

Ito S, Cherubini E (1991) Strychnine-sensitive glycine responses of neonatal rat hippocampal neurones. J Physiol (Lond) 440:67–83
Jacobs KM, Donoghue JP (1991) Reshaping the cortical motor map by unmasking latent intracortical connections. Science 251:944–947
Jagadeesh B, Wheat HS, Ferster D (1993) Linearity of summation of synaptic potentials underlying direction selectivity in simple cells of the cat visual cortex. Science 262:1901–1904
Jahr CE, Nicoll RA (1980) Dendrodendritic inhibition: demonstration with intracellular recording. Science 207:1473–1475
Jahr CE, Nicoll RA (1982) An intracellular analysis of dendrodendritic inhibition in the turtle in vitro olfactory bulb. J Physiol (Lond) 326:213–234
Jo Y-H, Schlichter R (1999) Synaptic corelease of ATP and GABA in cultured spinal neurons. Nature Neurosci 2:241–245
Johnston D, Magee JC, Colbert CM, Christie BR (1996) Active properties of neuronal dendrites. Ann Rev Neurosci 19:165–186
Jonas P, Bischofberger J, Sandkuhler J (1998) Corelease of two fast neurotransmitters at a central synapse. Science 281:419–424
Jones EG (1993) GABAergic neurons and their role in cortical plasticity in primates. Cereb Cortex 3:361–372
Jones KA, Borowsky B, Tamm JA, Craig DA, Durkin MM, Dai M, Yao W-J, Johnson M, Gunwaldsen C, Huang L-Y, Tang C, Shen Q, Salon JA, Morse K, Laz T, Smith KE, Nagarathnam D, Noble SA, Branchek TA, Gerald C (1998) $GABA_B$ receptors function as a heteromeric assembly of the subunits $GABA_BR1$ and $GABA_BR2$. Nature 396:674–679
Kaas JH (1991) Plasticity of sensory and motor maps in adult mammals. Ann Rev Neurosci 14:137–167
Kaila K (1994) Ionic basis of GABAA receptor channel function in the nervous system. Prog Neurobiol 42:489–537
Kaila K, Lamsa K, Smirnov S, Taira T, Voipio J (1997) Long-lasting GABA-mediated depolarization evoked by high-frequency stimulation in pyramidal neurons of rat hippocampal slice is attributable to a network-driven, bicarbonate-dependent K^+ transient. J Neurosci 17:7662–7672
Kaila K, Voipio J (1987) Postsynaptic fall in intracellular pH induced by GABA-activated bicarbonate conductance. Nature 330:163–165
Kano M, Rexhausen U, Dreessen J, Konnerth A (1992) Synaptic excitation produces a long-lasting rebound potentiation of inhibitory synaptic signals in cerebellar Purkinje cells. Nature 356:601–604
Kano M (1996) Long-lasting potentiation of GABAergic inhibitory synaptic transmission in cerebellar Purkinje cells: its properties and possible mechanisms. Behav Brain Sci 19:354–361
Kapur A, Lytton WW, Ketchum KL, Haberly LB (1997a) Regulation of the NMDA component of EPSPs by different components of postsynaptic GABAergic inhibition: computer simulation analysis in piriform cortex. J Neurophysiol 78:2546–2559
Kapur A, Pearce RA, Lytton WW, Haberly LB (1997b) $GABA_A$-mediated IPSCs in piriform cortex have fast and slow components with different properties and locations on pyramidal cells. J Neurophysiol 78:2531–2545
Katona I, Sperlagh B, Sik A, Kafalvi A, Vizi ES, Mackie K, Freund TF (1999) Presynaptically located CB1 cannabinoid receptors regulate GABA release from axon terminals of specific hippocampal interneurons. J Neurosci 19:4544–4558
Kaupmann K, Huggel K, Heid J, Flor PJ, Bischoff S, Mickel SJ, McMaster G, Angst C, Bittiger H, Froestl W, Bettler B (1997) Expression cloning of GABAB receptors uncovers similarity to metabotropic glutamate receptors. Nature 386:239–246
Kaupmann K, Malitschek B, Schuler V, Heid J, Froestl W, Beck P, Mosbacher J, Bischoff S, Kulik A, Shigemoto R, Karschin A, Bettler B (1998) $GABA_B$-receptor subtypes assemble into functional heteromeric complexes. Nature 396:683–687
Kawaguchi Y (1995) Physiological subgroups of nonpyramidal cells with specific morphological characteristics in layer II/III of rat frontal cortex. J Neurosci 15:2638–2655

Kawaguchi Y, Kubota Y (1996) Physiological and morphological identification of somatostatin- or vasoactive intestinal polypeptide-containing cells among GABAergic cell subtypes in rat frontal cortex. J Neurosci 16:2701–2715
Kawaguchi Y, Kubota Y (1997) GABAergic cell subtypes and their synaptic connections in rat frontal cortex. Cereb Cortex 7:476–486
Kim U, Sanchez-Vives MV, McCormick DA (1997) Functional dynamics of GABAergic inhibition in the thalamus. Science 278:130–134
Kirischuk S, Veselovsky N, Grantyn R (1999) Relationship between presynaptic calcium transients and postsynaptic currents at single gamma-aminobutyric acid (GABA)ergic boutons. Proc Natl Acad Sci USA 96:7520–7525
Kirsch J, Betz H (1998) Glycine-receptor activation is required for receptor clustering in spinal neurons. Nature 392:717–720
Klinke R, Kral A, Heid S, Tillein J, Hartmann R (1999) Recruitment of the auditory cortex in congenitally deaf cats by long-term cochlear electrostimulation. Science 285:1729–1733
Komatsu Y (1994) Age-dependent long-term potentiation of inhibitory synaptic transmission in rat visual cortex. J Neurosci 14:6488–6499
Komatsu Y (1996) GABAB receptors, monoamine receptors, and postsynaptic inositol trisphosphate-induced Ca2+ release are involved in the induction of long-term potentiation at visual cortical inhibitory synapses. J Neurosci 16:6342–6352
Konopacki J, Golebiewski H, Eckersdorf B, Blaszczyk M, Grabowski R (1997) Theta-like activity in hippocampal formation slices: the effect of strong disinhibition of $GABA_A$ and $GABA_B$ receptors. Brain Res 775:91–98
Kopysova IL, Debanne D (1998) Critical role of axonal A-type K^+ channels and axonal geometry in the gating of action potential propagation along CA3 pyramidal cell axons: a simulation study. J Neurosci 18:7436–7451
Krnjevic K (1981) Desensitization of GABA receptors. In Costa E, Di Chiara G, Gessa GL (eds) GABA and benzodiazepine receptors. Raven Press, New York, 111–120
Kudo N, Nishimaru H (1998) Reorganization of locomotor activity during development in the prenatal rat. Ann N Y Acad Sci 860:306–317
Lambert NA, Harrison NL, Teyler TJ (1991) Baclofen-induced disinhibition in area CA1 of rat hippocampus is resistant to extracellular Ba2+. Brain Res 547:349–352
Lambert NA, Wilson WA (1993a) Discrimination of post- and presynaptic GABAB receptor-mediated responses by tetrahydroaminoacridine in area CA3. J Neurophysiol 69:630–635
Lambert NA, Wilson WA (1993b) Heterogeneity in presynaptic regulation of GABA release from hippocampal inhibitory neurons. Neuron 11:1057–1067
Lambert NA, Wilson WA (1996) High-threshold Ca^{2+} currents in rat hippocampal interneurones and their selective inhibition by activation of $GABA_B$ receptors. J Physiol (Lond) 492:115–127
Lane RD, Killackey HP, Rhoades RW (1997) Blockade of GABAergic inhibition reveals reordered cortical somatotopic maps in rats that sustained neonatal forelimb removal. J Neurophysiol 77:2723–2735
Lang EJ, Sugihara I, Llinas R (1996) GABAergic modulation of complex spike activity by the cerebellar nucleoolivary pathway in rat. J Neurophysiol 76:255–275
Larkum ME, Zhu JJ, Sakmann B (1999) A new cellular mechanism for coupling inputs arriving at different cortical layers. Nature 398:338–341
Laurent G (1996) Dynamical representation of odors by oscillating and evolving neural assemblies. Trends Neurosci 19:489–496
Le Beau FEN, Alger BE (1998) Transient suppression of GABAA-receptor-mediated IPSPs after epileptiform burst discharges in CA1 pyramidal cells. J Neurophysiol 79:659–669
Leinekugel X, Tseeb V, Ben-Ari Y, Bregestovski P (1995) Synaptic GABAA activation induces Ca2+ rise in pyramidal cells and interneurons from rat neonatal hippocampal slices. J Physiol (Lond) 487:319–329
Lenz RA, Pitler TA, Alger BE (1997) High intracellular Cl^- concentrations depress G-protein-modulated ionic conductances. J Neurosci 17:6133–6141

Lenz RA, Wagner JJ, Alger BE (1998) N- and L-type calcium channel involvement in depolarization-induced suppression of inhibition in rat hippocampal CA1 cells. J Physiol (Lond) 512:61–73

Leung L-WS, Yim CY (1986) Intracellular records of theta rhythm in hippocampal CA1 cells of the rat. Brain Res 367:323–327

Lewis CA, Faber DS (1996a) Inhibitory synaptic transmission in isolated patches of membrane from cultured rat spinal cord and medullary neurons. J Neurophysiol 76:461–470

Lewis CA, Faber DS (1996b) Properties of spontaneous inhibitory synaptic currents in cultured rat spinal cord and medullary neurons. J Neurophysiol 76:448–460

Lim R, Alvarez FJ, Walmsley B (1999) Quantal size is correlated with receptor cluster area at glycinergic synapses in the rat brainstem. J Physiol (Lond) 516:505–512

Ling DSF, Benardo LS (1995) Recruitment of GABAA inhibition in rat neocortex is limited and not NMDA-dependent. J Neurophysiol 74:2329–2335

Lisman JE, Idiart MAP (1995) Storage of 7 ± 2 short-term memories in oscillatory subcycles. Science 267:1512–1515

Liu G, Choi S, Tsien RW (1999) Variability of neurotransmitter concentration and nonsaturation of postsynaptic AMPA receptors at synapses in hippocampal cultures and slices. Neuron 22:395–409

Liu Y-B, Disterhoft JF, Slater NT (1993) Activation of metabotropic glutamate receptors induces long-term depression of GABAergic inhibition in hippocampus. J Neurophysiol 69:1000–1004

Llano I, Leresche N, Marty A (1991) Calcium entry increases the sensitivity of cerebellar Purkinje cells to applied GABA and decreases inhibitory synaptic currents. Neuron 6:565–574

Llano I, Gerschenfeld HM (1993) Inhibitory synaptic currents in stellate cells of rat cerebellar slices. J Physiol (Lond) 468:177–200

Llano I, Marty A (1995) Presynaptic metabotropic glutamatergic regulation of inhibitory synapses in rat cerebellar slices. J Physiol (Lond) 486:163–176

Llinas R, Ribrary U (1993) Coherent 40-Hz oscillation characterizes dream state in humans. Proc Natl Acad Sci USA 90:2078–2081

Llinas R, Sugimori M (1980) Electrophysiological properties of in vitro Purkinje cell dendrites in mammalian cerebellar slices. J Physiol (Lond) 305:197–213

Llinas RR (1988) The intrinsic electrophysiological properties of mammalian neurons: insights into central nervous system function. Science 242:1654–1664

Lopantsev V, Schwartzkroin PA (1999) $GABA_A$-dependent chloride influx modulates $GABA_B$-mediated IPSPs in hippocampal pyramidal cells. J Neurophysiol 82: 1218–1223

Luhmann HJ, Prince DA (1991) Postnatal maturation of the GABAergic system in rat neocortex. J Neurophysiol 65:247–263

Luscher C, Jan LY, Stoffel M, Malenka RC, Nicoll RA (1997) G protein-coupled inwardly rectifying K^+ channels (GIRKs) mediate postsynaptic but not presynaptic transmitter actions in hippocampal neurons. Neuron 19:687–695

Lux HD, Heinemann U, Dietzel I (1986) Ionic changes and alterations in the size of the extracellular space during epileptic activity. In Delgado-Escueta AV, Ward AA Jr, Woodbury DM, Porter RJ (eds), Advances in neurology, vol 44: Basic mechanisms of the epilepsies. Molecular and cellular approaches. Raven Press, New York, 619–639

Maccaferri G, Toth K, McBain CJ (1998) Target-specific expression of presynaptic mossy fiber plasticity. Science 279:1368–1370

Maccaferri G, McBain CJ (1995) Passive propagation of LTD to stratum oriens-alveus inhibitory neurons modulates the temporoammonic input to the hippocampal CA1 region. Neuron 15:137–145

Maccaferri G, McBain CJ (1996) The hyperpolarization-activated current (Ih) and its contribution to pacemaker activity in rat CA1 hippocampal stratum oriens-alveus interneurones. J Physiol (Lond) 497:119–130

Macdonald RL, Olsen RW (1994) GABAA receptor channels. Ann Rev Neurosci 17:569–602
Mahanty NK, Sah P (1998) Calcium-permeable AMPA receptors mediate long-term potentiation in interneurons in the amygdala. Nature 394:683–687
Mann-Metzer P, Yarom Y (1999) Electrotonic coupling interacts with intrinsic properties to generate synchronized activity in cerebellar networks of inhibitory interneurons. J Neurosci 19:3298–3306
Martina M, Schultz JH, Ehmke H, Monyer H, Jonas P (1998) Functional and molecular differences between voltage-gated K^+ channels of fast-spiking interneurons and pyramidal neurons of rat hippocampus. J Neurosci 18:8111–8125
Martina M, Jonas P (1997) Functional differences in Na^+ channel gating between fast-spiking interneurones and principal neurones of rat hippocampus. J Physiol (Lond) 505:593–603
McBain CJ, Freund TF, Mody I (1999) Glutamatergic synapses onto hippocampal interneurons: precision timing without lasting plasticity. Trends Neurosci 22:228–235
McBain CJ, Dingledine R (1993) Heterogeneity of synaptic glutamate receptors on CA3 stratum radiatum interneurones of rat hippocampus. J Physiol (Lond) 462:373–392
McBain CJ, Maccaferri G (1997) Synaptic plasticity in hippocampal interneurons? A commentary. Can J Physiol Pharmacol 75:488–494
McCormick DA, Connors BW, Lighthall JW, Prince DA (1985) Comparative electrophysiology of pyramidal and sparsely spiny stellate neurons of the neocortex. J Neurophysiol 54:782–806
McCormick DA, Bal T (1997) Sleep and arousal: thalamocortical mechanisms. Ann Rev Neurosci 20:185–215
McIntyre DC, Racine RJ (1986) Kindling mechanisms: current progress on an experimental epilepsy model. Prog Neurobiol 27:1–12
McLean HA, Caillard O, Ben-Ari Y, Gaiarsa J-L (1996a) Bidirectional plasticity expressed by GABAergic synapses in the neonatal rat hippocampus. J Physiol (Lond) 496:471–477
McLean HA, Caillard O, Khazipov R, Ben-Ari Y, Gaiarsa J-L (1996b) Spontaneous release of GABA activates $GABA_B$ receptors and controls network activity in the neonatal rat hippocampus. J Neurophysiol 76:1036–1046
McMahon LL, Kauer JA (1997) Hippocampal interneurons express a novel form of synaptic plasticity. Neuron 18:295–305
McNamara JO, Bonhaus DW, Shin C, Crain BJ, Gellman RL, Giacchino JL (1984) The kindling model of epilepsy: a critical review. CRC Crit Rev Clin Neurobiol 1:341–392
Michelson HB, Wong RKS (1991) Excitatory synaptic responses mediated by GABAA receptors in the hippocampus. Science 253:1420–1423
Michelson HB, Wong RKS (1994) Synchronization of inhibitory neurones in the guinea-pig hippocampus in vitro. J Physiol (Lond) 477:35–45
Miledi R (1966) Strontium as a substitute for calcium in the process of transmitter release at the neuromuscular junction. Nature 5067:1233–1234
Miles R (1990a) Synaptic excitation of inhibitory cells by single CA3 hippocampal pyramidal cells of the guinea-pig in vitro. J Physiol (Lond) 428:61–77
Miles R (1990b) Variation in strength of inhibitory synapses in the CA3 region of guinea-pig hippocampus in vitro. J Physiol (Lond) 431:659–676
Miles R, Toth K, Gulyas AI, Hajos N, Freund TF (1996) Differences between somatic and dendritic inhibition in the hippocampus. Neuron 16:815–823
Miles R, Poncer J-C (1993) Metabotropic glutamate receptors mediate a post-tetanic excitation of guinea-pig hippocampal inhibitory neurones. J Physiol (Lond) 463:461–473
Miles R, Wong RKS (1984) Unitary inhibitory synaptic potentials in the guinea-pig hippocampus in vitro. J Physiol (Lond) 356:97–113

Miles R, Wong RKS (1987) Latent synaptic pathways revealed after tetanic stimulation in the hippocampus. Nature 329:724–726
Misgeld U, Bijak M, Jarolimek W (1995) A physiological role for GABAB receptors and the effects of baclofen in the mammalian central nervous system. Prog Neurobiol 46:423–462
Miura M, Yoshioka M, Miyakawa H, Kato H, Ito K-I (1997) Properties of calcium spikes revealed during $GABA_A$ receptor antagonism in hippocampal CA1 neurons from guinea pigs. J Neurophysiol 78:2269–2279
Mody I, De Koninck Y, Otis TS, Soltesz I (1994) Bridging the cleft at GABA synapses in the brain. Trends Neurosci 17:517–525
Morin F, Beaulieu C, Lacaille J-C (1996) Membrane properties and synaptic currents evoked in CA1 interneuron subtypes in rat hippocampal slices. J Neurophysiol 76:1–16
Morishita W, Kirov SA, Pitler TA, Martin LA, Lenz RA, Alger BE (1997) *N*-Ethylmaleimide blocks depolarization-induced suppression of inhibition and enhances GABA release in the rat hippocampal slice in vitro. J Neurosci 17:941–950
Morishita W, Kirov SA, Alger BE (1998) Evidence for metabotropic glutamate receptor activation in the induction of depolarization-induced suppression of inhibition in hippocampal CA1. J Neurosci 18:4870–4882
Morishita W, Alger BE (1997) Sr^{2+} supports depolarization-induced suppression of inhibition and provides new evidence for a presynaptic expression mechanism in rat hippocampal slices. J Physiol (Lond) 505:307–317
Morishita W, Sastry BR (1996) Postsynaptic mechanisms underlying long-term depression of GABAergic transmission in neurons of the deep cerebellar nuclei. J Neurophysiol 76:59–68
Mott DD, Lewis DV (1991) Facilitation of the induction of long-term potentiation by GABAB receptors. Science 252:1718–1720
Mott DD, Lewis DV (1994) The pharmacology and function of central $GABA_B$ receptors. Int Rev Neurobiol 36:97–223
Mulkey RM, Endo S, Shenolikar S, Malenka RC (1994) Involvement of a calcineurin/inhibitor-1 phosphatase cascade in hippocampal long-term depression. Nature 369:486–488
Nelson S, Toth L, Sheth B, Sur M (1994) Orientation selectivity of cortical neurons during intracellular blockade of inhibition. Science 265:774–777
Newberry NR, Nicoll RA (1984a) A bicuculline-resistant inhibitory post-synaptic potential in rat hippocampal pyramidal cells in vitro. J Physiol (Lond) 348:239–254
Newberry NR, Nicoll RA (1984b) Direct hyperpolarizing action of baclofen on hippocampal pyramidal cells. Nature 308:450–452
Newberry NR, Nicoll RA (1985) Comparison of the action of baclofen with gamma-aminobutyric acid on rat hippocampal pyramidal cells in vitro. J Physiol (Lond) 360:161–185
Nicoll RA, Eccles JC, Oshima T, Rubia F (1975) Prolongation of hippocampal inhibitory postsynaptic potentials by barbiturates. Nature 258:625–627
Nicoll RA, Alger BE (1979) Presynaptic inhibition: transmitter and ionic mechanisms. Int Rev Neurobiol 21:217–258
Nicoll RA, Jahr CE (1982) Self-excitation of olfactory bulb neurones. Nature 296:441–444
Nowycky MC, Mori K, Shepherd GM (1981) GABAergic mechanisms of dendrodendritic synapses in isolated turtle olfactory bulb. J Neurophysiol 46:639–648
Nurse S, Lacaille J-C (1997) Do $GABA_A$ and $GABA_B$ inhibitory postsynaptic responses originate from distinct interneurons in the hippocampus? Can J Physiol Pharmacol 75:520–525
Nusser Z, Cull-Candy S, Farrant M (1997) Differences in synaptic $GABA_A$ receptor number underlie variation in GABA mini amplitude. Neuron 19:697–709
Nusser Z, Hajos N, Somogyi P, Mody I (1998) Increased number of synaptic $GABA_A$ receptors underlies potentiation at hippocampal inhibitory synapses. Nature 395:172–177

Oda Y, Charpier S, Murayama Y, Suma C, Korn H (1995) Long-term potentiation of glycinergic inhibitory synaptic transmission. J Neurophysiol 74:1056–1074
Ohno-Shosaku T, Sawada S, Yamamoto C (1998) Properties of depolarization-induced suppression of inhibitory transmission in cultured rat hippocampal neurons. Pflugers Arch 435:273–279
Otis TS, Staley KJ, Mody I (1991) Perpetual inhibitory activity in mammalian brain slices generated by spontaneous GABA release. Brain Res 545:142–150
Otis TS, De Koninck Y, Mody I (1994) Lasting potentiation of inhibition is associated with an increased number of gamma-aminobutyric acid type A receptors activated during miniature inhibitory postsynaptic currents. Proc Natl Acad Sci USA 91:7698–7702
Otis TS, Mody I (1992) Differential activation of GABAA and GABAB receptors by spontaneously released transmitter. J Neurophysiol 67:227–235
Ouardouz M, Lacaille J-C (1997) Properties of unitary IPSCs in hippocampal pyramidal cells originating from different types of interneurons in young rats. J Neurophysiol 77:1939–1949
Pape H-C (1996) Queer current and pacemaker: the hyperpolarization-activated cation current in neurons. Ann Rev Physiol 58:299–327
Parra P, Gulyas AI, Miles R (1998) How many subtypes of inhibitory cells in the hippocampus? Neuron 20:983–993
Pearce RA (1993) Physiological evidence for two distinct GABAA responses in rat hippocampus. Neuron 10:189–200
Pearce RA, Grunder SD, Faucher LD (1995) Different mechanisms for use-dependent depression of two GABAA-mediated IPSCs in rat hippocampus. J Physiol (Lond) 484:425–435
Penttonen M (1998) Gamma frequency oscillation in the hippocampus of the rat: intracellular analysis in vivo. Eur J Neurosci 10:718–728
Perkins KL, Wong RKS (1996) Ionic basis of the postsynaptic depolarizing GABA response in hippocampal pyramidal cells. J Neurophysiol 76:3886–3894
Perreault P, Avoli M (1988) A depolarizing inhibitory postsynaptic potential activated by synaptically released gamma-aminobutyric acid under physiological conditions in rat hippocampal pyramidal cells. Can J Physiol Pharmacol 66:1100–1102
Perreault P, Avoli M (1989) Effects of low concentrations of 4-aminopyridine on CA1 pyramidal cells of the hippocampus. J Neurophysiol 61:953–970
Perreault P, Avoli M (1992) 4-Aminopyridine-induced epileptiform activity and a GABA-mediated long-lasting depolarization in the rat hippocampus. J Neurosci 12:104–115
Pitler TA, Alger BE (1992a) Cholinergic excitation of GABAergic interneurons in the rat hippocampal slice. J Physiol (Lond) 450:127–142
Pitler TA, Alger BE (1992b) Postsynaptic spike firing reduces synaptic GABAA responses in hippocampal pyramidal cells. J Neurosci 12:4122–4132
Pitler TA, Alger BE (1994a) Depolarization-induced suppression of GABAergic inhibition in rat hippocampal pyramidal cells: G protein involvement in a presynaptic mechanism. Neuron 13:1447–1455
Pitler TA, Alger BE (1994b) Differences between presynaptic and postsynaptic GABAB mechanisms in rat hippocampal pyramidal cells. J Neurophysiol 72:2317–2327
Pouzat C, Marty A (1999) Somatic recording of GABAergic autoreceptor current in cerebellar stellate and basket cells. J Neurosci 19:1675–1690
Racca C, Catania MV, Monyer H, Sakmann B (1996) Expression of AMPA-glutamate receptor B subunit in rat hippocampal GABAergic neurons. Eur J Neurosci 8:1580–1590
Rajendra S, Schofield PR (1995) Molecular mechanisms of inherited startle syndromes. Trends Neurosci 18:80–82
Reyes A, Lujan R, Rozov A, Burnashev N, Somogyi P, Sakmann B (1998) Target-cell-specific facilitation and depression in neocortical circuits. Nature Neurosci 1:279–285

Rivera C, Voipio J, Payne JA, Ruusuvuori E, Lahtinen H, Lamsa K, Pirvola U, Saarma M, Kaila K (1999) The K^+/Cl^- co-transporter KCC2 renders GABA hyperpolarizing during neuronal maturation. Nature 397:251–255
Roberts E (1991) Living systems are tonically inhibited, autonomous optimizers, and disinhibition coupled to variability generation is their major organizing principle: inhibitory command-control at levels of membrane, genome, metabolism, brain, and society. Neurochem Res 16:409–421
Rohrbacher J, Jarolimek W, Lewen A, Misgeld U (1997) $GABA_B$ receptor-mediated inhibition of spontaneous inhibitory synaptic currents in rat midbrain culture. J Physiol (Lond) 500:739–749
Ropert N, Miles R, Korn H (1990) Characteristics of miniature inhibitory postsynaptic currents in CA1 pyramidal neurones of rat hippocampus. J Physiol (Lond) 428:707–722
Salin PA, Prince DA (1996) Spontaneous $GABA_A$ receptor-mediated inhibitory currents in adult rat somatosensory cortex. J Neurophysiol 75:1573–1588
Scanziani M, Capogna M, Gahwiler BH, Thompson SM (1992) Presynaptic inhibition of miniature excitatory synaptic currents by baclofen and adenosine in the hippocampus. Neuron 9:919–927
Scheich H (1991) Auditory cortex: comparative aspects of maps and plasticity. Curr Opin Neurobiol 1:236–247
Schiller J, Schiller Y, Stuart G, Sakmann B (1997) Calcium action potentials restricted to distal apical dendrites of rat neocortical pyramidal neurons. J Physiol (Lond) 505:605–616
Schoppa NE, Kinzie JM, Sahara Y, Segerson TP, Westbrook GL (1998) Dendrodendritic inhibition in the olfactory bulb is driven by NMDA receptors. J Neurosci 18:6790–6802
Schreiner CE (1992) Functional organization of the auditory cortex: maps and mechanisms. Curr Opin Neurobiol 2:516–521
Schwartzkroin PA, Mathers LH (1978) Physiological and morphological identification of a nonpyramidal hippocampal cell type. Brain Res 157:1–10
Shepherd GM (1994) Discrimination of molecular signals by the olfactory receptor neuron. Neuron 13:771–790
Shin C, Pedersen HB, McNamara JO (1985) γ-Aminobutyric acid and benzodiazepine receptors in the kindling model of epilepsy: a quantitative radiohistochemical study. J Neurosci 5:2696–2701
Sik A (1997) Interneurons in the hippocampal dentate gyrus: an in vivo intracellular study. Eur J Neurosci 9:573–588
Sik A, Hajos N, Gulacsi A, Mody I, Freund TF (1998) The absence of a major Ca^{2+} signaling pathway in GABAergic neurons of the hippocampus. Proc Natl Acad Sci USA 95:3245–3250
Sillito AM (1984) Functional considerations of the operation of GABAergic inhibitory processes in the visual cortex. In: Jones EG, Peters A (eds) Cerebral cortex. Functional properties of cortical cells. Plenum, New York, pp 91–117
Silver RA (1998) Neurotransmission at synapses with single and multiple release sites. In: Faber DS, Korn H, Redman SJ, Thompson SM, Altman JS (eds) Workshop IV. Central Synapses. Quantal Mechanisms and Plasticity. Human Frontier, Strasbourg, pp 130–139
Singer JH, Berger AJ (1999) Contribution of single-channel properties to the time course and amplitude variance of quantal glycine currents recorded in rat motoneurons. J Neurophysiol 81:1608–1616
Singer W, Gray CM (1995) Visual feature integration and the temporal correlation hypothesis. Ann Rev Neurosci 18:555–586
Soltesz I, Bourassa J, Deschenes M (1993) The behavior of mossy cells of the rat dentate gyrus during theta oscillations in vivo. Neuroscience 57:555–564
Soltesz I, Mody I (1994) Patch-clamp recordings reveal powerful GABAergic inhibition in dentate hilar neurons. J Neurosci 14:2365–2376

Staley KJ, Soldo BL, Proctor WR (1995) Ionic mechanisms of neuronal excitation by inhibitory GABAA receptors. Science 269:977–981
Staley KJ, Mody I (1992) Shunting of excitatory input to dentate gyrus granule cells by a depolarizing GABAA receptor-mediated postsynaptic conductance. J Neurophysiol 68:197–212
Staubli U, Lynch G (1990) Stable depression of potentiated synaptic responses in the hippocampus with 1–5 Hz stimulation. Brain Res 513:113–118
Stelzer A (1992) GABAA receptors control the excitability of neuronal populations. Int Rev Neurobiol 33:195–287
Stelzer A, Simon G, Kovacs G, Rai R (1994) Synaptic disinhibition during maintenance of long-term potentiation in the CA1 hippocampal subfield. Proc Natl Acad Sci USA 91:3058–3062
Stelzer A, Shi H (1994) Impairment of GABAA receptor function by *N*-methyl-D-aspartate-mediated calcium influx in isolated CA1 pyramidal cells. Neuroscience 62:813–828
Stuart G, Spruston N, Sakmann B, Hausser M (1997) Action potential initiation and backpropagation in neurons of the mammalian CNS. Trends Neurosci 20:125–131
Stuart G (1999) Voltage-activated sodium channels amplify inhibition in neocortical pyramidal neurons. Nature Neurosci 2:144–150
Suga N, Zhang Y, Yan J (1997) Sharpening of frequency tuning by inhibition in the thalamic auditory nucleus of the mustached bat. J Neurophysiol 77:2098–2114
Svoboda KR, Adams CE, Lupica CR (1999) Opioid receptor subtype expression defines morphologically distinct classes of hippocampal interneurons. J Neurosci 19:85–95
Tamas G, Buhl EH, Somogyi P (1997a) Massive autaptic self-innervation of GABAergic neurons in cat visual cortex. J Neurosci 17:6352–6364
Tamas G, Buhl EH, Somogyi P (1997b) Fast IPSPs elicited via multiple synaptic release sites by different types of GABAergic neurone in the cat visual cortex. J Physiol (Lond) 500:715–738
Tamas G, Somogyi P, Buhl EH (1998) Differentially interconnected networks of GABAergic interneurons in the visual cortex of the cat. J Neurosci 18:4255–4270
Tang C-M, Margulis M, Shi Q-Y, Fielding A (1994) Saturation of postsynaptic glutamate receptors after quantal release of transmitter. Neuron 13:1385–1393
Thalmann RH (1988) Blockade of a late inhibitory postsynaptic potential in hippocampal CA3 neurons in vitro reveals a late depolarizing potential that is augmented by pentobarbital. Neurosci Lett 95:155–160
Thiels E, Barrionuevo G, Berger TW (1994) Excitatory stimulation during postsynaptic inhibition induces long-term depression in hippocampus in vivo. J Neurophysiol 72:3009–3016
Thompson SM, Deisz RA, Prince DA (1988) Relative contributions of passive equilibrium and active transport to the distribution of chloride in mammalian cortical neurons. J Neurophysiol 60:105–124
Thompson SM, Capogna M, Scanziani M (1993) Presynaptic inhibition in the hippocampus. Trends Neurosci 16:222–227
Thompson SM (1994) Modulation of inhibitory synaptic transmission in the hippocampus. Prog Neurobiol 42:575–609
Thompson SM, Poncer J-C, Capogna M, Gahwiler BH (1997) Properties of spontaneous miniature $GABA_A$ receptor mediated synaptic currents in area CA3 of rat hippocampal slice cultures. Can J Physiol Pharmacol 75:495–499
Thompson SM, Gahwiler BH (1989) Activity-dependent disinhibition III. Desensitization and GABAB receptor-mediated presynaptic inhibition in the hippocampus in vitro. J Neurophysiol 61:524–533
Thompson SM, Gahwiler BH (1992) Comparison of the actions of baclofen at pre- and postsynaptic receptors in the rat hippocampus in vitro. J Physiol (Lond) 451:329–345

Thomson AM, Deuchars J, West DC (1993) Single axon excitatory postsynaptic potentials in neocortical interneurons exhibit pronounced paired pulse facilitation. Neuroscience 54:347–360
Tomasulo RA, Ramirez JJ, Steward O (1993) Synaptic inhibition regulates associative interactions between afferents during the induction of long-term potentiation and depression. Proc Natl Acad Sci USA 90:11,578–11,582
Tong G, Jahr CE (1994) Multivesicular release from excitatory synapses of cultured hippocampal neurons. Neuron 12:51–59
Toth K, McBain CJ (1998) Afferent-specific innervation of two distinct AMPA receptor subtypes on single hippocampal interneurons. Nature Neurosci 1:572–578
Traub RD, Whittington MA, Stanford IM, Jefferys JGR (1996) A mechanism for generation of long-range synchronous fast oscillations in the cortex. Nature 383:621–624
Tsubokawa H, Ross WN (1996) IPSPs modulate spike backpropagation and associated $[Ca^{2+}]_i$ changes in the dendrites of hippocampal CA1 pyramidal neurons. J Neurophysiol 76:2896–2906
Turrigiano GG (1999) Homeostatic plasticity in neuronal networks: the more things change, the more they stay the same. Trends Neurosci 22:221–227
Van der Loos H, Glaser EM (1972) Autapses in neocortex cerebri: synapses between a pyramidal cell's axon and its own dendrites. Brain Res 48:355–360
Varela JA, Song S, Turrigiano GG, Nelson SB (1999) Differential depression at excitatory and inhibitory synapses in visual cortex. J Neurosci 19:4293–4304
Vautrin J, Schaffner AE, Barker JL (1994) Fast presynaptic GABAA receptor-mediated Cl^- conductance in cultured rat hippocampal neurones. J Physiol (Lond) 479:53–63
Verheugen JAH, Fricker D, Miles R (1999) Noninvasive measurements of the membrane potential and GABAergic action in hippocampal interneurons. J Neurosci 19:2546–2555
Vincent P, Armstrong CM, Marty A (1992) Inhibitory synaptic currents in rat cerebellar Purkinje cells: modulation by postsynaptic depolarization. J Physiol (Lond) 456:453–471
Vincent P, Marty A (1993) Neighboring cerebellar Purkinje cells communicate via retrograde inhibition of common presynaptic interneurons. Neuron 11:885–893
Vincent P, Marty A (1996) Fluctuations of inhibitory postsynaptic currents in Purkinje cells from rat cerebellar slices. J Physiol (Lond) 494:183–199
Vizi ES, Kiss JP (1998) Neurochemistry and pharmacology of the major hippocampal transmitter systems: synaptic and nonsynaptic interactions. Hippocampus 8:566–607
Voogd J, Glickstein M (1998) The anatomy of the cerebellum. Trends Neurosci 21: 370–375
Wagner JJ, Alger BE (1995) GABAergic and developmental influences on homosynaptic LTD and depotentiation in rat hippocampus. J Neurosci 15:1577–1586
Wagner JJ, Alger BE (1996a) Homosynaptic LTD and depotentiation: do they differ in name only? Hippocampus 6:24–29
Wagner JJ, Alger BE (1996b) Increased neuronal excitability during depolarization-induced suppression of inhibition. Soc Neurosci Abstr 22:1519
Wall PD (1995) Do nerve impulses penetrate terminal arborizations? A pre-synaptic control mechanism. Trends Neurosci 18:99–103
Wang J-H, Stelzer A (1996) Shared calcium signaling pathways in the induction of long-term potentiation and synaptic disinhibition in CA1 pyramidal cell dendrites. J Neurophysiol 75:1687–1701
Weinberger NM (1995) Dynamic regulation of receptive fields and maps in the adult sensory cortex. Ann Rev Neurosci 18:129–158
Weiser M, Bueno E, Sekirnjak C, Martone ME, Baker H, Hillman D, Chen S, Thornhill W, Ellisman M, Rudy B (1995) The potassium channel subunit KV3.1b is localized to somatic and axonal membranes of specific populations of CNS neurons. J Neurosci 15:4298–4314

White JH, Wise A, Main MJ, Green A, Fraser NJ, Disney GH, Barnes AA, Emson P, Foord SM, Marshall FH (1998) Heterodimerization is required for the formation of a functional $GABA_B$ receptor. Nature 396:679–682
Whittington MA, Traub RD, Jefferys JGR (1995) Synchronized oscillations in interneuron networks driven by metabotropic glutamate receptor activation. Nature 373:612–615
Whittington MA, Traub RD, Faulkner HJ, Stanford IM, Jefferys JGR (1997) Recurrent excitatory postsynaptic potentials induced by synchronized fast cortical oscillations. Proc Natl Acad Sci USA 94:12,198–12,203
Williams SR, Toth TI, Turner JP, Hughes SW, Crunelli V (1997) The 'window' component of the low threshold Ca^{2+} current produces input signal amplification and bistability in cat and rat thalamocortical neurones. J Physiol (Lond) 505:689–705
Wong RKS, Prince DA (1979) Dendritic mechanisms underlying penicillin-induced epileptiform activity. Science 204:1228–1231
Wong RKS, Watkins DJ (1982) Cellular factors influencing GABA response in hippocampal pyramidal cells. J Neurophysiol 48:938–951
Wong RKS, Prince DA, Basbaum AI (1979) Intradendritic recordings from hippocampal neurons. Proc Natl Acad Sci USA 76:986–990
Wu L-G, Saggau P (1995) $GABA_B$ receptor-mediated presynaptic inhibition in guinea-pig hippocampus is caused by reduction of presynaptic Ca^{2+} influx. J Physiol (Lond) 485:649–657
Wu L-G, Saggau P (1997) Presynaptic inhibition of elicited neurotransmitter release. Trends Neurosci 20:204–212
Xiang Z, Huguenard JR, Prince DA (1998) Cholinergic switching within neocortical inhibitory networks. Science 281:985–988
Xie X, Smart TG (1991) A physiological role for endogenous zinc in rat hippocampal synaptic neurotransmission. Nature 349:521–524
Xu T-L, Li J-S, Jin Y-H, Akaike N (1999) Modulation of the glycine response by Ca^{2+}-permeable AMPA receptors in rat spinal neurones. J Physiol (Lond) 514:701–711
Ylinen A, Bragin A, Nadasdy Z, Jando G, Szabo I, Sik A, Buzsaki G (1995a) Sharp wave-associated high-frequency oscillation (200 Hz) in the intact hippocampus: network and intracellular mechanisms. J Neurosci 15:30–46
Ylinen A, Soltesz I, Bragin A, Penttonen M, Sik A, Buzsaki G (1995b) Intracellular correlates of hippocampal theta rhythm in identified pyramidal cells, granule cells, and basket cells. Hippocampus 5:78–90
Yokoi M, Mori K, Nakanishi S (1995) Refinement of odor molecule tuning by dendrodendritic synaptic inhibition in the olfactory bulb. Proc Natl Acad Sci USA 92:3371–3375
Yoon K-W, Rothman SM (1991) The modulation of rat hippocampal synaptic conductances by baclofen and gamma-aminobutyric acid. J Physiol (Lond) 442:377–390
Zhang C-L, Messing A, Chiu SY (1999) Specific alteration of spontaneous GABAergic inhibition in cerebellar Purkinje cells in mice lacking the potassium channel Kv1.1. J Neurosci 19:2852–2864
Zhang DX, Levy WB (1993) Bicuculline permits the induction of long-term depression by heterosynaptic, translaminar conditioning in the hippocampal dentate gyrus. Brain Res 613:309–312
Zhang L, Spigelman I, Carlen PL (1991) Development of GABA-mediated, chloride-dependent inhibition in CA1 pyramidal neurones of immature rat hippocampal slices. J Physiol (Lond) 444:25–49
Zhang L, McBain CJ (1995a) Potassium conductances underlying repolarization and after-hyperpolarization in rat CA1 hippocampal interneurones. J Physiol (Lond) 488:661–672
Zhang L, McBain CJ (1995b) Voltage-gated potassium currents in stratum oriens-alveus inhibitory neurones of the rat CA1 hippocampus. J Physiol (Lond) 488:647–660

Zhang SJ, Jackson MB (1993) GABA-activated chloride channels in secretory nerve endings. Science 259:531–534
Zhang SJ, Jackson MB (1995) GABAA receptor activation and the excitability of nerve terminals in the rat posterior pituitary. J Physiol (Lond) 483:583–595
Zhang W, Vazquez L, Apperson M, Kennedy MB (1999) Citron binds to PSD-95 at glutamatergic synapses on inhibitory neurons in the hippocampus. J Neurosci 19:96–108
Zhang Y, Perez Velazquez JL, Tian GF, Wu C-P, Skinner FK, Carlen PL, Zhang L (1998) Slow oscillations ($\leq$1 Hz) mediated by GABAergic interneuronal networks in rat hippocampus. J Neurosci 18:9256–9268
Zheng W, Knudsen EI (1999) Functional selection of adaptive auditory space map by $GABA_A$-mediated inhibition. Science 284:962–965
Ziemann U, Corwell B, Cohen LG (1998) Modulation of plasticity in human motor cortex after forearm ischemic nerve block. J Neurosci 18:1115–1123

Section II
Pharmacology of the GABA System

$GABA_A$ Receptors

CHAPTER 2

The Molecular Architecture of GABA$_A$ Receptors

E. A. BARNARD

A. Repertoire of Subunit Types

I. Structural Diversity and Uniformity

The structure of the GABA$_A$ receptor was unknown until 1987, when its subunits were revealed by cDNA cloning. Much general information on its molecular properties had accrued from biochemical and pharmacological analyses prior to then (reviewed by STEPHENSON 1988). Starting from purification on a benzodiazepine affinity column of a protein preparation which retained the multiple types of binding site previously identified in the native receptors (SIGEL and BARNARD 1984), followed by peptide sequencing, cDNA cloning led to the structure of the first 2 subunit types, $\alpha 1$ and $\beta 1$ (SCHOFIELD et al. 1987). The topology of the subunits in the cell membrane (Fig. 1) was thus deduced and a superfamily of transmitter-gated ion channels became apparent (BARNARD et al. 1987). This, the "Cys-loop" superfamily (see Fig. 1) (COCKROFT et al. 1990; KARLIN and AKABAS 1995) is now known to contain five related receptor families (BARNARD 1996): acetylcholine (nicotinic), 5-hydroxytryptamine$_3$, GABA, glycine and glutamate (anion channel). The latter three, a set of anion channels, are more homologous to each other, sharing up to 27% amino acid sequence identity.

Based on those first two GABA$_A$ receptor sequences, homology screening led to the $\alpha 1$–3 and $\beta 1$–3 homologous subunits (LEVITAN et al. 1988) and subsequently to all of the others now known. These comprise a total of 19 related mammalian subunits (Fig. 2), each encoded by a different gene. Each of these polypeptides contains four deduced transmembrane hydrophobic segments (TM1–4). Figure 2 illustrates the eight different sequence sub-families into which these fall structurally and their relationships. The amino acid sequence identity shared between different sub-families is mostly about 35%, but can be as low as 23%, or as high as 47% ($\alpha 1/\gamma 2$, $\varepsilon/\gamma 3$). Within each sub-family the members, termed isoforms ($\alpha 1$, $\alpha 2$, . . .), generally share about 65%–80% sequence identity (but see Sect. A.II).

This high degree of heterogeneity is further increased by alternative exon splicing of the pro-mRNA, which generates from one gene two forms of the

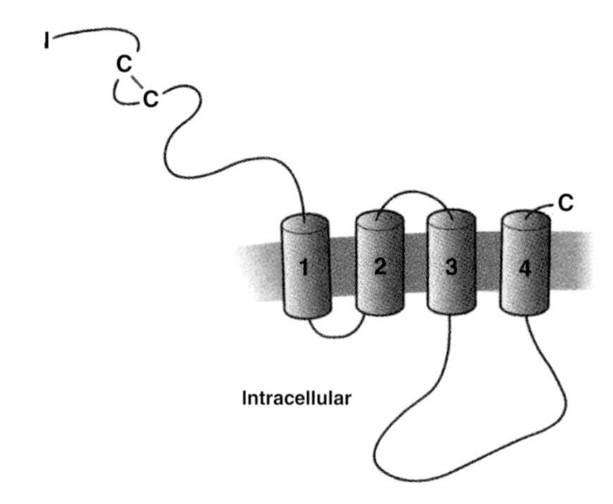

Fig. 1. The topology of the subunits of the $GABA_A$ receptors. This topology, deduced from hydropathy plots, has been confirmed in the case of the ACh receptor of this superfamily by the mapping of regions exposed to the extracellular or to the intracellular medium (Karlin and Akabas 1995) and by direct structural analysis (Miyazama et al. 1999). Since all of the subunits of both families have essentially the same pattern of hydrophobic sequences along the chain it is assumed that this topology is the same in both. Five of these subunits form one channel molecule. The two cysteines which form the Cys-loop structure are shown by *C-C*. The transmembrane domains are *numbered*; TM2 is selected from each of the five assembled subunits to form the major lining of the ion channel. The large *intracellular loop* shown starts at an approximately equivalent position along the chain in all of the subunits, but is very variable between subunits, both in sequence and in length. The *C-terminal tail* beyond TM4 is in only some of the subunits as shown, being limited to only about one or two residues in most

$\gamma 2$ subunit (Whiting et al. 1990; Kofuji et al. 1991) which can have different tissue distributions. This occurs likewise for the (avian) $\beta 2$ and $\beta 4$ subunits (Bateson et al. 1991; Harvey et al. 1994) and it is not excluded that these variants also occur in mammals, especially in the known mammalian $\beta 2$. In each case two products, longer and shorter, are expressed, designated "L" and "S" and differing by a short peptide at some point in the long intracellular loop between TM3 and TM4. Alternative transcripts of the $\beta 3$ and $\alpha 5$ subunits can also occur, but these would vary only the signal peptide or the 5′-untranslated region. Another product of alternative splicing deletes a short sequence at the N-terminus of the $\alpha 6$ subunit (Korpi et al. 1994), but this abolishes the receptor activity in the combinations of it so far tested. These latter three variants and the avian forms are not included in the enumerations of isoforms here.

The structural plan (Fig. 1) of the subunits is invariably conserved, with a cleaved signal peptide, an N-terminal extracellular, *N*-glycosylated domain of ~220 residues, near-constant locations of TM1–3, linkage of TM3 to TM4 by a long intracellular loop which is very variable between isoforms in both

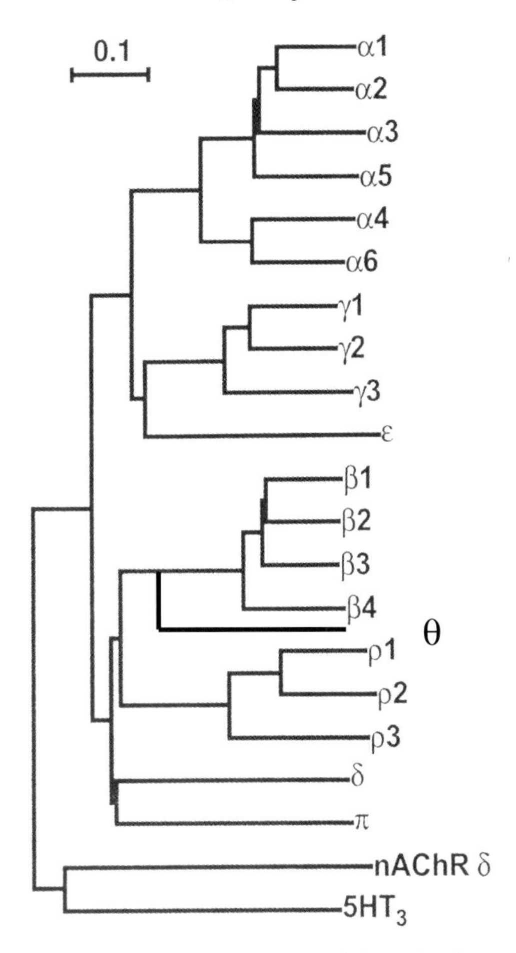

Fig. 2. A dendrogram depicting the relatedness of the subunit types of the $GABA_A$ receptors. The *scale bar* represents 10% sequence divergence on the horizontal axis. (The vertical distances are arbitrary). The eight functionally distinct subunit types form eight sub-families. Alignments are compared by a computer program which uses all sequence homology features. All sequences are from the rat except ε, π, θ (human) and $\beta 4$ (chicken, shown for comparison with θ). The mature amino acid sequences are used, after signal peptide removal was predicted by a uniform method. The tree was generated using as outgroup representatives rat nicotinic acetylcholine receptor δ and 5-HT3A subunits, which illustrate the degree of homology within the superfamily. Modified from Fig 3 of BARNARD et al. (1998), where the methods used and database accession numbers are given

size and sequence, and a small or vanishing C-terminal tail. Fifty-nine amino acid positions show complete constancy throughout all known mammalian (and indeed vertebrate) subunits, of which only 16 are in the four TM segments (13 in TM1 plus TM2).

The "Cys loop", a hallmark of the superfamily, is a 15-residue disulphide-bridged loop, with a constant central Tyr/Phe Pro X Asp motif, present in every subunit and ending at close to 70 residues before the start of TM1 (Fig. 1). In the centre of TM2 there is an octet sequence, which was thought to be fully conserved as a signature of the $GABA_A$ receptors, Thr Thr Val Leu Thr Met Thr Thr, and to be directly involved in the channel conduction. However, in the more recently discovered subunits (often missed in earlier cloning because that sequence was included as a basis for consensus probes) this constancy is in fact absent and six of these eight positions can show variation, four of them in the same subunit (θ). Nevertheless, certain residues in TM2 are involved in ion permeation, as discussed elsewhere in this volume and by BARNARD (2000). Two of those six natural changes are of Ser for Thr (at the first or seventh places in the octet), and this may function similarly. The θ subunit requires a γ subunit to be present also for channel function, and suppresses the channel function of $\alpha\beta$ heteromers (BONNERT et al. 1999), so it may not need all of the structural determinants of channel opening.

II. Subfamilies of Subunits

The size of the mature subunit varies noticeably, from about 420 amino acids (γ, π) to 609 (θ). The differences are largely due to very variable extension of the second intracellular loop; there is also an insertion in some cases (especially ε and θ) at the N-terminus. These differences have a considerable effect on estimating the true degrees of similarity of the subunits. Parts of those extensions may have only a low effect in differentiating function, as is suggested by the species differences in a given sequence being largely concentrated in them. As yet, this issue has been little probed by their truncation or peptide exchange. The percentage identity comparisons used here (and generally) and the nodal positions in the dendrograms that can be constructed are both influenced by it and by its ambiguities in aligning the sequences.

Four of the subfamilies contain (so far) only one member each, δ, ε, π, θ (Fig. 2). These play a different role to the other subunits in the composition of the receptors, as discussed in Sect. E. The separation into a one-member subfamily is less clear for θ, since its sequence identity with β subunits (about 50%: BONNERT et al. 1999) is distinctly higher than that *within* another subfamily, the α subunits. $\alpha 4$ and $\alpha 6$ are only about 38% identical to $\alpha 1$ or $\alpha 2$ (all compared in the human). This shows that the classification into sub-families, while generally stated to be by sequence, is partly based on function. $\alpha 4$ and $\alpha 6$ form a sub-group which is more distinct in sequence from the other sub-group of four α subunits seen in Fig. 2. That separation has turned out to be greater than, e.g. that of θ or ε from a neighbouring subfamily (ε having 42%–47% identity with γ1–3, all human: WHITING et al. 1997). This well reflects their roles in receptor composition: all of the α subunits can function in combination with a β and a γ subunit, but only $\alpha 4$ and $\alpha 6$ then confer insensitivity to benzodiazepines. Neither θ or ε function in combination with $\alpha\beta$ alone.

Another β subunit was cloned from the chicken (Bateson et al. 1991), termed β4 because it shares a maximum of 77% identity with mammalian β1–3 mature subunits, and often significantly less. That contrasts with the 92% or higher identity between the chicken and mammalian β3 orthologues (Bateson et al. 1990), which is typical of the very high species conservation in the $GABA_A$ receptors. Since the most recently cloned θ subunit is closest in sequence to β subunits (Bonnert et al. 1999) (Fig. 2) that raises the question whether θ and β4 are orthologues. This appears not to be the case, since human θ shares only about 51% identity with both human β1 and avian β4 (less with human β2 and β3). Occurrence of a mammalian β4 has not as yet been investigated, but cannot be excluded. Although the sequence divergence of θ from others is less than is the case for any other subfamily, since it does not act as a β subunit when expressed (Bonnert et al. 1999), it is assigned to a separate subfamily.

B. The Subunit Number per Receptor Molecule

To understand the construction of $GABA_A$ receptor subtypes from this repertoire of subunits, it is necessary first to establish the total number of subunits in each receptor molecule, and then to know whether this number is constant for all the native compositions. It has become clear that – with the possible exception of the ρ subunits, considered in Sect. F below – this number will be made up by several subunit types in each molecule, i.e. the receptors are in general heteromeric (discussed in Sect. C.II). Hence, the ultimate goal must be to know the stoichiometries of the subunit types within that number, for the range of $GABA_A$ receptors in situ.

Regarding the number of subunits per receptor, the prediction has often been made that this will be the same (five subunits) as for another transmitter-gated ion channel where the composition has been unequivocally established: the $GABA_A$ receptor subunits show a low but definite sequence homology with the subunits of the nicotinic acetylcholine receptors. Both are in the same superfamily within the transmitter-gated ion channels (Schofield et al. 1987; Barnard 1996, 2000). The muscle type of that receptor occurs in the *Torpedo* electric organ at such a high density in post-synaptic membrane sheets that it is possible to prepare membranes containing a surface lattice of the receptors which form into tubular crystals. From these a three-dimensional structure of the molecule could be obtained by cryo-electron microscopy and image analysis by N. Unwin and colleagues, now at 4.6 A resolution (Miyazawa et al. 1999). Those studies provide absolute proof that this receptor is pentameric, with the ion channel located in the centre of five homologous transmembranous subunits. A wealth of other studies had established that these are of four types, α, β, γ and δ, with two copies of the agonist-binding α subunit per molecule (reviewed by Devillers-Thiery et al. 1993; Karlin and Akabas 1995).

For the $GABA_A$ receptors, the situation is necessarily more complex, since the unique situation in the *Torpedo* post-synaptic membranes does not recur and since there are several classes of subunits involved in the receptor population in highly variable ways. It is, therefore, preferable to use the natural $GABA_A$ receptor population from the brain, rather than a selected expressed recombinant composition which may or may not be representative of that population, and to make direct analyses thereon, since these will not be limited by an assumption of the subunit classes to be taken as co-assembling. These requirements have been met using purified $GABA_A$ receptors from pig brain cortex and symmetry analysis on the electron microscope images of the dispersed molecules of the receptors. This method yields a power spectrum for each particle with a peak at its dominant symmetry. This symmetry is five-fold, over the population of particles analysed (NAYEEM et al. 1994). Further, the negatively-stained images obtained for all of the receptor particles indicated a central pore of the pentameric rosette, corresponding to the image observed similarly with negatively-stained *Torpedo* receptor particles, due in the latter (MIYAZAWA et al. 1999) to a central channel in the membrane enclosed within the receptor. The particles isolated from the brain will comprise a variety of $GABA_A$ receptor subtypes. We can only say that at least the majority are pentameric, since a deviating small minority with an atypical subunit composition would not be detected in the experimental noise. Evidence from independent methods has supported this conclusion: molecular weights in solution, when determined by rigorous hydrodynamic methods, of native brain receptors (MAMALAKI et al. 1989) or recombinant $\alpha 1$ $\beta 1$ $\gamma 2$ receptors (KNIGHT et al. 1998) agree with a pentamer, Estimates of the ratios of α, β and γ subunits in their recombinant receptors, noted below, also fit best with a subunit total of five. In view of the convergence in these diverse cases, and the concurrence with other receptors in the same superfamily (BARNARD et al. 1996), it is presumed that the pentameric structure holds for all of the $GABA_A$ receptors. This has not been studied, however, specifically for the subtypes containing δ, ε, π or ρ subunits.

C. Subunits Within the Pentamer

I. Two Subunit Pools for Receptor Assembly

Some of the ionotropic GABA receptors in the retina exhibit a highly distinctive pharmacology, these alone being insensitive to bicuculline and to barbiturates and neurosteroids (for details see BORMANN and FEIGENSPAN, Chap. 10, this volume). They are also sensitive to the agonist cis-4-aminocrotonic acid and are unaffected by benzodiazepines (although those latter properties are also found in minorities of other $GABA_A$ receptors). These have been described previously as "$GABA_C$ receptors". This type contains the $\rho 1$, $\rho 2$ or $\rho 3$ subunits, all three being predominantly in the retina (CUTTING et al. 1991, 1992; WANG et al. 1994; ENZ et al. 1995; ZHANG et al. 1995; OGURUSU et al. 1997) but some are expressed in other brain regions also (ENZ et al. 1995; WEGELIUS et al. 1998; ENZ and CUTTING 1999; OGURUSU et al. 1999). On the retinal bipolar cells, the ρ subunits occur in synaptic receptor clusters on cells which are separate from the clusters of non-ρ subunits at other synapses (FLETCHER et al. 1998; KOULEN et al. 1998). In recombinant co-expressions so far investigated, the ρ subunits do not participate in combinations with the aforementioned α, β or γ types (SHIMADA et al. 1992; KUSAMA et al. 1993a; ENZ et al. 1995; HACKAM et al. 1998). Hence, on present information a pool of at least 16 subunit types (plus at least 2 splice variants) is used in forming the main class of mammalian $GABA_A$ receptors, plus a second pool of at least 3 ρ subunits which are used separately. Since the ρ subunits are homologous to the other subunits and form similar anion channels, the International Union of Pharmacology (IUPHAR) places them within the $GABA_A$ receptors and discontinues the term "$GABA_C$ receptors", noting that it is illogical to place within the GABA receptors the metabotropic B class (a long-established designation) between A and C ion channel classes (BARNARD et al. 1998). The former C sub-class is termed now the A0r sub-class (r denoting ρ-containing and the zero the absence of the main pharmacological properties of the others), i.e. the $GABA_{A0r}$ receptors.

II. A Constrained Combinatorial System for the Receptor Compositions

We therefore start from the situation that a repertoire of at least 20 mammalian subunits (including the $\gamma 2$ splice variant) is available. The total is 21 if the splice variant of $\beta 2$, found to be expressed in the chicken (HARVEY et al. 1994), is included, its origin in the TM3/TM4 loop being similar to the splicing there of $\gamma 2$, which is known (see above) to occur both in birds and mammals; indeed, two polypeptide forms of $\beta 2$ have also been found (although not yet sequenced) in the mammal (BENKE et al. 1994). This set is drawn upon to a total of five for each receptor molecule. The selection for this produces a combinatorial system for constructing these receptors. This could in principle generate an impossibly large number of subtypes: constraints which reduce this exist at several levels. The first is the separation into two pools for co-assembly as described above. This removes from the potential total the combinations of the ρ subunits with any of the others. The second is that, so far as is known, all normal $GABA_A$ receptor molecules other than A_{0r} forms (i) require both α and β subunits; (ii) require in addition one or more of the γ, δ, ε, π or θ subunit types, which do not occur otherwise; (iii) usually contain either three or four different subunit types (which may include dual isoforms of one type, e.g. $\alpha 1 \alpha 6$). A possible exception to (ii) might be receptors containing only α and β subunits: in vitro these can robustly co-express functional receptors in all host cell types tried (SCHOFIELD et al. 1987; SHIVERS et al. 1989; PUIA et al. 1991; SHINGAI et al. 1991; ANGELOTTI and MACDONALD 1993; ATKINSON et al. 1994; HARTNETT et al. 1996), which behave as pentamers (KNIGHT et al. 1999) and which can be maintained long-term in stable cell lines (Moss et al. 1990; HADINGHAM et al. 1992). It is unknown, however, whether such $\alpha\beta$ receptors exist in vivo, though there is some evidence that they might in the case of an $\alpha 4\ \beta$ receptor (BENCSITS et al. 1999).

Binary combinations other than $\alpha\beta$ pairs, or even single subunits, can, for most of the α, β or γ isoforms, also be expressed functionally, in oocytes and in some but not all (ANGELOTTI et al. 1993) transfected mammalian cell types (BLAIR et al. 1988; PRITCHETT et al. 1988; SHIVERS et al. 1989; SIGEL et al. 1990; VERDOORN et al. 1990; SANNA et al. 1995; KRISHEK et al. 1996). This expression is in most cases weak, depending on the subtype, species or host cell, and it is always much increased when supplemented to give an $\alpha\beta\gamma$ combination. It is not considered, therefore, to give an exception to the aforementioned rules operating in vivo. The selection for $\alpha\beta\gamma$ is such that even the robustly expressed $\alpha\beta$ form disappears when a γ subunit is added (ANGELOTTI and MACDONALD 1993).

Considering rule (ii), in the great majority of brain receptors it is a γ subunit that complements α and β, as deduced from immunocytochemical and co-immunoprecipitation evidence (for references see BARNARD et al. 1998). That evidence also shows that $\gamma 2$ is by far the most abundant and ubiquitous of the $GABA_A$ receptor subunits in the CNS; by immunogold labelling the $\gamma 2$ subunit is very commonly seen localised at the same synaptic junction as α and β subunits (SOMOGYI et al. 1996; NUSSER et al. 1998). The dominance of $\alpha\beta\gamma$ types is also shown by the high percentage of the native $GABA_A$ receptors sensitive to benzodiazepine (BZ) drugs, for which an $\alpha\beta\gamma$ combination is required. The far-reaching effects on the receptor population of the deletion of γ subunits in transgenic mice are described by H. MÖHLER (Chap. 3, this volume).

In the limit of rule (iii), the theoretical maximum of different subunit types or isoforms combined in one molecule is five; analysis of extracted cerebellar $GABA_A$ receptors using several isoform-specific antibodies in turn (JECHLINGER et al. 1998) gave results that were compatible with this maximum of five types occurring in certain very limited cases. These rules are derived from a large body of observations on the formation of functional receptors in heterologous expression or on analyses of co-occurrence of subunits in receptors in or from native tissues.

That set of requirements arises, of course, from the types of interaction which can occur between the surfaces of different subunits. The interactions of the subunits which are thus selected must be energetically favourable for the assembly, the correct targeting and the stability of the active receptor. Mostly the structural barriers to that correct assembly must be low, since any ternary combination of the $\alpha\beta\gamma$ (i.e. $\alpha_i+\beta_j+\gamma_k$) form tested so far can interact in some or other host cell to produce a functional receptor, apparently self-directed to a single type (examples in SIGEL et al. 1990; HADINGHAM et al. 1992; ANGELOTTI and MACDONALD 1993; SAXENA and MACDONALD 1994; DUCIC et al.

1995; KIRSCH et al. 1995; SIEGHART 1995; WAFFORD et al. 1996; NEELANDS et al. 1999). This denotes high complementarity of the tertiary structures of three diverse subunit types, since (as reviewed above) homomers are strongly disfavoured. The exception is the ρ class of subunits, and these differ from the others (as tested in α and β) in a determinant in the N-terminal domain (HACKAM et al. 1998) which directs the interactions for separate assembly from the aforementioned two pools.

If the only constraint on assembly when α, β and γ subunits are present is that all those three types must co-assemble, then a possible total of 96 $\alpha\beta\gamma$-containing mammalian receptors could be created in this sub-class. The evidence on the native BZ-sensitive receptors suggests that their multiplicity, although considerable, is well below this. Obviously a further constraint is the local gene expression program, since some theoretical partners will not co-occur in the same cells. For example, $\alpha6$, $\alpha4$, δ or $\gamma3$ have not been found with certain others. A second level of constraint here is that of the targeting or chaperone or anchoring mechanisms, which can direct subunit selection in the targeting or localisation or synaptic clustering (e.g. via gephyrin) of $GABA_A$ receptors (CRAIG et al. 1996; ESSRICH et al. 1998; KNEUSSEL et al. 1999). Intermediate complexes which are not permissive for a preferred path of receptor assembly become degraded. That topic cannot be reviewed here, but it should be noted that for the $GABA_A$ receptors such processing in heterologous expression may not be a guide to its path in the neurones and may also differ between neuronal types, determined by the availabilities of specific controlling factors (as just noted). In vitro it has been found to vary for some $GABA_A$ receptor subunits even between different host cells. An example of more selective pairing in situ than in recombinant expression is given by the set of $\alpha1$, $\alpha6$ and δ subunits. The recombinant $\alpha1$ and δ subunits assemble well with β subunits to form functional receptors in each of three host systems used (SAXENA and MACDONALD 1994; DUCIC et al. 1995; KRISHEK et al. 1996). However, although those three subunit types co-exist in the same cerebellar granule cell, δ is replaced by $\gamma2$ in the receptors there which contain $\alpha1$ (alone) plus a β subunit, δ always being combined instead with $\alpha6$. This was shown by a variety of approaches: comprehensive immunogold localisations (NUSSER et al. 1998), co-immunopurifications (QUIRK et al. 1994b; JECHLINGER et al. 1998), an immuno/freeze-fracture technique (CARUNCHO and COSTA 1994) and again by $\alpha6$ truncation through gene targeting, which is found to deplete cerebellar $\alpha6$ and δ subunits together (JONES et al. 1997). This illustrates the additional level of constraint on the receptor compositions that can be exerted by processing in situ.

The α, β and γ subunits are used together, therefore, in a combinatorial selection, greatly limited by the specific constraints described here. The roles of the "alternative" subunits δ, ε, π and θ will be reviewed below.

D. Stoichiometry Within the Pentamer

I. Co-occurrence of Two Isoforms of One Subunit Type

The majority of $GABA_A$ receptors contain, therefore, α, β and γ subunits, while the total of the subunits per molecule is five (Fig. 2). Hence the receptors in this set can obviously have one of three general compositions: $(\alpha)_2\ (\beta)_2\ \gamma$ *or* $(\alpha)_2\ \beta\ (\gamma)_2$ *or* $\alpha\ (\beta)_2\ (\gamma)_2$. (Here, parentheses are used to indicate that the subscript numeral shown here represents counting of the isoforms present in one molecule and not the isoform identity). Such additional cases as $(\alpha)_3\ \beta\ \gamma$ *or* $\alpha\ \beta\ (\gamma)_3$ would have been theoretically possible, but measurements of an electrophysiological property determined quantitatively by the number of tagged recombinant subunits of each type forming the channel (BACKUS et al. 1993), in the case of co-expression of the $\alpha3\ \beta2\ \gamma2$ subunits, have excluded (at least in that case) the presence of three of any of those types in one receptor molecule.

The next logical step, therefore, in evaluating the potential combinations of subunits is to ask whether two isoforms of α *or* of β *or* of γ can occur in one receptor molecule, e.g. to produce compositions of the type $(\alpha1\ \alpha2)\ (\beta)_2\ \gamma$.

In the case of the α subunits, there is a variety of evidence for such co-occurrence of isoforms, in a minority of $GABA_A$ receptors. This has come first from antibody detection of (for example) an α isoform, when a brain-derived population of $GABA_A$ receptors is purified using an antibody specific for a different α isoform or from subtractive immuno-depletions of those two isoforms. Receptors containing at least the pairs $\alpha1\alpha2$, $\alpha1\alpha3$, $\alpha1\alpha5$, $\alpha2\alpha3$, $\alpha3\alpha5$, $\alpha4\alpha1$, $\alpha4\alpha2$ and $\alpha4\alpha3$ have been detected thus (each in a minority of the population containing the respective individual isoforms) (ENDO and OLSEN 1993; POLLARD et al. 1993; EBERT et al. 1994; MCKERNAN and WHITING 1996; BENKE et al. 1997; BENCSITS et al. 1999). The $\alpha6$ subunit can also pair in this manner. $\alpha6$ occurs (in the mature brain) only in the cerebellar granule cells (LAURIE et al. 1992; THOMPSON et al. 1994) and in the similar granule cells of the cochlear nucleus (VARECKA et al. 1994). In the cerebellum immuno-purification has shown $\alpha1$ and $\alpha6$ co-occurring in one receptor in a minority of cases (POLLARD et al. 1995; KHAN et al. 1996; JECHLINGER et al. 1998); a second approach, using antibody labelling with electron microscopy, has likewise shown that $\alpha6$ can co-localise with $\alpha1$ (NUSSER et al. 1998), although not for all of the $\alpha6$ subunits there. In those analyses most of the receptors also contained $\gamma2$ subunits.

Immunopurification analysis has also indicated such co-occurrence of two β isoforms in $GABA_A$ receptors in brain extracts (LI and DE BLAS 1997; Jechlinger et al. 1998). For the γ subunits, again the similar use of isoform-specific antibodies has, on brain extracts or purified receptor preparations, shown evidence for the co-occurrence of $\gamma2$ with $\gamma3$ and also of $\gamma2L$ with $\gamma2S$ (KHAN et al. 1994; QUIRK et al. 1994). However, results to the contrary, i.e. with only $\gamma1$, $\gamma3$ or $\gamma2$ (2L and 2S not being tested) separately immunopurified from brain receptors, have also been reported (BENCSITS et al. 1999).

II. Possibilities for Subunit Stoichiometry

Since two isoforms of the α subunit can sometimes occur in one receptor, as reviewed above, the $GABA_A$ receptors are considered as having two α places in the pentamer. Likewise, since there is evidence that two γ isoforms can co-occur, it must be considered that there could also be two γ places in the pentamer. Yet two β isoforms have also been reported to co-exist, as noted above, creating ambiguity.

This ambiguity has also been probed in recombinant combinations, BACKUS et al. (1993) showed, in the aforementioned study of a $\alpha3\ \beta2\ \gamma2$ receptor in HEK293 cells, that the $(\alpha)_2\ \beta\ (\gamma)_2$ composition best fitted the properties observed there, whereas CHANG et al. (1996), using a similar principle (but in oocyte expression, and employing $\alpha1$, not $\alpha3$), found that the evidence favours the $(\alpha)_2\ (\beta)_2\ \gamma$ composition. The latter stoichiometry was also derived for $\alpha1\ \beta3\ \gamma2$ receptors expressed in HEK 293 cells, from the staining ratios of those subunits seen when separated in Western blots (TRETTER et al. 1997).

We do not know if any of these statements hold for the whole native population of $GABA_A$ receptors of the $\alpha\beta\gamma$ type. If all of the findings are correct, then there is not a single stoichiometry for that type in vivo, and either $(\alpha)_2\ (\beta)_2\ \gamma$ or $(\alpha)_2\ \beta\ (\gamma)_2$ can exist, depending on the isoforms involved. This question is as yet unsettled.

Nevertheless, for any particular subunit set which will form one receptor, we can assume that there will only be one stoichiometry and arrangement in the native pentamer in situ. This is found to be so with all other heteromeric proteins which contain tightly-bound subunits. For example, there is only one cyclic order of the subunits $(\alpha)_2\ \beta\ \gamma\ \delta$ present in the population of *Torpedo* acetylcholine receptors (TOYOSHIMA and UNWIN 1990; MIYAZAWA et al. 1999) and further, using those subunits, one does not find that the same receptor type, in a variety of skeletal muscles, can be expressed in another stoichiometry. As a general principle of protein chemistry, for each composition the stoichiometry and the circular order of subunits around the channel will be fixed, due to optimisation of the interactions at the interfaces of the different subunits. This considerably reduces the total number of theoretically possible subtypes, particularly so for the receptors containing two isoforms of, e.g. the α subunit.

E. $GABA_A$ Receptors Containing Other Types of Subunits

As noted earlier, at least four other types, δ, ε, π and θ occur, each with α and β subunits also in the molecule so far as our present limited knowledge goes. None of those can replace an α or β subunit in expression studies. Each of those four appears to function, therefore, by either replacing or complementing the γ subunit in the receptor. A further type, the ρ subunits, form a separate pool. There is no indication so far that any of those five types ever co-exist in a receptor.

I. The δ Subunit

The δ subunit has a restricted distribution in the rat brain, expression being highest in the cerebellar granule cells, next in the thalamus and olfactory bulb and very low or absent in many areas (SHIVERS et al. 1989; BENKE et al. 1991; LAURIE et al. 1992). In confirmation, δ was found, by antibody reaction, in only 11% of all the $GABA_A$ receptors extracted from rat brain but in 27% in rat cerebellum (QUIRK et al. 1995).

Recombinant $\alpha\beta\delta$ combinations can form GABA-gated channels, which are insensitive to diazepam (SAXENA and MACDONALD 1994, 1996). In the cerebellum δ has been detected in $\alpha6\ \beta\ \delta$ or $\alpha6\ \alpha1\ \beta\ \delta$ combinations only (see Sect. C.II) and in the thalamus in $\alpha4\ \beta\ \delta$ only (SUR et al. 1999). The latter is the main $\alpha4$-containing subtype in the thalamus, although $\alpha4$ is also present there in $\alpha4\ \beta\ \gamma2$ receptors (SUR et al. 1999). $\alpha4$ is found in the latter subtype in some other forebrain areas, too, and in other receptors there paired with another α subunit, or without a γ or δ subunit, all at very low abundance and all diazepam-insensitive (BENKE et al. 1997; BENCSITS et al. 1999). The brain receptors containing δ are also all BZ-insensitive and it is generally found that native δ and γ subunits are mutually exclusive (CARUNCHO and COSTA 1994; QUIRK et al. 1994b, 1995; JECHLINGER et al. 1998; ARAUJO et al. 1998; BENCSITS et al. 1999). Despite this, recombinant $\alpha\ \beta\ \gamma2\ \delta$ receptors can assemble and are functional in vitro, with distinctive properties (SAXENA and MACDONALD 1994). It is unclear at present to what extent, outside the cerebellum and thalamus, δ also occurs in receptors having α subunits other than $\alpha6$ or $\alpha4$.

II. The ε Subunit

This has a very restricted distribution, its mRNA and protein showing clearly in situ (in the adult primate) only in the hypothalamus and in the dentate gyrus hilar and CA3 regions of the hippocampal formation (WHITING et al. 1997). It is also present in spinal cord and the heart.

In transfected HEK 293 cells studies of $\alpha2\ \beta1\ \varepsilon$ or $\alpha1\ \beta3\ \varepsilon$ (DAVIES et al. 1997) or $\alpha1\ \beta1\ \varepsilon$ combinations (WHITING et al. 1997: also in oocytes) showed that in all of them ternary receptors activated by GABA can be formed. These are not modulated by BZ drugs and desensitise much more rapidly than $\alpha1\ \beta1$ or $\alpha1\ \beta1\ \gamma2$ receptors. In L929 fibroblasts, NEELANDS et al. (1999) expressed the $\alpha1\ \beta3\ \varepsilon$ combination and found that its chloride channel is both spontaneously active and gated by GABA. Again BZ-insensitive, this receptor has also acquired inhibition by furosemide, otherwise seen (KORPI et al. 1995; WAFFORD et al. 1996) only with $\alpha6\ \beta\ \gamma2$ or $\alpha4\ \beta\ \gamma2$ receptors. The channel conductance of the ε-containing receptors is as for $\alpha1\ \beta3\ \gamma2$ and $\alpha1\ \beta3\ \delta$ and not $\alpha\beta$ channels (NEELANDS et al. 1999).

III. The π Subunit

The π (for "peripheral") subunit has been found present in several human peripheral organs, principally the uterus, and in very low levels in hippocampus and cortex (HEDBLOM and KIRKNESS 1997). When co-expressed in HEK 293 cells, it could remove BZ binding of $\alpha1\ \beta1\ \gamma2$ receptors, indicating complex formation (HEDBLOM and KIRKNESS 1997). In the L929 cell as host, it was deduced by NEELANDS and MACDONALD (1999) that π could combine to form $\alpha5\ \beta3\ \pi$ functional receptors, on the basis of the changes in several functional properties compared to $\alpha5\ \beta3$. Further, formation of $\alpha5\ \beta3\ \gamma3\ \pi$ receptors when the four subunits were co-expressed was inferred in a similar way. The π-containing receptors were BZ-insensitive and had a channel conductance as large as that of the $\alpha\beta\gamma$ receptors and unlike that of the $\alpha\beta$ receptors.

The neuronal precursor cell line NT2 expresses native mRNAs for the same π, $\alpha5$, $\beta3$ and $\gamma3$ subunits but does not appear to form any π-containing receptors (NEELANDS and MACDONALD 1999). The function of π in vivo is still uncertain.

IV. The θ Subunit

As noted earlier (Sect. A.II), θ is relatively close to the β subunits in sequence (Fig. 2), but not in functional properties. All of the data so far on θ come from the study of BONNERT et al. (1999). The mRNA and protein for this subunit were discovered by those authors to be prominent in certain regions of primate brain, particularly in the substantia nigra and the striatum, and absent in many others, including the cerebellum. In regions rich in dopaminergic or in noradrenergic neurones, θ co-localises with these.

Exceptionally, θ assembles (so far as was detectable from co-immunoprecipitation of rat striatal extracts) with one α isoform only, $\alpha2$, and with $\gamma1$ and not with $\gamma2$, $\gamma3$, δ nor ε. BONNERT et al. (1999) concluded that the preferred θ combination is $\alpha2\ \beta1\ \gamma1\ \theta$. It is interesting, in view of the structural closeness of θ to β subunits, that in assembly at the cell surface (in the case of co-expression in HEK293 cells) θ was found to act as a β subunit. This would be compatible with a $(\alpha)_2\ \beta\theta\ \gamma$ composition.

Functional heterologous expression of θ required a quaternary set, $\alpha\ \beta\ \gamma\ \theta$; $\alpha2$ or $\alpha1$, and $\gamma1$ or $\gamma2$ were active in this. Modulation by BZ agonists or inverse agonists, or by pentobarbital or pregnanolone, were all unchanged by θ incorporation.

The observations with θ and with ε show that some receptors containing four different subunit classes in the molecule may be needed, to refine the properties to fit particular functional niches.

F. The ρ Subunits

The properties of the $GABA_{A0r}$ receptors containing this subunit type are covered in detail by BORMANN and FEIGENSPAN (Chap. 10, this volume). Here their structure should be noted. As reviewed above (Sect. C.I), three isoforms are known, $\rho 1$, $\rho 2$ and $\rho 3$. The heterologous expression of each of these alone can give strong functional expression, such that native ρ homo-oligomers have been assumed. However, this view must be modified, since the rat $\rho 2$ subunit, unlike the previously studied human $\rho 2$, does not form functional receptors alone in the oocyte, but requires the rat $\rho 1$ to do so (Zhang et al. 1995). The pharmacology of $\rho 1$ is thereby changed (to picrotoxin resistance); hence $\rho 1 \rho 2$ heteromeric receptors exist. Although the human $\rho 1$ or $\rho 2$ subunits can each assemble alone to functional receptors after transfection into HEK 293 cells, after their co-transfection detailed analysis of the properties has disclosed the formation of human $\rho 1 \rho 2$ heteromers also (Enz and Cutting 1999).

The rat $\rho 1$, $\rho 2$ and $\rho 3$ mRNAs all occur in the retina (for details see Bormann and Feigenspan, Chap. 10, this volume). The responses found on rat retinal bipolar cells (which express both $\rho 1$ and $\rho 2$ subunits: Enz et al. 1995) do not match those of $\rho 1$ receptors or $\rho 2$ receptors but can correspond to the $\rho 1 \rho 2$ receptor (Zhang et al. 1995). Hence, that hetero-oligomer appears to be functional in situ. Rat $\rho 3$ subunits can also form hetero-oligomeric receptors with $\rho 2$, as well as homomeric receptors (Ogurusu et al. 1999). Cells which express both $\rho 2$ and $\rho 3$ may therefore contain the $\rho 2 \rho 3$ receptor, but there is as yet little information on this.

While the ρ subunits were previously regarded as purely retinal, accumulating evidence has shown that all three also occur in several brain regions, although in different distributions. Thus, rat $\rho 3$ is expressed more in the hippocampus than in the retina or in other brain regions (Wegelius et al. 1998) and is several times more abundant in the embryonic (day 16, non-retinal) brain than in the adult, unlike the other two isoforms (Ogurusu et al. 1999). Those studies and the work of Enz et al. (1995) also showed that $\rho 2$ is present in the hippocampus and $\rho 1$ is very low or undetectable there and in other brain regions, except the superior colliculus, where $\rho 1$ and $\rho 2$ co-occur. Hence there are no general associations of these three, and there are regions where only $\rho 2$ out of these is detectable. Since rat $\rho 2$ does not express alone, in the oocyte (Zhang et al. 1995), it seems probable that some other unknown pairing of it occurs in vivo. This could be with a fourth, as yet unknown, ρ isoform. However, we cannot exclude the alternative, that $\rho 2$ is sometimes complexed with non-ρ subunits, which are always present. The tests reported to exclude this with $\rho 2$ or $\rho 3$ have not been exhaustive; it might require testing with more than one other partner, or some chaperone or other trafficking factor only found in native neurones. If this occurs, then the segregation of the ρ pool would break down in special cases.

G. Conclusions on the Subtypes

Within the constraints summarised above, it appears that a considerable number of $GABA_A$ receptor subtypes can exist in vivo, more than for any other of the transmitter-gated channels. Each different combination could in principle generate an individual pharmacology. In the $\alpha\beta\gamma$ sub-class, these subtypes can be recognised mainly by the great variation in the responses to different "BZ/ω" drugs, i.e. a very wide range of benzodiazepines and many unrelated structures that are active at a single modulatory site which is a characteristic of that sub-class. They range through modulatory agonists, partial agonists and inverse agonists to antagonists, and members can be selected therefrom to discriminate between the $\alpha\beta\gamma$ combinations. Those structures are reviewed elsewhere (Barnard et al. 1998), with a list (Table 4 therein) illustrating 17 cases of potential $GABA_A$ receptor subtypes defined thus. In particular, each change at the α position(s) or the γ position in the combination creates a different pharmacology within the scope of that wide range of modulators.

Where γ is replaced by δ, ε, π or θ, that series cannot be used (see Sect. E). However, the $GABA_A$ receptors have a wealth of other modulatory sites (reviewed in several other chapters in this volume) which could be exploited similarly to recognise subtypes. The pharmacology of the receptors containing these alternative subunits is in its infancy, but there are already indications that those subunits introduce differences at such sites on the receptor as those for neurosteroids or for loreclezole (Davies et al. 1997; Neelands et al. 1999; Neelands and Macdonald 1999).

All of the discriminations discussed here are made in the first instance in recombinant co-expression. In some cases we can seek to relate these to actual native combinations, as recognised from co-immunopurification results or co-localisations of subunits in situ or differing channel characteristics. In a very few favourable cases at present this may allow us to define functional native subtypes. For each of these particular cases strong evidence exists, based on a concurrence of all three of those approaches (with the co-localisations made at the EM level). Thus, they include the combinations $\alpha1\beta2\gamma2$ (Benke et al. 1994; Quirk et al. 1994b; Brickley et al. 1996; Somogyi et al. 1996; Nusser et al. 1998), $\alpha6\beta\gamma2$ and $\alpha6\beta\delta$ (Caruncho and Costa 1994; Quirk et al. 1994b; Saxena and Macdonald 1994; Ducic et al. 1995; Brickley et al. 1996; Jechlinger et al. 1998; Nusser et al. 1998). Much caution is required in pursuing this: even in those favourable cases the native isoform of β or of $\gamma2$ is often not established and nor is the stoichiometry within the molecule. Major barriers to absolute identifications are inherent, first in the special combinatorial system of the $GABA_A$ receptors: many more subtypes can in this case (but not for most other receptors) be created in vitro than are likely to occur in vivo. Other barriers arise from the complexity of the brain circuitry, and from the co-occurrence therein of multiple subtypes of GABA receptors in small regions or within a single neurone. To recognise all of the native $GABA_A$ receptors is a challenge for the long term.

References

Angelotti TP, Macdonald RL (1993) Assembly of $GABA_A$ receptor subunits: $\alpha 1\beta 1$ and $\alpha_1\beta_1$ γ_{2S} subunits produce unique ion channels with dissimilar single-channel properties. J Neurosci 13:1429–1440

Angelotti TP, Uhler MD, Macdonald RL (1993) Assembly of $GABA_A$ receptor subunits: analysis of transient single-cell expression utilizing a fluorescent substrate/marker gene technique. J Neurosci 13:1418–1428

Araujo F, Ruano D, Vitorica J (1998) Absence of association between δ and $\gamma 2$ subunits in native $GABA_A$ receptors from rat brain. Eur J Pharmacol 347:347–353

Atkinson AE, Bermudez I, Darlison MG, Barnard EA, Earley FGP, Possee RD, Beadle DJ, King LA (1993) Assembly of functional $GABA_A$ receptors in insect cells using baculovirus expression vectors. Neuroreport 3:597–600

Backus KH, Arigoni M, Drescher U, Scheurer L, Malherbe P, Möhler H, Benson A (1993) Stoichiometry of a recombinant $GABA_A$-receptor deduced from mutation-induced rectification. Neuroreport 5:285–288

Barnard EA (1992) Receptor classes and the transmitter-gated ion channels. Trends Biochem Sci 17:368–374

Barnard EA (1996) The transmitter-gated channels: a range of receptor types and structures. Trends Pharmacol Sci 17:305–308

Barnard EA (1998) The range of structures of the transmitter-gated channels. In: Endo M (ed) Pharmacology of ionic channel function (Handbook of Experimental Pharmacology). Springer, Berlin Heidelberg New York, pp 363–390

Barnard EA, Skolnick P, Olsen RW, Mohler H, Sieghart W, Biggio G, Braestrup C, Bateson AN, Langer SZ (1998) International Union of Pharmacology XV. Subtypes of γ-aminobutyric acid-A receptors: classification on the basis of subunit structure and receptor function. Pharmacol Rev 50:291–314

Baude A, Sequier JM, Mckernan RM, Oliver KR, Somogyi P (1992) Differential subcellular distribution of the $\alpha 6$ subunit versus the $\alpha 1$ and $\beta 2/3$ subunits of the $GABA_A$/benzodiazepine receptor complex in granule cells of the cerebellar cortex. Neuroscience 51:739–748

Bateson AN, Lasham A, Darlison MG (1991) γ-Aminobutyric acid-A receptor heterogeneity is increased by alternative splicing of a novel β-subunit gene transcript. J Neurochem 56:1437–1440

Bencsits E, Ebert V, Tretter V, Sieghart W (1999) A significant part of native γ-aminobutyric $acid_A$ receptors containing $\alpha 4$ subunits do not contain γ or δ subunits. J Biol Chem 274:19613–19616

Benke D, Fritschy JM, Trzeciak A, Bannwarth W, Mohler H (1994) Distribution, prevalence, and drug binding profile of γ-aminobutyric acid type A receptor subtypes differing in the β-subunit variant. J Biol Chem 269:27100-27107

Benke D, Michel C, Mohler H (1997) GABA(A) receptors containing the $\alpha 4$-subunit: prevalence, distribution, pharmacology, and subunit architecture in situ. J Neurochem 69:806–814

Blair LAC, Levitan ES, Dionne VE, Barnard EA (1988) Single subunits of the $GABA_A$ receptor form ion channels with properties characteristics of the native receptor. Science 242:577–579

Bonnert TP, McKernan RM, Farrar S, le Bourdelles B, Heavens RP, Smith DW, Hewson L, Rigby MR, Sirinathsinghji DJ, Brown N, Wafford KA, Whiting PJ (1999) Theta, a novel γ-aminobutyric acid type A receptor subunit. Proc Natl Acad Sci USA 96:9891–9896

Brickley SG, Cull-Candy SG, Farrant M (1996) Development of a tonic form of synaptic inhibition in rat cerebellar granule cells resulting from persistent activation of $GABA_A$ receptors. J Physiol (Lond) 497:753-759

Caruncho HJ, Costa E (1994) Double-immunolabeling analysis of $GABA_A$ receptor subunits in label-fracture replicas of cultured cerebellar granule cells. Receptors Channels 2:143–153

Chang Y, Wang R, Barot S, Weiss DS (1996) Stoichiometry of a recombinant $GABA_A$ receptor. J Neurosci, 16:5415–5424
Cockroft VB, Ostedgaard DJ, Barnard EA, Lunt GG (1990) Modelling of agonist binding to the ligand-gated ion channel superfamily of receptors. Proteins 8:386–397
Craig AM, Banker G, Chang W, McGrath ME, Serpinskaya AS (1996) Clustering of gephyrin at GABAergic but not glutamatergic synapses in cultured rat hippocampal neurons. J Neurosci 16:3166–3177
Cutting GR, Lu L, O'Hara BF, Kasch LM, Montrose Rafizadeh C, Donovan DM, Shimada S, Antonarakis SE, Guggino WB, Uhl GR, Azazian HH (1991) Cloning of the γ-aminobutyric acid (GABA) ρ_1 cDNA: a GABA receptor subunit highly expressed in the retina. Proc Natl Acad Sci 88:2673–2677
Cutting GR, Curristin S, Zoghbi H, O'Hara B, Seldin MF, Uhl GR (1992) Identification of a putative γ-aminobutyric acid (GABA) receptor $\rho 2$ cDNA and colocalization of the genes encoding rho2 (GABRR2) and rho1 (GABRR1) to human chromosome 6q14–q21 and mouse chromosome 4. Genomics 12:801–806
Davies PA, Hanna MC, Hales TG, Kirkness EF (1997) A novel class of GABA-A receptor subunit confers insensitivity to anaesthetic agents. Nature 385:820–823
Devillers-Thiery A, Galzi JL, Eisele JL, Bertrand S, Bertrand D, Changeux JP (1993) Functional architecture of the nicotinic acetylcholine receptor: a prototype of ligand-gated ion channels. J Membr Biol 136:97–112
Ducic I, Caruncho HJ, Zhu WJ, Vicini S, Costa E (1995) γ-Aminobutyric acid gating of Cl^- channels in recombinant $GABA_A$ receptors. J Pharmacol Exp Ther 272:438–445
Ebert B, Wafford KA, Whiting PJ, Krogsgaard-Larsen P, Kemp JA (1994) Molecular pharmacology of γ-aminobutyric acid type A receptor agonists and partial agonists in oocytes injected with different α, β, and γ receptor subunit combinations. Mol Pharmacol 46:957–963
Endo S, Olsen RW (1993) Antibodies specific for α subunit subtypes of $GABA_A$ receptors reveal brain regional heterogeneity. J Neurochem 60:1388–1398
Enz R, Brandstatter JH, Hartveit E, Wassle H, Bormann J (1995) Expression of GABA receptor $\rho 1$ and $\rho 2$ subunits in the retina and brain of the rat. Eur J Neurosci 7:1495–1501
Enz R, Cutting GR (1999) $GABA_C$ receptor ρ subunits are heterogeneously expressed in the human CNS and form homo- and hetero-oligomers with distinct physical properties. Eur J Neurosci 11:41–50
Essrich C, Lorez M, Benson JA, Fritschy JM, Luscher B (1998) Postsynaptic clustering of major $GABA_A$ receptor subtypes requires the $\gamma 2$ subunit and gephyrin. Nature Neurosci 1:563–571
Fletcher EL, Koulen P, Wassle H (1998) $GABA_A$ and $GABA_C$ receptors on mammalian rod bipolar cells. J Comp Neurol 396:351–365
Hackam AS, Wang TL, Guggino WB, Cutting GR (1998) Sequences in the amino termini of GABA ρ and $GABA_A$ subunits specify their selective interaction in vitro. J Neurochem 70:40–46
Hadingham KL, Harkness PC, McKernan RM, Quirk K, Le Bourdelles B, Horne AL, Kemp JA, Barnard EA, Ragan CI, Whiting P (1992) Stable expression of mammalian $GABA_A$ receptors in mouse cells: demonstration of functional assembly of benzodiazepine-responsive sites. Proc Natl Acad Sci USA 89:6378–6382
Hadingham KL, Wafford KA, Thompson SA, Palmer KJ, Whiting PJ (1995) Expression and pharmacology of human $GABA_A$ receptors containing $\gamma 3$ subunits. Eur J Pharmacol 291:301–309
Hadingham KL, Wafford KA, Bain C, Garrett EM, Heavens RP, Sirinathsinghji DJS, Whiting PJ (1996) Cloning of cDNA encoding the human γ-aminobutyric $acid_A$ receptor $\alpha 6$ subunit and characterization of the pharmacology of $\alpha 6$-containing receptors. Mol Pharmacol 49:253–259

Harvey RJ, Chinchetru MA, Darlison MG (1994) Alternative splicing of a 51-nucleotide exon that encodes a putative protein kinase C phosphorylation site generates two forms of the chicken γ-aminobutyric acid-A receptor $\beta 2$ subunit. J Neurochem 62:10–16
Heblom E, Kirkness EF (1997) A novel class of $GABA_A$ receptor subunit in tissues of the reproductive system. J Biol Chem 272:15346–15350
Jechlinger M, Pelz R, Tretter V, Klausberger T, Sieghart W (1998) Subunit composition and quantitative importance of hetero-oligomeric receptors: $GABA_A$ receptors containing $\alpha 6$ subunits. J Neurosci 18:2449–2457
Jones A, Korpi ER, McKernan RM, Pelz R, Nusser Z, Makela R, Mellor JR, Pollard S, Bahn S, Stephenson FA, Randall AD, Sieghart W, Somogyi P, Smith AJ, Wisden W (1997) Ligand-gated ion channel subunit partnerships: $GABA_A$ receptor $\alpha 6$ subunit gene inactivation inhibits δ subunit expression. J Neurosci 17:1350–1362
Karlin A, Akabas MH (1995) Toward a structural basis for the function of nicotinic acetylcholine receptors and their cousins. Neuron 15:1231–1244
Khan ZU, Gutierrez A, De Blas A (1994) Short and long forms of $\gamma 2$ subunits of GABA/benzodiazepine receptor from rat cerebellum. J Neurochem 63:371–374
Khan ZU, Gutierrez A, De Blas AL (1996): The $\alpha 1$ and $\alpha 6$ subunits can co-exist in the same cerebellar $GABA_A$ receptor maintaining their individual benzodiazepine-binding affinities. J Neurochem 66:685–691
Kirsch T, Kuhse I, Betz H (1995) Targeting of glycine receptor subunits to gephyrin-rich domains in transfected human embryonic kidney cells. Mol Cell Neurosci 6:450–462
Kneussel M, Brandstatter JH, Laube B, Stahl S, Muller U, Betz H (1999) Loss of post-synaptic $GABA_A$ receptor clustering in gephyrin-deficient mice. J Neurosci 19:9289–9297
Knight AR, Hartnett C, Marks C, Brown M, Gallager D, Tallman J, Ramabhadran TV (1998) Molecular size of recombinant $\alpha 1\beta 1$ and $\alpha 1\beta 1\gamma 2$ $GABA_A$ receptors expressed in Sf9 cells. Receptors Channels 6:1–18
Kofuji P, Wang JB, Moss SJ, Huganir RL, Burt DR (1991) Generation of two forms of the γ-aminobutyric acid-A receptor $\gamma 2$-subunit in mice by alternative splicing. J Neurochem 56:713–715
Korpi ER, Kuner T, Kristo P, Kohcer M, Herb A, Luddens H, Seeburg PH (1994) Small N-terminal deletion by splicing in cerebellar $\alpha 6$ subunit abolishes $GABA_A$ receptor function. J Neurochem 63:1167–1170
Korpi ER, Kuner T, Seeburg PH, Luddens H (1995) Selective antagonist for the cerebellar granule cell-specific γ-aminobutyric acid type A receptor. Mol Pharmacol 47:283–289
Koulen P, Brandstatter JH, Enz R, Bormann J, Wassle H (1998) Synaptic clustering of $GABA_C$ receptor ρ-subunits in the rat retina. Eur J Neurosci 10:115–127
Krishek BJ, Amato A, Connolly CN, Moss SJ, Smart TG (1996) Proton sensitivity of the $GABA_A$ receptor is associated with the receptor subunit composition. J Physiol (Lond) 492:431–443
Kusama T, Spivak CE, Whiting P, Dawson VL, Schaeffer JC, Uhl GR (1993) Pharmacology of GABAρ1 and GABAα/β receptors expressed in *Xenopus* oocytes and COS cells. Br J Pharmacol 109:200–206
Laurie DJ, Seeburg PH, Wisden W (1992) The distribution of 13 $GABA_A$ receptor subunit mRNAs in the rat brain. 2. Olfactory bulb and cerebellum. J Neurosci 12:1063–1076
Levitan ES, Blair LAC, Dionne VE, Barnard EA (1988a) Biophysical and pharmacological properties of cloned $GABA_A$ receptor subunits expressed in *Xenopus* oocytes. Neuron 1:773–781
Levitan ES, Schofield PR, Burt DR, Rhee LM, Wisden W, Kohler M, Fujita N, Rodriguez HF, Stephenson FA, Darlison MG, Barnard EA, Seeburg PH (1988b) Structural and functional basis for $GABA_A$ receptor heterogeneity. Nature 336:76–79

Li M, De Blas AL (1997) Coexistence of two β subunit isoforms in the same γ-aminobutyric acid type A receptor. J Biol Chem 272:16564–16569
Mamalaki C, Barnard EA, Stephenson FA (1989) Molecular size of the γ-aminobutyric $acid_A$ receptor purified from mammalian cerebral cortex. J Neurochem 52:125–134
Mckernan RM, Whiting PJ (1996) Which $GABA_A$-receptor subtypes really occur in the brain? Trends Neurosci 19:139–143
Miyazawa A, Fujiyoshi Y, Stowell M, Unwin N (1999) Nicotinic acetylcholine receptor at 4.6 Å resolution: transverse tunnels in the channel wall. J Mol Biol 288:765–786
Moss S, Smart TG, Porter N, Nayeem N, Devine J, Stephenson FA, Macdonald RL, Barnard EA (1990) Cloned GABA receptors are maintained in a stable cell line: allosteric and channel properties. Mol Brain Research 189:77–88
Nayeem N, Green TP, Martin IL, Barnard EA (1994) Quaternary structure of the native $GABA_A$ receptor determined by electron microscope image analysis. J Neurochem 62:815–818
Neelands TR, Fisher JL, Bianchi M, Macdonald RL (1999) Spontaneous and γ-aminobutyric acid activated $GABA_A$ receptor channels formed by epsilon subunit-containing isoforms. Mol Pharmacol 55:168–178
Neelands TR, Macdonald RL (1999) Incorporation of the π subunit into functional γ-aminobutyric acid-A receptors. Mol Pharmacol 56:598–610
Nusser Z, Sieghart W, Somogyi P (1998) Segregation of different $GABA_A$ receptors to synaptic and extrasynaptic membranes of cerebellar granule cells. J Neurosci 18:1693–1703
Ogurusu T, Eguchi G, Shingai R (1997) Localisation of GABA receptor $\rho 3$ subunit in rat retina. Neuroreport 8:925–927
Ogurusu T, Yanagi K, Watanabe M, Fukaya M, Shingai R (1999) Localisation of GABA receptor $\rho 2$ and $\rho 3$ subunits in rat brain and functional expression of homo-oligomeric $\rho 3$ receptors and hetero-oligomeric $\rho 2 \rho 3$ receptors. Receptors Channels 6:463–476
Pollard S, Duggan MJ, Stephenson FA (1993) Further evidence for the existence of alpha subunit heterogeneity within discrete γ-aminobutyric $acid_A$ receptor subpopulations. J Biol Chem 268:3753–3757
Pollard S, Thompson CL, Stephenson FA (1995) Quantitative characterization of $\alpha 6$ and $\alpha 1 \alpha 6$ subunit-containing native γ-aminobutyric acid-A receptors of adult rat cerebellum demonstrates two α subunits per receptor oligomer. J Biol Chem 270:3753–3757
Pritchett DB, Sontheimer H, Gorman CM, Kettenmann H, Seeburg PH, Schofield PR (1988) Transient expression shows ligand gating and allosteric potentiation of $GABA_A$ receptor subunits. Science 242:1306–1308
Puia G, Vicini S, Seeburg PH, Costa E (1991) Influence of recombinant γ-aminobutyric acid-A receptor subunit composition on the action of allosteric modulators of γ-aminobutyric acid-gated Cl^- currents. Mol Pharmacol 39:691–696
Quirk K, Gillard NP, Ragan CI, Whiting PJ, Mckernan RM (1994a) γ-Aminobutyric acid type A receptors in the rat brain can contain both $\gamma 2$ and $\gamma 3$ subunits, but $\gamma 1$ does not exist in combination with another γ subunit. Mol Pharmacol 45:1061–1070
Quirk K, Gillard NP, Ragan CI, Whiting PJ, Mckernan RM (1994b) Model of subunit composition of γ-aminobutyric acid A receptor subtypes expressed in rat cerebellum with respect to their α and γ/δ subunits. J Biol Chem 269:16020–16028
Quirk K, Whiting PJ, Ragan CI, Mckernan RM (1995) Characterisation of δ-subunit containing $GABA_A$ receptors from rat brain. Eur J Pharmacol 290:175–181
Sanna E, Garau F, Harris RA (1995) Novel properties of homomeric beta 1 γ-aminobutyric acid type A receptors: actions of the anesthetics propofol and pentobarbital. Mol Pharmacol 46:213–217
Saxena NC, Macdonald RL (1994) Assembly of $GABA_A$ receptor subunits: role of the δ-subunit. J Neurosci 14:7077–7086
Saxena NC, Macdonald RL (1996) Properties of putative cerebellar γ-aminobutyric acid A receptor isoforms. Mol Pharmacol 49:567–579

Schofield PR, Darlison MG, Fujita N, Burt DR, Stephenson FA, Rodriguez H, Rhee LM, Ramachandran J, Reale V, Glencorse TA, Seeburg PH, Barnard EA (1987) Sequence and functional expression of the $GABA_A$ receptor shows a ligand-gated receptor superfamily. Nature 328:221–227

Shimada S, Cutting GR, Uhl GR (1992) γ-Aminobutyric acid A or C receptor? γ-Aminobutyric acid $\rho 1$ receptor RNA induces bicuculline-, barbiturate-, and benzodiazepine-insensitive γ-aminobutyric acid responses in *Xenopus* oocytes. Mol Pharmacol 41:683–687

Shingai R, Sutherland ML, Barnard EA (1991) Effects of subunit types of the cloned $GABA_A$ receptor on the response to a neurosteroid. Eur J Pharmacol 206:77–80

Shingai R, Yanagi K, Fukushima T, Sakata K, Ogurusu T (1996) Functional expression of rat GABA receptor $\rho 3$ subunit. Neurosci Res 26:387–390

Shivers BD, Killisch I, Sprengel R, Sontheimer H, Kohler M, Schofield PR, Seeburg PH (1989) Two novel $GABA_A$ receptor subunits exist in distinct neuronal subpopulations. Neuron 3:327–337

Sieghart W (1995) Structure and pharmacology of γ-aminobutyric $acid_A$ receptor subtypes. Pharmacol Rev 47:181–234

Sigel E, Barnard EA (1984) A γ-aminobutyric acid/benzodiapzepine receptor complex from bovine cerebral cortex. Improved purification with preservation of regulatory sites and their regulations. J Biol Chem 259:7129–7223

Sigel E, Baur R, Trube G, Mohler H, Malherbe P (1990) The effect of subunit composition of rat brain $GABA_A$ receptors on channel function. Neuron 5:703–711

Somogyi P, Fritschy JM, Benke D, Roberts JD, Sieghart W (1996) The $\gamma 2$ subunit of the $GABA_A$ receptor is concentrated in synaptic junctions containing the $\alpha 1$ and $\beta 2/3$ subunits in hippocampus, cerebellum and globus pallidus. Neuropharmacology 35:1425–1444

Sur C, Farrar SJ, Kerby J, Whiting PJ, Atack JR, McKernan RM (1999) Preferential coassembly of $\alpha 4$ and δ subunits of the γ-aminobutyric acid-A receptor in rat thalamus. Mol Pharmacol 56:110–1155

Toyoshima C, Unwin N (1990) Three-dimensional structure of the acetylcholine receptor by cryoelectron microscopy and helical image reconstruction. J Cell Biol 111:2623–2635

Tretter V, Ehya N, Fuchs K, Sieghart W (1997) Stoichiometry and assembly of a recombinant $GABA_A$ receptor subtype. J Neurosci 17:2728–2737

Varecka L, Wu CH, Rotter A, Frostholm A (1994) $GABA_A$ benzodiazepine receptor $\alpha 6$ subunit mRNA in granule cells of the cerebellar cortex and cochlear nuclei: expression in developing and mutant mice. J Comp Neurol 339:341–352

Verdoorn TA (1994) Formation of heteromeric γ-aminobutyric acid type A receptors containing two different α subunits. Mol Pharmacol 45:475–480

Verdoorn TA, Draguhn A, Ymer S, Seeburg PH, Sakmann B (1990) Functional properties of recombinant rat $GABA_A$ receptors depend upon subunit composition. Neuron 4:919–928

Wafford KA, Thompson SA, Thomas D, Sikela J, Wilcox AS, Whiting PJ (1996) Functional characterization of human γ-aminobutyric $acid_A$ receptors containing the $\alpha 4$ subunit. Mol Pharmacol 50:670–678

Wang TL, Guggino WB, Cutting GR (1994) A novel γ-aminobutyric acid receptor subunit ($\rho 2$) cloned from human retina forms bicuculline-insensitive homo-oligomeric receptors in *Xenopus* oocytes. J Neurosci 14:6524–6531

Wegelius K, Pasternack M, Hiltunen JO, Rivera C, Kaila K, Saarma M, Reeben M (1998) Distribution of GABA receptor ρ-subunit transcripts in the rat brain. Eur J Neurosci 10:350–357

Whiting P, Mckernan RM, Iversen LL (1990) Another mechanism for creating diversity in gamma-aminobutyrate type A receptors: RNA splicing directs expression of two forms of gamma 2 subunit, one of which contains a protein kinase C phosphorylation site. Proc Natl Acad Sci USA 87:9966–9970

Whiting PJ, McAllister G, Vassilatis D, Bonnert TP, Heavens RP, Smith DW, Hewson L, O'Donnell R, Rigby MR, Sirinathsinghji DJS, Marshall G, Thompson SA, Wafford KA (1997) Neuronally restricted RNA splicing regulates the expression of a novel $GABA_A$ receptor subunit conferring atypical functional properties. J Neurosci 17:5027–5037

Zhang D, Pan ZH, Zhang X, Brideau AD, Lipton SA (1995) Cloning of a γ-aminobutyric acid type C receptor subunit in rat retina with a methionine residue critical for picrotoxinin channel block. Proc Natl Acad Sci 92:11756–11760

CHAPTER 3
Functions of $GABA_A$-Receptors: Pharmacology and Pathophysiology

H. MÖHLER

A. Introduction

Based on the diversity of constituent subunits the structual heterogenity of $GABA_A$-receptors is well established (see BARNARD Chap. 2, this volume). The functional significance of $GABA_A$-receptors subtypes in vivo, however, has largely remained unknown. It is only through genetic means – gene inactivation, reduction of gene dosage, point mutations – that the functional role of $GABA_A$-receptor subtypes is beginning to be identified. The present chapter summarizes these attempts with regard to the pharmacology and patholphysiology of $GABA_A$-receptor subtypes.

B. Pharmacology of $GABA_A$-Receptor Subtypes

I. Benzodiazepine Actions at $GABA_A$-Receptor Subtypes

1. Distinction of Receptor Subtypes by Point Mutations

$GABA_A$ receptors are molecular substrates for the regulation of vigilance, anxiety, muscle tension, epileptogenic activity and anterograde amnesia, which is evident from the spectrum of actions elicited by clinically effective drugs acting at their modulatory benzodiazepine (BZ) binding site (for review see MÖHLER et al. 1997a,b, 2000). BZ-sensitive $GABA_A$ receptors are characterized by the subunits $\alpha1$, $\alpha2$, $\alpha3$, or $\alpha5$ (Fig. 1). Their opening frequency is enhanced by agonists of the BZ site, which is the basis of their therapeutic effectiveness in the treatment of anxiety disorders, sleep disturbances, muscle spasms, and epilepsy but also of their undesired side effects. The classical benzodiazepines such as diazepam interact indiscriminately with all BZ-sensitive $GABA_A$ receptor subtypes ($\alpha1$, $\alpha2$, $\alpha3$, and $\alpha5$) with comparable affinity (MÖHLER and OKADA 1977; BRAESTRUP et al. 1977) whereby a conserved histidine residue is critical for ligand binding at the BZ site (WIELAND et al. 1992; BENSON et al. 1998). In contrast, the BZ-insensitive receptor subtypes in the brain display an arginine residue in the corresponding position. Recombinant diazepam-sensitive receptors have previously been shown to be rendered diazepam-insensitive by replacing this histidine residue by arginine without

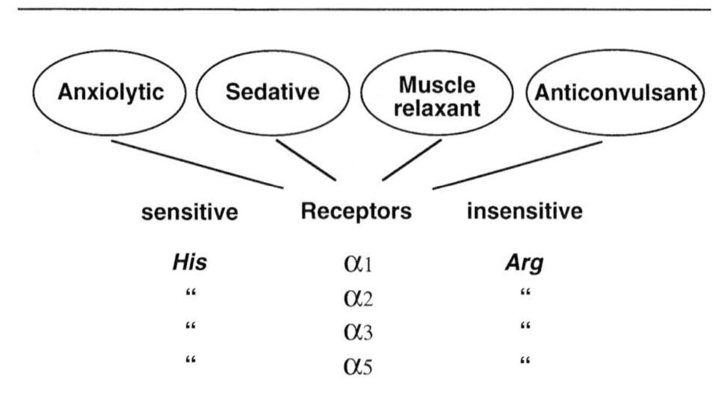

Fig. 1. Attribution of benzodiazepine actions to $GABA_A$-receptor subtypes ($\alpha 1$, $\alpha 2$, $\alpha 3$, $\alpha 5$) by generating mouse lines in which selected receptor subtypes are rendered diazepam-insensitive by a point mutation (replacement of a histidine by an arginine residue)

altering the GABA sensitivity as shown for the $\alpha 1$ subunit (WIELAND et al. 1992; KLEINGOOR et al. 1993) and the $\alpha 2$, $\alpha 3$, and $\alpha 5$ subunits (BENSON et al. 1998). In the brain, the predominant $GABA_A$ receptor subtype contains the $\alpha 1$ subunit (FRITSCHY and MÖHLER 1995; FRITSCHY et al. 1992, 1998). Its pharmacological significance was therefore evaluated by introducing the $\alpha 1$(H101R) point mutation into the germline of mice by gene targeting (Fig. 1) (RUDOLPH et al. 1999). The replacement vector contained not only the desired point mutation in exon 4 but also a loxP-flanked neomycin resistance marker in intron 4. This procedure permitted breeding of the mice carrying the mutant allele to Ella-cre mice (LAKSO et al. 1996) to eliminate the neomycin resistance cassette. The pharmacological analysis of the point mutated mice was therefore free of any potential interference which may have resulted from the presence of the neomycin marker.

The receptors from $\alpha 1$(H101R) mice displayed a ligand binding profile consistent with that of physiologically diazepam insensitive $GABA_A$ receptors, i.e., a virtual lack of affinity for diazepam, clonazepam, and zolpidem (Fig. 2). In sections of $\alpha 1$(H101R) mutant brain, the diazepam-insensitive sites were visualized autoradiographically in all regions known to express the $\alpha 1$-subunit, i.e., in particular in olfactory bulb, cerebral cortex, thalamus, pallidum, mid-brain, and cerebellum. Most importantly, gating of the point-mutated receptor by GABA remained unaltered as shown in Purkinje cells, in which $\alpha 1$ receptors predominate. The response to GABA was indistinguishable between cells from wild type and $\alpha 1$(H101R) mice. It was only the potentiation by diazepam which was strongly reduced in cells from $\alpha 1$(H101R) mice with the remaining diazepam effect being attributed to diazepam-sensitive receptors other than $\alpha 1$ in these cells. Thus, the repertoire of BZ actions in $\alpha 1$(H101R) mice was expected to be based exclusively on receptors containing $\alpha 2$, $\alpha 3$, and $\alpha 5$ subunits. The drug responses mediated by $GABA_A$ $\alpha 1$ receptors were expected to be silenced in the $\alpha 1$(H101R) mice (Fig. 1).

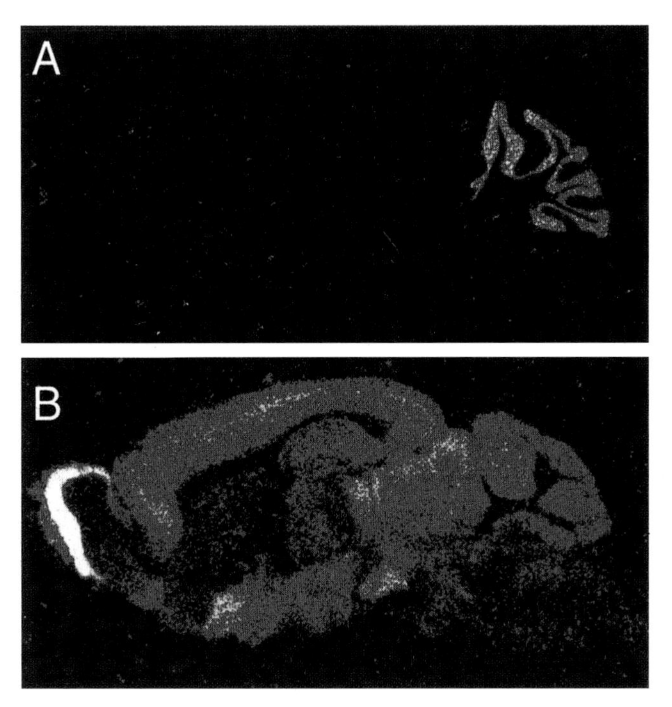

Fig. 2. Autoradiographic visualization of diazepam-insensitive benzodiazepine binding sites in brain slices from: **A** wildtype, **B** α1(H101R) point mutated mice. The sections were incubated with 3H-Ro 15–4513 in the presence of 100 μmol/l diazepam. In wild-type brain diazepam-insensitive GABA$_A$-receptors are represented by the small population of α4 and α6 receptors. In the mutant brain diazepam-insensitive sites are additionally present in all areas expressing the α1-subunit (RUDOLPH et al. 1999)

2. Sedation and Receptor Subtypes

The α1(H101R) mice were resistant to the sedative effect of diazepam (depression of motor activity) as tested up to a dose of 30 mg/kg i.p. The selectivity of this effect was underlined by the unaltered responsiveness of α1(H101R) mice to the sedative/hypnotic effects of drugs other than ligands of the BZ site such as the neurosteroid 3α-hydroxy-5β-pregnan-20-one or sodium pentobarbital which remained as effective as in wild-type mice in reducing motor activity or inducing a loss of righting reflex, respectively. These results support the view that diazepam-induced sedation is mediated via the α1-GABA$_A$ receptor (Table 1).

3. Amnesia and Receptor Subtypes

The memory impairing effect of diazepam, analysed in a step-through passive avoidance paradigm, was strongly reduced in the α1(H101R) mice as shown by the shortened latency for re-entering the dark compartment 24 h after training compared to the wild-type. The ability of the α1(H101R) mice to exhibit amnesia induced by a muscarinic antagonist remained unaffected. The

Table 1. Proposed roles of $GABA_A$-receptor subtypes in benzodiazepine actions

	$\alpha 1$	$\alpha 2$ $\alpha 3$ $\alpha 5$
Sedation	+	–
Amnesia	+	–
Seizure protection	+	+
Anxiolysis	–	+
Myorelaxation	–	+
Motor impairment	–	+
Ethanol potentiation	–	+

From Rudolph et al. 1999.

memory impairment induced by scopolamine was apparent to the same extent in both $\alpha 1$(H101R) mice and wild-type mice. These results demonstrate that the diazepam-induced anterograde amnesia is mediated via $GABA_A$ $\alpha 1$ receptors.

4. Anticonvulsant Activity and Receptor Subtypes

The anticonvulsant activity of diazepam, assessed by its protection against pentylenetetrazole-induced tonic convulsions, was reduced in $\alpha 1$(H101R) mice compared to wild-type mice. The partial anticonvulsant effect of diazepam which remained in $\alpha 1$(H101R) mice was due to $GABA_A$ receptors other than $\alpha 1$, since it was antagonized by the BZ antagonist flumazenil (Hunkeler et al. 1981). However, sodium phenobarbital was fully effective as anticonvulsant in $\alpha 1$(H101R) mice with a dose-response relationship similar to that of wild-type mice. These results show that the anticonvulsant activity of BZ-site ligands is largely – but not fully – mediated by $GABA_A$ $\alpha 1$ receptors.

5. Myorelaxation, Potentiation and Receptor Subtypes

The myorelaxant, motor impairing and ethanol potentiating properties of diazepam were not impaired in the $\alpha 1$(H101R) mice. Diazepam induced myorelaxation to the same extent in wild-type and $\alpha 1$(H101R) mice (horizontal wire test). In the rotarod test, both $\alpha 1$(H101R) and wild-type mice displayed a dose-dependent motor impairment. This muscle relaxant effect may be mediated by the $\alpha 2$- and $\alpha 5$-receptors present on motoneurons (Fritschy and Möhler 1995). Furthermore, diazepam potentiated in a dose-dependent manner the sedative effect of ethanol by increasing the duration of the loss of the righting reflex in both wild-type and $\alpha 1$(H101R) mice. Thus, the myorelaxant and ethanol potentiating activity of BZ site ligands are exclusively mediated by $GABA_A$ receptors of the $\alpha 2$, $\alpha 3$, and/or $\alpha 5$ type but not the $\alpha 1$ type.

6. Anxiolytic Activity and Receptor Subtypes

The anxiolytic activity of diazepam was unaltered in the $\alpha 1$(H101R) mice as assessed in two paradigms, the light-dark choice test, as well as in the elevated

X-maze. These results demonstrate that the anxiolytic actions of diazepam can be attributed to the small populations of neurons expressing the $\alpha 2$, $\alpha 3$, and/or $\alpha 5$ receptors (Table 1). They include parts of the limbic system ($\alpha 2$, $\alpha 5$) and the reticular activating system (noradrenergic and serotonergic neurons; $\alpha 3$) (Fritschy et al. 1992; Fritschy and Möhler 1995), supporting their role in the drug-induced regulation of anxiety (Gray 1995; Iversen 1984; File and Pellow 1987).

7. Strategies for Drug Design

Strategies for the design of a new generation of BZ site ligands acting selectively on GABA$_A$-receptor subtypes are apparent (Table 1). For instance, agonists acting on $\alpha 2$, $\alpha 3$, and/or $\alpha 5$ receptors are expected to include non-sedative and non-amnesic anxiolytics for the treatment of anxiety disorders and anxious depression. Furthermore, in schizophrenia, BZ monotherapy has not been fully evaluated despite reports on their antipsychotic effects (Wolkowitz and Pickar 1991; Delini-Stula et al. 1992) and their use as co-medication. Since only selected parts of the GABA system are affected in schizophrenia (Woo et al. 1998; Huntsman et al. 1998; Benes 1995; Akbarian et al. 1995) and the dopamine system is linked to particular populations of GABA neurons (Mrzljak et al. 1996), subtype-specific BZ-site ligands may provide a new focus for the treatment of schizophrenia. Finally, the point-mutated mice will be valuable in defining the relevance of receptor subtypes for the sequelae of chronic BZ treatment such as tolerance and dependence. For instance, ligands acting on particular receptor subtypes would not be expected to induce dependence liability to the same extent as ligands acting on all GABA$_A$ receptors. This opens the prospect for tailor-made subtype-specific drugs that may lack dependence liability. By applying the point mutation strategy to the $\alpha 2$, $\alpha 3$, and $\alpha 5$ subunits, it will be possible to refine the dissection of the pharmacological spectrum of drug effects elicited through the BZ site of GABA$_A$ receptor subtypes. Recently, the anxiolytic action was attributed to the α_2-receptor subtype (Löw et al. 2000).

II. Ethanol and GABA$_A$ Receptor Subtypes

The mechanism of action of ethanol has been analysed using different mutant mice. It had been demonstrated earlier that mice lacking the γ-isoform of protein-kinase C show a reduced response to ethanol (Harris et al. 1995). This result supported the view that the phosphorylation of GABA$_A$-receptors at sites of the large cytoplasmic loop of the $\gamma 2$L-subunit may be critical for mediating the effect of ethanol. In order to test this hypothesis mice were generated in which a 24 bp exon was deleted which distinguishes the $\gamma 2$L splice variant from by the $\gamma 2$S-variant (Homanics et al. 1999a). However these animals showed the same sensitivity to ethanol as control mice. There was no difference in the potentiation of GABA currents by ethanol observed in neurons from wildtype or $\gamma 2L^{-/-}$ mice. Furthermore, several behavioural effects of ethanol were likewise unchanged such as the ethanol-induced sleep-time, anxiolysis, acute tolerance, chronic withdrawal hyperexcitability, and hyperlocomotor activity (Homanics et al. 1999a). Thus, γ2L does not appear to be required for the ethanol-induced modulation of $GABA_A$-receptors and whole animal behaviour (Homanics et al. 1999b). The mechanism of action of ethanol was further analysed in animals with mutations affecting the α6-subunit. A naturally occurring point mutation in the α6-subunit gene was earlier shown to cosegregate with a phenotype which was more sensitive than controls to the motor impairing effect of alcohol (Korpi et al. 1993; Hellevno et al. 1989). However, α6 null mutant mice failed to display altered responses to ethanol (Homanics et al. 1997b). In particular, ethanol-induced motor impairment, tolerance and withdrawal hyperexcitability were not different between genotypes ($\alpha 6^{+/+}$, $\alpha 6^{-/-}$) (Homanics et al. 1998; Korpi et al. 1998). Thus, the $GABA_A$ receptors containing α6-subunit do not appear to be critically involved in the behavioural response to ethanol (Homanics et al. 1997b).

III. Anaesthetics and Pentobarbital

The role of $GABA_A$-receptors in mediating the action of anaesthetics was genetically assessed by targeting the $GABA_A$-receptor β3- and α6-subunit genes. Although mice lacking the β3-subunit gene generally die as neonates, some survive with abnormal behaviour (hyperactivity, incoordination, epilepsy) (Homanics et al. 1997a). In these animals the effectiveness of pentobarbital, enflurane, and halothane to induce a loss of righting reflex remained unaltered while midazolam and etomidate were less effective (Quinlan et al. 1998). The latter agents were therefore postulated to produce hypnosis by different molecular mechanisms. However, in contrast to the unaltered effectiveness of the volatile anaesthetics enflurane and halothane in inducing a loss of the righting reflex, their immobilizing effect (tail clamp test) was impaired in the β3 null mutant mice. Absence of the α6-subunit did not change the response to pentobarbital and general anaesthetics (Homanics et al. 1997b), a result which is somewhat surprising since at least pentobarbital can directly activate α6 but not α1-receptors at concentrations of $100\,\mu$mol/l (Hadingham et al. 1996; Thompson et al. 1996). However, a naturally occurring point mutation in the α6-subunit gene enhanced the ataxic effects of volatile anaesthetics and the loss of righting reflex by pentobarbital (Korpi et al. 1993; Hellevno et al. 1989).

C. $GABA_A$-Receptor Mutants as Models for Disease

I. Anxiety-Behaviour and Bias for Threat Cues

It is widely accepted that pathological anxiety has a neurobiological and genetic underpinning. A crucial role has been delineated for the amygdala and its array of connections to higher cortical, subcortical areas in particular the hippocampus and brainstem structures in the acquisition and retention of conditioned fear in animals. These connections facilitate acquisition of the sensory and interpretive information needed to select fear responses according to context and allow the coordinated expression of cognitive, affective, motor and autonomic components of anxiety. The locus coeruleus and the brainstem respiratory centres have reciprocal connections to the amygdala and may contribute to the processing of stimuli and the expression of anxiety via descending pathways (Roy-Burne and Cowley 1998). Key roles are attributed to excitatory circuits from the cortex to the amygdala and to the inhibitory GABAergic local-circuit neurons, the latter being consistent with the efficacy of benzodiazepine anxiolytics. Thus, the $GABA_A$-receptor system provides a fruitful molecular target for a pathophysiological inquiry of anxiety.

1. Genetically Defined Animal Model of Anxiety

$GABA_A$-receptor deficits have been identified in patients with anxiety disorders. In patients suffering from panic attacks, a deficit of $GABA_A$-receptors has been identified in the hippocampus, parahippocampus and orbitofrontal cortex in 11C-flumazenil PET-studies (Schlegel et al. 1994; Kaschka et al. 1995; Malizia et al. 1998). A $GABA_A$-receptor deficit has also been implicated in generalized anxiety disorders (Tiikonen et al. 1997) although only in particular areas (Abadie et al. 1999). The hypothesis was therefore tested, whether an impairment of $GABA_A$-receptor function is sufficient to induce a state of anxiety characterized by behavioural inhibition and hypersensitivity to negative associations in an animal model. Since the $\gamma 2$-subunit is required for synaptic clustering and normal single channel conductance of most $GABA_A$-receptors (Günther et al. 1995; Essrich et al. 1998), mice heterozygous for the $\gamma 2$-subunit of $GABA_A$-receptors were expected to provide a limited reduction of $GABA_A$-receptor function. The $\gamma 2^{+/0}$ mice were analysed with regard to the presence of both behavioural inhibition and hypersensitivity to negative associations as characteristic features of anxiety states in humans.

By generating mice that are heterozygous mutant for the $\gamma 2$-subunit gene, a limited reduction of $GABA_A$-receptor function was implemented. The $GABA_A$-receptor dysfunction in $\gamma 2^{+/0}$ mice, visualized by decreased benzodiazepine binding and receptor clustering, was most pronounced in brain areas that are also known to be affected in anxiety disorders in man. In patients with panic disorder, tested in the interepisode state, the cerebral blood flow is increased in the parahippocampal-hippocampal area (Reimann et al. 1984; Nordahl et al. 1990). The same brain region has been shown to display decreased benzodiazepine binding in patients with generalized anxiety disorder (Tiikonen et al. 1997) or panic disorder (Schlegel et al. 1994; Kaschka et al. 1995; Malizia et al. 1998), in line with the pronounced hippocampal and cortical $GABA_A$-receptor impairment in $\gamma 2^{+/0}$.

2. Enhanced Reactivity to Natural Aversive Stimuli

The deficit in $GABA_A$-receptor function resulted in an enhanced reactivity of $\gamma 2^{+/0}$ mice to natural aversive stimuli, as demonstrated by the aversion to novelty, exposed space, and brightly illuminated areas. This behavioural inhibition represents anxiety-related responses that are generally thought to include the activity of the septo-hippocampal system in both animals and humans (Gray and McNaughton 1996; Rogan and LeDoux 1996; Blanchard and Blanchard 1988; Knight 1996). Thus, the pronounced impairment of receptor clustering, notably in cerebral cortex and hippocampus, appears to contribute to the anxiety-related behaviour of the $\gamma2^{+/0}$ mice. The diazepam-induced reversal of the behavioural inhibition of $\gamma2^{+/0}$ mice corresponded to that in the human condition. Subjects with high anxiety scores are more sensitive to the anxiolytic action of benzodiazepines than the controls (O'Boyle et al. 1986; Glue et al. 1995).

3. Learned Aversive Stimuli

In humans, anxiety states are characterized not only by harm avoidance behaviour but also by a heightened responsiveness to negative associations in assessing the emotional quality of a situation (Eysenck 1992). This includes a bias for interpreting ambiguous scenarios as threatening, an attentional bias favouring the selective processing of threat cues and a bias of explicit memory for threat (McNally 1996). Such features of anxiety found a correspondence in the behaviour of the $\gamma2^{+/0}$ mice. Trace conditioning was enhanced in $\gamma2^{+/0}$ mice (Fig. 3a) indicating that these animals displayed a heightened sensitivity to negative associations in this fear conditioning variation. It is important to note that the acquisition and retention of the classical conditioned response to the context or to a cue (Fig. 3c,d) were unaltered in $\gamma2^{+/0}$ suggesting that implicit forms of learning were not affected. It appears to be rather the perception of the temporal contingency of negative stimuli that was enhanced in $\gamma2^{+/0}$.

Similarly, in cue discrimination learning (Fig. 3b) the $\gamma2^{+/0}$ mice displayed a heightened fear response in assessing the negative association of an ambiguous stimulus. In this test, the $\gamma2^{+/0}$ mice perceived the partial stimulus to be as threatening as the fully conditioned stimulus. This behaviour has previously been attributed to a hyperactivity of the hippocampus (McNaughton 1997) and would be in line with a heightened sensitivity to negative associations in the $\gamma2^{+/0}$ mice. The enhanced reactivity in both trace conditioning and cue discrimination learning suggests that the $\gamma2^{+/0}$ mice represent a model of anxiety behaviour which includes a hypersensitivity to negative associations.

4. Pathophysiology of Anxiety Disorders

Human anxiety disorders arise from a combination of genetic vulnerability and traumatic experience. Mice with the $GABA_A$-receptor $\gamma2$-subunit heterozygosity overreact to various specific anxiety-provoking situations. The $\gamma2$ mutant mice therefore represent a valid genetic model of at least some forms of anxiety. Such genetic models (Crestani et al. 1999; Heisler et al. 1998) are important in furthering the study of innate contributors to anxiety disorders.

Selective anxiety responses in $\gamma_2^{+/0}$ mice

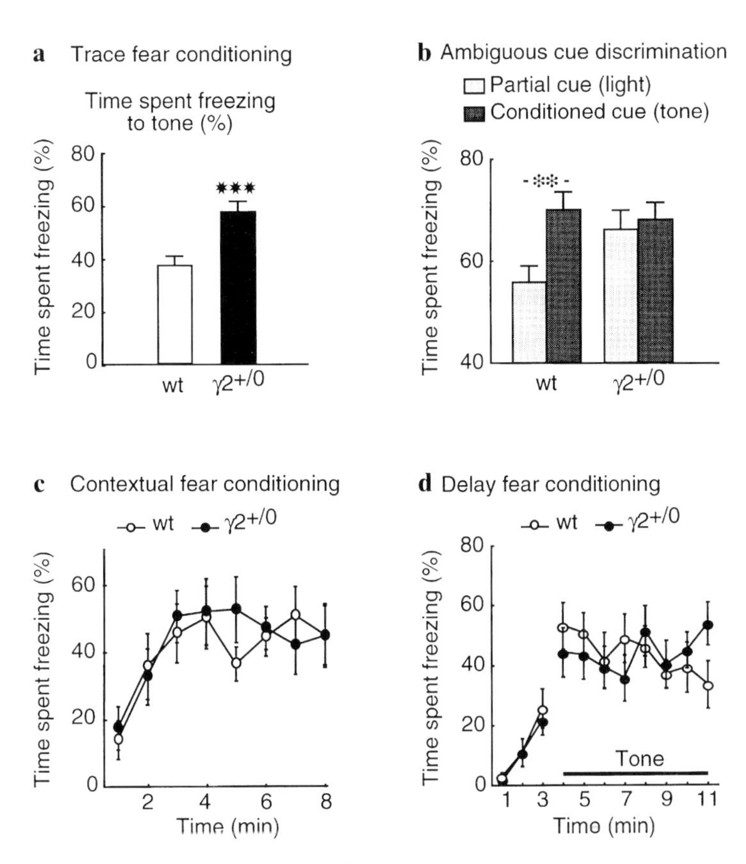

Fig. 3a–d. Behavioural responses of $\gamma 2^{+/0}$ and wildtype mice to learned aversive stimuli. In contrast to: **c** contextual, **d** delay fear conditioning, hightened sensitivity of $\gamma 2^{+/0}$ to negative associations is apparent in: **a** trace fear conditioning, **b** assessing ambiguous stimuli. (CRESTANI et al. 1999)

First, the mice offer the promise of a genetic model of the anxiety-predisposed human, which may be useful in improving drug discovery. Rather than examining the effects of novel anxiolytics on normal rats, one may examine genetic models of anxiety. Second, these mice offer easily testable predictions about mutations that may be found in anxiety patients. Finally, although the identification of genetic predisposing factors would certainly be a major advance, it is clear that genes alone will not explain human anxiety. These mutant mice should therefore be a valuable model for testing ideas about how genes and the environment interact to produce this condition.

II. Craniofacial Development

Mice which are devoid of the $\beta 3$-subunit (Homanics et al. 1997a) mostly die as neonates, displaying only half the normal density of $GABA_A$-receptors in brain. Some of the $\beta 3$-deficient neonatal mortality, but not all, is accompanied by the development of cleft palate. A role of $GABA_A$-receptors in craniofacial development is supported by the emergence of the neonatally lethal cleft palate in mice homozygous for the p4THO-II deletion which includes the $\alpha 5$, $\gamma 3$ and the $\beta 3$-subunit gene (Culiat et al. 1993, 1994). Since the cleft palate phenotype could be rescued by introducing a $\beta 3$-subunit transgene into the p4THO-II homozygous mutants the $\beta 3$-$GABA_A$-receptors appear to play an essential role in craniofacial development (Culiat et al. 1995) (see Kim and Olsen, Chap. 9, this volume).

III. Angelman's Syndrome

The $\beta 3$-subunit null mutants are considered to be a model of the genetic disorder Angelman's syndrome in humans (Homanics et al. 1997a; DeLorey et al. 1998). The $\beta 3$ null mutants which survive the neonatal period show the four hallmarks of this disease in man: cognitive deficits, motor impairment, hyperactivity (including sleep disorders) and spontaneous seizures. Most patients have a deletion in material chromosome 15 that encompasses several genes including three $GABA_A$-receptor subunits ($\alpha 5$, $\beta 3$, $\gamma 3$) and the major candidate gene UBE3A (see Kim and Olsen, Chap. 9, this volume).

IV. Desynchrony of Neuronal Oscillations

Mice lacking the $GABA_A$-receptor $\beta 3$-subunit largely show neonatal lethality due to cleft palate (see above). The few $\beta 3$-deficient mice that survive eventually reach normal body size although with reduced life span. They display many neurological impairments including deficits in neuronal inhibition in spinal cord and higher cortical centres as shown by their hyperresponsiveness to sensory stimuli, their strong motor impairment and frequent myoclonus and occasional epileptic seizures (Homanics et al. 1997a). In particular, in the reticular nucleus of the thalamus, which normally acts as "desynchronizer", recurrent GABA-mediated inhibitions were abolished in brain slices of $\beta 3$ null mutants. Since $\beta 3$-receptors are present in the reticular nucleus but not in principal neurons of thalamic relay cells, oscillatory synchrony was dramatically intensified in the mutant tissue (Huntsman et al. 1999). This may explain the occurrence of spontaneous seizures in $\beta 3$ homozygous null mutants, pointing to a crucial role of $\beta 3$ GABA-receptors in the responsiveness to sensory stimuli and seizure control.

D. Limitations of the Gene Inactivation Approach

Inactivation or alteration of a $GABA_A$-receptor subunit gene can result in a functional impairment of the receptor and thereby provide information on the mechanism of particular neuroanatomical circuits and human disease. However, the road from the genotype to the phenotype can be circuitous and the phenotype may result from multiple changes including developmental aberrations, functional deficits in adult brain as well compensatory adaptations (for review see RUDOLPH and MÖHLER 1999).

I. Adaptation

GABA$_A$-receptors in adult cerebellar granule cells are predominantly of the α6-type. Mutants which lack a functional α6 subunit gene displayed a grossly normal cerebellar cytoarchitecture, while the number of cerebellar GABA$_A$-receptors appeared normal and no differences in motor function or motor learning were identified (HOMANICS et al. 1997b). Furthermore, the affinity for muscimol was reduced, which points to an upregulation of α1-receptors as adaptive mechanism (HOMANICS et al. 1997b). It is however unclear whether this is a general phenomenon since no upregulation was apparent in another α6 null mutant (JONES et al. 1997). In mice lacking the γ2L subunit variant an upregulation of the γ2S subunit variant (about 2.4-fold) has been observed in immunoprecipitation studies (HOMANICS et al. 1999b).

II. Severity of Impairment

The γ2 and β3 null mutants are neonatally lethal although some animals survive with neurological deficits. In these cases the molecular and cellular phenotype can be studied in primary cultures of embryonic brain or in tissue slices (ESSRICH et al. 1998; HUNTSMAN et al. 1999). However, the behavioural phenotype of the few animals which survive to adolescence or even adulthood is not representative for the mutation but rather reflects a fortuitous constellation of genetic and other factors.

III. Marker Genes

The presence of selectable marker genes in the mutant animals expressing neomycin phosphotransferase and herpes simplex virus thymidine kinase can also interfere with the phenotype. This became apparent in mice in which the γ2L subunit variant was mutated into the γ2S variant (deletion of a 24 bp exon) (HOMANICS et al. 1999a). The $\gamma 2L^{+/-}$ male mutants showed a reduced fertility or were partly sterile. In addition, the modified γ2 allele was transmitted at a reduced frequency. Although it cannot be excluded that this effect is due to the γ2L deletion it is most likely attributed to the presence of the selectable markers, in particular herpes simplex virus thymidine kinase (HOMANICS et al. 1999b). Expression of viral thymidine kinase in spermatids can be lethal to these cells (BRAUN et al. 1990).

IV. Strain Differences

The strain-specific effects on behaviour can be greater than the contributions made by a single gene, i.e. mouse strain differences can sometimes confound the results of a gene knock-out experiment. For instance, in the $\alpha 6$ null mutants, withdrawal hyperexcitability following chronic ethanol was markedly enhanced in the mutant 129/SvJ strain compared to controls but was unaltered in the mutant C57BL/6J (HOMANICS et al. 1998). Thus, significant differences in tests of withdrawal hyperexcitability maybe confounded by the influence of genes that cosegregate with the targeted allele.

Frequently, mutants of mixed genetic background are generated and F2–F4 generations are behaviourally tested which may retain a bias of the genetic background (JONES et al. 1997; HOMANICS et al. 1997b). For instance, an $\alpha 6$ null mutant of mixed background (129/SvJ $\times$ C57BL/6J) showed a stronger response to diazepam (10 mg/kg and 20 mg/kg) in the rotarod test than all control lines (mixed background, 129/SvJ or C57BL/6J). However, the different types of control mice differed among themselves in their drug response and thereby influenced the quantitative impact of the mutation (KORPI et al. 1998).

To minimize the influence of the genetic background, it is recommended to generate two different pure mutant lines by backcrossing for at least five, better ten or more, generations (GERLAI 1996; BANBURY CONFERENCE 1997), followed by testing both strains separately or subsequent F1 hybrids. This procedure has been followed for the behavioural assessment of $\gamma 2^{+/-}$ mice (CRESTANI et al. 1999). Finally, many of the shortcomings of the gene inactivation approach to probe $GABA_A$-receptor function can be avoided when the expression of the gene remains unaltered and the functional impairment is introduced by a point mutation. This strategy has been very successfully employed to attribute the benzodiazepine pharmacology to distinct $GABA_A$-receptor subtypes (see Sect. A) (RUDOLPH et al. 1999).

Acknowledgement. I wish to express my gratitude to the collaborators who contributed to the genetic analysis of $GABA_A$-receptor function: S. Balsiger, R. Baer, D. Benke, J. Benson, H. Bluethmann, I. Brünig, F. Crestani, C. Essrich, J. M. Fritschy, R. Keist, K. Löw, M. Lorez, B. Lüscher, I. R. Martin, C. Michel and U. Rudolph.

References

Abadie P, Boulenger JP, Benali K, Barre L, Zarifian E, Baron JC (1999) Relationships between trait and state anxiety and the central benzodiazepine receptor: a PET study. Eur J Neurosci 11:1470–1478

Akbarian S, Kim JJ, Potkin SG, Hagman JO, Tafazzoli A, Bunney WE Jr, Jones EG (1995) Gene expression for glutamic acid decarboxylase is reduced without loss of neurons in prefrontal cortex of schizophrenics. Arch Gen Psychiatry 52:258–266

Banbury Conference on the Genetic Background in mice (1997) Mutant mice and neuroscience: recommendations concerning genetic background. Neuron 19:755–759

Benes FM (1995) Altered glutamatergic and GABAergic mechanisms in the cingulate cortex of the schizophrenic brain. Arch Gen Psychiatry 52:1015–1018

Benson JA, Löw K, Keist R, Möhler H, Rudolph U (1998) Pharmacology of recombinant γ-aminobutyric acid A receptors rendered diazepam-insensitive by point-mutated α-subunits. FEBS Lett 431:400–404

Blanchard DC, Blanchard RJ (1988) Ethopharmacological approaches to the biology of emotion. Ann Rev Psychol 39:43–68

Braestrup C, Albrechtsen R, Squires RF (1977) High densities of benzodiazepine receptors in human cortical areas. Nature 269:702–704

Braun RE, Lo D, Pinkert CA, Videra G, Flavell RA, Palmiter RD, Wrinster RL (1990) Infertility in male transgenic mice: disruption of sperm development by HSV-tk expression in postmeiotic germ cells. Biol Reprod 43:684–693

Crestani F, Lorez M, Baer K, Essrich C, Benke D, Laurent JP, Belzung C, Fritschy JM, Luscher B, Möhler H (1999) Impairment of $GABA_A$-receptor clustering results in enhanced anxiety responses and a bias for threat cues. Nature Neurosci 2:833–839 (see also News and View, Nature Neurosci 2:780–782 [1999] and Nature Medicine 5:1131–1132 [1999])

Culiat CT, Stubbs L, Nicholls RD, Montgomery CS, Russel LB, Johnson DK, Rinchik EM (1993) Concordance between isolated cleft palate in mice and alterations within a region including the gene encoding the $\beta 3$ subunit of the type A γ-aminobutyric acid receptor. Proc Natl Acad Sci USA 90:5105–5109

Culiat CT, Stubbs LJ, Montgomery CS, Russell LB, Rinchik EM (1994) Phenotypic consequences of deletion of the $\gamma 3$ $\alpha 5$ or $\beta 3$ subunit of the type A γ-aminobutyric acid receptor in mice. Proc Natl Acad Sci USA 91:2815–2818

Culiat CT, Stubbs LJ, Woychik RP, Russell LB, Johnson DK, Rinchik EM (1995) Deficiency of the $\beta 3$ subunit of the type A γ-aminobutyric acid receptor causes cleft palate in mice. Nature Genetics 11:344–346

Delini-Stula A, Berdah-Tordjman D, Neumann N (1992) Partial benzodiazepine agonists in schizophrenia: expectations and present clinical findings. Clinical Neuropharmacology 15:405A–406A

DeLorey TM, Handforth A, Anagnostaras SG, Homanics GE, Minassian BA, Asatourian A, Fanselow MS, Delgado-Escueta A, Ellison GD, Olsen RW (1998) Mice lacking the $\beta 3$ subunit of the $GABA_A$ receptor have the epilepsy phenotype and many of the behavioural characteristics of Angelman syndrome. J Neurosci 18:8505–8514

Essrich C, Lorez M, Benson JA, Fritschy JM, Lüscher B (1998) Postsynaptic clustering of major $GABA_A$-receptor subtypes requires the $\gamma 2$-subunit and gephyrin. Nature Neurosci 1:563–571

Eysenck MW (1992) In: Gale A, Eysenck MW (eds) Handbook of individual differences: biological perspectives. John Wiley & Sons, New York, pp. 157-178

File SE, Pellow S (1987) Behavioral pharmacology of minor tranquilizers Pharmacol Ther 35:265–290

Fritschy JM, Möhler H (1995) $GABA_A$-receptor heterogeneity in the adult rat brain: differential regional and cellular distribution of seven major subunits. J Comp Neurol 359:154–194

Fritschy JM, Mertens S, Oertel WH, Bachi T, Möhler H (1992) Five subtypes of type A γ-aminobutyric acid receptors identified in neurons by double and triple immunofluorescence staining with subunit-specific antibodies. Proc Natl Acad Sci USA 89:6726–6730

Fritschy JM, Weinmann O, Wenzel A, Benke D (1998) Synapse-specific localization of NMDA- and $GABA_A$-receptor subunits revealed by antigen-retrieval immunohistochemistry. J Comp Neurol 390:194–210

Gerlai R (1996) Gene targeting studies of mammalian behaviour: is it the mutation or the background genotype? Trends Neurosci 19:177–181

Glue P, Wilson S, Coupland N, Ball D, Nutt D (1995) The relationship between benzodiazepine receptor sensitivity and neuroticism. J Anxiety Disord 9:33–45

Gray JA, McNaughton N (1996) In: Dienstbier RA, Hope DA (eds) Nebraska Symposium on Motivation, 43. University of Nebraska Press, London, pp 61–134
Gray JA (1995) A model of the limbic system and basal ganglia: applications to anxiety and schizophrenia In: Gazzaniga (ed) The cognitive neurosciences. MIT Press, Cambridge, Mass., pp 1165–1176
Günther U, Benson J, Benke D, Fritschy JM, Reyes G, Knoflach F, Crestani F, Aguzzi A, Arigoni M, Lang Y et al (1995) Benzodiazepine-insensitive mice generated by targeted disruption of the $\gamma 2$ subunit gene of γ-aminobutyric acid type A receptors. Proc Natl Acad Sci USA 92:7749–7753
Hadingham KL, Garrett EM, Wafford KA, Bain C, Heavens RP, Sirinathsinghji DJ, Whiting PJ (1996) Cloning of cDNAs encoding the human gamma-aminobutyric acid type A receptor alpha 6 subunit and characterization of the pharmacology of alpha 6-containing receptors. Mol Pharmacol 49:253–259
Harris RA, McQuilkin SJ, Paylor R, Abeliovich A, Tonegawa S, Wehner JM (1995) Mutant mice lacking the γ isoform of protein kinase C show decreased behavioral actions of ethanol and altered function of γ-aminobutyrate type A receptors. Proc Natl Acad Sci USA 92:3658–3662
Heisler LK, Chu HM, Brennan TJ, Danao JA, Bajwa P, Parsons LH, Tecott LH (1998) Elevated anxiety and antidepressant-like responses in serotonin 5-HT1A receptor mutant mice. Proc Natl Acad Sci USA 95:15049–15054
Hellevno K, Kiianmaa K, Korpi ER (1989) Effect of GABAergic drugs on motor impairment from ethanol barbital and lorazepam in rat lines selected for differential sensitivity to ethanol. Pharm Biochem Behav 34:399–404
Homanics GE, Delorey TM, Firestone LL, Quinlan JJ, Handforth A, Harrison NL, Krasowski MD, Rick CEM, Korpi ER, Makela R, Brilliant MH, Hagiwara N, Ferguson C, Snyder K, Olsen RW (1997a) Mice devoid of γ-aminobutyrate type A receptor $\beta 3$ subunit have epilepsy cleft palate and hypersensitive behavior. Proc Natl Acad Sci USA 94:4143–4148
Homanics GE, Ferguson C, Quinlan JJ, Daggett J, Snyder K, Lagenaur C, Mi ZP, Wang XH, Grayson DR, Firestone LL (1997b) Gene knockout of the $\alpha 6$ subunit of the γ-aminobutyric acid type A receptor: lack of effect on responses to ethanol pentobarbital and general anaesthetics. Mol Pharmacol 51:588–596
Homanics GE, Le NQ, Kist F, Mihalek R, Hart AR, Quinlan JJ (1998) Ethanol tolerance and withdrawal responses in $GABA_A$ receptor alpha 6 subunit null allele mice and in inbred C57BL/6J and strain 129/SvJ mice. Alcohol Clin Exp Res 22:259–265
Homanics GE, Harrison NL, Quinlan JJ, Krasowski MD, Rick CEM, de Blas AL, Mehta AK, Kist F, Mihalek RM, Aul JJ, Firestone LL (1999a) Normal electrophysiological and behavioral responses to ethanol in mice lacking the long splice variant of the $\gamma 2$ subunit of the γ-aminobutyrate type A receptor. Neuropharmacol 38:253–265
Homanics GE, Harrison NL, Quinlan JJ, Krasowski MD, Rick CEM, de Blas AL, Mehta AK, Kist F, Mihalek RM, Aul JJ, Firestone LL (1999b) Mice lacking the long splice variant of $\gamma 2$ subunit of γ-aminobutyric acid type A receptor demonstrate increased anxiety and enhanced behavioral responses to benzodiazepine receptor agonists but not to ethanol JPET (in press)
Hunkeler W, Mohler H, Pieri L, Polc P, Bonetti EP, Cumin R, Schaffner R, Haefely W (1981) Selective antagonists of benzodiazepines. Nature 290:514–516
Huntsman MM, Tran BV, Potkin SG, Bunney WE Jr, Jones EG (1998) Altered ratios of alternatively spliced long and short $\gamma 2$ subunit mRNAs of the γ-amino butyrate type A receptor in prefrontal cortex of schizophrenics. Proc Natl Acad Sci USA 95:15066–15071
Huntsman MM, Porcello DM, Homanics GE, DeLorey TM, Huguenard JR (1999) Reciprocal inhibitory connections and network synchrony in the mammalian thalamus. Science 283:541–543
Iversen SD (1984) 5-HT and anxiety. Neuropharmacol 23:1553–1560

Jones A, Korpi ER, McKernan RM, Pelz R, Nusser Z, Makela R, Mellor JR, Pollard S, Bahn S, Stephenson FA, Randall AD, Sieghart W, Somogyi P, Smith AJH, Wisden W (1997) Ligand-gated ion channel subunit partnerships: GABA$_A$ receptor $\alpha 6$ subunit gene inactivation inhibits δ subunit expression. J Neurosci 17:1350–1362
Kaschka W, Feistel H, Ebert D (1995) Reduced benzodiazepine receptor binding in panic disorders measured by iomazenil SPECT. J Psychiat Res 29:427–434
Kleingoor C, Wieland HA, Korpi ER, Seeburg PH, Kettenmann H (1993) Current potentiation by diazepam but not GABA sensitivity is determined by a single histidine residue. Neuroreport 4:187–190
Knight RT (1996) Contribution of human hippocampal region to novelty detection. Nature 383:256–259
Korpi ER, Kleingoor C, Kettenmann H, Seeburg PH (1993) Benzodiazepine-induced motor impairment linked to point mutation in cerebellar GABA$_A$ receptor. Nature 361:356–359
Korpi ER, Koikkalainen P, Vekovischeva OY, Makela R, Kleinz R, Uusi-Oukari M, Wisden W (1998) Cerebellar granule-cell-specific GABA$_A$ receptors attenuate benzodiazepine-induced ataxia: evidence from $\alpha 6$-subunit-deficient mice. Eur J Neurosci 11:233–240
Lakso M, Pichel JG, Gorman JR, Sauer B, Okamoto Y, Lee E, Alt FW, Westphal H (1996) Efficient in vivo manipulation of mouse genomic sequences at the zygote stage. Proc Natl Acad Sci USA 93:5860–5865
Lōw K, Crestani F, Keist R, Benke D, Brūnig I, Benson FA, Fritschy FM, Rūlicke T, Bluethmann H, Mōhler H, Rudolph K (2000) Molecular and neuronal substrate for the selective attenuation of anxiety. Science (in press)
Malizia AL, Cunningham VJ, Bell CJ, Liddle PF, Jones T, Nutt DJ (1998) Decreased brain GABA$_A$-benzodiazepine receptor binding in panic disorder. Arch Gen Psychiat 55:715–720
McNally RJ (1996) In: Dienstbier RA, Hope DA (eds), Nebraska Symposium on Motivation, 43. University of Nebraska Press, London, pp 211–250
McNaughton N (1997) Cognitive dysfunction resulting from hippocampal hyperactivity – a possible cause of anxiety disorder? Pharmacol Biochem Behav 56:603–611
Möhler H, Okada T (1977) Benzodiazepine receptor: demonstration in the central nervous system. Science 198:849–851
Möhler H, Benke D, Benson J, Lüscher B, Rudolph U, Fritschy JM (1997a) Diversity in structure pharmacology and regulation of GABA$_A$-receptors In: Enna SJ, Bowery NG (eds) GABA-receptors. Humana Press, Totowa, pp 11–36
Möhler H, Benke D, Fritschy JM (1997b) The GABA$_A$-receptors. In: Avanzini G et al (eds) Molecular and cellular targets for antiepileptic drugs: John Libbey, pp 39–53
Möhler H, Benke D, Fritschy JM, Benson J (2000) The GABA$_A$-receptor benzodiazepine site. In: Martin DL, Olsen RW (eds) GABA in the nervous system. Lippincott Williams & Wilkins (in press)
Mrzljak L, Bergson C, Pappy M, Huff R, Levenson R, Goldman-Rakic PS (1996) Localization of dopamine D4 receptors in GABAergic neurons of the primate brain. Nature 381:245–248
Nordahl TE et al (1990) Cerebral glucose metabolic differences in patients with panic disorder. Neuropsychopharmacology 3:261–272
O'Boyle CA, Harris D, Barry H, Cullen JH (1986) Differential effects of benzodiazepine sedation in high and low anxious patients in a real life stress setting. Psychopharmacology 88:226–229
Quinlan JJ, Homanics GE, Firestone LL (1998) Anesthesia sensitivity in mice that lack the $\beta 3$ subunit of the γ-aminobutyric acid type A receptor. Anaesthesiology 88:775–780
Reiman EM et al (1984) A focal brain abnormality in panic disorder a severe form of anxiety. Nature 310:683–685
Rogan MT, LeDoux JE (1996) Emotion: systems, cells, synaptic plasticity. Cell 85:469–475

Roy-Burne P, Cowley DB (1998) Search for pathophysiology of panic disorder. Lancet 352:1646–1647

Rudolph U, Crestani F, Benke D, Brünig I, Benson I, Fritschy IM, Martin IR, Bluethmann H, Möhler H (1999) Benzodiazepine actions mediated by specific $GABA_A$-receptor subtypes. Nature 401:797–800 (see also News and Views, Nature 401:751–752 [1999])

Rudolph U, Möhler H (1999) Genetically modified animals in pharmacological research: future trends. Europ J Pharmacol 375:327–337

Schlegel S et al (1994) Decreased benzodiazepine receptor binding in panic disorder measured by iomazenil SPECT. Eur Arch Psychiatry Clin Neurosci 244:49–51

Thompson SA, Whiting PJ, Wafford KA (1996) Barbiturate interactions at the human $GABA_A$ receptor: dependence on receptor subunit combination Br J Pharmacol 117:521–527

Tiihonen J, Kuikka J, Rasanen P, Lepola U, Koponen H, Liuska A, Lehmusvaara A, Vainio P, Kononen M, Bergstrom K, Yu M, Kinnunen I, Akerman K, Karhu J (1997) Cerebral benzodiazepine receptor binding and distribution in generalized anxiety disorder: a fractal analysis. Mol Psychiatry 2:463–471

Wieland HA, Lüddens H, Seeburg PH (1992) A single histidine in $GABA_A$-receptors is essential for benzodiazepine agonist binding. J Biol Chem 267:1426–1429

Wolkowitz OM, Pickar D (1991) Benzodiazepines in the treatment of schizophrenia: a review and reappraisal. Am J Psychiatry 148:714–726

Woo TU, Whitehead RE, Melchitzky DS, Lewis DA (1998) A subclass of prefrontal γ-aminobutyric acid axon terminals are selectively altered in schizophrenia. Proc Natl Acad Sci USA 95:5341–5346

CHAPTER 4

Steroid Modulation of $GABA_A$ Receptors

J.J. LAMBERT, J.A. PETERS, S.C. HARNEY, and D. BELELLI

A. Introduction

In 1984, Harrison and Simmonds demonstrated the synthetic steroidal anaesthetic alphaxalone (5α-pregnan-3α-ol-11,20-dione) to enhance potently and selectively the interaction of γ-aminobutyric acid (GABA) with the $GABA_A$ receptor (HARRISON and SIMMONDS 1984). In the same year, the steroid hormone androsterone (5α-androstan-3α-ol-17-one) was shown to share this activity, albeit with reduced potency (SIMMONDS et al. 1984). Alphaxalone and androsterone are closely related structurally to some endogenously occurring metabolites of progesterone (i.e. 5α- or 5β-pregnan-3α-ol-20-one) and deoxycorticosterone (5α-pregnane-3α,21-diol-20-one) which led logically to the evaluation of such steroids as allosteric modulators of $GABA_A$ receptor function. In electrophysiological, tracer-flux and radioligand binding studies, such steroids were found to be more potent than alphaxalone in potentiating the action of agonists at the $GABA_A$ receptor and allosteric interactions with established binding sites for other modulators (e.g. benzodiazepines) were revealed (MAJEWSKA et al. 1986, CALLACHAN et al. 1987; HARRISON et al. 1987a; GEE et al. 1987, 1988). In addition, at relatively high concentrations, the steroids exerted a direct GABA-mimetic effect (CALLACHAN et al. 1987; COTTRELL et al. 1987).

The rapidity of modulatory and agonist effects of the steroid in single cell studies, and their activity in radioligand binding studies performed on membrane homogenates, obviously precluded a traditional genomic mechanism of action. Instead, the potency and stereoselectivity of the modulatory effect, combined with the results of drug interaction studies strongly suggested the presence of a novel steroid binding site on the $GABA_A$ receptor protein (LAMBERT et al. 1995). Recently, this concept has been greatly strengthened by a comparison of the $GABA_A$ receptor modulatory activity of the enantiomers of endogenous and synthetic steroids and steroid analogues (see Sect. B.I). It is now generally accepted that the $GABA_A$ receptor harbours perhaps multiple steroid binding sites that are one major molecular target underlying the non-genomic effects of steroids upon neurones. Collectively, steroids acting in this manner have been coined 'neuroactive-steroids' with the term 'neurosteroid' being reserved for those steroids actually synthesised *de novo* from cholesterol, or formed by metabolism of blood-borne precursors, within the CNS (ROBEL and BAULIEU 1994).

Consistent with their actions on the $GABA_A$ receptor, neuroactive steroids have anxiolytic, anticonvulsant and sedative properties including, at relatively high doses, inducing a state of general anaesthesia (LAMBERT et al. 1995; GASIOR et al. 1999; RUPPRECHT and HOLSBOER 1999). At present, synthetic derivatives of 5α-pregnan-3α-ol-20-one are undergoing clinical trials for the treatment of epilepsy, anxiety and insomnia (GASIOR et al. 1999). Clearly, their potential in the clinical arena will be influenced not only by their behavioural efficacy, but additionally by whether they exhibit a reduced propensity to induce side-effects when compared to currently available $GABA_A$ receptor modulators such as the benzodiazepines (GASIOR et al. 1999). Rather than administer steroids per se, an alternative therapeutic strategy may be to develop drugs which interfere with the synthesis or metabolism of the endogenous neurosteroids.

Endocrine glands such as the adrenal cortex and ovaries are established endogenous sources of neuroactive steroids (PURDY et al. 1991; PAUL and PURDY 1992). However, it is now recognised that within the brain itself, certain glial cells and neurones contain the enzymatic machinery necessary for the local synthesis of neurosteroids (BAULIEU and SCHUMACHER 1996). Some of these enzymes play a ubiquitous role in steroid synthesis and hence drugs targeted to these proteins may have non-selective actions. Of particular interest is the NADH/NADPH-dependent enzyme 3α-hydroxysteroid dehydrogenase, which reduces 5α-pregnane 3,20-dione to 5α-pregnan-3α-ol-20-one or, indeed, can operate in the reverse direction to reform the genomically active 5α-pregnane-3,20-dione (RUPPRECHT et al. 1993). The enzymic regulatory mechanisms that determine whether oxidation or reduction of the steroid predominates remains to be determined. However, recent evidence has emerged that the antidepressant fluoxetine may influence the activity of this enzyme to favour the production of the $GABA_A$ receptor-active 5α-pregnan-3α-ol-20-one (UZUNOV et al. 1996; GUIDOTTI and COSTA 1998). This action of fluoxetine appears to be independent of the established effects of this antidepressant on the uptake of 5-hydroxytryptamine. In a clinical study, patients with unipolar major depression were reported to have relatively low cerebrospinal fluid levels of 5α-pregnan-3α-ol-20-one, an imbalance that was addressed by treatment with fluoxetine (UZUNOVA et al. 1998; GUIDOTTI and COSTA 1998). Furthermore, the improvement in patient symptomatology was correlated with the increase in neurosteroid levels (UZUNOVA et al. 1998). Hence, given the known behavioural effects of 5α-pregnan-3α-ol-20-one, it is conceivable that an effect on neurosteroid synthesis may contribute to the alleviation by fluoxetine of the anxiety and dysphoria associated with conditions such as premenstrual syndrome and certain forms of depression (UZUNOVA et al. 1998). These findings broaden the potential clinical utility of the neurosteroids and suggest the 3α-hydroxysteroid dehydrogenase enzyme family as a new drug target.

It is now established that steroids such as 5α-pregnan-3α-ol-20-one are potent and selective GABA$_A$ receptor modulators that act in vivo to produce clear behavioural effects consistent with the enhancement of inhibitory synaptic transmission. However, the fundamental question remains as to whether the endogenous levels of such steroids are sufficient to regulate GABA$_A$ receptor function under physiological, or pathophysiological, conditions. In female rats, 5α-pregnan-3α-ol-20-one is estimated to be present within the brain at low nanomolar concentrations which, in vitro, would produce a modest enhancement of GABA$_A$ receptor function. However, GABA-modulatory activity may be more pronounced during stress, or in the later stages of pregnancy, during which substantially raised neurosteroid levels have been reported (PURDY et al. 1991; PAUL and PURDY 1992; CONCAS et al. 1999). Furthermore, it is now evident that the synthesis of 5α-pregnan-3α-ol-20-one within the brain is not uniform. Such regional dependency may render consideration of whole brain levels of the steroid misleading (CHENEY et al. 1995; GUIDOTTI et al. 1996; GUIDOTTI and COSTA 1998).

That endogenous concentrations of 5α-pregnan-3α-ol-20-one present physiologically are indeed sufficient to enhance neural inhibition, is strongly suggested by recent studies investigating the influence of inhibitors and promoters of neurosteroid synthesis on the loss of the righting reflex induced by pentobarbitone in mice (MATSUMOTO et al. 1999). Pretreatment with a 5α-reductase inhibitor considerably decreased the cortical content of 5α-pregnan-3α-ol-20-one and concomitantly reduced the duration of the barbiturate induced "anaesthesia". By contrast, fluoxetine raised cortical neurosteroid levels and the central depressant effects of pentobarbitone were enhanced (MATSUMOTO et al. 1999). As pentobarbitone and neuroactive steroids act synergistically at the GABA$_A$ receptor (CALLACHAN et al. 1987; PETERS et al. 1988) these data are consistent with the presence of steroids at facilitating concentrations under physiological conditions (MATSUMOTO et al. 1999).

In summary, a potent, selective and stereospecific interaction of certain synthetic and endogenous neuroactive steroids with the GABA$_A$ receptor is now firmly established. When administered to animals, such steroids exhibit a behavioural profile consistent with the enhancement of neuronal inhibition, including anxiolytic, anticonvulsant, sedative/hypnotic and general anaesthetic activities. Synthetic steroid analogues are currently undergoing clinical assessment in an attempt to exploit this behavioural profile. The demonstration that the brain can synthesise 5α-pregnan-3α-ol-20-one raises the exciting prospect that the activity of the major inhibitory neurotransmitter in the central nervous system may be finely tuned by this locally produced modulator. Furthermore, the centrally located enzymes that synthesise, or metabolise, 5α-pregnan-3α-ol-20-one could present novel therapeutic targets. Indeed, some of the behavioural effects of established psychotherapeutic agents such as fluoxetine may, in part, be due to an effect upon the metabolism of neurosteroids.

Irrespective of whether or not these findings result in novel drugs, there is a burgeoning literature that indicates a physiological/pathophysiological role for neurosteroids. Hence, their study may provide a better understanding of some forms of epilepsy and psychiatric disorders where a perturbation of neurosteroid homeostasis is suspected (e.g. premenstrual tension and postnatal depression).

The present review focuses upon the effects of neuroactive steroids upon $GABA_A$ receptor function at the molecular and cellular levels, commencing with a description of the structural elements of the steroid molecule essential for activity. Thereafter, the influence of the subunit composition of the $GABA_A$ receptor upon the steroidal modulation is considered, along with altered sensitivity to such regulation as a potential consequence of the differential expression of subunit isoforms in response to changing levels of endogenous steroids. The remainder of the chapter describes the mechanistic aspects of neurosteroid action, including their influence upon the kinetics of $GABA_A$ receptor single channel activity under steady-state and non-equilibrium conditions. The latter underlies the modulatory activity of neurosteroids on $GABA_A$ receptor-mediated inhibitory postsynaptic currents, the subsequent modification of the integrative capacity of central neurones and, ultimately, behaviour.

B. Structure Activity Relationship for Steroids at the $GABA_A$ Receptor

Early studies of the structural requirements for potent modulation of $GABA_A$ receptor activity by steroids emphasised the requirement for a 5α- or 5β-reduced pregnane (or androstane) skeleton, an α-hydroxyl substituent at C3 of the steroid A ring, and a keto group at either C20 of the pregnane steroid side chain or C17 of the androstane ring system (HARRISON and SIMMONDS 1984; MAJEWSKA et al. 1986; CALLACHAN et al. 1987; HARRISON et al. 1987a;GEE et al. 1987, 1988; PETERS et al. 1988; see Fig. 1). Inevitably, subsequent investigations have led to refinement and extension of this simple scheme. It is now probably an oversimplification to attempt to define a single structure activity relationship for steroids at the $GABA_A$ receptor. Complications arise from the heterogeneity of $GABA_A$ receptors within the nervous system and the fact that the GABA-modulatory and GABA-mimetic activities of the steroids can be differentially influenced by the subunit composition of the receptor (see Sect. C). Furthermore, certain sulphated steroids act as negative allosteric modulators of $GABA_A$ receptor activity, though this activity may be mediated at a site distinct from that recognising positive steroidal modulators (see Sect. B.III). The following summary of the structure activity relationship for steroid interaction with the $GABA_A$ receptor should be read with these limitations in mind.

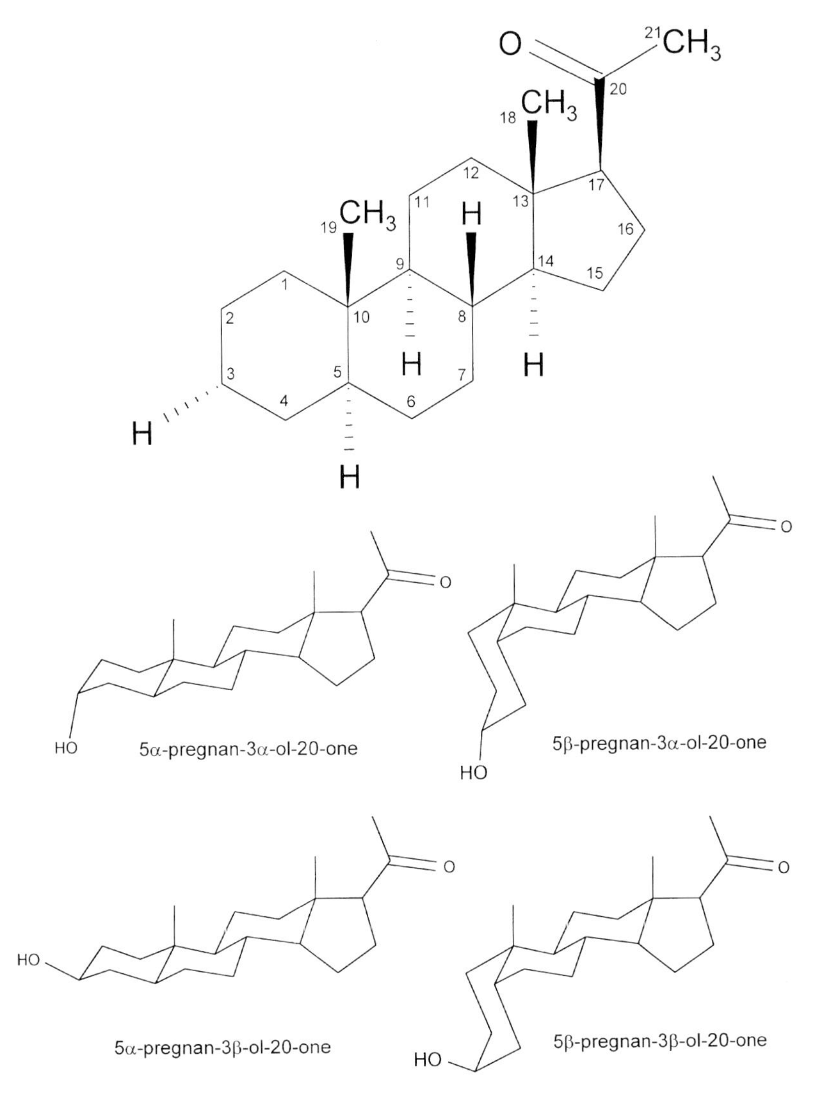

Fig. 1. The numbering of the carbon atoms in the steroid 5α-pregnan-3α-ol-20-one and perspective drawings of the stereoisomeric pairs of compounds resulting from the configuration of the C5- and C3- hydroxyl groups. By convention, substituents projecting below (*broken wedges*) and above (*solid wedges*) the plane of the steroid ring system are in the α- and β-configurations respectively. Note that the orientation of the C5 hydroxyl determines the configuration of the steroid A and B ring fusion (5α-pregnane series – *trans*; 5β-pregnane series – *cis*)

I. Enantioselectivity of Steroid Action

Studies demonstrating a differential effect of enantiomeric pairs of pregnane and androstane steroids and the structurally related benz[*e*]indenes (see Sect. B.II and Fig. 2) upon $GABA_A$ receptor function provide the most convincing evidence that such compounds act directly upon the receptor. This is so because enantioselectivity is only manifest in a chiral (e.g. protein) environment. The endogenous eutomer (+)-5α-pregnan-3α-ol-20-one (which differs from alphaxalone in that C11 is unsubstituted; see Fig. 2) acts as a potent positive allosteric modulator of the $GABA_A$ receptor and is an anaesthetic in animal studies (MAJEWSKA et al. 1986; HARRISON et al. 1987a; PETERS et al. 1988). In comparison, the distomer (–)-5α-pregnan-3α-ol-20-one has much reduced GABA-modulatory and anaesthetic potencies in tadpoles and mice (WITTMER et al. 1996; ZORUMSKI et al. 1996). A similar correlation between GABA-modulatory and anaesthetic potency exists for androstane enantiomers bearing a 17β-carbonitrile substituent (see Sect. B.VII) (WITTMER et al. 1996) and for the enantiomers of the benz[e]indene BI-1 (ZORUMSKI et al. 1996). These observations strongly support the concept of (a) distinct binding site(s) for steroids on the $GABA_A$ receptor and reinforce observations suggesting alphaxalone and 5β-pregnan-3α-ol-20-one to be effective modulators of GABA only when applied extracellularly (LAMBERT et al. 1990; POISBEAU et al. 1997).

II. The Ring System

Recent studies have demonstrated that a saturated ring system is not an absolute requirement for positive allosteric modulation of the $GABA_A$ receptor activity by steroids. In several assays, 4-pregnen-3α-ol-20-one exhibits a potency and efficacy comparable to that of 5α-pregnan-3α-ol-20-one (HAWKINSON et al. 1994). Similarly, 5α-preg-9(11)-en-3-ol-20-one retains some activity (HAWKINSON et al. 1994). Furthermore, the steroid A ring per se is not essential for activity because certain benz[*e*]indene compounds (e.g. BI-1, Fig. 2) which may be viewed as tricyclic steroid analogues in which the steroid A-ring is partially opened and removed (ZORUMSKI et al. 1996), retain the ability to potentiate and activate $GABA_A$ receptors in an enantioselective fashion (RODGERS-NEAME et al. 1992; WITTMER et al. 1996; ZORUMSKI et al. 1996) (see Sect. B.I also). The introduction of a double bond within the steroid D ring between C16 and C17 reduces, but does not abolish, the activity of some naturally occurring and synthetic pregnanes (BOLGER et al. 1996) (see Sect. B.VI also).

III. C2 Substitution

Modulation of $GABA_A$ receptor activity by pregnane steroids rendered water soluble by the introduction of a 2β-morpholinyl group has recently been

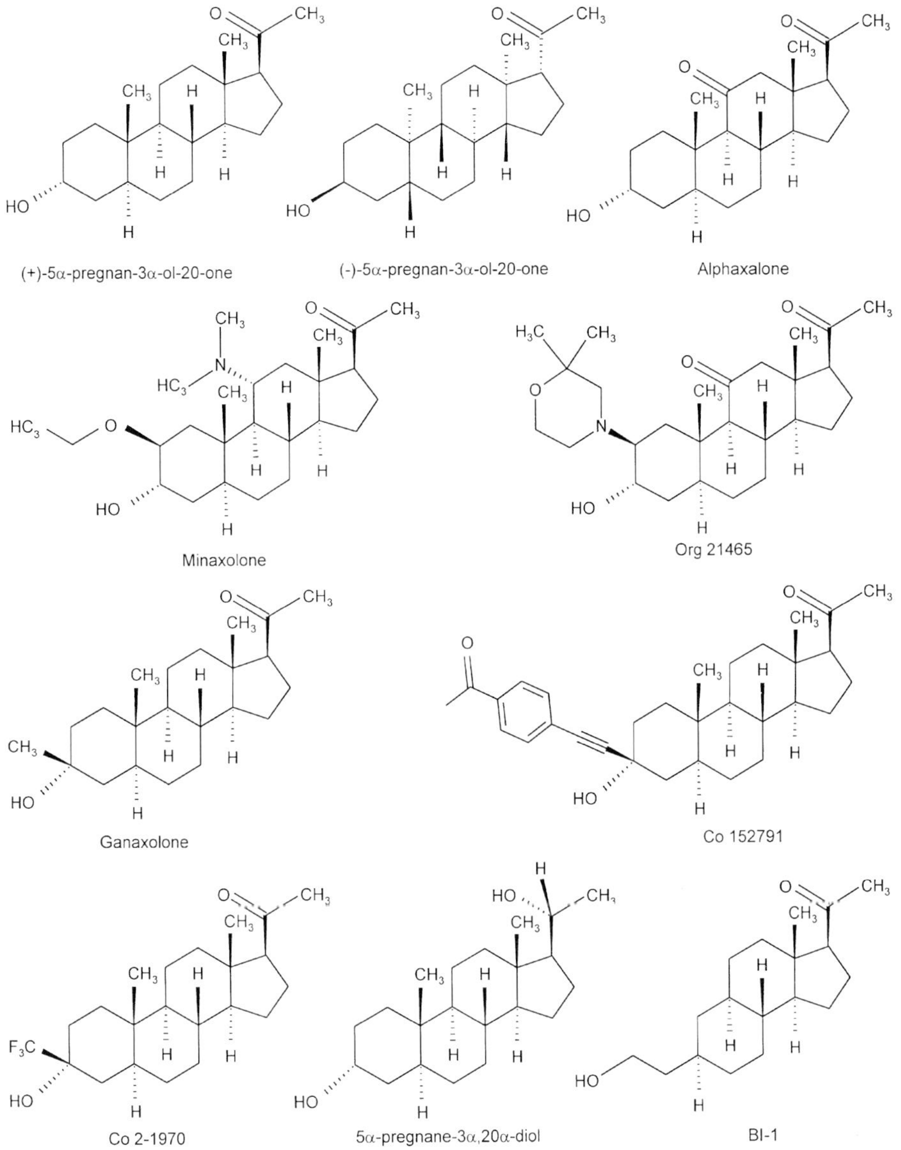

(+)-5α-pregnan-3α-ol-20-one

(-)-5α-pregnan-3α-ol-20-one

Alphaxalone

Minaxolone

Org 21465

Ganaxolone

Co 152791

Co 2-1970

5α-pregnane-3α,20α-diol

BI-1

Fig. 2. Chemical structures of selected naturally occurring and synthetic neuroactive steroids and a neuroactive steroid analogue. The structures depicted show: the naturally occurring (+) and synthetic (–) enantiomers of 5α-pregnan-3α-ol-20-one; the 11-keto substituted compound alphaxalone; the water-soluble derivatives minaxolone and Org 21465; the 3β-substituted compounds ganaxolone and Co 152791; the partial agonists Co-1970 and 5α-pregnane-3α,20α-diol and the benz[*e*]indene, BI-1. See text for further details.

described in detail (HILL-VENNING et al. 1996; ANDERSON et al. 1997). It is clear that the steroid binding site(s) of the $GABA_A$ receptor can tolerate rather bulky substituents at the 2β-position, since even alkylated 2β-morpholinyl derivatives of alphaxalone (e.g. Org 21465, Fig. 2) can be accommodated without loss of potency (ANDERSON et al. 1997). Similarly, the modulatory activity of the anaesthetic steroid, minaxolone (Fig. 2), at the $GABA_A$ receptor is not adversely affected by structural modifications to the parent compound alphaxalone (2β-ethoxy and 11α-dimethylamino substitutions) that confer solubility in water (SHEPHERD et al. 1996).

IV. C3 Substitution

Numerous studies have demonstrated that 5α- and 5β-reduced pregnane (and androstane) steroids are essentially equally potent as modulators of the $GABA_A$ receptor. Thus, despite the substantial conformational difference introduced by the stereochemistry of the steroid A/B ring fusion (*trans* and *cis* in the 5α- and 5β series of compounds respectively, see Fig. 1), binding within the $GABA_A$ receptor is accommodated. By contrast, the nature and configuration of the substituent at the C3 position of the steroid A ring is an extremely important determinant of steroid action at the $GABA_A$ receptor. For example, epimerization of the 3-hydroxyl group of the anaesthetic steroid alphaxalone to the β-configuration, yields betaxalone (5α-pregnane-3β-ol-11,20-dione), which is neither an anaesthetic, nor a positive allosteric modulator of the receptor (HARRISON and SIMMONDS 1984; COTTRELL et al. 1987). The 3β-hydroxy epimers of the naturally occurring steroids 5α-pregnan-3α-ol-20-one, 5β-pregnan-3α-ol-20-one and 5α-pregnan-3α,11β,21-triol-20-one, are similarly ineffective in potentiating GABA (HARRISON et al. 1987a; GEE et al. 1988; PETERS et al. 1988; KOKATE et al. 1994). However, 5α-pregnan-3α-ol-20-one and 5α-pregnan-3β-ol-20-one, when utilized at relatively high concentrations, do share the ability to increase the rate of desensitization of current responses mediated by the $GABA_A$ receptor, indicating that this aspect of their action is not diastereoselective (WOODWARD et al. 1992).

Oxidation of the 3-hydroxyl group to the ketone (CALLACHAN et al. 1987; HARRISON et al. 1987a; PURDY et al. 1990; HAWKINSON et al. 1994), markedly attenuates, or abolishes, positive allosteric modulation by 5α- and 5β-pregnanes. Similarly, pregnenes (e.g. progesterone) and androstenes (e.g. testosterone and androstenedione) wherein a C3 ketone substituent is present within an unsaturated (C4-C5 double bond) steroid A-ring exert only a limited activity even when utilized at very high concentrations (PARK-CHUNG et al. 1999). In addition, the substitution of oxime, acetate or methyl groups at the C3 position greatly diminishes activity (PURDY et al. 1990; HAWKINSON et al. 1994; UPSANI et al. 1997). It seems likely that the free hydroxyl group at C3, via hydrogen bond donatation, is an important determinant of the primary docking of the steroid molecule to the positive allosteric regulator site(s) of $GABA_A$ receptor (UPSANI et al. 1997).

The formation of a sulphate ester at the 3α-hydroxyl group of 5α- or 5β-pregnan-3α-ol-20-one, or 5α-androstan 3α-ol-17-one (androsterone), results in compounds that, at sub-micromolar concentrations, have minimal $GABA_A$ receptor activity (NILSSON et al. 1998; PARK-CHUNG et al. 1999). However, at higher concentrations, the sulphated pregnanes and androstanes inhibit $GABA_A$ receptor activity (NILSSON et al. 1998; PARK-CHUNG et al. 1999) in a manner qualitatively similar to that documented for other endogenous sulphated steroids including pregnenolone sulphate and dehydroepiandrosterone sulphate (MAJEWSKA et al. 1988, 1990a,b; MIENVILLE and VICINI 1989; LE FOLL et al. 1997). The inhibitory action of pregnenolone sulphate and 5β-pregnane-3α-ol-20-one sulphate does not demonstrate enantioselectivity, unlike the potentiating effect of, for example, 5α-pregnan 3α-ol-20-one (WITTMER et al. 1996; NILSSON et al. 1998) (see Sect. B.I). This might indicate that sulphated and un-sulphated steroids bind to distinct sites to produce their opposing effects, a suggestion consistent with the results of interaction studies between the two classes of compounds (PARK-CHUNG et al. 1999). Adding further complexity, blockade of $GABA_A$ receptor mediated currents by the unnatural enantiomer of dehydroepiandrosterone sulphate is clearly less potent than for the naturally occurring steroid, which may suggest differences in the nature of the site(s) that recognise specific sulphated compounds (NILSSON et al. 1998). Hereafter, we confine the discussion to steroids that act as positive allosteric modulators of $GABA_A$ receptor activity.

The potential therapeutic utility of pregnane steroids (other than as short acting intravenous general anaesthetic agents) is limited by their rapid metabolism via conjugation or oxidation of the crucial 3-hydroxyl group. It is possible to retard such reactions by substitution at the 3β-position. Thus, the 3β-methyl substituted analogue of 5α-pregnan-3α-ol-20-one (i.e. ganaxolone) (Fig. 2) retains potency and efficacy as a modulator of the $GABA_A$ receptor and, unlike the parent compound, demonstrates anticonvulsant activity against chemically induced seizures in rats when administered orally (CARTER et al. 1997). Within the 5α-pregnane series, the introduction of simple alkyl 3β-substituents larger than a methyl group results in a reduction in both potency and efficacy (the latter being inferred from incomplete displacement of the binding of [^{35}S]TBPS to the receptor complex in radioligand binding assays). The reduction in potency does not correlate simply with the size of the substituted alkyl group (HOGENKAMP et al. 1997). By contrast, the reduction in potency produced by 3β-substitution with either ethers or alkyl halides tends to increase with size. An interesting example of the latter group of compounds is the 3β-trifluoromethyl derivative of 5α-pregnan-3α-ol-20-one (i.e. Co 2–1970) (Fig. 2) which acts in a manner consistent with partial agonism in both radioligand binding and electrophysiological assays of allosteric regulation of the $GABA_A$ receptor (HAWKINSON et al. 1996). Steroids with limited efficacy could, in principle, offer advantages over full-agonists in certain clinical settings.

In contrast to the deleterious effect of 3β-alkyl substitutions, the incorporation of alkene and alkyne groups at this position is generally well tolerated in both the 5α- and 5β-pregnane series if the unsaturated bond is immediately adjacent to the steroid A ring (HOGENKAMP et al. 1997; HAWKINSON et al. 1998). Indeed, certain 3β-phenylethynyl analogues of 5α- and 5β-pregnan 3α-ol-20-one (e.g. Co 152791) (Fig. 2) retain not only the full agonist character of the parent steroid, but in addition demonstrate a marked increased in potency (UPSANI et al. 1997; HAWKINSON et al. 1998). Optimal activity is associated with the ethynyl spacer unit, which is postulated to place the phenyl ring at an appropriate distance from the steroid nucleus, and the presence of hydrogen bond acceptors (e.g. acetyl or carbethoxy groups) at the *para*-position of the phenyl ring. The enhanced potency of, for example, the 3β-(*p*-acetylmethylphenylethynyl) derivatives of 5α- and 5β-pregnan-3α-ol-20-one, has been interpreted as evidence for the existence of an auxiliary docking site at the GABA receptor. The latter is proposed to be accessed via the rigid spacer extending from the 3β-position and binding is subsequently stabilized by hydrogen bond formation (UPSANI et al. 1997; HAWKINSON et al. 1998).

V. C5, C10 or C11 Substitution

The substitution of the C5 hydrogen atom by a methyl group in the α-orientation (i.e. projecting below the plane of the steroid ring system) greatly reduces, or abolishes, potentiation of $GABA_A$ receptor activity by pregnane and androstane steroids. By contrast, 5β-methyl substitution is better tolerated, indicating that steric restrictions exist in the region of space below the steroid A ring (HAN et al. 1996). The 19-Nor steroids generated by the replacement of the C19 methyl group at C10 by H exhibit activities more closely related to their parent compounds, suggesting steric hindrance to be less pronounced above the plane of the steroid ring (HAN et al. 1996). At the C11 position, the introduction a ketone group into 5α-pregnanes (e.g. alphaxalone; 5α-pregnan-3α-ol-11,20-dione) (Fig. 2) causes some loss of activity at the $GABA_A$ receptor, whereas introducing an hydroxyl function at this, or the adjacent C12, position abolishes activity (HAWKINSON et al. 1994; ANDERSON et al. 1997).

VI. The C17 Side Chain

For all pregnane steroids examined to date, the side chain at C17 must be in the β-configuration for activity (PURDY et al. 1990; HAWKINSON et al. 1994) Similarly, whilst substitution of the acetyl side chain with a carbonitrile moiety produces a compound with an activity similar to that of 5α-pregnan-3α-ol-20-one, the β orientation of the substituent is once again crucial. The insertion of a double bond between C16 and C17 of the pregnane steroid D ring (see also Sect. B.I) produces 16-ene analogues whose reduced potency is thought to result from changes in the conformation of the side chain that place the C20 ketone group (see Sect. B.VII) in an unfavourable orientation (BOLGER et al. 1996).

VII. C20 Substitution

The presence of a ketone group at C20 of the acetyl side chain was initially deemed essential to the activity of pregnane steroids at the $GABA_A$ receptor (HARRISON et al. 1987a). It is postulated that the C20 ketone acts as a hydrogen bond acceptor, which, together with the 3α-hydroxyl-group (see Sect. IV), serves to anchor the steroid in the primary binding pocket of the $GABA_A$ receptor (UPSANI et al. 1997; HAWKINSON et al. 1998). However, subsequent studies have revealed 20-keto reduced analogues of 5α- and 5β-pregnan-3α-ol-20-one (e.g. 5α-pregnane-3α,20α-diol) (Fig. 2) to modulate $GABA_A$ receptor activity in a manner consistent with partial agonism. The potency and efficacy of such pregnanediols are dependent upon structural determinants that include *cis* or *trans* fusion of the A and B rings and the orientation (α or β) or the 20-hydroxyl moiety, which, in contrast to a C20 ketone substituent, might function as a hydrogen bond donor (McCAULEY et al. 1995; BELELLI et al. 1996).

VIII. C21 Substitution

The presence of a hydroxyl group at C21 (as the in naturally-occurring 5α-pregnan-3α,21-diol-20-one) or its esterification to the acetate or mesylate produces only modest reductions in activity (HAWKINSON et al 1994). Similarly, from studies conducted with a series of 2β-morpholinyl substituted steroids (see Sect. B.III), it appears that the steroid binding site of the $GABA_A$ receptor can accept functional groups that include hydroxyl, chloride, acetate, thioacetate, thiocyanate and azide moieties (ANDERSON et al. 1997). A hemisuccinate group can also be tolerated (GASIOR et al.1999). However, unlike 5β-pregnan-3α-ol-20-one, 5β-pregnan-3α,21-diol-20-one is reported to act as a partial agonist, suggesting some interaction between C21 substituents and the orientation (*cis* or *trans* – see Sect. B.I) of the steroid (XUE et al. 1997).

IX. Summary

An α-hydroxyl group at C3 and ketone moiety at C20 most probably serve, by donating and accepting hydrogen bonds respectively, as points of attachment of neurosteroids within the primary binding pocket of the $GABA_A$ receptor. However, the energy provided by such interactions would clearly be insufficient to account for the high apparent affinity of many pregnane steroids. Additional important stabilizing influences most probably include hydrophobic interactions between the steroid ring system and receptor protein. In this respect, the area immediately beneath the A/B ring fusion appears to present a forbidden volume, but the configuration of rings appears to be of little importance. Substitutions in the β-orientation at C2, and at C21 are well tolerated, whereas the effect of chemical modification at C11 is dependent upon the precise substituent. The metabolism of neurosteroids can be retarded by substitutions at the 3β position which, in the case of phenylethynyl dervatives, may also contribute to potency by contacting an auxiliary binding pocket.

C. Neurosteroid Binding Site Heterogeneity and the Influence of $GABA_A$ Receptor Subunit Composition upon Neurosteroid Action

There is considerable indirect evidence from radioligand binding and chloride flux studies with native $GABA_A$ receptors to suggest that neuroactive steroids can differentiate between $GABA_A$ receptor isoforms. As a consequence, the effects of the steroids may be brain region dependent (GEE et al. 1988; PRINCE and SIMMONDS 1993; OLSEN and SAPP 1995). However, studies investigating the dependence of neurosteroid action on the subunit composition of the $GABA_A$ receptor have not provided a consistent picture (LAMBERT et al. 1995). For clarity, the findings presented here will be restricted to those obtained in electrophysiological assays.

I. α-Subunits

The benzodiazepine pharmacology of the $GABA_A$ receptor is highly dependent upon the isoform of the α subunit (α_{1-6}) present with the hetero-oligomer (LÜDDENS et al. 1995; SMITH and OLSEN 1995). By contrast, differences in neuroactive steroid potency across the α isoforms are relatively modest (SHINGAI et al. 1991; PUIA et al. 1993; LAMBERT et al. 1995). Indeed, the presence of the α subunit is not a prerequisite for modulation by neuroactive steroids, because recombinant receptors assembled solely from β_1 and γ_2 subunits are sensitive to 5α-pregnan-3α-ol-20-one and alphaxalone. At such receptors, steroids exhibit a similar EC_{50} to that found for α, β and γ subunit combinations, albeit with a reduced maximum effect (MAITRA and REYNOLDS 1999). Utilizing the *Xenopus laevis* oocyte expression system, we have recently investigated the influence of the α isoform ($\alpha_X\beta_1\gamma_2$ where x = 1–6) on the potency (EC_{50}) and maximal (E_{max}) GABA-modulatory effects of 5α-pregnan-3α-ol-20-one (see Table 1). Essentially, and in agreement with previous studies, inspection of Table 1 confirms that the neurosteroid does not discriminate clearly between the α isoforms. Hence, apart the receptor assembled from α_6, β_1 and γ_2 subunits, the E_{max}. varies little (i.e. a seven- to ninefold increase of the current induced by an EC_{10} concentration of GABA) for receptors containing the different α isoforms (Table 1). Evaluation of the neurosteroid EC_{50} reveals, at most, only a three- to fourfold difference (Table 1).

The effects of the neurosteroids on the α_4-subunit containing receptor are of particular interest given the recent reports on the increased expression of this subunit in the hippocampus upon progesterone withdrawal (SMITH et al. 1998a,b). Hippocampal neurones, isolated from progesterone-withdrawn rats,

Table 1. The influence of $GABA_A$ receptor subunit composition upon the modulatory effects of 5α-pregnan-3α-ol-20-one

Subunit combination	EC_{50}[a]	E_{MAX}[b]
$\alpha_1\beta_1$	380 ± 10 nmol/l	143 ± 2%
$\alpha_1\beta_1\gamma_{2L}$	89 ± 6 nmol/l	69 ± 4%
$\alpha_1\beta_2\gamma_{2L}$	177 ± 2 nmol/l	75 ± 4%
$\alpha_1\beta_3\gamma_{2L}$	195 ± 36 nmol/l	72 ± 4%
$\alpha_2\beta_1\gamma_{2L}$	146 ± 11 nmol/l	66 ± 6%
$\alpha_3\beta_1\gamma_{2L}$	74 ± 1 nmol/l	67 ± 7%
$\alpha_4\beta_1\gamma_{2L}$	317 ± 25 nmol/l	72 ± 6%
$\alpha_5\beta_1\gamma_{2L}$	302 ± 38 nmol/l	81 ± 2%
$\alpha_6\beta_1\gamma_{2L}$	220 ± 12 nmol/l	131 ± 6%
$\alpha_6\beta_2\gamma_{2L}$	350 ± 29 nmol/l	108 ± 5%
$\alpha_6\beta_3\gamma_{2L}$	264 ± 33 nmol/l	90 ± 9%

All parameters are calculated from steroid concentration-effect relationships obtained from a minimum of 4 oocytes expressing combinations of human $GABA_A$ receptor subunits. Data are collated from LAMBERT et al. (1999) and the unpublished observations of D. Belelli.

[a] The EC_{50} is defined as the concentration of steroid which causes the GABA (EC_{10}) evoked current to be enhanced to 50% of the maximum potentiation that can be produced by the steroid.

[b] The E_{MAX} is defined as the maximum potentiation produced by the steroid (expressed as a percentage of the peak current evoked by a saturating concentration of GABA alone).

express $GABA_A$ receptors with physiological and pharmacological properties consistent with those reported for α_4 subunit-containing recombinant receptors. In particular, GABA-evoked currents recorded from such neurones are brief in duration, insensitive to lorazepam, and characteristically are enhanced by benzodiazepine antagonists and inverse agonists (WAFFORD et al. 1996; SMITH et al. 1998a,b, 1999). This alteration of the hippocampal $GABA_A$ receptors appears to be in response to the withdrawal of the progesterone metabolite 5α-pregnan-3α-ol-20-one, rather than progesterone itself (SMITH et al. 1998a,b). In addition to expressing an altered benzodiazepine pharmacology, the hippocampal $GABA_A$ receptors of these treated animals are insensitive to "physiological" (10 nmol/l) levels of 5α-pregnan-3α-ol-20-one (SMITH et al. 1998b). This feature would appear inconsistent with the properties of α_4-containing receptors, which are reported to be neurosteroid-sensitive (WAFFORD et al. 1996) (Table 1) although given the fourfold difference in the EC_{50} value for 5α-prenan-3α-ol-20-one acting at α_1- vs α_4-subunit-containing receptors, the latter would be expected to be less sensitive to physiological levels of the steroid.

II. β Subunits

Alphaxalone, 5α-pregnan-3α-ol-20-one and 5α-pregnane-3α,21-diol-20-one do not discriminate between the β-subunit isoforms when expressed in hetero-oligomeric receptors of the composition $\alpha_1\beta_X\gamma_2$ (where x = 1, 2 or 3) (HADINGHAM et al. 1993; SANNA et al. 1997) (see also Table 1). In this respect, the neuroactive steroids differ from the anaesthetic etomidate and the anticonvulsant loreclezole, which preferentially modulate β_2- and β_3- over β_1-subunit containing receptors (WINGROVE et al. 1994; BELELLI et al. 1997).

III. γ Subunits

The presence of a γ subunit in a heteromeric $GABA_A$ receptor complex is a prerequisite for a consistent allosteric modulation by benzodiazepines (LÜDDENS et al. 1995). Furthermore, the nature of the benzodiazepine interaction with the receptor is additionally influenced by the γ subunit isoform present within the receptor complex (LÜDDENS et al. 1995). However, in contrast to the benzodiazepines, the presence of a γ subunit is not required for steroid modulation of GABA-evoked currents (PUIA et al. 1990; SHINGAI et al.1991). In a recent electrophysiological study utilizing oocytes, the identity of γ subunit had little effect on the potency with which alphaxalone, or 5α-pregnan-3α-ol-20-one, modulated GABA-evoked currents, although the maximal effect of the steroids was greater at γ_3- vs γ_1- or γ_2-subunit-containing receptors (MAITRA and REYNOLDS 1999).

IV. The δ Subunit

The expression of a δ subunit, in combination with α and β subunits, dramatically reduces the GABA-modulatory effects of 5α-pregnane-3α,21-diol-20-one, but has little effect on the GABA-mimetic effects of this steroid (ZHU et al. 1996). The potential physiological importance of this observation is illustrated by experiments performed on cerebellar granule cells, in which $GABA_A$ receptor mediated responses exhibit a reduced responsiveness to neurosteroid modulation with development. Analysis of the potential subunit composition of granule cell $GABA_A$ receptors by single cell PCR techniques suggests that the loss of neurosteroid sensitivity may be due to increased incorporation of the δ subunit into the receptor complex (ZHU et al. 1996).

V. The ε Subunit

As found for the δ subunit, the incorporation of the ε subunit into α- and β-subunit-containing $GABA_A$ receptors dramatically reduces the GABA-modulatory effects of 5α-pregnan-3α-ol-20-one and the general anaesthetics propofol and pentobarbitone (DAVIES et al. 1997). Although the direct effects of such agents are little affected by the ε subunit (DAVIES et al. 1997), interpretation of this finding is complicated because receptors embodying an ε subunit exhibit spontaneous channel openings in the absence of GABA (NEELANDS et al. 1999). In addition, an independent study found the ε subunit to exert little effect upon the GABA-modulatory properties of the neurosteroids (WHITING et al. 1997). These contradictory findings are currently inexplicable.

VI. Summary

The identity of the α and β subunits has little, or no, effect upon neurosteroid action and the γ subunit is not required for their activity. Substitution of a γ subunit by a δ subunit clearly suppresses the GABA-modulatory activity of the neurosteroids, but the influence of the ε subunit remains to be clarified. Recent studies have revealed that synaptic $GABA_A$ receptors are differentially sensitive to neurosteroids (see Sect. E), but the molecular basis of such diversity remains to be elucidated.

D. Molecular Mechanism of Neurosteroid Action

Experiments investigating the influence of alphaxalone on GABA-induced current fluctuations recorded from mouse spinal neurones suggested that the steroid acts primarily to prolong the mean open time of the $GABA_A$ receptor ion channel (BARKER et al. 1987). In agreement, single channel recordings made from membrane patches excised from bovine chromaffin cells clearly demonstrated 5α- or 5β-pregnan-3α-ol-20-one to prolong the open time of channels activated by GABA with no effect on the single channel conductance (CALLACHAN et al. 1987; LAMBERT et al. 1987). Additionally, these studies established that at concentrations greater than those required for GABA modulation, these steroids in the absence of GABA directly activated the receptor complex. Similar actions have recently been noted for 5β-pregnan-3α-ol-20-one acting on the $GABA_A$ receptor(s) expressed by frog pituitary melanotrophs (LE FOLL et al. 1997).

The $GABA_A$ receptor of the chromaffin cell exhibits multiple interconverting conductance states which prevents a quantitative analysis on the effect of these neurosteroids on GABA-gated ion channel kinetics. However, the $GABA_A$ receptors of mouse spinal neurones often exhibit one predominant conductance state (MACDONALD et al. 1989; MACDONALD and OLSEN 1994). By restricting analysis to such data segments, three kinetically distinct open states of the GABA-gated ion channel were revealed and the depressant neuroactive steroids were shown primarily to promote the occurrence of the open states of intermediate and long duration at the expense of openings of brief duration (TWYMAN and MACDONALD 1992; MACDONALD and OLSEN 1994). The anaesthetic barbiturates act in a similar way to perturb channel kinetics (MACDONALD et al. 1989), but the neuroactive steroids additionally increase the frequency of single channel openings (TWYMAN and MACDONALD 1992). Whether this latter effect is caused by the direct activation of the receptor channel complex by the neuroactive steroid is not known.

The aforementioned kinetic studies were performed with relatively low concentrations of GABA. However, for at least some central GABA-ergic synapses, it appears that the concentration of synaptically-released GABA is sufficient to saturate briefly a small number of postsynaptic $GABA_A$ receptors (MODY et al. 1994; EDWARDS 1995). Hence, an examination of the effects of the neuroactive steroids on GABA-evoked currents induced by rapidly applied saturating concentrations of the agonist may be more instructive in understanding how the steroid-induced perturbation of channel kinetics modifies synaptic transmission. Rapid (200 μs) and brief (1 ms) applications of a saturating concentration of GABA to nucleated membrane patches excised from cerebellar granule cells induces currents that decay with a biphasic time course consisting of fast and slow components (ZHU and VICINI 1997). The decay of some miniature inhibitory postsynaptic currents (mIPSCs – the result of the activation of synaptically located $GABA_A$ receptors by a single vesicle of GABA) also exhibit a bi-exponential decay (EDWARDS 1995; ZHU and VICINI 1997). In both instances, the fast time component is thought to originate from channels oscillating between GABA bound open and closed confirmations, whereas the slower phase is proposed to be caused by receptors visiting, and exiting, various desensitized states (JONES and WESTBROOK 1996). Hence, $GABA_A$ receptors exiting desensitization could re-enter conducting states and by this mechanism, effectively prolong the GABA-evoked current. The neurosteroid 5α-pregnane-3α,21-diol-20-one has been shown to prolong the slow time constant of decay of GABA-evoked currents recorded from nucleated patches (ZHU and VICINI 1997). This effect is postulated to result from the steroid acting to slow the recovery of receptors from desensitization (ZHU and VICINI 1997). Consistent with this proposal, 5α-pregnane-3α,21-diol-20-one, in the presence of a saturating concentration of GABA, increases the probability of the channel being in the open state, by increasing the number of late channel openings (ZHU and VICINI 1997). This mechanism is thought to underlie the neurosteroid-induced prolongation of GABA-mediated synaptic events (see below).

E. Neurosteroid Effects on Synaptic Transmission

The effects 5α-pregnan-3α-ol-20-one and 5α-pregnane-3α,21-diol-20-one on evoked inhibitory postsynaptic currents (IPSCs) were first examined in voltage-clamp studies performed on rat hippocampal neurones in cell culture (HARRISON et al. 1987a,b). The neurosteroids were found to prolong the decay of the GABA-mediated synaptic current with little, or no, effect upon IPSC amplitude or rise time. Surprisingly, given the interest in neurosteroids, it is only recently that their effects on inhibitory synaptic transmission have been investigated further.

Evoked, or spontaneous, IPSCs are thought to result from the asynchronous release of GABA from multiple release sites (Mody et al. 1994; Williams et al. 1998), making the interpretation of the effects of neuroactive steroids complex. However, synaptic events recorded in the presence of the voltage-activated sodium channel blocker tetrodotoxin, which prevents release due to local presynaptic action potential discharge and isolates miniature inhibitory postsynaptic currents (mIPSCs), are thought to arise from the release of a single vesicle of GABA. The latter most probably briefly saturates a relatively small number of postsynaptic $GABA_A$ receptors with neurotransmitter (Mody et al. 1994). A number of studies utilizing the in vitro brain slice preparation, or acutely dissociated neurones with adherent synaptic terminals, have reported nanomolar concentrations of 5α- or 5β-pregnan-3α-ol-20-one and 5α-pregnane-3α,21diol-20-one to prolong the mIPSC decay time recorded from neurones of the medial preoptic nucleus, cerebellar Purkinje neurones, hippocampal dentate granule and CA1 pyramidal neurones (Cooper et al. 1996; Harney et al. 1999; Lambert et al. 1999; Haage and Johansson 1999). Interestingly, mIPSCs recorded from dentate granule cells within slices prepared from 20-day-old animals appear relatively insensitive to neurosteroid modulation, compared to those of Purkinje and CA1 hippocampal neurones. By contrast, the mIPSCs of dentate granule cells of 10-day-old animals are neurosteroid-sensitive (Cooper et al. 1996). The physiological and pharmacological properties of dentate granule $GABA_A$ receptors are reported to undergo considerable developmental changes (Hollrigel and Soltesz 1997; Kapur et al. 1999), presumably reflecting changes in $GABA_A$ receptor subunit composition that are known to occur at this time (Fritschy et al. 1994). Hence, the neurosteroid sensitivity of the dentate granule neurones may be developmentally regulated by changes in the subunit complement of the $GABA_A$ receptor.

That the neurosteroid sensitivity of $GABA_A$ receptors can be both a dynamic and plastic property is demonstrated by recent studies on spontaneous IPSCs (sIPSCs) recorded, in the absence of tetrodotoxin, from hypothalamic magnocellular oxytocin neurones during the reproductive cycle of the rat (Brussaard et al. 1997, 1999). Such neurones secrete oxytocin during parturition and lactation. Acting upon the neurones of virgin animals, and animals one day prior to parturition, 5α-pregnan-3α-ol-20-one produces a concentration-dependent prolongation of the sIPSC decay time with no effect on sIPSC amplitude. However, upon parturition, which is coincident with a dramatic decrease of endogeneous 5α-pregnan-3α-ol-20-one levels, the sIPSCs become insensitive to the neurosteroid and exhibit a prolonged decay (Brussaard et al. 1997, 1999). The altered synaptic decay and neurosteroid insensitivity of sIPSCs is long-lived, with their properties only reverting to those of pre-pregnancy several weeks after the end of lactation (Brussaard et al. 1999). The inhibitory input to these neurones plays an important regulatory role and these changes in the properties of the $GABA_A$ receptors may underlie the timed release of oxytocin required for parturition and lactation. Coincident with the altered synaptic decay and neurosteroid insensitivity, the ratio of α_2 to α_1

mRNA is increased in these neurones (BRUSSAARD et al. 1997, 1999). If such changes are mirrored at the level of the expressed protein, an altered subunit composition of the synaptic $GABA_A$ receptors might underlie the changes in sIPSC kinetics. However, for recombinant $GABA_A$ receptors, the effect of 5α-pregnan-3α-ol-20-one is little influenced by the nature of the α isoform (see Sect. C), although a reduced metabolite (5α-pregnane-3α,20α-diol) is less potent at α_2- vs α_1- or α_3-subunit-containing receptors (BELELLI et al. 1996). Whether the properties of synaptic α_2 subunit-containing receptors are functionally distinct from recombinant receptors that incorporate an α_2 subunit, or whether these neurones express additional subunits that might explain the neurosteroid-insensitivity (e.g. δ or ε), remains to be determined.

Finally, some studies have reported that neurosteroids, in addition to influencing the mIPSC time course, may additionally increase the frequency of mIPSCs, implying a presynaptic effect of the steroid (POISBEAU et al. 1997; REITH and SILLAR 1997; HAAGE and JOHANSSON 1999).

F. Concluding Remarks

The stereoselectivity and potency of the neurosteroid interaction with the $GABA_A$ receptor is indicative of the presence of a high affinity binding site on the receptor protein. However, although genetically modified recombinant $GABA_A$ receptors have been successfully utilised to identify key amino acids, or domains, of the protein that contribute to the benzodiazepine and GABA binding sites (SIGEL and BAUR 1997), to date this approach has had limited success for the neurosteroids (RICK et al. 1998). Irrespective of the nature of the interaction with the $GABA_A$ receptor, the more accurate estimation of the likely synaptic concentrations of 5α-pregnan-3α-ol-20-one, coupled with the demonstration of its central synthesis, strongly suggests that this potent steroid-receptor interaction could subserve an important physiological/pathophysiological role. Clearly, the identification of a selective neurosteroid antagonist, analogous to the benzodiazepine receptor antagonist flumazenil, would be invaluable in evaluating an endogenous function.

Therapeutically, synthetic steroids are currently undergoing clinical trials as anticonvulsants, anxiolytics and in the treatment of sleep disorders (GASIOR et al. 1999). It will be of interest not only to determine their clinical efficacy, but to establish whether compounds which are based on the structure of an endogeneous modulator offer any advantages, particularly regarding side-effects, over currently available $GABA_A$ receptor ligands such as the benzodiazepines. Finally, the discovery of novel compounds which selectively interact with the brain enzymes that synthesise or metabolise the neurosteroids may offer a new therapeutic avenue.

Acknowledgements. Some of the work reported here was supported by an EC Bioscience and Health Grant BMH4-CT97–2359. Dr. D. Belelli is an MRC Senior Fellow. Ms. S.C. Harney is supported by an MRC Studentship.

References

Anderson A, Boyd AC, Byford A, Campbell AC, Gemmell DK, Hamilton NM, Hill DR, Hill-Venning C, Lambert JJ, Maidment MS, May V, Marshall RJ, Peters JA, Rees DC, Stevenson D, Sundaram H (1997) Anaesthetic activity of novel water-soluble 2β-morpholinyl steroids and their modulatory effects at $GABA_A$ receptors. J Med Chem 40:1668–1681

Barker JL, Harrison NL, Lange GD, Owen DG (1987) Potentiation of γ-aminobutyric-acid-activated chloride conductance by a steroid anaesthetic in cultured rat spinal neurones. J Physiol 386:485–501

Baulieu E-E, Schumacher M (1996) Synthesis and functions of neurosteroids. In: Genazzani AR, Petraglia F, Purdy RH (eds) The brain: source and target for sex steroid hormones. Parthenon, New York, pp 5–24

Belelli D, Lambert JJ, Peters JA, Gee KW, Lan NC (1996) Modulation of human $GABA_A$ receptor by pregnanediols. Neuropharmacology 35:1223–1231

Belelli D, Lambert JJ, Peters JA, Wafford KA, Whiting PJ (1997) The interaction of the general anesthetic etomidate with the γ-aminobutyric acid type A receptor is influenced by a single amino acid. Proc Natl Acad Sci USA. 94:11031–11036

Bolger M, Wieland S, Hawkinson J, Xia H, Upsani R, Lan NC (1996) In vitro and in vivo activity of 16,17-dehydro-epipregnanolones: 17,20-bond torsional energy analysis and D-ring conformation. Pharm Res 13:1488–1494

Brussaard AB, Kits KS, Baker RE, Willems WP, Leyting-Vermeulen JW, Voorn P, Smit AB, Bicknell RJ, Herbison AE (1997) Plasticity in fast synaptic inhibition of adult oxytocin neurons caused by switch in $GABA_A$ receptor subunit expression. Neuron 19:1103–1114

Brussaard AB, Devay P, Leyting-Vermeulen JL, Kits KS (1999) Changes in properties and neurosteroid regulation of GABAergic synapses in the supraoptic nucleus during the mammalian female reproductive cycle. J Physiol 516:513–524

Callachan H, Cottrell GA, Hather NY, Lambert JJ, Nooney JM, Peters JA (1987) Modulation of the $GABA_A$ receptor by progesterone metabolites. Proc R Soc Lond B Biol Sci 231:359–369

Carter RB, Wood PL, Weiland S, Hawkinson JE, Belelli D, Lambert JJ, White HS, Wolf HF, Mirsadeghi S, Tahir SH, Bolger MB, Lan NC, Gee KW (1997) Characterization of the anticonvulsant properties of ganaxolone (CCD 1042; 3α-hydroxy-3β-methyl-5α-pregnan-20-one), a selective, high-affinity, steroid modulator of the γ-aminobutyric $acid_A$ receptor. J Pharmacol Exp Ther 280:1284–1295

Cheney DL, Uzunov D, Costa E, Guidotti A (1995) Gas chromatographic-mass fragmentographic quantitation of 3α-hydroxy-5α-pregnan-20-one (allopregnanolone) and its precursors in blood and brain of adrenalectomized and castrated rats. J Neurosci 15:4641–4650

Cooper EJ, Johnston GAR, Edwards FA (1996) Developmental differences in synaptic GABA-ergic currents in hippocampal and cerebellar cells of male rats. Soc Neurosci Abs 22:810

Concas A, Follesa P, Barbaccia ML, Purdy RH, Biggio G (1999) Physiological modulation of $GABA_A$ receptor plasticity by progesterone metabolites. Eur J Pharmacol 375:225–235

Cottrell GA, Lambert JJ, Peters JA (1987) Modulation of $GABA_A$ receptor activity by alphaxalone. Br J Pharmacol 90:491–500

Davies PA, Hannah MC, Hales TG, Kirkness EF (1997) Insensitivity to anaesthetic agents conferred by a class of $GABA_A$ receptor subunit. Nature 385:820–823

Edwards FA (1995) Anatomy and electrophysiology of fast central synapses lead to a structural model for long-term potentiation. Physiol Rev 75:759–87

Fritschy JM, Paysan J, Enna A, Mohler H (1994) Switch in the expression of rat $GABA_A$-receptor subtypes during postnatal development: an immunohistochemical study. J Neurosci 14:5302–5324

Gasior N, Carter RB, Witkin JM (1999) Neuroactive steroids: potential therapeutic use in neurological and psychiatric disorders. Trends Pharmacol Sci 20:107–112

Gee KW, Chang WC, Brinton RE, McEwen BS (1987) GABA-dependent modulation of the Cl-ionophore by steroids in rat brain. Eur J Pharmacol. 136:419–423

Gee KW, Bolger MB, Brinton RE, Coirini H, McEwen BS (1988) Steroid regulation of the chloride ionophore in rat brain: structure activity requirements, regional dependence and mechanism of action. J Pharmacol Exp Ther 241:346–353

Guidotti A, Uzunov DP, Auta J, Costa E (1996) Application of gas chromatography mass fragmentography with negative ion chemical ionization technology to measure neurosteroids and their biosynthesis rate in the rat brain. In: Genazzani AR, Petraglia F, Purdy RH (eds) The brain: source and target for sex steroid hormones. Parthenon, New York, pp 25–41

Guidotti A, Costa E (1998) Can the antidysphoric and anxiolytic profiles of selective serotonin reuptake inhibitors be related to their ability to increase brain 3α, 5α-tetrahydroprogesterone (allopregnanolone) availability? Biol Psychiatry 44:865–873

Haage D, Johansson S (1999) Neurosteroid modulation of synaptic and GABA-evoked currents in neurons from the rat medial preoptic nucleus. J Neurophysiol 82:143–151

Hadingham KL, Wingrove PB, Wafford KA, Bain C, Kemp JA, Palmer KJ, Wilson AW, Wilcox AS, Sikela JM, Ragan CI, Whiting PJ (1993) Role of the β subunit in determining the pharmacology of human γ-aminobutyric acid type A receptors. Mol Pharmacol 44:1211–1218

Han M, Zorumski CF, Covey DF (1996) Neurosteroid analogues. 4. The effect of methyl substitution at the C-5 and C-10 positions of neurosteroids on electrophysiological activity at $GABA_A$ receptors. J Med Chem 39:4218–4232

Harney S, Frenguelli BG, Lambert JJ (1999) Neurosteroid modulation of $GABA_A$ receptor-mediated miniature inhibitory postsynaptic currents in the rat hippocampus. Br J Pharmacol 126:7P

Harrison NL, Simmonds MA (1984) Modulation of the $GABA_A$ receptor complex by a steroid anaesthetic. Brain Res 323:287–292

Harrison NL, Majewska MD, Harrington JW, Barker JL (1987a) Structure activity relationships for steroid interaction with the γ-amino-butyric $acid_A$ receptor complex. J Pharmacol Exp Ther 241:346–353

Harrison NL, Vicini S, Barker JL (1987b) A steroid anesthetic prolongs inhibitory postsynaptic currents in cultured rat hippocampal neurons. J Neurosci 7:604–609

Hawkinson JE, Kimbrough CL, Belelli D, Lambert JJ, Purdy RH, Lan NC (1994) Correlation of neuroactive steroid modulation of [^{35}S]t-butylbicyclophosphorothionate and [^{3}H]flunitrazepam binding and γ-aminobutyric $acid_A$ receptor function. Mol Pharmacol 46:977–985

Hawkinson JE, Drew JA, Kimbrough CL, Chen J-S, Hogenkamp DJ, Lan NC, Gee KW, Shen K-Z, Whittemore ER, Woodward RM (1996) 3α-hydroxy-3β-trifluoromethyl-5α-pregnan-20-one (Co 2–1970): A partial agonist at the neuroactive steroid site of the γ-aminobutyric $acid_A$ receptor. Mol Pharmacol 49:897–906

Hawkinson JE, Acosta-Burruel M, Yang KC, Hogenkamp DJ, Chen JS, Lan NC, Drewe JA, Whittemore ER, Woodward RM, Carter RB, Upsani RB (1998) Substituted 3β-phenylethynyl derivatives of 3α-hydroxy-5β-pregnan-20-one: remarkably potent neuroactive steroid modulators of gamma-aminobutyric $acid_A$ receptors. J Pharmacol Exp Ther 287:198–207

Hill-Venning C, Peters JA, Callachan H, Lambert JJ, Gemmell DK, Anderson A, Byford A, Hamilton N, Hill DR, Marshall RJ, Campbell AC (1996) The anaesthetic action and modulation of $GABA_A$ receptor activity by the novel water soluble aminosteroid Org 20599. Neuropharmacology 35:1209–1222

Hogenkamp DJ, Tahir SH, Hawkinson JE, Upsani RB, Alauddin M, Kimbrough CL, Acosta-Burreul M, Whittemore ER, Woodward RM, Lan NC, Gee KW, Bolger MB (1997) Synthesis and in vitro activity of 3β-substituted-3α-hydroxypregnan-20-ones: allosteric modulators of the $GABA_A$ receptor. J Med Chem 40:61–72

Hollrigel GS, Soltesz I (1997) Slow kinetics of miniature IPSCs during early postnatal development in granule cells of the dentate gyrus. J Neurosci 17:5119–5128

Jones MV, Westbrook GL (1996) The impact of receptor desensitization on fast synaptic transmission Trends Neurosci 19:96–101

Kapur J, Haas KF, Macdonald RL (1999) Physiological properties of $GABA_A$ receptors from acutely dissociated rat dentate granule cells. J Neurophysiol 81: 2464–2471

Kokate G, Svensson BE, Rogawski MA (1994) Anticonvulsant activity of neurosteroids: correlation with γ-aminobutyric acid-evoked chloride current potentiation. J Pharmacol Exp Ther 270:1223–1229

Lambert JJ, Peters JA, Cottrell GA (1987) Actions of synthetic and endogenous steroids on the $GABA_A$ receptor. Trends Pharmacol Sci 8:224–227

Lambert JJ, Peters JA, Sturgess NC, Hales TG (1990) Steroid modulation of the $GABA_A$ receptor complex: electrophysiological studies. In: Chadwick D, Widdows K (eds) Steroids and neuronal activity, CIBA Foundation Symposium 153. Wiley, Chichester, pp 56–82

Lambert JJ, Belelli D, Hill-Venning C, Peters JA (1995) Neurosteroids and $GABA_A$ receptor function. Trends Pharmacol Sci 16:295–303

Lambert JJ, Belelli D, Shepherd SE, Pistis M, Peters JA (1999) The selective interaction of neurosteroids with the $GABA_A$ receptor. In: Baulieu E-E, Robel P, Schumacher M (eds) Neurosteroids: a new regulatory function in the nervous system. Humana Press, Totowa, pp 125–142

Le Foll F, Castel H, Louiset E, Vaudry H, Cazin L (1997) Multiple modulatory effects of neuroactive steroid pregnanolone on $GABA_A$ receptor in frog pituitary melanotrophs. J Physiol 504:387–400

Lüddens H, Korpi ER, Seeburg PH (1995) $GABA_A$/benzodiazepine receptor heterogeneity: neurophysiological implications. Neuropharmacology 34:245–254

MacDonald RL, Rogers CJ, Twyman RE (1989) Barbiturate modulation of kinetic properties of the $GABA_A$ receptor channel of mouse spinal neurones in culture. J Physiol 417:483–500

MacDonald RL, Olsen RW (1994) $GABA_A$ receptor channels. Ann Rev Neurosci 17:569–602

Maitra R, Reynolds JN (1999) Subunit dependent modulation of $GABA_A$ receptor function by neuroactive steroids. Brain Res 819:75–82

Majewska MD, Harrison NL, Schwartz RD, Barker, JL, Paul SM (1986) Steroid hormones are barbiturate-like modulators of the GABA receptor. Science 323:1004–1007

Majewska MD, Mienville JM, Vicini S (1988) Neurosteroid pregnenolone sulfate antagonizes electrophysiological responses to $GABA_A$ in neurons. Neurosci Lett 90:279–284

Majewska MD, Demirgoren S, London ED (1990a) Binding of pregnenolone sulfate to rat brain membranes suggests multiple sites of steroid action at the $GABA_A$ receptor. Eur J Pharmacol 189:307–315

Majewska MD, Demirgoren S, Spivak CE, London ED (1990b) The neurosteroid dehydroepiandrosterone sulfate is an allosteric antagonist of the $GABA_A$ receptor. Brain Res 526:143–146

Matsumoto K, Uzunova V, Pinna G, Taki K, Uzunov DP, Watanabe H, Mienville JM, Guidotti A, Costa E (1999) Permissive role of brain allopregnanolone content in the regulation of pentobarbital-induced righting reflex loss. Neuropharmacology 38:955–963

McCauley LD, Liu V, Chen J.-S, Hawkinson JE, Lan NC, Gee KW (1995) Selective actions of certain neuroactive pregnanediols at the γ-aminobutyric acid type A receptor complex in rat brain. Mol Pharmacol 47:354–362

Mienville JM, Vicini S (1989) Pregnenolone sulfate antagonizes $GABA_A$ receptor-mediated currents via a reduction of channel opening frequency. Brain Res 489:190–194

Mody I, DeKoninck Y, Otis TS, Soltesz I (1994) Bridging the cleft at GABA synapses in the brain. Trends Neurosci 17:517–525

Neelands TR, Fisher JL, Bianchi M, Macdonald RL (1999) Spontaneous and gamma-aminobutyric acid (GABA)-activated $GABA_A$ receptor channels formed by epsilon subunit-containing isoforms. Mol Pharmacol 55:168–178

Nilsson KR, Zorumski CF, Covey DF (1998) Neurosteroid analogues. 6. The synthesis and $GABA_A$ receptor pharmacology of enantiomers of dehydroepiandrosterone sulfate, pregnenolone sulfate, and ($3\alpha,5\beta$)-3-hydroxypregnan-20-one sulfate. J Med Chem 41:2604–2613

Olsen RW, Sapp DW (1995) Neuroactive steroid modulation of $GABA_A$ receptors. In: Biggio G, Sanna E, Serra M, Costa E (eds) $GABA_A$ receptors and anxiety; from neurobiology to treatment. Advances in Biochemical Psychopharmacology 48. Raven, New York, pp 57–74

Park-Chung M, Malayev A, Purdy RH, Gibbs TT, Farb DH (1999) Sulfated and unsulfated steroids modulate gamma-aminobutyric acid A receptor function through distinct sites. Brain Res 830:72–87

Paul SM, Purdy RH (1992) Neuroactive steroids FASEB J 6:2311–2322

Peters JA, Kirkness EF, Callachan H, Lambert JJ, Turner AJ (1988) Modulation of the $GABA_A$ receptor by depressant barbiturates and pregnane steroids. Br J Pharmacol 94:1257–1269

Poisbeau P, Feltz P, Schlichter R (1997) Modulation of $GABA_A$ receptor-mediated IPSCs by neuroactive steroids in a rat hypothalamo-hypophyseal co-culture model. J Physiol 500:475–485

Prince RJ, Simmonds MA (1993) Differential antagonism by epipregnanolone of alphaxalone and pregnanolone potentiation of [^{3}H] flunitrazepam binding suggests more than one class of binding site for steroids at $GABA_A$ receptors. Neuropharmacology 32:59–63

Puia G, Santi MR, Vicini S, Pritchett DB, Purdy RH, Paul SM, Seeburg PH, Costa E (1990) Neurosteroids act on recombinant human $GABA_A$ receptors. Neuron 4:759–765

Puia G, Ducic I, Vicini S, Costa E (1993) Does neurosteroid modulatory efficacy depend on $GABA_A$ receptor subunit composition? Receptors and Channels 1:135–142

Purdy RH, Morrow AL, Blinn JR, Paul SM (1990) Synthesis, metabolism and pharmacological activity of 3α-hydroxy steroids which potentiate GABA-receptor mediated chloride ion uptake in rat cerebral cortical synaptosomes. J Med Chem 33:1572–1581

Purdy RH, Morrow AL, Moore PH, Paul SM (1991) Stress-induced elevations of γ-aminobutyric acid type A receptor active steroids in the rat brain. Proc Natl Acad Sci USA 88:4553–4557

Reith CA, Sillar KT (1997) Pre- and postsynaptic modulation of spinal GABAergic neurotransmission by the neurosteroid, 5β-pregnan-3α-ol-20- one. Brain Res 770:202–212

Rick CE, Ye Q, Finn SE, Harrison NL (1998) Neurosteroids act on the $GABA_A$ receptor at sites on the N-terminal side of the middle of TM2. Neuroreport 9:379–383

Robel P, Baulieu E-E (1994) Neurosteroids: biosynthesis and function. In: de Kloet R, Sutanto W (eds) Neurobiology of steroids: methods in neurosciences 22. Academic Press, San Diego, pp 36–50

Rodgers-Neame NT, Covey DF, Hu Y, Isenberg KE, Zorumski CF (1992) Effects of a benz[*e*]indene on GABA-gated chloride currents in cultured post-natal rat hippocampal neurons. Mol Pharmacol 42:952–957

Rupprecht R, Reul JM, Trapp T, van Steensel B, Wetzel C, Damm K, Ziegelgansberger W, Holsboer F (1993) Progesterone receptor-mediated effects of neuroactive steroids. Neuron 11:523–530

Rupprecht R, Holsboer F (1999) Neuroactive steroids: mechanisms of action and neuropsychopharmacological consequences. Trends Neurosci 22:410–416

Sanna E, Murgia A, Casula A, Biggio G (1997) Differential subunit dependence of the actions of the general anesthetics alphaxalone and etomidate at γ-aminobutyric acid type A receptors expressed in *Xenopus laevis* oocytes. Mol Pharmacol 51:484–490

Shepherd SE, Peters JA, Lambert JJ (1996) The interaction of intravenous anaesthetics with rat inhibitory and excitatory amino acid receptors expressed in *Xenopus laevis* oocytes. Br J Pharmacol 119:364P

Shingai R, Sutherland ML, Barnard EA (1991) Effects of subunit types of cloned $GABA_A$ receptor on the response to a neurosteroid. Eur J Pharmacol 206:77–80

Sigel E, Baur A (1997) The benzodiazepine binding site of $GABA_A$ receptors. Trends Pharmacol Sci 18:425–429

Simmonds MA, Turner JP, Harrison NL (1984) Interactions of steroids with the GABA-A receptor complex. Neuropharmacology 23:877–878

Smith GB, Olsen RW (1995) Functional domains of $GABA_A$ receptors. Trends Pharmacol Sci 16:162–168

Smith SS, Gong QH, Hsu FC, Markowitz RS, Ffrench-Mullen MH, Li X (1998a) $GABA_A$ receptor α_4 subunit suppression prevents withdrawal properties of an endogenous neurosteroid. Nature 392:926–930

Smith SS, Gong QH, Li X, Moran MH, Bitran D, Frye CA, Hsu FC (1998b) Withdrawal from 3α-OH-5α-pregnan-20-one using a pseudopregnancy model alters the kinetics of hippocampal $GABA_A$-gated current and increases the $GABA_A$ receptor α_4 subunit in association with increased anxiety. J Neurosci 18:5275–5284

Smith SS, Gong QH, Li X, Markowitz RS (1999) Neurosteroid effects on $GABA_A$ receptor subunit expression. J Physiol 518P:11S

Twyman RE, MacDonald RL (1992) Neurosteroid regulation of $GABA_A$ receptor single channel kinetic properties of mouse spinal cord neurones in culture. J Physiol 456:215–224

Upsani RB, Yang KC, Acosta-Burruel M, Konkoy CS, McLellan JA, Woodward RM, Lan NC, Carter RB, Hawkinson JE (1997) 3α-hydroxy-3β-(phenylethynyl)-5β-pregnan-20-ones: synthesis and pharmacological activity of neuroactive steroids with high affinity for $GABA_A$ receptors. J Med Chem 40:73–84

Uzunov DP, Cooper TB, Costa E, Guidotti A (1996) Fluoxetine-elicited changes in brain neurosteroid content measured by negative mass ion fragmentography. Proc Natl Acad Sci USA 93:12599–12604

Uzunova V, Sheline Y, Davis JM, Rasmusson A, Uzunov DP, Costa E, Guidotti A (1998) Increase in the cerebrospinal fluid content of neurosteroids in patients with unipolar major depression who are receiving fluoxetine or fluvoxamine. Proc Natl Acad Sci USA 95:3239–3244

Wafford KA, Thompson SA, Thomas D, Sikela J, Wilcox AS, Whiting PJ (1996) Functional characterization of human gamma-aminobutyric acid (A) receptors containing the alpha 4 subunit Mol Pharmacol, 50:670–678

Whiting PJ, McAllister G, Vasilatis D, Bonnert TP, Heavens RP, Smith DW, Hewson L, O'Donnell R, Rigby MR, Sirinathsinghji DJS, Marshall G, Thompson SA, Wafford KA (1997) Neuronally restricted RNA splicing regulates the expression of a novel $GABA_A$ receptor subunit conferring atypical functional properties. J Neurosci 17:5027–5037

Williams SR, Buhl EH, Mody I (1998) The dynamics of synchronised neurotransmitter release determined from compound spontaneous IPSCs in rat dentate granule neurones in vitro. J Physiol 510:477–497

Wingrove PB, Wafford KA, Bain C, Whiting PJ (1994) The modulatory action of loreclezole at the γ-aminobutyric acid type A receptor is determined by a single amino acid in the β_2 and β_3 subunit. Proc Natl Acad Sci USA 91:4569–4573

Wittmer LL, Hu Y, Kalkbrenner M, Evers AS, Zorumski CF, Covey DF (1996) Enantioselectivity of steroid-induced γ-aminobutyric $acid_A$ receptor modulation and anesthesia. Mol Pharmacol 50:1581–1586

Woodward RM, Polenzani L, Miledi R (1992) Effects of steroids on γ-aminobutyric acid receptors expressed in *Xenopus* oocytes by poly $(A)^+$RNA from mammalian brain and retina. Mol Pharmacol 41:89–103
Xue BG, Whittemore ER, Park CH, Woodward RM, Lan NC, Gee KW (1997) Partial agonism by 3α, 21-dihydroxy-5β-pregnan-20-one at the γ-aminobutyric acid A receptor neurosteroid site. J Pharmacol Exp Ther 281:1095–1101
Zhu WJ, Wang JF, Krueger KE, Vicini S (1996) δ Subunit inhibits neurosteroid modulators of $GABA_A$ receptors. J Neurosci 16:6648–6656
Zhu WJ, Vicini S (1997) Neurosteroid prolongs $GABA_A$ channel deactivation by altering kinetics of desensitized states. J Neurosci 17:4032–4036
Zorumski CF, Wittmer LL, Isenberg KE, Hu Y, Covey DF (1996) Effects of neurosteroid and benz[*e*]indene enantiomers on $GABA_A$ receptors in cultured hippocampal neurones and transfected HEK-293 cells. Neuropharmacology 35:1161–1168

CHAPTER 5

Allosteric Modulation of $GABA_A$ Receptor Function by General Anesthetics and Alcohols

M.D. KRASOWSKI, R.A. HARRIS, and N.L. HARRISON

A. Introduction

Since their introduction into clinical practice nearly 150 years ago, general anesthetics have become some of the most widely used and important therapeutic agents. Alcohol, specifically ethanol, is arguably the most important non-prescription drug in most Western countries. Despite over a century of research, the molecular mechanisms of action of general anesthetics and alcohols in the central nervous system (CNS) have remained elusive. Ligand-gated ion channels have emerged as promising molecular targets to mediate the CNS effects of both classes of drug. In this review, we aim to describe the actions of general anesthetics and alcohols on γ-aminobutyric $acid_A$ ($GABA_A$) receptors. We will begin by summarizing the chemical classes of anesthetics. We will briefly examine contemporary experimental methodology and review the pharmacological criteria that can help define proteins that represent plausible molecular targets for general anesthetics and alcohols. We will then describe the actions of these agents on the $GABA_A$ receptors. The last decade has witnessed an explosion of such studies, and we will focus in particular on recent work which utilizes recombinant chimeric and mutated receptors to identify regions of the $GABA_A$ receptors that are important for the modulatory actions of general anesthetics and alcohols.

B. What is a General Anesthetic?

General anesthetics include a startling range of structurally diverse molecules that can be, somewhat arbitrarily, divided into volatile anesthetics, anesthetic gases, alcohols, and intravenous anesthetics (Fig. 1). The observation that a spectrum of chemically dissimilar agents produces general anesthesia greatly influenced the thinking of early investigators seeking to explain mechanisms of anesthetic action. A landmark series of experiments reported independently by Hans Meyer and Charles Ernest Overton around the turn of the century determined that the potencies of general anesthetic molecules correlated well with their oil/water partition coefficients (MEYER 1899, 1901; OVERTON 1901). The so-called "Meyer-Overton correlation" was later extended to embrace the

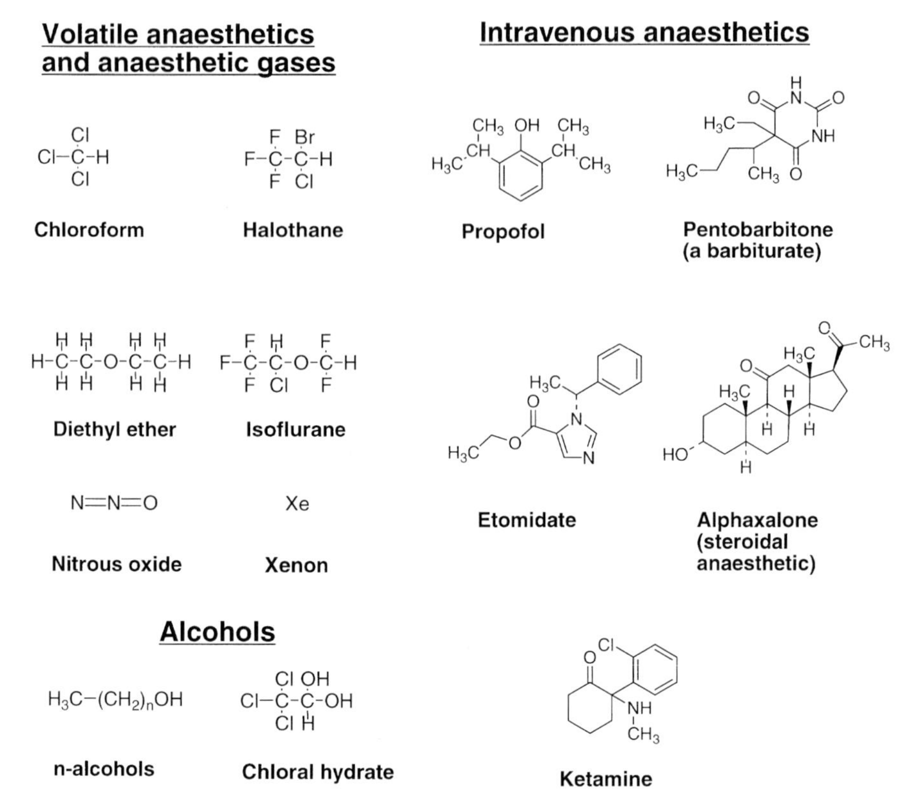

Fig. 1. Chemical structures of selected general anesthetics

concept that certain molecules produce general anesthesia by a non-specific mechanism. Non-specific theories of anesthesia usually include some notion that general anesthetics perturb membrane lipids within the CNS to reduce neuronal excitability and thereby produce anesthesia (MEYER 1937; MULLINS 1954; SEEMAN 1972). Research within the last several decades has demonstrated numerous inconsistencies between experimental observations and non-specific theories of general anesthesia. The main problems including the following (FRANKS and LIEB 1994; HARRISON and FLOOD 1998):

1. Some chemical compounds are predicted by non-specific theories to be anesthetics but do not, in fact, produce anesthesia.
2. Non-specific theories of anesthesia cannot account for the stereoselectivity demonstrated by some anesthetic isomers.
3. Anesthetic effects on lipids (such as alterations in membrane bilayer fluidity), when measured experimentally, are often negligible at clinically relevant concentrations, and are easily reproduced by very small *increases* in ambient temperature. In contrast, *decreases* in body temperature mimic

the behavioral effects of general anesthetics (FRANKS and LIEB 1986, 1994; TOMLIN et al. 1998).

Some prescient investigators recognized a number of decades ago that anesthetics may act instead on specific targets. For example, Sir John Eccles and colleagues studied spinal synaptic reflexes in animals under pentobarbitone anesthesia (ECCLES and MALCOLM 1946; ECCLES et al. 1963) and raised the possibility of anesthetic actions at neurotransmitter receptors, important in synaptic transmission.

C. Special Considerations for Alcohol

Although the *n*-alcohols are general anesthetics at high doses, they are not used clinically as such. The real interest is in the pharmacology of sub-anesthetic alcohol doses, as well as the chronic effects of alcohol. Effects of ethanol commonly associated with mild intoxication, such as relaxation, reduced anxiety and behavioral disinhibition occur (in mice, rats, and humans) at blood alcohol concentrations of 5–20 mmol/l, whereas general anesthesia requires 100–200 mmol/l ethanol (DEITRICH and HARRIS 1996). Demonstrating reliable effects of 5 mmol/l ethanol on defined receptors (proteins) has proven difficult, and it is unlikely that actions on lipid properties can account for actions of ethanol at non-toxic concentrations. In this review, we focus on acute actions of anesthetics and alcohols, but it should be noted that continuous exposure to ethanol may result in tolerance and dependence. There is some evidence that the chronic neuronal adaptations in response to ethanol are related to changes in the initial targets of the drug (e.g., $GABA_A$ receptors, NMDA receptors), but molecular mechanisms of alcohol tolerance and dependence remain to be defined.

D. Overview of Ligand-Gated Ion Channels

This review summarizes recent progress in the understanding of general anesthetic and alcohol actions on the $GABA_A$ receptors. A number of excellent reviews over the last decade have summarized work on the molecular and cellular actions of general anesthetics (WEIGHT et al. 1992; FRANKS and LIEB 1993, 1994, 1996a; TANELIAN et al. 1993; HARRIS et al. 1995b; LAMBERT et al. 1995, 1996; MIHIC et al. 1995; SMITH and OLSEN 1995; WHITING et al. 1995; LOVINGER 1997; HARRISON and FLOOD 1998; PEARCE 1999). Ligand-gated ion channels are certainly not the only possible molecular targets for general anesthetics; other neuronal proteins such as voltage-gated ion channels and G-protein coupled receptors may also play a role in the overall spectrum of behavioral actions of some of the general anesthetics. However, extensive research has arrived at an almost universal consensus; voltage-gated ion channels are, in general, relatively insensitive to clinically relevant concentrations of general anesthetics (FRANKS and LIEB 1994). Detailed studies of general anesthetic actions on G-protein-coupled receptors are scarce, and it can be difficult to distinguish effects on the receptor *per se* from general anesthetic perturbations of second messengers or effector molecules such as protein kinases and phospholipases. Receptors for the neurotransmitters glutamate, GABA, glycine, serotonin (5-HT), and acetylcholine (ACh) are currently strong candidates as molecular mediators of the CNS effects of general anesthetics (FRANKS and LIEB 1994, 1996a; HARRIS et al. 1995b). The ligand-gated ion channels include the $GABA_A$, glycine, serotonin-3 (5-HT_3), and nicotinic ACh receptors, along with the AMPA-, kainate-, and NMDA-sensitive subtypes of ion-otropic glutamate receptors. (Note: GABA, glutamate, 5-HT, and ACh also act on 'slow' neurotransmitter receptors, e.g., $GABA_B$, muscarinic acetylcholine, and metabotropic glutamate receptors, which are coupled to second messenger systems.) $GABA_A$, glycine, 5-HT_3, and nicotinic ACh receptors form part of an evolutionarily related ligand-gated ion channel gene superfamily (ORTELLS and LUNT 1995). Ionotropic glutamate receptors were originally thought to be part of this superfamily but are now thought to belong to a distinct ion channel class.

E. $GABA_A$ and Glycine Receptors

$GABA_A$ and glycine receptors are chloride-selective ion channels. These are generally considered to be inhibitory neurotransmitter receptors, since in most cells, opening of chloride channels results in membrane hyperpolarization and/or stabilization of the membrane potential away from the threshold for firing action potentials (MCCORMICK 1989). GABA and glycine are the primary fast inhibitory neurotransmitters in the CNS, with glycine abundant in the spinal cord and brainstem (KUHSE et al. 1995; ZAFRA et al. 1997) and GABA predominant in higher brain regions (MCCORMICK 1989). It has been estimated that one-third of all synapses in the CNS are GABA-ergic (BLOOM and IVERSEN 1971).

$GABA_A$ and glycine receptors, like the other members of the ligand-gated ion channel superfamily to which they belong, appear to share a common subunit topology, with a large N-terminal extracellular domain, four putative membrane-spanning regions (TM1–TM4), a heterogeneous intracellular loop between TM3 and TM4, and a short extracellular C-terminal domain. Residues within the extracellular N-terminal domain form the agonist binding domains (KUHSE et al. 1995; SMITH and OLSEN 1995) while amino acid residues within TM2 line the ion channel pore (XU and AKABAS 1993; AKABAS et al. 1994; see Fig. 2). Native receptors are composed of pentameric arrangements of individual receptor subunits (LANGOSCH et al. 1988; COOPER et al. 1991).

Subunit heterogeneity creates extensive diversity among the inhibitory ligand-gated ion channels. Multiple subunits have been cloned for $GABA_A$ (α_{1-6}, β_{1-4}, γ_{1-4}, δ, ε, and π) (reviewed in: MACDONALD and OLSEN 1994; RABOW et al. 1995; MCKERNAN and WHITING 1996; DAVIES et al. 1997; HEDBLOM and

KIRKNESS 1997; WHITING et al. 1997; BARNARD et al. 1998) and glycine (α_{1-4}, β) (BETZ 1991, 1992; KUHSE et al. 1995; ZAFRA et al. 1997) receptors. GABA$_A$ receptors *in vivo* predominantly consist of α, β, and γ subunits with a proposed stoichiometry of 2α:2β:1γ (CHANG et al. 1996; TRETTER et al. 1997). The existence of six α subunit isoforms enables considerable anatomical and functional diversity of GABA$_A$ receptors (FRITSCHY and MOHLER 1995; SIEGHART 1995; NUSSER et al. 1996). In particular, the α subunit isoform may

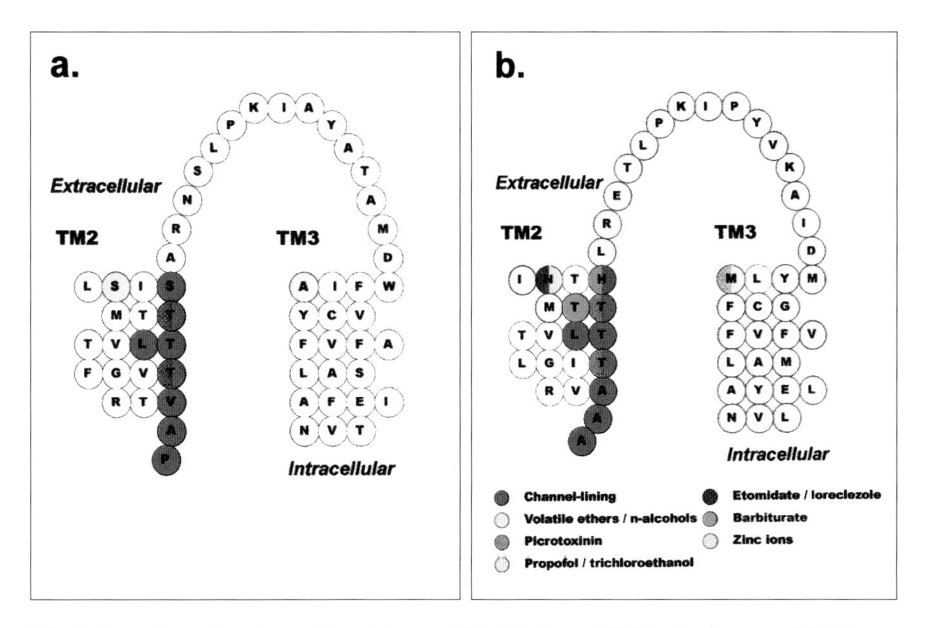

Fig. 2. Location of amino acid residues within TM2 and TM3 of: **a** human GABA$_A$ α_1 (HADINGHAM et al. 1993a), **b** human GABA$_A$ β_2 (HADINGHAM et al. 1993b) receptor subunits that are critical for general anesthetic modulation or block by the non-competitive antagonists picrotoxinin and Zn^{2+}, in addition to amino acid residues which are thought to line the ion channel pore. GABA$_A$ α_1 and β_2 subunit isoforms are chosen since they represent the most common neuronal α and β subunit isoforms (MCKERNAN and WHITING 1996; BARNARD et al., 1998). The residue positions are from published studies: channel-lining residues (XU and AKABAS 1993, 1996), volatile ethers [enflurane (MIHIC et al. 1997) and isoflurane (MIHIC et al. 1997; KRASOWSKI et al. 1998b)], *n*-alcohols (MIHIC et al. 1997), picrotoxinin (GURLEY et al. 1995), propofol (KRASOWSKI et al. 1998b), trichloroethanol (KRASOWSKI et al. 1998a), etomidate (BELELLI et al. 1997; MCGURK et al. 1998), loreclezole (WINGROVE et al. 1994), barbiturate (pentobarbitone) (BIRNIR et al., 1997), and zinc ions (HORENSTEIN and AKABAS 1998). Note that some of the residue positions highlighted were actually uncovered in α or β subunit isoforms different from α_1 or β_2. To date, detailed three-dimensional structural information about TM2, the TM2-TM3 linker, and TM3 is lacking. The spatial relationship between TM2 and TM3 in the functional GABA$_A$ receptor complex is currently unknown

influence agonist potency (LEVITAN et al. 1988; SIGEL et al. 1990), agonist efficacy (EBERT et al. 1994), regulation by benzodiazepines (WAFFORD et al. 1991), and channel kinetics (TIA et al. 1996; LAVOIE et al. 1997). The most common neuronal subunit combination is $\alpha_1\beta_2\gamma_2$ (MCKERNAN and WHITING 1996; BARNARD et al. 1998). $GABA_A$ receptors are blocked competitively by bicuculline and non-competitively by picrotoxinin and Zn^{2+} (see Fig. 2).

Strychnine-sensitive glycine receptors *in vivo* consist of both α homomers and $\alpha\beta$ heteromeric receptors, with a switch from homomeric α_2 to heteromeric $\alpha_1\beta$ receptors occurring during development (BETZ 1991, 1992; KUHSE et al. 1995). The best described physiological role for glycine receptors is in Renshaw cell inhibition of motor neurones in the spinal cord; however, glycine receptors are also widely expressed in the brainstem and throughout higher regions of the neuraxis (BETZ 1991, 1992).

$GABA_C$ receptors are formed from ρ subunits (ρ_{1-3}) (CUTTING et al. 1991, 1992; JOHNSTON 1996). $GABA_C$ receptors show greatest expression in the retina but are also found in other areas of the brain (WEGELIUS et al. 1998). The designation of '$GABA_C$' for ρ subunits, while potentially confusing (BARNARD et al. 1998), follows from their extensive pharmacological differences from $GABA_A$ and $GABA_B$ receptors, including insensitivity to the classical $GABA_A$ competitive antagonist bicuculline (CUTTING et al. 1991, 1992; JOHNSTON 1996).

F. Pharmacological Criteria for a Reasonable General Anesthetic/Alcohol Target Site

Before discussing the actions of specific agents on ligand-gated ion channels, it is worthwhile to define specific criteria that a target molecule (receptor protein or otherwise) must fulfill in order to qualify as a candidate in mediating the behavioral actions of the general anesthetics (FRANKS and LIEB 1994; HARRISON and FLOOD 1998):

1. The general anesthetic (or alcohol) must alter the function of the receptor at behaviorally relevant concentrations.
2. The receptor must be expressed in the appropriate anatomical locations to mediate the specific behavioral effects of the anesthetic or alcohol.
3. If an anesthetic molecule shows stereoselective effects *in vivo*, these should be mirrored by the *in vitro* actions at the receptor.
4. The hydrophobicity of a compound within a homologous series of anesthetics or alcohols should correlate with potency at the receptor and with *in vivo* anesthetic potency.
5. If a target molecule exhibits a 'cutoff' phenomenon, e.g., in the homologous series of *n*-alcohols, this should reflect the cutoff for the biological effect under consideration.

I. What is the "Clinically Relevant Concentration" for a General Anesthetic?

For an inhaled anesthetic such as isoflurane, one 'minimum alveolar concentration' (MAC) conventionally refers to the concentration of inhaled anesthetic that produces immobility in 50% of patients or animals studied (EGER et al. 1965; QUASHA et al. 1980). Immobility, a lack of purposeful response to a noxious stimulus, represents an easily determined endpoint across a large variety of different animal species. The use of immobility as an experimental endpoint is helpful in that, for most general anesthetics, anesthetic concentrations two- to four-fold above the EC_{50} for producing immobility are invariably lethal (FRANKS and LIEB 1994). The anesthetic concentrations that produce significant inhibition of cognitive functions and cortical activity, assessed using EEG-derived indicators, are lower than those required for producing immobility (CHORTKOFF et al. 1995a,b; ISELIN-CHAVES et al. 1998). *Thus, anesthetic concentrations several-fold greater than those that produce immobility define the upper boundary of the concentration range that is clinically relevant.* For a target to have any relevance for anesthesia, it must at least be sensitive to sub-lethal but immobilizing concentrations of anesthetics. This issue of relevant concentrations alone poses a severe challenge to the plausibility of 'lipid' theories of anesthetic action, since 'non-specific' effects of general anesthetics (e.g., disruption of lipid bilayer fluidity) appear to be negligible at clinically relevant concentrations (FRANKS and LIEB 1986, 1994; TOMLIN et al. 1998).

While the issue of relevant concentrations is obviously of paramount importance to molecular studies of general anesthetics, the physicochemical and pharmacokinetic properties of the various anesthetic drugs pose some obstacles to the determination of relevant concentrations. Volatile anesthetic potency is usually quantified in terms of MAC (EGER et al. 1965; QUASHA et al. 1980). MAC values (often expressed in the operating room in terms of % anesthetic gas by volume) can be converted to 'aqueous MAC equivalent concentrations' by use of the appropriate water/gas (or blood/gas) partition coefficients (FRANKS and LIEB 1993, 1996b). This provides an estimate for the concentration of anesthetic in the blood that is in equilibrium with the inspired partial pressure of anesthetic in the gas phase. Aqueous MAC equivalents are useful for *in vitro* experiments which involve the study of volatile anesthetics in aqueous solution (FRANKS and LIEB 1993, 1994, 1996b).

The issue of clinically relevant concentrations for the intravenous anesthetics and the alcohols in mammals is complicated by pharmacokinetic aspects of these drugs and the difficulty of ascertaining steady-state drug concentrations in the brain (FRANKS and LIEB 1994). In some cases (e.g., for propofol and the barbiturates), detailed pharmacokinetic studies have addressed these issues, and reasonable free anesthetic concentrations in brain can be estimated (FRANKS and LIEB 1994). In other cases (e.g., ketamine and the steroid anesthetic alphaxalone), only total anesthetic concentrations in blood are known, thus invariably overestimating brain concentrations and therefore underestimating the potency of this class of anesthetics, often by as much as one to two orders of magnitude (Cohen et al. 1973; Sear and Prys-Roberts 1979). The reader is referred to Franks and Lieb (1994) and to an extensive tabulation of anesthetic concentrations recently published elsewhere (Krasowski et al. 1999).

Although the *n*-alcohols are useful research tools for scientists interested in anesthetic mechanisms, where it is perfectly appropriate to study effects of concentrations corresponding to 1 or 2 MAC, the effects of sub-anesthetic 'recreational' alcohol concentrations are more relevant to its social consumption! The MAC for ethanol in mice, rats and tadpoles is 100–200 mmol/l (Deitrich and Harris 1996) and there are reports of chronic alcoholics actually achieving these levels without loss of consciousness (for example, the driver of the car in which Princess Diana was killed had a blood ethanol level of about 40 mmol/l). However, subtle behavioral effects of ethanol (e.g., anti-anxiety effects) are demonstrable at concentrations as low as 5 mmol/l. It has proven remarkably difficult to demonstrate reliable effects of low concentrations of ethanol on isolated brain receptors, channels, transporters or enzymes.

II. Anatomical Location

This is a more difficult issue to discuss since there is considerable debate about precisely which synaptic circuits are responsible for the various reflexes and complex behaviors that are perturbed by general anesthetics. The immobility produced by general anesthetics, perhaps not surprisingly, appears to involve depression of spinal reflex pathways, since it is independent of drug actions in the brain (Antognini and Schwartz 1993; Rampil et al. 1993; Collins et al. 1995). Receptors such as $GABA_A$ and AMPA receptors are promising general anesthetic targets due to their ubiquitous distribution and essential physiological roles as the major fast transmitters of the CNS. However, given the uncertainty concerning the exact anatomy of the synapses that are disrupted to produce the constellation of behavioral effects seen during general anesthesia, receptors with more limited distribution (e.g., 5-HT_3 receptors) may conceivably play major roles as molecular mediators of specific components of the general anesthetic state.

III. Stereoselectivity

Stereoselectivity represents one of the most powerful tests for the relevance of a putative anesthetic target (Franks and Lieb 1994; Harrison 1998). A number of general anesthetic molecules possess a chiral carbon atom, and some pairs of stereoisomers exert different anesthetic potencies in vivo. Stereoselectivity for producing immobility has been documented for the isomers of etomidate (Heykants et al. 1975; Tomlin et al. 1998) (see Fig. 3), the barbiturates (Andrews and Mark 1982), isoflurane (Harris et al. 1992; Lysko et al.

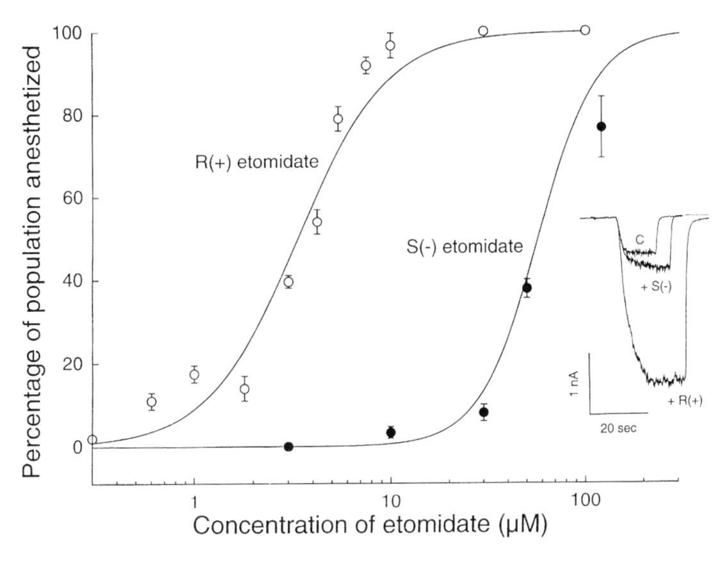

Fig. 3. The selectivity of etomidate optical isomers for producing general anesthesia in tadpoles mirrors the selectivity for potentiation of $GABA_A$ receptor function. The main graph illustrates the concentration-response curves for immobility produced by etomidate stereoisomers in *Rana temporaria* tadpoles. Note that the in vivo potency of R(+) etomidate is approximately one order of magnitude greater than that of S(–) etomidate (HEYKANTS et al. 1975). *The inset* depicts electrophysiological traces from GABA responses at bovine $GABA_A$ $\alpha_1\beta_1\gamma_{2L}$ receptors stably transfected in mouse L-cell fibroblast cells. Co-application of R(+) etomidate produces a vastly greater enhancement of the control submaximal GABA response (C) than co-application of S(–) etomidate. [From Tomlin SL, Jenkins A, Lieb WR, Franks NP (1998) Stereoselective effects of etomidate optical isomers on gamma-aminobutyric acid type A receptors and animals. Anesthesiology 88:708–717. Reproduced in adapted form with permission of the authors and Lippincott-Raven Publishers, 227 East Washington Square, Philadelphia, PA 19106–3708 USA]

1994; although see EGER et al. 1997), ketamine (RYDER et al. 1978; WHITE et al. 1985), and steroid anesthetics (ATKINSON et al. 1965). The formulation of these anesthetics is usually based on the racemic mixture due to the difficulty of separating enantiomers in large quantities (an exception is etomidate, which is prepared by a chiral synthesis (HEYKANTS et al. 1975). Production of pure enantiomers perhaps would improve the clinical profile for other general anesthetics (MOODY et al. 1994), although cost considerations probably preclude such a development.

General anesthetic stereoselectivity poses the most severe challenge to traditional lipid theories of anesthetic action. The optical isomers of isoflurane (DICKINSON et al. 1994) and etomidate (TOMLIN et al. 1998), despite significant differences in their *in vivo* potency (see Fig. 3), behave identically with respect to their ability to disorder lipid bilayers. In contrast, stereoselectivity supports the plausibility of the $GABA_A$ receptor as a target in mediating the actions of etomidate (Tomlin et al. 1998), pentobarbitone (Huang and Barker 1980), isoflurane (Jones and Harrison 1993; Hall et al. 1994), and the steroid anesthetics (Atkinson et al. 1965; Wittmer et al. 1996), since *in vivo* potency and activity at the $GABA_A$ receptor display identical trends. The *in vivo* stereoselectivity of ketamine stereoisomers is paralleled by the inhibitory action of the isomers at the NMDA receptor (Lodge et al. 1982).

Despite the rewards of studying general anesthetic stereoisomers, exemplified by the etomidate work outlined above (Tomlin et al. 1998) (see Fig. 3), the stereoselectivity approach has been under-utilized, mainly due to the limited supply and expense of purified stereoisomers (Moody et al. 1994). Furthermore, only limited anesthetic endpoints (mainly immobility) have been assessed for the anesthetic stereoisomers. It would be quite interesting to know whether the additional neurobiological actions of anesthetics (e.g., amnesia, analgesia) display similar patterns of stereoselectivity.

IV. Hydrophobicity

The so-called 'Meyer-Overton hypothesis,' which led to the adoption of the traditional dogma concerning lipid mechanisms of anesthesia, arose from the fundamental observation that the *in vivo* potency of general anesthetics rises in parallel with increasing hydrophobicity of the anesthetic molecules. This trend is most noticeable with the homologous series of *n*-alcohols but also holds true for diverse anesthetic molecules with oil/water partition coefficients varying over numerous orders of magnitude (Meyer 1899, 1901; Overton 1901). General anesthetic actions at a plausible receptor target should, therefore, exhibit similar trends. The Meyer-Overton correlation was traditionally interpreted to suggest non-specific mechanisms of action for general anesthetics in membrane lipids; however, an alternative explanation is that anesthetics bind to hydrophobic domains of receptor proteins (Franks and Lieb 1984, 1994). A major problem for traditional theories arose with the discovery of hydrophobic compounds which disobey the Meyer-Overton hypothesis (Koblin et al. 1994). These 'non-anesthetics' or 'non-immobilizers' can provide additional clues to which receptor targets might underlie the behavioral actions of general anesthetics.

V. Alcohol Cutoff

Another useful property of series of anesthetics, particularly the *n*-alkane and *n*-alkanol series, is the cutoff effect. The potencies of *n*-alcohols increase with increasing carbon chain length, to some length ("the cutoff") where there is no increase in potency with further lengthening of the carbon chain. In fact, longer chain length alcohols may be inactive, but this is often difficult to determine, because very long alcohols (e.g., C13, C14) have a very low water solubility and are therefore difficult to deliver *in vitro* or *in vivo*. One common assumption is that the cutoff occurs because the anesthetics occupy a site or cavity of finite dimensions and that long chain length compounds cannot enter the site. If this is true, then cutoff is a powerful tool for studying anesthetic sites and can be applied in many experimental systems. Glycine $\alpha 1$ receptors have an alcohol cutoff at decanol, while the cutoff for the related GABA $\rho 1$ receptors is at heptanol (WICK et al. 1998). If the alcohol cutoff reflects a limiting size of an alcohol binding site, the shorter cutoff in this GABA receptor is consistent with a smaller alcohol binding site than that in the glycine receptor.

G. Experimental Approaches to Studying General Anesthetic and Alcohol Actions at the $GABA_A$ Receptors

General anesthetic actions at ligand-gated ion channels have been studied using a variety of methodologies, including protein chemistry, radioligand binding, ion flux studies, and electrophysiology (TANELIAN et al. 1993; FRANKS and LIEB 1994; HARRIS et al. 1995b). We will focus mainly on electrophysiological studies since these, in general, provide superior time resolution and also offer the possibility of analyzing isolated cells or even single ion channels. The general anesthetics have properties that limit the utility of other experimental techniques. For example, specific binding of radiolabeled general anesthetics to ligand-gated ion channels has proven exceedingly difficult to demonstrate due to the low affinity of the interactions and the high degree of non-specific binding to neuronal membranes (TANELIAN et al. 1993; FRANKS and LIEB 1994; HARRIS et al. 1995b). Allosteric effects of general anesthetics have been monitored using radioligand binding of drugs to other sites on the ligand-gated ion channels (e.g., OLSEN and SNOWMAN 1982; HARRIS et al. 1995a). In addition, limited progress has been made in developing anesthetic congeners useful for photoaffinity labeling or other covalent modification of receptors (although see ECKENHOFF 1996). These limitations contrast starkly with the studies of other classes of agents at ligand-gated ion channels. For instance, the high-affinity benzodiazepine binding site on the $GABA_A$ receptor has been mapped out in some detail due to the ability to perform both specific radioligand binding and photoaffinity labeling (SIGEL and BUHR 1997; MCKERNAN et al. 1998), which powerfully complements the extensive body of literature on electrophysiological actions of benzodiazepines at $GABA_A$ receptors (SIGEL and BUHR 1997).

Another exciting tool in the quest to establish the *in vivo* significance of a putative anesthetic target is the use of targeted gene manipulations in mice (HOMANICS et al. 1998). A variety of manipulations are possible, including introducing a gene not normally present (transgenic mice), removing an endogenous gene ('knock-out mice'), or replacing an endogenous gene with an altered copy ('knock-in mice') (HOMANICS et al. 1998). Gene targeting in mice has already been very valuable for elucidating the mechanism of action for some drugs. Knock-out of the $GABA_A$ γ_2 receptor subunit gene resulted in mice that were insensitive to the sedative/hypnotic actions of benzodiazepines such as diazepam (GUNTHER et al. 1995). The γ_2 subunit gene knock-out, in conjunction with the *in vitro* dependence of benzodiazepine modulation of the $GABA_A$ receptor on the presence of a γ subunit (PRITCHETT et al. 1989), effectively demonstrates the $GABA_A$ receptor as the major target mediating the sedative/hypnotic actions of benzodiazepines. Mice homozygous for a deletion of the $GABA_A$ receptor β_3 subunit gene exhibit cleft palate, absence seizures, hyperexcitability (HOMANICS et al. 1997; DELOREY et al. 1998), and some resistance to the immobilizing actions of intravenous and volatile anesthetics (QUINLAN et al. 1998). Another gene targeting experiment in mice involved the replacement of the α_{2a}-adrenoreceptor with a dysfunctional receptor mutant. These 'knock-in' mice failed to show analgesic and sedative responses to α_{2a}-adrenoreceptor agonists such as dexmedetomidine and clonidine (LAKHLANI et al. 1997). Additional elegant examples of 'knock-in' mouse experiments may be found elsewhere in this volume, (RUDOLPH et al. 1999; MCKERNAN et al. 2000).

H. Actions of General Anesthetics at $GABA_A$ Receptors

General anesthetics act as *positive or negative allosteric modulators* of agonist actions at ligand-gated ion channels. Among the ligand-gated ion channels, there is no known case in which the anesthetic competes for the same binding site as the endogenous neurotransmitter. The most extensively examined ligand-gated ion channel target for general anesthetics has been the $GABA_A$ receptor (TANELIAN et al. 1993; FRANKS and LIEB 1994; HARRIS et al. 1995b). Virtually every general anesthetic tested enhances the function of the $GABA_A$ receptor at clinically relevant concentrations (FRANKS and LIEB 1994; ZIMMERMAN et al. 1994; HARRIS et al. 1995b) The exceptions are ketamine (SIMMONDS and TURNER 1987), xenon (FRANKS et al. 1998), and possibly nitrous oxide (DZOLJIC and VAN DUJIN 1998; JEVTOVIC-TODOROVIC et al. 1998; MENNERICK et al. 1998). General anesthetic enhancement of $GABA_A$ receptor function is evident in single cell electrophysiological experiments as potentiation of a submaximal GABA response (see Fig. 4) or, at the synaptic level, as prolongation of inhibitory post-synaptic potentials (NICOLL et al. 1975; SCHOLFIELD 1980) or currents (HARRISON et al. 1987b; MACIVER et al. 1991; JONES and HARRISON 1993; BANKS and PEARCE 1999) (Fig. 5). Potentiation of submaximal GABA-induced currents remains the most popular paradigm for electrophysiological experiments, since it is easily reproducible and can be used to study native $GABA_A$ receptors in dissociated neurones or recombinant receptors expressed in mammalian cell lines or *Xenopus* oocytes (TANELIAN et al. 1993; FRANKS and LIEB 1994; HARRIS et al. 1995b).

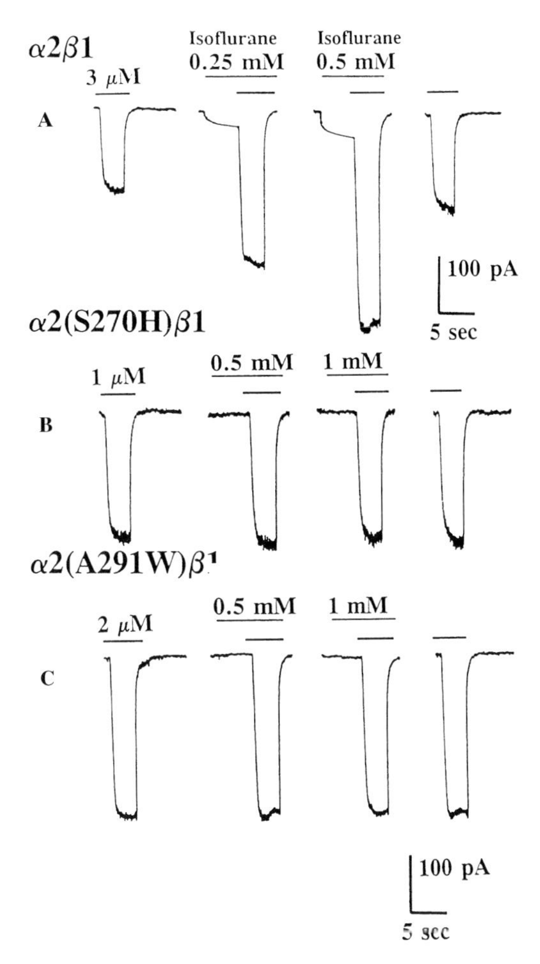

Fig. 4A–C. Specific mutations in TM2 or TM3 of the human GABA$_A$ receptor α_2 subunit abolish positive allosteric modulation by the volatile anesthetic isoflurane at GABA$_A$ $\alpha_2\beta_1$ receptors. **A** Submaximal GABA currents in wild-type GABA$_A$ $\alpha_2\beta_1$ receptors are strongly enhanced (i.e., potentiated) by co-application of clinically relevant concentrations of isoflurane (0.25 and 0.5 mmol/l = 0.5–1.0 MAC). **B,C** In contrast, submaximal GABA currents in α_2(S270H)β_1 or α_2(A291 W)β_1 mutant receptors are *not* enhanced by co-application of isoflurane concentrations up to 1 mmol/l (2 MAC). Thus, these mutant receptors are insensitive to GABA potentiation by isoflurane even at supra-anesthetic concentrations. Individual whole-cell voltage-clamp recordings from human embryonic kidney 293 cells transfected with cDNAs encoding the indicated subunit combination. [From Krasowski MD, Koltchine VV, Rick CE, Ye Q, Finn SE, Harrison NL (1998) Propofol and other intravenous anesthetics have sites of action on the γ-aminobutyric acid type A receptor distinct from that for isoflurane, Mol Pharmacol 53:530–538. Reproduced with permission of the authors and the American Society for Pharmacology and Experimental Therapeutics, 9650 Rockville Pike, Bethesda, MD 20814–3995 USA]

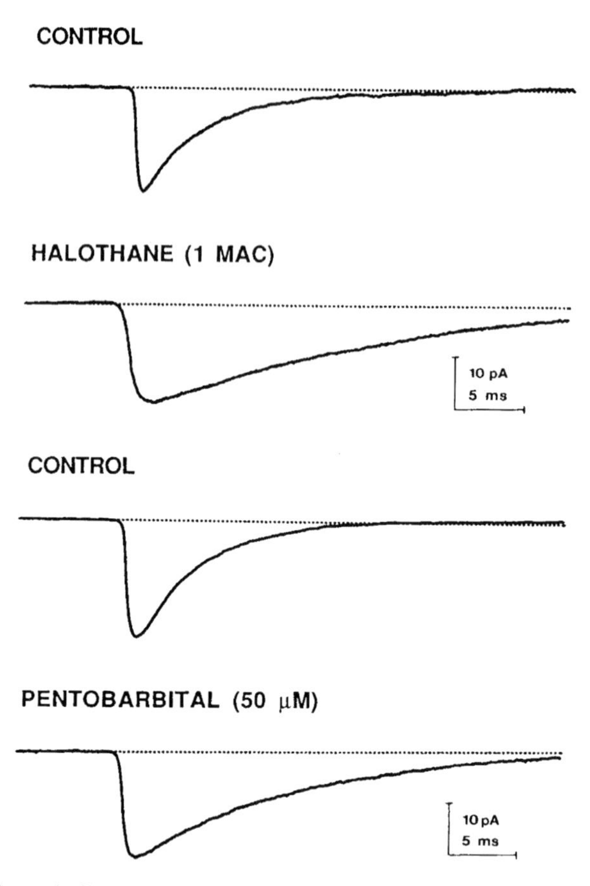

Fig. 5. Both the volatile anesthetic halothane and the intravenous anesthetic pentobarbitone prolong inhibitory post-synaptic currents (IPSCs) mediated by $GABA_A$ receptors. Data obtained from whole-cell patch-clamp recordings of rat hippocampal neurones from brain slices. Average records from whole-cell voltage-clamp recordings in hippocampal neurones of 100 individual spontaneous IPSCs for each trace, showing the prolongation of the decay phase of the IPSC produced by halothane and pentobarbitone. Data from halothane and pentobarbitone are from different neurones and preparations. [From MacIver MB, Tanelian DT, Mody I (1991) Two mechanisms for anesthetic-induced enhancement of $GABA_A$-mediated neuronal inhibition. Ann NY Acad Sci 625:91–96. Reproduced with permission of the authors and the Annals of the New York Academy of Sciences, 655 Madison Avenue, New York, NY 10021 USA]

Some anesthetics, particularly the intravenous agents, open the $GABA_A$ receptor chloride channel in the absence of agonist (BARKER and RANSOM 1978; CALLACHAN et al. 1987; ROBERTSON 1989; HALES and LAMBERT 1991; YANG et al. 1992; HARA et al. 1993; ADODRA and HALES 1995; JONES et al. 1995; BELELLI et al. 1996; RHO et al. 1996; HILL-VENNING et al. 1997; KRASOWSKI et al. 1997, 1998b; SANNA et al. 1997;). This 'direct activation' by general anesthetics involves a binding site completely distinct from that for classical $GABA_A$ receptor agonists such as GABA and muscimol (AMIN and WEISS 1993). Although direct activation usually occurs at supra-clinical concentrations, direct activation effects do sometimes occur at lower concentrations for some anesthetics (e.g., propofol) suggesting possible clinical relevance. Direct activation by anesthetics has been observed in other ligand-gated ion channels (e.g., for the anesthetic isoflurane at the strychnine-sensitive glycine receptor) (DOWNIE et al. 1996) but is most pronounced at the $GABA_A$ receptor.

The advent of cloning and recombinant expression techniques has greatly accelerated and facilitated attempts to classify ligand-gated ion channel sensitivity to general anesthetics. Molecular biology techniques may now be used to determine which regions of ligand-gated ion channels are critical for anesthetic modulation. Sensitivity to general anesthetics varies considerably, sometimes even among closely related receptors, and this forms the basis for the use of ‘chimeric’ receptors to isolate regions of a receptor essential for anesthetic modulation. Chimeric receptors are created by joining together, at the cDNA level, complementary fragments of receptor subunits, in which the parental subunits exhibit markedly different responses to anesthetic. The analysis of chimeric receptors can be used to delimit a region of a receptor essential for general anesthetic modulation, after which site-directed mutagenesis can be used to identify key residues. Chimeric receptors constructed to date include panels of $GABA_A$/glycine (KOLTCHINE et al. 1996), $GABA_A$/$GABA_C$ (LU and HUANG 1998), and glycine/$GABA_C$ (MIHIC et al. 1997; WICK et al. 1998) receptors.

Several problems may accompany the study of such chimeric receptors, including: (1) lack of functional expression (greatly reduced or absent responses to agonist), (2) chimeric receptor function differing radically from the constituent parent receptors, and/or (3) ambiguous pharmacological data. The first problem has substantially reduced the utility of $GABA_A$/$GABA_C$ (HACKAM et al. 1998) chimeras. Lack of functional chimeric receptor responses could potentially be due to protein assembly problems, very low single channel conductance, and/or a minuscule probability of opening following agonist binding (i.e., a defect in ion channel gating). Assembly problems are especially likely when blending heteromeric and homomeric receptors (e.g., $GABA_A$ with $GABA_C$ receptors). Despite these potential pitfalls, the use of chimeric receptors has already helped define putative sites of general anesthetic action on some of the ligand-gated ion channels (see below).

I. Volatile Anesthetics and Anesthetic Gases

Volatile anesthetics (e.g., halogenated ethers and alkanes) alter the function of many ligand-gated ion channels at reasonable concentrations. In general, agonist responses at $GABA_A$ and glycine receptors are positively modulated by volatile anesthetics. The low potency and physicochemical properties of the volatile anesthetics pose some technical challenges for *in vitro* experiments (FRANKS and LIEB 1993, 1994, 1996b; HARRIS et al. 1995b). Nevertheless, recent years have witnessed a steady increase in the number of careful studies of volatile anesthetic actions on $GABA_A$ receptors.

Considerable progress has been made in determining the amino acid residues within $GABA_A$ and glycine receptors that are critical for volatile anesthetic potentiation of agonist-induced currents. The use of a panel of glycine α_1/$GABA_C$ ρ_1 chimeric receptors allowed the identification of a 45 amino acid region encompassing TM2 and TM3 of the glycine α_1 receptor as both *necessary* and *sufficient* for potentiation of agonist-induced currents by the volatile ether enflurane (MIHIC et al. 1997). Extensive site-directed mutagenesis of glycine α_1 and $GABA_A$ α_2 and β_1 subunits determined that specific amino acid positions within TM2 and TM3 are also critical for agonist potentiation by isoflurane (MIHIC et al. 1997; KRASOWSKI et al. 1998b) (see Fig. 2 and 4), *n*-alcohols (including ethanol) (MIHIC et al. 1997; WICK et al. 1998; YE et al. 1998) and trichloroethanol (KRASOWSKI et al. 1998a) (see Fig. 2).

Most halogenated alkanes and ethers containing six or fewer carbons have anesthetic properties, but some notable exceptions to this rule exist. The work of Eger, Koblin, and colleagues has demonstrated that certain highly lipid-soluble halogenated cyclobutanes and alkanes are *unable* to produce immobility at concentrations predicted by the Meyer-Overton correlation to be in the anesthetic range (KOBLIN et al. 1994). These compounds, originally called non-anesthetics, are now more properly referred to as non-immobilizers, since although they do not produce immobility (KOBLIN et al. 1994) or analgesia (SONNER et al. 1998), they may interfere with learning and memory (KANDEL et al. 1996). The non-immobilizers, which are often heavily halogenated compounds (e.g., 1,2-dichlorohexafluorocyclobutane), elicit convulsions at higher concentrations (KOBLIN et al. 1994). The non-immobilizers have no modulatory actions at $GABA_A$ (MIHIC et al. 1994), glycine (MASCIA et al. 1996) or $GABA_C$ (MIHIC and HARRIS 1996) receptors. These results would seem to support the feasibility of $GABA_A$ and glycine receptors as viable molecular targets for producing immobility.

The anesthetic gases nitrous oxide and xenon have a pattern of action on the ligand-gated ion channels different from the volatile ethers and alkanes. This is perhaps not surprising since the clinical effects of xenon and nitrous oxide vary from that of the ethers and alkanes; for instance, unlike the ethers and alkanes, nitrous oxide is a potent analgesic with only weak immobilizing activity (MARSHALL and LONGNECKER 1996). Nitrous oxide inhibits agonist responses at NMDA receptors (JEVTOVIC-TODOROVIC et al. 1998; MENNERICK et al. 1998) but has only weak potentiating actions at $GABA_A$ receptors (DZOLJIC and VAN DUJIN 1998; JEVTOVIC-TODOROVIC et al. 1998; MENNERICK et al. 1998). Very recently, xenon has been demonstrated to inhibit NMDA receptors at clinically relevant concentrations but does not modulate the function of $GABA_A$ or AMPA receptors (FRANKS et al. 1998). This pharmacological profile, with NMDA receptor inhibition and a lack of potent actions on $GABA_A$ receptors, is shared by the 'dissociative anesthetic' ketamine.

J. Intravenous Anesthetic Agents

Etomidate and propofol both appear to be relatively selective modulators of the $GABA_A$ receptor. The $GABA_A$ receptor fulfills all of the above criteria for a plausible target underlying the anesthetic actions of these compounds. Propofol and etomidate do not modulate other ligand-gated ion channels at clinically relevant concentrations with the exception of propofol actions at the strychnine-sensitive glycine receptor (HALES and LAMBERT 1991; MASCIA et al. 1996; PISTIS et al. 1997). Amino acid residues within the β subunit of the $GABA_A$ receptor have been identified that are essential for potentiation of $GABA_A$ receptor function by etomidate (BELELLI et al. 1997; MOODY et al. 1997; MCGURK et al. 1998) and propofol (KRASOWSKI et al. 1998b) (see Fig. 2), consistent with previous studies suggesting that the β subunit of the $GABA_A$ receptor was likely to contain binding sites for these compounds (SANNA et al. 1995a,b).

Many steroid anesthetics such as alphaxalone are relatively selective for the $GABA_A$ receptor, although certain steroids have potent actions on other ligand-gated ion channels. For the steroid anesthetics, structure-activity studies comparing *in vivo* and *in vitro* potencies support a role for $GABA_A$ receptors in the actions of these compounds (HARRISON et al. 1987a; HU et al. 1993; LAMBERT et al. 1995, 1996; RUPPRECHT et al. 1996). For example, the non-anesthetic structural isomer betaxalone does not modulate the $GABA_A$ receptor (HARRISON and SIMMONDS 1984; COTTRELL et al. 1987). Critical residues for modulation by alphaxalone or other steroid anesthetics have not yet been identified within any ligand-gated ion channel, although studies of $GABA_A$/glycine chimeric receptors suggest a contribution of the N-terminal half of the $GABA_A$ receptor to GABA potentiation by alphaxalone (RICK et al. 1998).

Unlike propofol, etomidate, and the steroid anesthetics, the barbiturates are much less selective for the $GABA_A$ receptor. In addition to their actions at $GABA_A$ receptors, barbiturates also potently inhibit AMPA, kainate, and neuronal nicotinic ACh receptors. The optical isomers of pentobarbitone display the same order of potency for modulatory actions at the $GABA_A$ receptor as for their *in vivo* anesthetic actions (HUANG and BARKER 1980; FRANKS and LIEB 1994). A residue within TM2 of the β_1 subunit of the $GABA_A$ receptor has been identified that is apparently necessary for GABA potentiation by pentobarbitone (BIRNIR et al. 1997) (see Fig. 2), although this is a conserved residue in this sub-family, so this finding does not explain the pharmacologic differences between $GABA_A$ and glycine receptors (which are strikingly insensitive to barbiturates) (KOLTCHINE et al. 1996). GABA potentiation by barbiturates is not abolished by mutations in $GABA_A$ receptors that abolish potentiation by volatile anesthetics, *n*-alcohols, propofol, or trichloroethanol (Mihic et al. 1997; Krasowski et al. 1998a,b).

Compared with the other intravenous anesthetic agents discussed above, the 'dissociative anesthetic' ketamine has a very different *in vivo* and *in vitro* profile of action. Ketamine and related arylcycloalkylamines such as phencyclidine produce an atypical anesthesia characterized by a state of sedation, immobility, amnesia, marked analgesia, and a feeling of dissociation from the environment without true unconsciousness (Winters et al. 1972). These compounds can also produce intense hallucinations, especially in adults, and this limits their clinical usefulness (Marshall and Longnecker 1996). In contrast to most other general anesthetics, ketamine does not potentiate $GABA_A$ receptor function at clinically relevant concentrations (Simmonds and Turner 1987). Ketamine appears instead likely to produce anesthesia by inhibition of NMDA receptors (Lodge et al. 1982; Anis et al. 1983; Harrison and Simmonds 1984; Zeilhofer et al. 1992; Orser et al. 1997).

K. Alcohols

In parallel with the anesthetic field, there has been a transition from studies of non-specific actions of ethanol on membrane lipids to a search for specific sites of action on neuronal proteins. Key questions are:

1. Which neuronal proteins (or functions) are sufficiently sensitive to account for the intoxicating action of ethanol?
2. What is the molecular mechanism by which ethanol affects these proteins?
3. Which neuronal functions determine specific behavioral actions of ethanol (e.g., activating, sedative, anxiolytic, ataxic).

Because of the data implicating low ethanol sensitivity (or "responsiveness") as a positive factor in susceptibility for development of alcoholism (Schuckit 1992, 1994; Schuckit and Smith 1996), it is critical to identify molecular sites of alcohol action in the brain. These targets provide candidate systems for possible therapeutic interventions, as well as suggesting candidate genes for evaluation in human alcoholism. At the molecular level, a key question is whether there is a common mechanism for the action of ethanol on multiple ligand-gated ion channels as well as specific voltage-gated channels. Molecular techniques make it feasible to pinpoint regions of proteins critical for alcohol action and, more importantly, to construct mutant animals that can tell us if these candidate proteins are indeed responsible for distinct behavioral actions of ethanol *in vivo*.

L. $GABA_A$ and Glycine Receptors and Ethanol Action

In the early 1980s, a number of laboratories found that drugs (e.g., $GABA_A$ agonists, uptake inhibitors) that augment GABAergic function enhance the behavioral actions of ethanol, while drugs (e.g., $GABA_A$ receptor antagonists, synthesis inhibitors) that inhibit GABAergic function reduce ethanol behaviors (MARTZ et al. 1983; DEITRICH et al. 1989). In addition, the Long-sleep/Short-sleep (LS/SS) mice, which differ in genetic sensitivity to ethanol, were found to differ in their behavioral sensitivity to GABAergic drugs (MARTZ et al. 1983). These studies suggested that ethanol may exert some of its effects by enhancing GABA-mediated inhibition. One early electrophysiological study also presented evidence supporting this idea (DAVIDOFF 1973), but it was not developed further until 1986 when three laboratories independently demonstrated that intoxicating concentrations (5–50 mmol/l) of ethanol enhance the function of $GABA_A$ receptors (ALLAN and HARRIS 1986; SUZDAK et al. 1986; TICKU et al. 1986). These studies used different tissue preparations (mouse cerebellar and cortical microsacs, rat cortical synaptoneurosomes, and cultured mouse spinal neurons, respectively), but all measured the uptake of $^{36}Cl^-$ stimulated by GABA agonists and all obtained similar potentiation of $GABA_A$ receptor function by ethanol. These observations stimulated numerous electrophysiological studies of ethanol action on $GABA_A$ receptor function, and the results were inconsistent. A detailed discussion of this literature is beyond the scope of this chapter but is covered in reviews (DEITRICH et al. 1989; MIHIC and HARRIS 1995). At the risk of oversimplification, the literature suggests that there are ethanol-sensitive and ethanol-resistant $GABA_A$ receptors in brain, and that this ethanol sensitivity is likely determined both by subunit composition and by post-translational processing. However, the molecular details that define an ethanol-sensitive $GABA_A$ receptor remain to be determined. It is of interest to note that one of the first publications in this area (ALLAN and HARRIS 1986) showed that $GABA_A$ receptors of brain membranes from SS mice were resistant to ethanol, whereas those from LS mice were sensitive. Thus, the existence of ethanol-sensitive and -insensitive receptors, as well as their genetic association with ethanol sensitivity *in vitro*, is not a new idea, but has yet to be proven rigorously.

The function of recombinant, as well as neuronal, $GABA_A$ receptors can be enhanced by short- and long-chain alcohols, but effects of pharmacologically relevant concentrations of ethanol itself have not been found in all studies (see SIEGHART 1995; MIHIC et al. 1995). Glycine receptors are also modulated by ethanol and longer chain alcohols (CELENTANO et al. 1988; AGUAYO and PANCETTI 1994; ENGBLOM et al. 1991; MASCIA et al. 1996). The extensive behavioral evidence implicating $GABA_A$ receptors in ethanol action will not be reviewed here but has been presented in detail elsewhere (KOOB 1995; DRASKI and DEITRICH 1995). A role for the strychnine-sensitive glycine receptor in alcohol action is supported by behavioral studies in which glycine and the glycine precursor serine were shown to enhance the depressant effects of ethanol; this action was blocked by strychnine (WILLIAMS et al. 1995).

As noted above, the structurally related homomeric glycine $\alpha 1$ and GABA $\rho 1$ receptors exhibit opposing effects of ethanol: enhancement of function is seen in the former (MASCIA et al. 1996), and inhibition in the latter (MIHIC and HARRIS 1996). Using a chimeragenesis and mutagenesis approach, researchers identified two amino acids, in transmembrane domains two (TM2) and three (TM3), of glycine and $GABA_A$ receptors that were required for ethanol enhancement of receptor function (MIHIC et al. 1997) (Fig. 2). Other amino acid residues of glycine and $GABA_A$ receptors also affect ethanol enhancement of receptor function. Quantitative differences in ethanol enhancement of homomeric glycine $\alpha 1$ and $\alpha 2$ receptor function have been attributed to a difference in amino acid 52 (MASCIA et al. 1996). Gly-R $\alpha 1$ receptors, with alanine at residue 52, are more sensitive to ethanol than $\alpha 2$ receptors which have a serine residue at the homologous position. Furthermore, $\alpha 1$ subunits mutated from Ala to Ser at residue 52 have the same ethanol sensitivity as wild-type $\alpha 2$ receptors (MASCIA et al. 1996).

A major problem in this area is that not all $GABA_A$ receptors are sensitive to sub-anesthetic (<100 mmol/l) concentrations of ethanol, and the exact determinants of ethanol sensitivity remain to be defined. There is increasing support for the idea that activation of PKC is important for ethanol actions on $GABA_A$ and glycine receptors (WEINER et al. 1997b; MASCIA et al. 1998). In hippocampus, there are recent reports that ethanol sensitivity depends on the population of $GABA_A$ receptors studied (WEINER et al. 1997a), the activity of protein kinase C (PKC) (WEINER et al. 1994, 1997b), and even the degree of activation of $GABA_B$ receptors (WAN et al. 1996). Another study used null mutant mice lacking PKCγ to link the behavioral and neurochemical observations by showing that this mutation reduces sensitivity to ethanol *in vivo* and abolishes the action of ethanol on the function of cerebellar $GABA_A$ receptors (HARRIS et al. 1995c). One speculative synthesis of recent results is that ethanol binds directly to GABA or glycine receptors (perhaps between TM2 and TM3, perhaps elsewhere) and that phosphorylation of these receptors or associated proteins alters the affinity of alcohol binding.

M. Cutoff

We have followed previous suggestions (FRANKS and LIEB 1994; WICK et al. 1998) in defining cutoff as the point at which the potency of the *n*-alcohol no longer increases with increasing carbon chain length. As with stereoselectivity, alcohol cut-off severely challenges non-specific theories of anesthetic action, since there appears to be no cut-off for the disordering actions of *n*-alcohols on lipid bilayers (FRANKS and LIEB 1986). In general, the immobilizing actions of *n*-alcohols show a cut-off around dodecanol (C12) (MCCREERY and HUNT 1978; LYON et al. 1981; ALIFIMOFF et al. 1989), although the limited aqueous solubility of long-chain alcohols complicates matters (DILDY-MAYFIELD et al. 1996). The alcohol cut-off for the ligand-gated ion channels varies between receptors, and this is useful in implicating or eliminating the involvement of various receptors in the biological effects of the alcohols. Alcohol cut-off has recently been applied to the study of glycine and $GABA_C$ ρ_1 receptors harboring mutations in TM2 and TM3. It was first noted that mutation of a smaller to a larger amino acid residue in TM2 of the glycine α_1 subunit reduced the alcohol cut-off for the glycine receptor from dodecanol to propanol (WICK et al. 1998). In contrast, a double mutation of larger to smaller residues in TM2 and TM3 of the $GABA_C$ ρ_1 receptor extended the alcohol cut-off from heptanol to beyond dodecanol (WICK et al. 1998). This provides evidence that mutation of selected residues within TM2 and TM3 of glycine and $GABA_C$ receptors may actually alter the dimensions of the binding pocket for *n*-alcohols.

N. Discussion and Future Directions

Recent advances in the molecular biology of GABA receptors have provided tremendous opportunities for understanding actions of anesthetics and alcohol on these receptors. The availability of cDNAs encoding the receptor subunits, combined with expression systems and methods for the rapid introduction of mutations, has allowed rapid advances toward the 'Holy Grail' of anesthesia research: defining molecular sites of anesthetic action in the brain. There are tantalizing suggestions for an anesthetic binding site within GABA receptor subunits, but the low affinity of anesthetic binding makes it difficult to prove rigorously that the anesthetic is indeed binding at that site. Despite the obstacles it is likely that the site(s) of anesthetic and alcohol action on $GABA_A$ receptors will be defined to the satisfaction of many within a few years. These advances in molecular analysis will allow researchers to address the bigger question of which aspects of anesthetic and alcohol action are due to enhancement of GABAergic function. This will be accomplished by constructing mice with mutations in $GABA_A$ receptor subunits.

Targeted gene manipulations in mice will provide hypothesis-driven tests of the *in vivo* roles of certain ligand-gated ion channels in mediating the diverse behavioral actions of general anesthetics. Researchers over the last 5 years have created 'global knock-out mice' for various subunits of the ligand-gated ion channels. With the emergence of ligand-gated ion channel knock-out mice (and the commercial availability of some of these knock-outs), it should prove useful to test anesthetic sensitivity in these mice. Although these knock-out mice may provide initial clues as to the nature of anesthetic targets, some mice will be difficult to analyze for anesthetic sensitivity if they exhibit abnormal behavior, lethality, or gross alterations in neural development. These problems with knock-out mice may be circumvented by 'conditional' gene knock-outs where the gene of interest is disrupted only in limited brain regions and/or specified developmental time periods (HOMANICS et al. 1998). Another elegant example of gene targeting is the 'knock-in mouse.' One possibility is the introduction of the gene encoding a mutated receptor subunit that is insensitive to anesthetic modulation, in place of the normal endogenous gene

(Lakhlani et al. 1997). Knock-in mouse experiments potentially provide an elegant bridge between *in vitro* experiments and whole animal behavior. Ideally, the mutated receptor subunit would differ from the normal subunit only in terms of general anesthetic modulation (i.e., agonist response, voltage-dependence, kinetics, etc. of the receptor would be relatively normal) (Rudolph et al. 1999; McKernan et al. 2000). Recently described mutations within TM2 and TM3 of $GABA_A$ (see Figs. 2, 4) and glycine receptors, which confer insensitivity to volatile ether anesthetics (Mihic et al. 1997; Krasowski et al. 1998b), *n*-alkanols (Mihic et al. 1997; Wick et al. 1998; Ye et al. 1998), propofol (Krasowski et al. 1998b), trichloroethanol (Krasowski et al. 1998a), pentobarbitone (Birnir et al. 1997), and etomidate (Belelli et al. 1997; McGurk et al. 1998) essentially fit this qualification. A complication to gene targeting experiments is the presence of multiple subunit isoforms for the $GABA_A$ receptor subunits; if some or all of these isoforms play a role in general anesthesia, targeting of multiple genes may be required to obtain a clear alteration in anesthetic sensitivity.

There is now ample evidence that clinical concentrations of most volatile or intravenous general anesthetics, including the *n*-alcohols, enhance the function of $GABA_A$ receptors and we are on the verge of a molecular understanding of the sites of action of these drugs on $GABA_A$ receptors. However, there is still little information, or at least agreement, about the consequence of actions of these agents on $GABA_A$ receptors. This is particularly true for ethanol, where pharmacological interest is focused on the actions of sub-anesthetic doses, yet concentrations corresponding to these doses have small and variable effects on $GABA_A$ receptor function. This problem reflects our basic ignorance of how the brain works, in that we have no idea how small changes in channel function will influence behavior. We can be optimistic that construction of mice with mutant GABA receptors that differ in these subtle effects of anesthetics and alcohols will indeed address the fundamental question of how specific receptors influence specific behaviors. Indeed, recent work with the benzodiazepines suggests this era has already dawned.

Acknowledgments. We thank Dr. Caroline Rick for many helpful suggestions on this manuscript. Funding has been generously provided by NIH grants GM45129, GM56850, and GM00623 to N.L.H., by AA 03699 to R.A.H., by GM 47818 to R.A.H., N.L.H., and E. I. Eger II, and by NIMH training grant MH11504 to M.D.K.

References

Adodra S, Hales TG (1995) Potentiation, activation and blockade of $GABA_A$ receptors of clonal murine hypothalamic GT1–7 neurones by propofol. Br J Pharmacol 115:953–960

Aguayo, LG, Pancetti FC (1994) Ethanol modulation of the γ-aminobutyric $acid_A$- and glycine-activated Cl^- current in cultured mouse neurons. J Pharmacol Exp Ther 270:61–69

Akabas MH, Kaufmann C, Archdeacon P, Karlin A (1994) Identification of acetylcholine receptor channel-lining residues in the entire M2 segment of the α subunit. Neuron 13:919–927

Alifimoff JK, Firestone LL, Miller KW (1989) Anaesthetic potencies of primary alkanols: implications for the molecular dimensions of the anaesthetic site. Br J Pharmacol 96:9–16

Allan AM, Harris RA (1986) Gamma-aminobutyric acid and alcohol actions: Neurochemical studies of long sleep and short sleep mice. Life Sci 39:2005–2015

Amin J, Weiss DS (1993) $GABA_A$ receptor needs two homologous domains of the β-subunit for activation by GABA but not by pentobarbital. Nature 366:565–569

Andrews PR, Mark LC (1982) Structural specificity of barbiturates and related drugs. Anesthesiol 57:314–320

Anis NA, Berry SC, Burton NR, Lodge D (1983) The dissociative anaesthetics, ketamine and phencyclidine, selectively reduce excitation of central mammalian neurones by *N*-methyl-aspartate. Br J Pharmacol 79:565–575

Antognini JF, Schwartz K (1993) Exaggerated anesthetic requirements in the preferentially anesthetized brain. Anesthesiol 79:1244–1249

Atkinson RM, Davis B, Pratt MA, Sharpe HM, Tomich EG (1965) Action of some steroids on the central nervous system of the mouse. J Med Chem 8:426–432

Banks MI, Pearce RA (1999) Dual actions of volatile anesthetics on $GABA_A$ IPSCs: dissociation of blocking and prolonging effects. Anesthesiol 90:120–134

Barker JL, Ransom BR (1978) Pentobarbitone pharmacology of mammalian central neurones grown in tissue culture. J Physiol 280:355–372

Barnard EA, Skolnick P, Olsen RW, Mohler H, Sieghart W, Biggio G, Braestrup C, Bateson AN, Langer SZ (1998) International union of pharmacology XV. Subtypes of γ-aminobutyric $acid_A$ receptors: classification on the basis of subunit structure and receptor function. Pharmacol Rev 50:291–313

Belelli D, Callachan H, Hill-Venning C, Peters JA, Lambert JJ (1996) Interaction of positive allosteric modulators with human and Drosophila recombinant GABA receptors expressed in *Xenopus laevis* oocytes. Br J Pharmacol 118:563–576

Belelli D, Lambert JJ, Peters JA, Wafford K, Whiting PJ (1997) The interaction of the general anesthetic etomidate with the γ-aminobutyric acid type A receptor is influenced by a single amino acid. Proc Natl Acad Sci USA 94:11031–11036

Betz H (1991) Glycine receptors: heterogeneous and widespread in the mammalian brain. Trends Neurosci 14:458–461

Betz H (1992) Structure and function of inhibitory glycine receptors. Q Rev Biophys 25:381–394

Birnir B, Tierney ML, Dalziel JE, Cox GB, Gage PW (1997) A structural determinant of desensitization and allosteric regulation by pentobarbital of the $GABA_A$ receptor. J Membrane Biol 155:157–166

Bloom FE, Iversen LL (1971) Localizing [^{3}H]GABA in nerve terminals of cerebral cortex by electron microscopic autoradiography. Nature 229:628–630

Callachan H, Cottrell GA, Hather NY, Lambert JJ, Nooney JM, Peters JA (1987) Modulation of the $GABA_A$ receptor by progesterone metabolites. Proc R Soc Lond ser B Biol Sci 231:359–369

Celentano JJ, Gibbs TT, Farb DH (1988) Ethanol potentiates GABA- and glycine-induced chloride currents in chick spinal cord neurons. Brain Res 455:377–380

Chang Y, Wang R, Barot S, Weiss DS (1996) Stoichiometry of a recombinant $GABA_A$ receptor. J Neurosci 16:5415–5424

Chortkoff BS, Eger EI, Crankshaw DP, Gonsowski CT, Dutton RC, Ionescu P (1995a) Concentrations of desflurane and propofol that suppress response to command in humans. Anesth Analg 81:737–743

Chortkoff BS, Gonsowski CT, Bennett HL, Levinson B, Crankshaw DP, Dutton RC, Ionescu P, Block RI, Eger EI (1995b) Subanesthetic concentrations of desflurane and propofol suppress recall of emotionally charged information. Anesth Analg 81:728–736

Cohen ML, Chan SL, Way WL, Trevor AJ (1973) Distribution in the brain and metabolism of ketamine in the rat after intravenous administration. Anesthesiol 39:370–376

Collins JF, Kendig JJ, Mason P (1995) Anesthetic actions within the spinal cord: contributions to the state of general anesthesia. Trends In Neurosciences 18:549–553

Cooper E, Couturier S, Ballivet M (1991) Pentameric structure and subunit stoichiometry of a neuronal nicotinic acetylcholine receptor Nature 350:235–238

Cottrell GA, Lambert JJ, Peters JA (1987) Modulation of $GABA_A$ receptor activity by alphaxalone. Br J Pharmacol 90:491–500

Cutting GR, Curristin S, Zoghbi H, O'Hara B, Seldin MF, Uhl GR (1992) Identification of a putative γ-aminobutyric acid (GABA) receptor subunit rho_2 cDNA and colocalization of the genes encoding rho_2 (GABRR2) and rho_1 (GABRR1) to human chromosome 6q14-q21 and mouse chromosome 4. Genomics 12:801–806

Cutting GR, Lu L, O'Hara BF, Kasch LM, Montrose-Rafizadeh C, Donovan DM, Shimada S, Antonarakis SE, Guggino WB, Uhl GR, Kazazian HH (1991) Cloning of the γ-aminobutyric acid (GABA) rho 1 cDNA: a GABA receptor subunit highly expressed in the retina. Proc Natl Acad Sci USA 88:2673–2677

Davidoff RA (1973) Alcohol and presynaptic inhibition in an isolated spinal cord. Arch Neurol 28:60–63

Davies PA, Hanna MC, Hales TG, Kirkness EF (1997) Insensitivity to anaesthetic agents conferred by a class of $GABA_A$ receptor subunit. Nature 385:820–823

Deitrich RA, Dunwiddie TV, Harris RA, Erwin VG (1989) Mechanism of action of ethanol: initial central nervous system actions. Pharmacol Reviews 41:491–537

Deitrich RA, Harris RA (1996) How much alcohol should I use in my experiments? Alcohol Clin Exp Res 20:1–2

DeLorey TM, Handforth A, Anagnostaras SG, Homanics GE, Minassian BA, Asatourian A, Fanselow MS, Delgado-Escueta A, Ellison GD, Olsen RW (1998) Mice lacking the β_3 subunit of the $GABA_A$ receptor have the epilepsy phenotype and many of the behavioral characteristics of Angelman Syndrome. J Neurosci 18:8505–8514

Dickinson R, Franks NP, Lieb WR (1994) Can the stereoselective effects of the anesthetic isoflurane be accounted for by lipid solubility? Biophys J 66:2019–2023

Dildy-Mayfield JE, Mihic SJ, Liu Y, Deitrich RA, Harris RA (1996) Actions of long chain alcohols on $GABA_A$ and glutamate receptors: relation to in vivo effects. Br J Pharmacol 118:378–384

Downie DL, Hall AC, Lieb WR, Franks NP (1996) Effects of inhalational general anaesthetics on native glycine receptors in rat medullary neurones and recombinant glycine receptors in *Xenopus* oocytes. Br J Pharmacol 118:493–502

Draski LJ, Deitrich RA (1995) Initial effects of ethanol on the central nervous system In: Deitrich, RA, Erwin, VG (eds) Pharmacological effects of ethanol on the nervous system. CRC Press, Boca Raton, FL, pp 227–250

Dzoljic M, Van Dujin B (1998) Nitrous oxide-induced enhancement of γ-aminobutyric $acid_A$-mediated chloride currents in acutely dissociated hippocampal neurons. Anesthesiol 88:473–480

Ebert B, Wafford KA, Whiting PJ, Krogsgaard-Larsen P, Kemp JA (1994) Molecular pharmacology of γ-aminobutyric acid type A receptor agonists and partial agonists in oocytes injected with different α, β, and γ receptor subunit combinations. Mol Pharmacol 46:957–963

Eccles JC, Malcolm JL (1946) Dorsal root potentials of the spinal cord. J Neurophysiol 9:139–160
Eccles JC, Schmidt R, Willis WD (1963) Pharmacological studies on presynaptic inhibition. J Physiol 168:500–530
Eckenhoff RG (1996) An inhalational anesthetic binding domain in the nicotinic acetylcholine receptor. Proc Natl Acad Sci USA 93:2807–2810
Eger EI, Koblin DD, Laster MJ, Schurig V, Juza M, Ionescu P, Gong D (1997) Minimum alveolar anesthetic concentration values for the enantiomers of isoflurane differ minimally. Anesth Analg 85:188–192
Eger EI, Saidman LJ, Brandstater B (1965) Minimum alveolar anesthetic concentration: a standard of anesthetic potency. Anesthesiol 26:756–763
Engblom AC, Akerman KEO (1991) Effect of ethanol on α-aminobutyric acid and glycine receptor-coupled Cl^- fluxes in rat brain synaptoneurosomes. J Neurochem 57:384–390
Franks NP, Dickinson R, de Sousa SLM, Hall AC, Lieb WR (1998) How does xenon produce anaesthesia? Nature 396:324
Franks NP, Lieb WR (1984) Do general anaesthetics act by competitive binding to specific receptors? Nature 310:599–601
Franks NP, Lieb WR (1994) Molecular and cellular mechanisms of general anaesthesia. Nature 367:607–614
Franks NP, Lieb WR (1986) Partitioning of long-chain alcohols into lipid bilayers: implications for mechanisms of general anesthesia. Proc Natl Acad Sci USA 83:5116–5120
Franks NP, Lieb WR (1993) Selective actions of volatile general anaesthetics at molecular and cellular levels. Br J Anaesth 71:65–76
Franks NP, Lieb WR (1996a) An anesthetic-sensitive superfamily of neurotransmitter-gated ion channels. J Clin Anesth 8:3S–7S
Franks NP, Lieb WR (1996b) Temperature dependence of the potency of volatile general anesthetics: implications for in vitro experiments. Anesthesiol 84:716–720
Fritschy JM, Mohler H (1995) $GABA_A$-receptor heterogeneity in the adult rat brain: differential regional and cellular distribution of seven major subunits. J Comp Neurol 359:154–194
Gunther U, Benson J, Benke D, Fritschy JM, Reyes G, Knoflach F, Crestani F, Aguzzi A, Arigoni M, Lang Y, Bluethmann H, Mohler H, Luscher B (1995) Benzodiazepine-insensitive mice generated by targeted disruption of the $\gamma2$ subunit gene of γ-aminobutyric acid type A receptors. Proc Natl Acad Sci USA 92:7749–7753
Gurley D, Amin J, Ross PC, Weiss DS, White G (1995) Point mutations in the M2 region of the α, β, or γ subunit of the $GABA_A$ channel that abolish block by picrotoxin. Receptors Channels 3:13–20
Hackam AS, Wang TL, Guggino WB, Cutting GR (1998) Sequences in the amino termini of GABA ρ and $GABA_A$ subunits specify their selective interaction in vitro. J Neurochem 70:40–46
Hadingham KL, Wingrove P, Le Bourdelles B, Palmer KJ, Ragan CI, Whiting PJ (1993a) Cloning of cDNA sequences encoding human $\alpha2$ and $\alpha3$ γ-aminobutyric $acid_A$ receptor subunits and characterization of the benzodiazepine pharmacology of recombinant $\alpha1$-, $\alpha2$-, $\alpha3$-, and $\alpha5$-containing human γ-aminobutyric $acid_A$ receptors. Mol Pharmacol 43:970–975
Hadingham KL, Wingrove PB, Wafford KA, Bain C, Kemp JA, Palmer KJ, Wilson AW, Wilcox AS, Sikela JM, Ragan CI, Whiting PJ (1993b) Role of the β subunit in determining the pharmacology of human γ-aminobutyric acid type A receptors. Mol Pharmacol 44:1211–1218
Hales TG, Lambert JJ (1991) The actions of propofol on inhibitory amino acid receptors of bovine adrenomedullary chromaffin cells and rodent central neurones. Br J Pharmacol 104:619–628
Hall AC, Lieb WR, Franks NP (1994) Stereoselective and non-stereoselective actions of isoflurane on the $GABA_A$ receptor. Br J Pharmacol 112:906–910

Hara M, Kai Y, Ikemoto Y (1993) Propofol activates $GABA_A$ receptor-chloride ionophore complex in dissociated hippocampal pyramidal neurons of the rat. Anesthesiol 79:781–788

Harris B, Moody E, Skolnick P (1992) Isoflurane anesthesia is stereoselective. Eur J Pharmacol 217:215–216

Harris BD, Wong G, Moody EJ, Skolnick P (1995a) Different subunit requirements for volatile and nonvolatile anesthetics at γ-aminobutyric acid type A receptors. Mol Pharmacol 47:363–367

Harris RA, Mihic SJ, Dildy-Mayfield JE, Machu TK (1995b) Actions of anesthetics on ligand-gated ion channels: role of receptor subunit composition. FASEB J 9:1454–1462

Harris RA, McQuilkin SJ, Paylor R, Abeliovich A, Tonegawa S, Wehner JM (1995c) Mutant mice lacking the isoform of protein kinase C show decreased behavioral actions of ethanol and altered function of γ-aminobutyrate type A receptors. Proc Natl Acad Sci USA 92:3658–3662

Harrison, NL (1998) Optical isomers open a new window on anesthetic mechanism. Anesthesiol 88:566–568

Harrison NL, Flood P (1998) Molecular mechanisms of general anesthetic action. Sci Med 5:18–27

Harrison NL, Simmonds MA (1984) Modulation of the GABA receptor complex by a steroid anaesthetic. Brain Res 323:287–292

Harrison NL, Majewska MD, Harrington JW, Barker JL (1987a) Structure-activity relationships for steroid interaction with the γ-aminobutyric $acid_A$ receptor complex. J Pharmacol Exp Ther 241:346–353

Harrison NL, Vicini S, Barker JL (1987b) A steroid anesthetic prolongs inhibitory postsynaptic currents in cultured rat hippocampal neurons. J Neurosci, 7:604–609

Hedblom E, Kirkness EF (1997) A novel class of $GABA_A$ receptor subunit in tissues of the reproductive system. J Biol Chem 272:15346–15350

Heykants JJ, Meuldermans WE, Michiels LJ, Lewi PJ, Janssen PA (1975) Distribution, metabolism and excretion of etomidate, a short-acting hypnotic drug, in the rat. Comparative study of (R)-(+) and S-(–)-etomidate. Arch Intl Pharmacodyn Ther 216:113–129

Hill-Venning C, Belelli D, Peters JA, Lambert JJ (1997) Subunit-dependent interaction of the general anaesthetic etomidate with the γ-aminobutyric acid type A receptor. Br J Pharmacol 120:749–756

Homanics GE, DeLorey TM, Firestone LL, Quinlan JJ, Handforth A, Harrison NL, Krasowski MD, Rick CEM, Korpi ER, Makela R, Brilliant MH, Hagiwara N, Ferguson C, Snyder K, Olsen RW (1997) Mice devoid of γ-aminobutyric type A receptor $\beta3$ subunit have epilepsy, cleft palate, and hypersensitive behavior. Proc Natl Acad Sci USA 94:4143–4148

Homanics GE, Quinlan JJ, Mihalek RM, Firestone LL (1998) Alcohol and anesthetic mechanisms in genetically engineered mice. Front Biosci 3:D548–D558

Horenstein J, Akabas MH (1998) Location of a high affinity Zn^{2+} binding site in the channel of $\alpha1\beta1$ $GABA_A$ receptors. Mol Pharmacol 53:870–877

Hu Y, Zorumski CF, Covey DF (1993) Neurosteroid analogues: structure-activity studies of benz[*e*]indene modulators of $GABA_A$ receptor function. 1 The effect of 6-methyl substitution on the electrophysiological activity of 7-substituted benz[*e*]indene-3-carbonitriles. J Med Chem 36:3956–3967

Huang LY, Barker JL (1980) Pentobarbital: stereospecific actions of (+) and (–) isomers revealed on cultured mammalian neurons. Science 207:195–197

Iselin-Chaves IA, Flaishon R, Sebel PS, Howell S, Gan TJ, Sigl J, Ginsberg B, Glass PSA (1998) The effect of the interaction of propofol and alfentanil on recall, loss of consciousness, and the bispectral index. Anesth Analg 87:949–955

Jevtovic-Todorovic V, Todorovic SM, Mennerick S, Powell S, Dikranian K, Benshoff N, Zorumski CF, Olney JW (1998) Nitrous oxide (laughing gas) is an NMDA antagonist, neuroprotectant, and neurotoxin. Nature Med 4:460–463

Johnston GA (1996) $GABA_C$ receptors: relatively simple transmitter-gated ion channels? Trends Pharmacol Sci 17:319–323
Jones MV, Harrison NL (1993) Effects of volatile anesthetics on the kinetics of inhibitory postsynaptic currents in cultured rat hippocampal neurons. J Neurophysiol 70:339–1349
Jones MV, Harrison NL, Pritchett DB, Hales TG (1995) Modulation of the $GABA_A$ receptor by propofol is independent of the γsubunit. J Pharmacol Exp Ther 274:962–968
Kandel L, Chortkoff BS, Sonner J, Laster MJ, Eger EI (1996) Nonanesthetics can suppress learning. Anesth Analg 82:321–326
Koblin DD, Chortkoff BS, Laster MJ, Eger EI Halsey, MJ Ionescu P (1994) Polyhalogenated and perfluorinated compounds that disobey the Meyer-Overton hypothesis. Anesth Analg 79:1043–1048
Koltchine VV, Ye Q, Finn SE, Harrison NL (1996) Chimeric $GABA_A$/glycine receptors: expression and barbiturate pharmacology. Neuropharmacol, 35:1445–1456
Koob GF (1995) The neuropharmacology of ethanol's behavioral action: new data, new paradigms, new hope. In: Deitrich RA, Erwin VG (eds) Pharmacological effects of ethanol on the nervous system. CRC Press, Boca Raton, FL, pp 1–12
Krasowski MD, Finn SE, Ye Q, Harrison NL (1998a) Trichloroethanol modulation of recombinant $GABA_A$, glycine, and GABA r1 receptors. J Pharmacol Exp Ther 284:934–942
Krasowski MD, Koltchine VV, Rick CE, Ye Q, Finn SE, Harrison NL (1998b) Propofol and other intravenous anesthetics have sites of action on the γ-aminobutyric $acid_A$ receptor distinct from that for isoflurane. Mol Pharmacol 53:530–538
Krasowski MD, O'Shea SM, Rick CEM, Whiting PJ, Hadingham KL, Czajkowski C, Harrison NL (1997) α subunit isoform influences $GABA_A$ receptor modulation by propofol. Neuropharmacol 36:941–949
Krasowski MD, Harrison NL (1999) General anaesthetic actions on ligand-gated ion channels. Cell Mol Life Sci 55:1278–1303
Kuhse J, Betz H, Kirsch J (1995) The inhibitory glycine receptor: architecture, synaptic localization and molecular pathology of a postsynaptic ion-channel complex. Curr Opin Neurobiol 5:318–323
Lakhlani PP, MacMillan LB, Guo TZ, McCool BA, Lovinger DM, Maze M, Limbird LE (1997) Substitution of a mutant α_{2a}-adrenergic receptor via "hit and run" gene targeting reveals the role of this subtype in sedative, analgesic, and anesthetic-sparing responses in vivo. Proc Natl Acad Sci USA 94.9950–9955
Lambert JJ, Belelli D, Hill-Venning C, Callachan H, Peters JA (1996) Neurosteroid modulation of native and recombinant $GABA_A$ receptors. Cell Mol Neurobiol 16:155–174
Lambert JJ, Belelli D, Hill-Venning C, Peters JA (1995) Neurosteroids and $GABA_A$ receptor function. Trends Pharmacol Sci 16:295–303
Langosch D, Thomas L, Betz H (1988) Conserved quaternary structure of ligand-gated ion channels: the postsynaptic glycine receptor is a pentamer. Proc Natl Acad Sci USA 85:7394–7398
Lavoie AM, Tingey JJ, Harrison NL, Pritchett DB, Twyman RE (1997) Activation and deactivation rates of recombinant $GABA_A$ receptor channels are dependent on α-subunit isoform. Biophys J 73:2518–2526
Levitan ES, Blair LA, Dionne VE, Barnard EA (1988) Biophysical and pharmacological properties of cloned $GABA_A$ receptor subunits expressed in *Xenopus* oocytes. Neuron 1:773–781
Lodge D, Anis NA, Burton NR (1982) Effects of optical isomers of ketamine on excitation of cat and rat spinal neurones by amino acids and acetylcholine. Neurosci Lett 29:281–286
Lovinger DM (1997) Alcohols and neurotransmitter gated ion channels: past, present and future. Naunyn-Schmiedebergs Arch Pharmacol 356:267–282
Lu L, Huang Y (1998) Separate domains for desensitization of GABA ρ_1 and β_2 subunits expressed in *Xenopus* oocytes. J Membrane Biol 164:115–124

Lyon RC, McComb JA, Schreurs J, Goldstein DB (1981) A relationship between alcohol intoxication and the disordering of brain membranes by a series of short-chain alcohols. J Pharmacol Exp Ther 218:669–675
Lysko GS Robinson JL, Casto R, Ferrone RA (1994) The stereospecific effects of isoflurane isomers in vivo. Eur J Pharmacol 263:25–29
Macdonald RL, Olsen RW (1994) $GABA_A$ receptor channels. Annu Rev Neurosci 17:569–602
MacIver MB, Tanelian DL, Mody I (1991) Two mechanisms for anesthetic-induced enhancement of $GABA_A$-mediated neuronal inhibition. Ann NY Acad Sci 625:91–96
Marshall BE, Longnecker DE (1996) General anesthetics In: Hardman JG, Limbird LE, Molinoff PB, Ruddon RW, Gilman AG (eds) The pharmacological basis of therapeutics. McGraw-Hill, New York, pp 307–330
Martz A, Deitrich RA, Harris RA Behavioral evidence for the involvement of γ-aminobutyric acid in the actions of ethanol. Eur J Pharmacol 89:53–62
Mascia MP, Machu TK, Harris RA (1996) Enhancement of homomeric glycine re-ceptor function by long-chain alcohols and anaesthetics. Br J Pharmacol 119:1331–1336
Mascia MP, Wick MJ, Martinez LD, Harris RA (1998) Enhancement of glycine receptor function by ethanol: role of phosphorylation. Br J Pharmacol 125:263–270
McCormick DA (1989) GABA as an inhibitory neurotransmitter in human cerebral cortex. J Neurophysiol 62:1018–1027
McCreery MJ, Hunt WA (1978) Physico-chemical correlates of alcohol intoxication. Neuropharmacol 17:451–461
McGurk KA, Pistis M, Belelli D, Hope AG, Lambert JJ (1998) The effect of a transmembrane amino acid on etomidate sensitivity of an invertebrate GABA receptor. Br J Pharmacol 124:13–20
McKernan RM, Farrar S, Collins I, Emms F, Asuni A, Quirk K, Broughton H (1998) Photoaffinity labeling of the benzodiazepine binding site of $\alpha1\beta3\gamma2$ γ-aminobutyric $acid_A$ receptors with flunitrazepam identifies a subset of ligands that interact directly with His102 of the α subunit and predicts orientation of these within the benzodiazepine pharmacophore. Mol Pharmacol 54:33–43
McKernan RM, Whiting PJ (1996) Which $GABA_A$-receptor subtypes really occur in the brain? Trends Neurosci 19:139–143
McKernan RM et al. (2000) Sedative but not anxiolytic properties of benzodiazepines are mediated by the GABA-A receptor alpha-1 subtype. Nature Neurosci 3: 587–592
Mennerick S, Jevtovic-Todorovic V, Todorovic SM, Shen WX, Olney JW, Zorumski CF (1998) Effect of nitrous oxide on excitatory and inhibitory synaptic transmission in hippocampal cultures. J Neurosci 18:9716–9726
Meyer H (1899) Welche Eigenschaft der Anasthetica bedingt ihre narkotische Wirkung? Naunyn-Schmiedebergs Arch Exp Path Pharmakol 42:109–118
Meyer H (1901) Zur Theorie der Alkolnarkose: der Einfluss wechselnder Temperatur auf Wirkungsstärke und Theilungscoefficient der Narcotica. Naunyn-Schmiedebergs Arch Exp Path Pharmakol 46:338–346
Meyer KH (1937) Contribution to the theory of narcosis. Trans Faraday Soc 33:1062–1068
Mihic SJ, Harris RA (1996) Inhibition of ρ_1 receptor GABAergic currents by alcohols and volatile anesthetics. J Pharmacol Exp Ther 277:411–416
Mihic SJ, McQuilkin SJ, Eger EI, Ionescu P, Harris RA (1994) Potentiation of γ-aminobutyric acid type A receptor-mediated chloride currents by novel halogenated compounds correlates with their abilities to induce general anesthesia. Mol Pharmacol 46:851–857
Mihic SJ, Sanna E, Whiting PJ, Harris RA (1995) Pharmacology of recombinant $GABA_A$ receptors. Adv Biochem Psychopharmacol, 48:17–40
Mihic SJ, Ye Q, Wick MJ, Koltchine VV, Krasowski MD, Finn SE, Mascia MP, Valenzuela CF Hanson KK Greenblatt, EP Harris RA Harrison NL (1997)

Sites of alcohol and volatile anaesthetic action on $GABA_A$ and glycine receptors. Nature 389:385–389
Mihic SJ, Harris RA (1995) Alcohol actions at the GABA-A receptor/chloride channel complex In: Deitrich RA, Erwin VG (eds) Pharmacological effects of ethanol on the nervous system. CRC Press, Boca Raton, FL, pp51–72
Moody EJ, Harris BD, Skolnick P (1994) The potential for safer anaesthesia using stereoselective anaesthetics. Trends Pharmacol Sci 15:387–391
Moody EJ, Knauer C, Granja R, Strakhova M, Skolnick P (1997) Distinct loci mediate the direct and indirect actions of the anesthetic etomidate at $GABA_A$ receptors. J Neurochem 69:1310–1313
Mullins LJ (1954) Some physical mechanisms in narcosis. Chem Rev 54:289–322
Nicoll RA, Eccles JC, Oshima T, Rubia F (1975) Prolongation of hippocampal inhibitory postsynaptic potentials by barbiturates. Nature 258:625–627
Nusser Z, Sieghart W, Benke D, Fritschy JM, Somogyi P (1996) Differential synaptic localization of two major γ-aminobutyric acid type A receptor α subunits on hippocampal pyramidal cells. Proc Natl Acad Sci USA 93:11939–11944
Olsen RW, Snowman AM (1982) Chloride-dependent enhancement by barbiturates of γ-aminobutyric acid receptor binding. J Neurosci 2:1812–1823
Orser BA, Pennefather PS, MacDonald JF (1997) Multiple mechanisms of ketamine blockade of *N*-methyl-D-aspartate receptors. Anesthesiol 86:903–917
Ortells MO, Lunt GG (1995) Evolutionary history of the ligand-gated ion-channel superfamily of receptors. Trends Neurosci 18:121–127
Overton E (1901) Studien über die Narkose, zugleich ein Beitrag zur allgemeiner Pharmakologie. Gustav Fischer, Jena, Switzerland
Pearce RA (1999) Effects of volatile anesthetics on $GABA_A$ receptors: electrophysiological studies In: Moody EJ, Skolnick P (eds) Molecular bases of anesthesia. CRC Press, Boca Raton, FL (in press)
Pistis M, Belelli D, Peters JA, Lambert JJ (1997) The interaction of general anaesthetics with recombinant $GABA_A$ and glycine receptors expressed in *Xenopus laevis* oocytes: a comparative study. Br J Pharmacol 122:1707–1719
Pritchett DB, Sontheimer H, Shivers BD, Ymer S, Kettenmann H, Schofield PR, Seeburg PH (1989) Importance of a novel $GABA_A$ receptor subunit for benzodiazepine pharmacology. Nature 338:582–585
Quasha AL, Eger EI, Tinker JH (1980) Determination and applications of MAC. Anesthesiol 53:315–334
Quinlan JJ, Homanics GE, Firestone LL (1998) Anesthesia sensitivity in mice that lack the $\beta 3$ subunit of the γ-aminobutyric acid type A receptors. Anesthesiol 88:775–780
Rabow LE, Russek SJ, Farb DH (1995) From ion currents to genomic analysis: recent advances in $GABA_A$ receptor research. Synapse 21:189–274
Rampil IJ, Mason P, Singh H (1993) Anesthetic potency (MAC) is independent of forebrain structures in the rat. Anesthesiol 78:707–712
Rho JM, Donevan SD, Rogawski MA (1996) Direct activation of $GABA_A$ receptors by barbiturates in cultured rat hippocampal neurons. J Physiol 497:509–522
Rick CE, Ye Q, Finn SE, Harrison NL (1998) Neurosteroids act on the $GABA_A$ receptor at sites on the N-terminal side of the middle of TM2. Neuroreport 9:379–383
Robertson B (1989) Actions of anaesthetics and avermectin on $GABA_A$ chloride channels in mammalian dorsal root ganglion neurones. Br J Pharmacol 98:167–176
Rudolph H et al. (1999) Benzodiazepine actions mediated by specific gamma-aminobutyric acid-A receptor subtypes. Nature 401:796–800
Rupprecht R, Berning B, Hauser CA, Holsboer F, Reul JM (1996) Steroid receptor-mediated effects of neuroactive steroids: characterization of structure-activity relationship. Eur J Pharmacol 303:227–234
Ryder S, Way WL, Trevor AJ (1978) Comparative pharmacology of the optical isomers of ketamine in mice. Eur J Pharmacol 49:15–23

Sanna E, Garau F, Harris RA (1995a) Novel properties of homomeric $\beta1$ γ-aminobutyric acid type A receptors: actions of the anesthetics propofol and pentobarbital. Mol Pharmacol 47:213–217
Sanna E, Mascia MP, Klein RL, Whiting PJ, Biggio G, Harris RA (1995b) Actions of the general anesthetic propofol on recombinant human $GABA_A$ receptors: influence of receptor subunits. J Pharmacol Exp Ther 274:353–360
Sanna E, Murgia A, Casula A, Biggio G (1997) Differential subunit dependence of the actions of the general anesthetics alphaxalone and etomidate at γ-aminobutyric acid type A receptors expressed in *Xenopus laevis* oocytes. Mol Pharmacol 51:484–490
Scholfield CN (1980) Potentiation of inhibition by general anaesthetics in neurones of the olfactory cortex in vitro. Pflügers Archiv – Eur J Physiol 383:249–55
Schuckit MA (1994) Low level of response to alcohol as a predictor of future alcoholism. Am J Psychiatry 151:184–189
Schuckit, MA (1992) Reaction to alcohol as a predictor of alcoholism. Alcohol Clin Exp Res 16:656
Schuckit MA, Smith TL (1996) An 8-year follow-up of 450 sons of alcoholic and control subjects. Arch Gen Psych 53:202–210
Sear JW, Prys-Roberts C (1979) Plasma concentrations of alphaxalone during continuous infusion of Althesin. Br J Anaesth 51:861–865
Seeman P (1972) The membrane actions of anesthetics and tranquilizers. Pharmacol Rev 24:583–655
Sieghart W (1995) Structure and pharmacology of γ-aminobutyric $acid_A$ receptor subtypes. Pharmacol Rev 47:181–234
Sigel E, Baur R, Trube G, Mohler H, Malherbe P (1990) The effect of subunit composition of rat brain $GABA_A$ receptors on channel function. Neuron 5:703–711
Sigel E, Buhr A (1997) The benzodiazepine binding site of the $GABA_A$ receptor. Trends Pharmacol Sci 18:425–429
Simmonds MA, Turner JP (1987) Potentiators of responses to activation of γ-aminobutyric acid ($GABA_A$) receptors. Neuropharmacol 26:923–930
Smith GB, Olsen RW (1995) Functional domains of $GABA_A$ receptors. Trends Pharmacol Sci 16:162–168
Sonner J, Li J, Eger EI (1998) Desflurane and nitrous oxide but not non-immobilizers, affect nociceptive responses. Anesth Analg 86:629–634
Suzdak PD, Schwartz RD, Skolnick P, Paul SM (1986) Ethanol stimulates γ-aminobutyric acid receptor-mediated chloride transport in rat brain synaptoneurosomes. Proc Natl Acad Sci 83:4071–4075
Tanelian DL, Kosek P, Mody I, MacIver MB (1993) The role of the $GABA_A$ receptor/chloride channel complex in anesthesia. Anesthesiol 78:757–776
Tia S, Wang JF, Kotchabhakdi N, Vicini S (1996) Developmental changes of inhibitory synaptic currents in cerebellar granule neurons: role of $GABA_A$ receptor $\alpha6$ subunit. J Neurosci 16:3630–3640
Tickue MK, Lowrimore P, LeHoullier P (1986) Ethanol enhances GABA-induced $^{36}Cl^-$ influx in primary spinal cord cultured neurons. Brain Res Bull 17:128–126
Tomlin SL, Jenkins A, Lieb WR, Franks NP (1998) Stereoselective effects of etomidate optical isomers on gamma-aminobutyric acid type A receptors and animals. Anesthesiol 88:708–717
Tretter V, Ehya N, Fuchs K, Sieghart W (1997) Stoichiometry and assembly of a recombinant $GABA_A$ receptor subtype. J Neurosci 17:2728–2737
Wafford KA, Burnett DM, Leidenheimer NJ, Burt DR, Wang JB, Kofuji P, Dunwiddie TV, Harris RA, Sikela JM (1991) Ethanol sensitivity of the $GABA_A$ receptor expressed in *Xenopus* oocytes requires 8 amino acids contained in the $\gamma2L$ subunit. Neuron 7:27–33
Wan FJ, Berton F, Madamba SG, Francesconi W, Siggins GR (1996) Low ethanol concentrations enhance GABAergic inhibitory postsynaptic potentials in hippocam-

pal pyramidal neurons only after block of GABA-B receptors. Proc Natl Acad Sci USA 93:5049–5054

Wegelius K, Pasternack M, Hitunen JO, Rivera C, Kaila K, Saarma M, Reeben M (1998) Distribution of GABA receptor ρ subunit transcripts in the rat brain. Eur J Neurosci 10:350–357

Weight FF, Aguayo LG, White G, Lovinger DM, Peoples RW (1992) GABA- and glutamate-gated ion channels as molecular sites of alcohol and anesthetic action. Adv Biochem Psychopharmacol 47:335–347

Weiner JL, Gu C, Dunwiddie TV (1997a) Differential ethanol sensitivity of subpopulations of GABA-A synapses onto rat hippocampal CA1 pyramidal neurons. J Neurophys 77:1306–1312

Weiner JL, Valenzuela CF, Watson PL, Frazier CJ, Dunwiddie TV (1997b) Elevation of basal protein kinase C activity increases ethanol sensitivity of $GABA_A$ receptors in rat hippocampal CA1 pyramidal neurons. J Neurochem 68:1949–1959

Weiner JL, Zhang L, Carlen PL (1994) Potentiation of $GABA_A$-mediated synaptic current by ethanol in hippocampal CA1 neurons: Possible role of protein kinase C. J Pharm Exp Ther 268:1388–1395

White PF, Schuttler J, Shafer A, Stanski DR, Horai Y, Trevor AJ (1985) Comparative pharmacology of the ketamine isomers Studies in volunteers. Br J Anaesth 57:197–203

Whiting PJ, McAllister G, Vasilatis D, Bonnert TP, Heavens RP, Smith DW, Hewson L O'Donnell R, Rigby MR, Sirinathsinghji DJS, Marshall G, Thompson SA, Wafford KA (1997) Neuronally restricted RNA splicing regulates the expression of a novel $GABA_A$ receptor subunit conferring atypical functional properties. J Neurosci 17:5027–5037

Whiting PJ, McKernan RM, Wafford KA (1995) Structure and pharmacology of vertebrate $GABA_A$ receptor subtypes. Intl Rev Neurobio 38:95–138

Wick MJ, Mihic SJ, Ueno S, Mascia MP, Trudell JR, Brozowski SJ, Ye Q, Harrison NL, Harris RA (1998) Mutations of γ-aminobutyric acid and glycine receptors change alcohol cutoff: evidence for an alcohol receptor? Proc Natl Acad Sci USA 95:6504–6509

Williams KL, Ferko AP, Barbieri EJ, Digregoria GJ (1995) Glycine enhances the central depressant properties of ethanol in mice. Pharmacol Biochem Behav 50:199–205

Wingrove PB, Wafford KA, Bain C, Whiting PJ (1994) The modulatory action of loreclezole at the γ-aminobutyric acid type A receptor is determined by a single amino acid in the $\beta 2$ and $\beta 3$ subunit. Proc Natl Acad Sci USA 91:4569–4573

Winters WD, Ferrar-Allado T, Guzman-Flores C, Alcaraz M (1972) The cataleptic state induced by ketamine: a review of the neuropharmacology of anesthesia. Neuropharmacol 11:303–315

Wittmer LL, Hu Y, Kalkbrenner M, Evers AS, Zorumski CF, Covey DF (1996) Enantioselectivity of steroid-induced γ-aminobutyric $acid_A$ receptor modulation and anesthesia. Mol Pharmacol 50:1581–1586

Xu M, Akabas MH (1993) Amino acids lining the channel of the γ-aminobutyric acid type A receptor identified by cysteine substitution. J Biol Chem 268:21505–21508

Xu M, Akabas MH (1996) Identification of channel-lining residues in the M2 membrane-spanning segment of the $GABA_A$ receptor $\alpha 1$ subunit. J Gen Physiol 107:195–205

Yang J, Isenberg KE, Zorumski CF (1992) Volatile anesthetics gate a chloride current in postnatal rat hippocampal neurons. FASEB J 6:914–918

Ye Q, Koltchine VV, Mihic SJ, Mascia MP, Wick M, Finn SE, Harrison NL, Harris RA (1998) Enhancement of glycine receptor function by ethanol is inversely correlated with molecular volume at position $\alpha 267$. J Biol Chem 273:3314–3319

Zafra F, Aragon C, Gimenez C (1997) Molecular biology of glycinergic neurotransmission. Mol Neurobiol 14:117–142

Zeilhofer HU, Swandulla D, Geisslinger G, Brune K (1992) Differential effects of ketamine enantiomers on NMDA receptor currents in cultured neurons. Eur J Pharmacol 213:155–158

Zimmerman SA, Jones MV, Harrison NL (1994) Potentiation of γ-aminobutyric $acid_A$ receptor Cl^- current correlates with in vivo anesthetic potency. J Pharmacol Exp Ther 270:987–991

CHAPTER 6

Anticonvulsants Acting on the GABA System

B.S. MELDRUM and P. WHITING

A. Introduction

I. Role of GABA and GABA Receptors in Epilepsy

Modification of activity at GABAergic synapses powerfully influences epileptic phenomena. These effects show significant differences according to the type of epilepsy involved. The predominant effect for focal motor and tonic-clonic seizures is that impairment or reduction of function at $GABA_A$ receptors facilitates epileptic discharges and motor seizure activity and enhancement of function diminishes epileptic activity. This is clearly a consequence of the role of GABAergic synapses in recurrent inhibitory systems in cortical and other structures, and their effect in limiting the excessive discharge of principal neurons in time and space. Compounds blocking the inhibitory action of GABA at $GABA_A$ receptors such as bicuculline and picrotoxin are powerful convulsants when given focally in the brain or systemically. Compounds inhibiting glutamic acid decarboxylase activity, thereby blocking GABA synthesis, such as pyridoxal phosphate antagonists, are convulsant (for a more extensive list of epilepsy syndromes and seizures caused by GABA-related mechanisms see Table 1). Compounds potentiating the action of GABA at $GABA_A$ receptors are anticonvulsant (see below).

Absence epilepsy in man, with a 2–3 Hz spike-and-wave discharge in the cortex, is dependent on a thalamo-cortical loop which involves several sets of GABAergic synapses in cortex and thalamus. The "waves" correspond to hyperpolarising activity resulting from synchronous firing of GABAergic neurons. The effects of GABA-related drugs are complex. Agonists at $GABA_B$ receptors, such as baclofen, exacerbate the spike-and-wave discharges in man and animals, $GABA_B$ antagonists suppress them. Compounds potentiating $GABA_A$ synaptic function commonly exacerbate the discharges although some benzodiazepines with subtype selective actions can decrease the spike-and-wave discharges (see below).

The question therefore arises as to which genetic or acquired syndromes of epilepsy are a consequence of altered GABAergic function (see Table 1), and whether such syndromes respond selectively to drugs acting on GABAer-

Table 1. Seizures and epilepsy syndromes related to altered GABAergic function

A. Animal models
Mouse KO of $\beta 3$ subunit of $GABA_A$ receptor (DeLorey et al. 1998)
Mouse KO of GAD 65 (Kash et al. 1997)
Mouse KO of TNAP (phosphatase involved in pyridoxal phosphate metabolism) GABA deficit, (seizures respond to pyridoxine) (Waymire et al. 1995)
Administration of pyridoxal phosphate antagonists and other GAD inhibitors (deoxypyridoxine, isoniazid, allylglycine)
Antagonists acting on $GABA_A$ receptor (Bicuculline, picrotoxin)
Inverse agonists at BZ receptor on $GABA_A$ receptors (DMCM)
Convulsant barbiturates
Kindled seizures in rats
Limbic seizures following pilocarpine-induced status epilepticus
Status epilepticus (secondary phase, drug unresponsive) (diazepam sensitivity reduced)
B. Human syndromes
Pyridoxine deficiency
Pyridoxine dependency
Angelman Syndrome (deletion on maternal Chr 15q11–13, loss includes GABRB3 gene)
Complex partial seizures

gic function. The simplest example is that of generalised seizures in infancy related to pyridoxine deficiency or dependency where the seizures are related to deficient synthesis of GABA and can be treated by moderate or high doses of pyridoxine. Multiple forms of epilepsy occur in a neurodevelopmental disorder, known as Angelman syndrome, which also shows mental retardation and facial dysmorphism. Genetic studies commonly reveal a major deletion on maternal chromosome 15q11–13 (Minassian et al. 1998). Two genes appear to contribute to the syndrome – one is UBE3A, encoding a ubiquitin ligase, the other is GABRB3 encoding the $\beta 3$ subunit of the $GABA_A$ receptor subunit. Mice deficient in the murine homolog of GABRB3 also show multiple seizure types (DeLorey et al. 1998).

1. Developmental Changes in $GABA_A$ Receptor Effects

During early development (in neonatal rats, but in mid-term primates) $GABA_A$ responses are often depolarising. This is due to an abnormal Cl^- gradient prior to the expression of the neuronal Cl^- extruding K^+/Cl^- cotransporter, KCC2 (Rivera et al. 1999). This does not appear to be a critical factor for neonatal seizures in man, which respond to barbiturates and benzodiazepines indicating that potentiating GABA at $GABA_A$ receptors is anticonvulsant at this developmental stage.

A cell type that is prominent in early development is the Cajal-Retzius cell. These cells normally disappear from cortex and hippocampus early in development. They persistently show depolarising responses to GABA

(MIENVILLE 1998). They appear to persist in the hippocampus of those patients with complex partial seizures secondary to a critical event early in life (such as a prolonged febrile convulsion or traumatic brain injury) (BLÜMCKE et al. 1996)

II. Mechanism of Action of Antiepileptic Drugs

The mechanisms of action of antiepileptic drugs currently in clinical use are only partially understood. An action on voltage-dependent Na^+ channels, involving a prolongation of the inactivated state, contributes importantly to the antiepileptic action of phenytoin, carbamazepine and lamotrigine and may be significant for topiramate, zonisamide, valproate and diazepam (MACDONALD and MELDRUM 1995). It is probable that the anti-absence action of ethosuximide and trimethadione can be explained by their action to decrease T-type voltage-dependent Ca^{++} currents.

Approximately half the antiepileptic drugs in clinical use are thought to owe their efficacy either totally or partially to potentiating GABAergic inhibitory effects (see Table 2). Three principal mechanisms of action on synaptic function can be distinguished (see Fig. 1):

1. Compounds, such as tiagabine, may decrease GABA uptake into neurons and glia and thereby prolong the synaptic action of GABA.
2. Compounds inhibiting the further metabolism of GABA, such as vigabatrin may increase the brain content of GABA and enhance its synaptic release.
3. Compounds may act at various sites on $GABA_A$ receptors to potentiate or mimic the action of GABA.

These mechanisms will be discussed in turn.

B. GABA Transporters and Tiagabine

GABA is cleared from the synaptic cleft by diffusion and by uptake into neurons and glia via specific carriers in the plasma membrane. Four such carriers have been identified, sequenced and cloned in the mammalian brain (see Chap. 14 by B. KANNER). In rats they are referred to as GAT-1, GAT-2, GAT-3 and BGT-1 (the latter may be primarily a betaine transporter). These transporters belong to a family of Na^+/Cl^- neurotransmitter transporters (NELSON 1998). They show marked differences in their regional and cellular expression (MINELLI et al. 1995, 1996). In the rat GAT-1 is the principal transporter in the cerebral and cerebellar cortices and in the hippocampus, where it is predominantly neuronal but is also present in astrocytic processes (MINELLI et al. 1995). GAT-2 is expressed principally in the leptomeninges. GAT-3 is found predominantly in astrocytic processes in midbrain and brain stem structures,

Table 2. Antiepileptic drugs acting on GABAergic function

Compound	Action on GABA function	Other actions	Anti-epileptic action
Diazepam	$GABA_A$ potentiation	Na^+ channel inactivated	Myoclonic epilepsy Status epilepticus
Clonazepam	$GABA_A$ potentiation		Absence seizures Atypical absences Complex partial seizures Tonic-clonic seizures Myoclonic seizures
Clobazam	$GABA_A$		Myoclonic epilepsy
Lorazepam	$GABA_A$		Status epilepticus
Loreclezole	$GABA_A$		Complex partial seizure
Chlormethiazole	$GABA_A$		Status epilepticus
Phenobarbital	$GABA_A$	AMPA receptor block	Generalised (t.c) seizures Partial seizures Neonatal seizures
Ganaxolone	$GABA_A$		
Tiagabine	GAT1 inhibition		Complex partial seizure
Vigabatrin	GABA-T inhibition		Complex partial seizure Infantile spasm
Felbamate	$GABA_A$ potentiation	NMDA receptor block	Complex partial seizure
Gabapentin	Altered GABA metabolism	Ca^{++} channels	Complex partial seizures
Valproate	Altered GABA metabolism	Na^+ channels	Absence seizures Generalised (t-c) seizure Complex partial seizure
Topiramate	$GABA_A$ potentiation	Na^+ channels	Complex partial seizure

GAT1, GABA transporter 1; GABA-T, GABA transaminase; t.c., tonic clonic; AMPA, α-amino-3-hydroxy-5-methyl-4-isoxazole propionic acid; NMDA, *N*-methyl-D-aspartate.

including the thalamus and superior and inferior colliculi (MINELLI et al. 1996; DEBIASI et al. 1998).

The three GABA transporters show differences in their substrate selectivity and in their sensitivity to different inhibitors. β-Alanine is a substrate for GAT-2 and GAT-3 and thus competes with GABA for uptake. Nipecotic acid, guvacine and their various derivatives preferentially inhibit GAT-1 (BORDEN et al. 1994). Nipecotic acid and guvacine penetrate the blood brain barrier poorly. The addition of a lipophilic side chain, however, provides compounds that penetrate the blood brain barrier, such as SKF 89976 A, SKF 100330 A, Cl-966, NNC-711 and NO328(= tiagabine) (see Fig. 2). These compounds have been shown to be anticonvulsant in a variety of animal models of epilepsy.

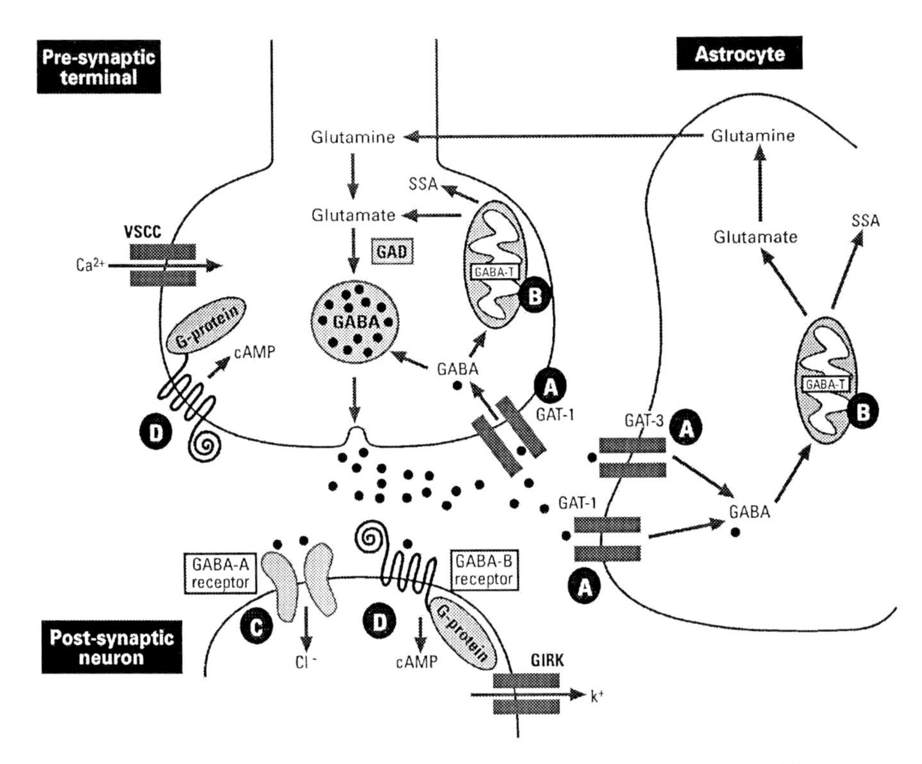

Fig. 1. GABAergic synapse showing sites of action of GABA-transport inhibitors (*A*), GABA-transaminase inhibitors (*B*) and drugs acting on $GABA_A$ (*C*) and $GABA_B$ (*D*) receptors in relation to the presynaptic terminal, postsynaptic neuron and an astrocytic process. GAD, glutamic acid decarboxylase; SSA, succinic semialdehyde; GAT-1 and GAT-3, GABA transporters; GABA-T, 4-aminobutyrate:2-oxoglutarate transaminase

Some of these compounds are neurotoxic in animals and man. Tiagabine has been shown to be antiepileptic in several animal models of epilepsy, being particularly potent in reflex epilepsy in rodents and photosensitive baboons and against kindled seizures in the rat (MORIMOTO et al. 1997; SMITH et al. 1995). It enhances spike-and-wave discharges, however, in rodent models of absence seizures, e.g. lethargic mice and WAG/Rij rats (HOSFORD and WANG 1997; COENEN et al. 1995). In accordance with its preclinical spectrum of activity it is clinically effective against complex partial seizures (KÄLVIÄINEN et al. 1998; RICHENS et al. 1995) but it may exacerbate absence seizures.

Recordings of inhibitory post-synaptic currents or potentials show that the effect of tiagabine is to prolong their duration (THOMPSON and GÄHWILER 1992; ROEPSTORFF and LAMBERT 1992) consistent with the concept that GABA-uptake serves to shorten the duration of inhibitory synaptic potentials.

GABA uptake inhibitors with a different selectivity for the transporter molecules have also been studied preclinically. These include compounds such

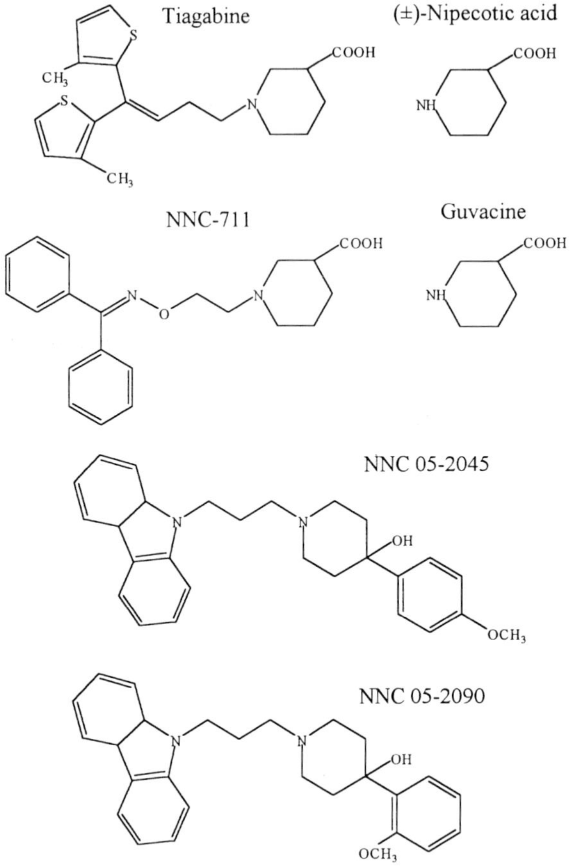

Fig. 2. Molecular structures of GABA-transport inhibitors, nipecotic acid, guvacine, tiagabine, NNC-711, NNC 05–2045 and NNC 05–2090

as NNC 05–2045 and NNC 05–2090 that are more potent inhibitors of GAT3 (= GAT4 in mice) than of GAT-1. These show little anticonvulsant activity against kindled seizures but significant activity against maximal electroshock seizures (Dalby et al. 1997), a result consistent with the dominant role of GAT1 in the limbic system and of GAT3 in the brain stem and midbrain.

I. Effects of Other Anti-Epileptic Drugs on GABA-Transporters

Although tiagabine is the only antiepileptic drug in current clinical use whose primary mechanism of action is via inhibition of GABA transport, it is possible that some other antiepileptic drugs can modify GABA transport. The evidence for this comes either from expression studies employing GABA transporters or from studies in cell cultures. Thus studies of [^{3}H]GABA uptake in oocytes expressing GAT-1 show inhibition of uptake with vigabatrin, 1 nmol/l and 0.5 mmol/l, gabapentin 50 μmol/l, and valproate 100 μmol/l (ECKSTEIN-LUDWIG et al. 1999). In cultures of human cortical astrocytes uptake of GABA is inhibited by valproate 1 mmol/l and by vigabatrin 0.1 mmol/l (compared with tiagabine 0.2 mmol/l). The possible role of in vivo actions on GABA transport in the antiepileptic actions of valproate, vigabatrin and gabapentin remains to be elucidated.

II. Changes in GABA Transporters in Epilepsy

Studies on the binding of [^{3}H]-nipecotic acid suggest that GABA transporter levels are reduced in the hippocampus of amygdala-kindled rats (DURING et al. 1995). Microdialysis data in patients with drug-resistant complex partial seizures have also been interpreted as showing impaired GABA-transporter function in the hippocampus on the epileptogenic hemisphere (DURING et al. 1995). This interpretation has been supported by electrophysiological studies in hippocampal slices from anterior temporal lobectomies in patients with hippocampal sclerosis (WILLIAMSON et al. 1995).

C. Vigabatrin and Inhibition of GABA-Transaminase

The synthesis of GABA from glutamate by the decarboxylase GAD is part of the so-called "GABA-shunt" that links the two TCA cycle intermediates 2-oxoglutarate and succinate. The further metabolism of GABA is provided by two mitochondrial enzymes, GABA-transaminase (4-aminobutyrate:2-oxoglutarate aminotransferase) and succinic semialdehyde dehydrogenase (succinate-semialdehyde:NAD(P)oxidoreductase). Inhibiting these enzymes can lead to a marked accumulation of GABA in the brain. In the 1970s it was shown that irreversible (catalytic) inhibitors of GABA-transaminase, such as ethanolamine-*O*-sulphate, γ-vinyl-GABA (vigabatrin) and γ-acetylenic GABA (Fig. 3) were anticonvulsant in some preclinical models of epilepsy (MELDRUM and HORTON 1978; METCALF 1979). In vivo and ex vivo experiments provide evidence that the synaptic release of GABA is enhanced in the cortex or hippocampus of rats given vigabatrin (ABDUL-GHANI et al. 1981; QUME and FOWLER 1997). Vigabatrin accumulates in neurons. In the CSF of patients treated for 3 months with vigabatrin the levels of GABA, homocarnosine and glycine are increased and those of glutamate decreased (KÄLVIÄINEN et al. 1993). There is also evidence that the turnover of 5-HT is reduced. All these changes can be interpreted as secondary to a primary effect on GABA.

Vigabatrin is effective in complex partial seizures and in some forms of primary generalised seizures, particularly infantile spasms (West's syndrome) (VIGEVANO and CILIO 1997). As add-on therapy in patients with partial seizures, in which the original therapy is predominantly drugs acting on voltage-

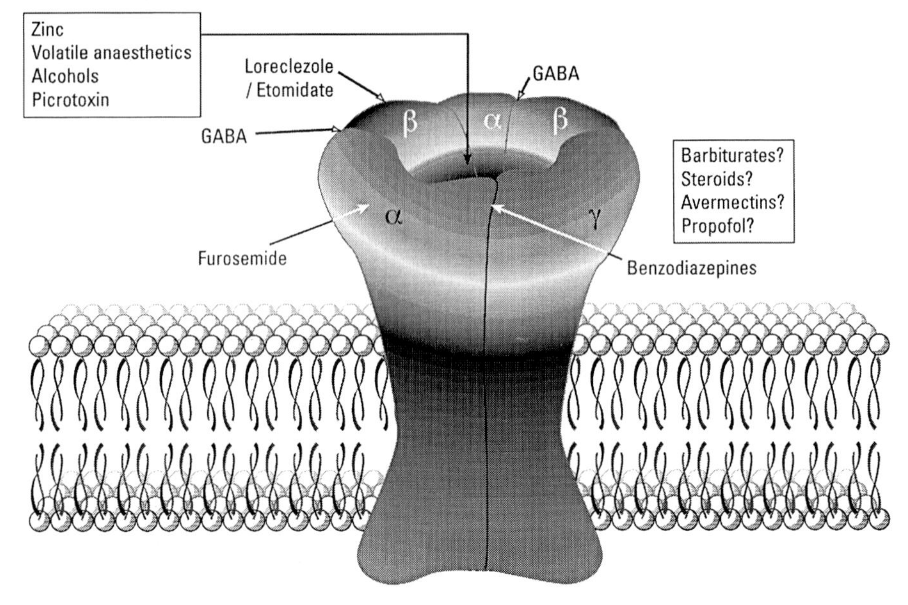

Fig. 3. Diagram of $GABA_A$ receptor showing sites of action of the various classes of anti-epileptic drug

sensitive sodium channels, vigabatrin appears to be more effective than several alternative novel agents (MARSON et al. 1997). As monotherapy in newly diagnosed patients with partial seizures, vigabatrin has less efficacy than carbamazepine (CHADWICK 1999). It is generally better tolerated than carbamazepine but is associated with a higher incidence of psychiatric symptoms (25% vs 15%). Visual field defects occur in a significant proportion of patients taking vigabatrin (KÄLVIÄINEN et al. 1999; Wild et al. 1999); this retinal toxicity is probably directly linked to inhibition of GABA-transaminase.

D. Anticonvulsants Acting Through the $GABA_A$ Receptor

The $GABA_A$ receptor is obviously central to GABAergic neurotransmission and as such is an important therapeutic target for anticonvulsants. The structure of $GABA_A$ receptor subtypes is discussed in detail elsewhere in this volume (see Chap. 2 by E.A. BARNARD). One of the most interesting, and therapeutically useful, aspects of $GABA_A$ receptors is their rich pharmacology (see Chap. 3 by H. MÖHLER and Fig. 3). There are a number of modulatory sites on the receptor (i.e. drug binding sites distinct from the GABA agonist site) through which various pharmacological agents act to potentiate or inhibit allosterically the action of GABA. A number of these sites have been, and are continuing to be, exploited to generate therapeutically useful drugs. Here we will discuss those pertinent to the treatment of convulsions: the benzodiazepine site, the barbiturate site, the loreclezole site and the steroid site (the latter is discussed in detail in Chap. 4 by J. LAMBERT et al.), and we will also discuss the activity of topiramate and chlormethiazole at the receptor.

I. Benzodiazepines

The most therapeutically useful of these modulatory sites on the $GABA_A$ receptor is the so-called benzodiazepine (BZ) binding site (WHITING et al. 1995). This is named after one class of compounds that act via this site. It is important to note that other compounds, which do not have a benzodiazepine chemical structure, also act at this site, e.g. zolpidem and various β-carbolines. (To date no endogenous agonist or inverse agonist has been identified.) Benzodiazepine agonists (which potentiate the action of GABA and thereby lead to increased hyperpolarisation of the postsynaptic membrane, see below) are widely prescribed for absence epilepsy (clonazepam) and status epilepticus (diazepam and lorazepam) and for myoclonic epilepsies. BZs act by increasing the frequency of channel opening (ROGERS et al. 1994). Receptors require both an α and γ subunit to have a BZ site (PRITCHETT et al. 1989); those containing a δ or ε subunit (i.e. $\alpha\beta\delta$ or $\alpha\beta\varepsilon$) do not have a BZ site (QUIRK et al. 1994; SAXENA and MACDONALD 1996; WHITING et al. 1997). $\alpha1$, $\alpha2$, $\alpha3$, $\alpha5$ coassembled with a β and $\gamma2$ subunit, have high affinity for so-called non-selective BZ ligands such as diazepam and clonazepam. In fact this is a misnomer, as $\alpha4$ and $\alpha6$ (coassembled with a β and $\gamma2$) containing receptors have very low affinity for these compounds (LUDDENS et al. 1990; WISDEN et al. 1991; WAFFORD et al. 1996; HADINGHAM et al. 1996). Similarly, while clonazepam and diazepam have high affinity at receptors containing $\gamma2$, their affinity at receptors containing $\gamma1$ and $\gamma3$ is considerably reduced (BENKE et al. 1996; WINGROVE et al. 1997). An additional complication is the concept of efficacy, i.e. the degree of modulation mediated by the BZ site ligand. Compounds can act as agonists (potentiating the GABA response, with an anticonvulsant effect) or inverse agonists (negatively modulating the GABA response, with a proconvulsant effect) according to the ligand and the type of α and γ subunit present in the receptor (see WHITING et al. 1995 for review). BZ site ligands (unlike barbiturates, see below) do not activate the receptor in the absence of GABA. A key point is that these observations hold out the possibility for the development of BZ site drugs targeted to a defined receptor subtype (i.e. a subtype known to be integral to the pathogenesis of seizures) through either selective affinity or selective efficacy.

II. Barbiturates

Barbiturates were first used as anticonvulsants in the USA in 1912. Phenobarbital is prescribed for both generalised and partial motor seizures. The primary site of action of barbiturates such as phenobarbital is via the $GABA_A$

receptor (EVANS 1979). Barbiturates act by increasing the channel open time (MATHERS and BARKER 1980). At low concentrations barbiturates positively modulate response to GABA via an allosteric mechanism, leading to hyperpolarisation of the postsynaptic membrane (THOMPSON et al. 1996). At higher concentrations barbiturates are GABAmimetic, i.e. they are able to activate the receptor in the absence of GABA, and this in part reflects the poorer safety profile of these drugs compared to benzodiazepines. The site of action of barbiturates on the $GABA_A$ receptor at the molecular level has yet to be defined. While there is certainly some selectivity of these compounds, e.g. pentobarbitone is more efficacious at $\alpha6$ containing receptors (THOMPSON et al. 1996), the selectivity is not absolute, and thus this class of compounds can be considered active at all subtypes so far examined.

Phenobarbital is widely used in primary generalised, tonic-clonic seizures. It is also effective in simple and complex partial seizures, but requires higher plasma concentrations for efficacy in this indication. It has been widely used in neonatal seizures, febrile convulsions and status epilepticus.

III. Steroids

The pharmacology of steroids acting at the $GABA_A$ receptor is discussed in detail elsewhere in this volume (see Chap. 4 by J. LAMBERT et al.). Like the barbiturates, their site of action on the $GABA_A$ receptor has yet to be defined at a molecular level. Similarly the receptor subtype selectivity of this class of compounds is not that profound, such that in general they can be considered active at all subtypes. None of this class of compound is currently prescribed, although some have been in clinical trial. Ganaxolone (3α-hydroxy-3β-methyl-5α-pregnan-20-one) is a non-subtype-selective positive allosteric modulator of the $GABA_A$ receptor (CARTER et al. 1997) which is in clinical trials for partial and generalised seizures. While ganaxolone is active in pentylenetetrazole-induced general seizures in animals (BEEKMAN et al. 1998) it appears to exacerbate the seizures seen in models of absence epilepsy (SNEAD 1998), thus resembling barbiturates. This phenomenon is discussed in more detail below.

IV. Loreclezole

Loreclezole is a potent anticonvulsant active in a number of animal seizure models (ASHTON et al. 1992). In clinical trials loreclezole was found to be active in the treatment of partial seizures (RENTMEESTER et al. 1991). Using recombinant $GABA_A$ receptor subtypes it has been possible to define the molecular target for this drug. It acts through a novel modulatory site on the β subunit of the receptor (WAFFORD et al. 1994). Loreclezole has a 300-fold higher affinity for $\beta2$- and $\beta3$-containing receptors compared to $\beta1$-containing receptors, and this selectivity is determined by amino acids in the transmembrane 2 domain of the β subunit (WINGROVE et al. 1994). Interestingly etomidate (Hypnomidate), the widely prescribed general anaesthetic, appears to act through the same site on the $GABA_A$ receptor as loreclezole (BELELLI et al. 1997).

V. Topiramate

Topiramate is a relatively new antiepileptic drug (WALKER and SANDER 1996) approved for adjunctive therapy in partial and secondarily generalised seizures. This drug is active in animal models of seizures, including pentylenetetrazole induced seizures (SHANK et al. 1994), in seizures in spontaneously epileptic rats and sound-induced seizures of DBA/2 mice (NAKAMURA et al. 1994), and amygdaloid kindling induced seizures in rats (AMANO et al. 1998). The mode of action through which this drug exerts its antiepileptic effect has yet to be definitively defined, but probably involves activity at sodium channels (TAVERNA et al. 1999), AMPA/kainate type glutamate receptors (SHANK 1995) and $GABA_A$ receptors (WHITE et al. 1997). The activity at the latter has yet to be clearly defined at the molecular level. It is clear however that topiramate enhances GABA mediated currents through a site on the receptor that is distinct from the BZ site (SHANK 1995; WHITE et al. 1997).

VI. Chlormethiazole

Chlormethiazole is a hypnotic and a sedative that in the past has been used in elderly patients, and also as treatment for acute alcohol withdrawal. Evidence has recently been presented for efficacy in stroke (GREEN 1998). More relevant to this discussion, it is also used in status epilepticus, though generally when patients have failed to respond to first line drugs (HARVEY et al. 1975; MILLER and KOVAR 1983; MARTIN and MILLAC 1994). Chlormethiazole does not appear to interact with glutamate receptors, calcium or sodium channels (GREEN et al. 1998). Its primary site of action is most likely through the $GABA_A$ receptor (CROSS et al. 1989; ZHONG and SIMMONDS 1997), at micromolar concentrations potentiating the GABA response and at millimolar concentrations directly activating the receptor through sites on the receptor distinct from the BZ site (HALES and LAMBERT 1992). Chlormethiazole has similar effects at the strychnine sensitive glycine receptor (HALES and LAMBERT 1992). It has been shown to interact with recombinant $GABA_A$ receptors, though any subunit or subtype selectivity remains to be determined (SLANY et al. 1995; ZEZULA et al. 1996).

E. Alterations in GABA Receptors in Epilepsy

I. Alterations in the Expression of $GABA_A$ Receptors in Animal Models of Seizure

A number of studies have attempted to identify changes in the expression of $GABA_A$ receptors (particularly changes in the expression of subunit mRNAs) in animal models of seizures. Interpretation and comparison of results from these studies is difficult due to the use of different animal models, different protocols, and the determination of the expression of different receptor subunit mRNAs.

An experimental model of status epilepticus can be generated by pilocarpine treatment (Walton and Treiman 1988). This results in loss of $GABA_A$-mediated inhibition in the CA1 region of the hippocampus (Kapur and Coulter 1995) and indeed a loss of receptors as measured by radioligand binding (Kapur et al. 1994). At the molecular level this loss of $GABA_A$ receptor is correlated with a loss of $\alpha5$ (Houser et al. 1995; Rice et al. 1996) and also $\alpha2$ $GABA_A$ receptor subunit mRNAs (Rice et al. 1996) in the CA1–3 region, with no change in $\alpha1$, $\beta2$ or $\gamma2$ (Rice et al. 1996). There is a small increase in $\alpha5$ mRNA in the dentate gyrus. Receptors containing $\alpha2$ and $\alpha5$ subunits are abundantly expressed in the pyramidal cells of CA1-CA3, while receptors containing the $\alpha1$ subunit are expressed in these cells and also abundantly in the interneurons throughout the hippocampal formation (Wisden et al. 1992; Sperk et al. 1997). The changes in mRNA level are relatively small (20%–30%) compared to the large decreases in receptor function which have been reported (Kapur et al. 1994). These mRNA changes need to be linked to changes in levels of receptor subtype to show cause and effect. Using the higher resolution approach of single cell polymerase reaction (PCR) combined with electrophysiology, Brooks-Kayal et al. (1998) have looked at changes in $GABA_A$ receptor subunit mRNAs in individual dentate gyrus granule cells from animals in which prolonged seizures had been induced with pilocarpine. They found significant changes in these chronically epileptic animals, particularly decreases in $\alpha1$ and $\beta1$, and increases in $\alpha4$, $\beta3$, δ and ε subunits. These changes correlate with changes in the pharmacological properties of the receptors. As before, the question arises as to whether these changes are compensatory or causative. They do, however, suggest receptor subtypes as possible therapeutic targets.

Hippocampal or amygdala kindling (where repeated high-frequency electrical stimulation leads to the gradual appearance of increasingly overt seizures) is used as an animal model of temporal lobe epilepsy and complex partial seizures. Hippocampal kindling has been shown to lead to a small decrease in $GABA_A$ receptors in the CA1 region of the hippocampus (up to 25%) and a more significant increase in $GABA_A$ receptors (up to 50% in the dentate), as measured by the binding of the GABA site ligand [^{3}H]-muscimol (Titulaer et al. 1994). This is correlated with changes in receptor function in these areas at the fully-kindled stage (24h after the last seizure), but these changes return to normal within a month (Titulaer et al. 1995). Nusser et al. (1998) have used a combination of electrophysiology and immunoelectronmicroscopy to show that the increase in size of the inhibitory postsynaptic potential of dentate gyrus granule cells in kindled animals is correlated with an increase in the number of $GABA_A$ receptors inserted at the synapse. This functional enhancement of GABAergic inhibition can be more than fully reversed by an increase of the effect of zinc to oppose GABA-mediated inhibition (BUHL et al. 1996; COULTER 1999).

Levels of receptor subunit mRNA have also been measured in the hippocampus of kindled animals (KOKAIA et al. 1994). Changes observed depend upon the number of stimulations. The most robust changes are observed in the dentate gyrus after 40 stimulations; 4 h after the last seizure there are significant decreases in both $\alpha 1$ and $\gamma 2$, while between 12 h and 48 h there are significant increases, with levels of mRNA returning to normal within 5 days. This is clearly a biphasic change in mRNA levels, and the increase in mRNA after 12 h correlates with changes in receptor measured by radioligand binding and electrophysiology, discussed above. One interpretation of these observations is that the changes in mRNA and receptor in the dentate gyrus are a response to stabilise granule cell excitability, and thus reduce the susceptibility to seizures. The changes appear, however, to be transient, arguing against a direct role in the more permanently increased excitability characteristic of kindling.

Selective breeding gives rise to fast- and slow-kindling strains of rats. Using subunit-specific antibodies it has been shown that in the fast-kindling strain there is reduced expression of the $\alpha 1$ subunit and an increase in the $\alpha 5$ subunit (POULTER et al. 1999). This suggests that epileptogenesis is enhanced when there is a failure of the normal developmental shift in subunit expression from $\alpha 5$ (in development) to $\alpha 1$ (in adulthood).

Kainic acid-induced seizures are an animal model of temporal lobe epilepsy, with spontaneous recurrent seizures and neuronal loss in the hippocampus. SPERT and colleagues have examined the changes in both $GABA_A$ receptor subunit polypeptide and mRNA in the hippocampus of lesioned animals (SCHWARZER et al. 1997; TSUNASHIMA et al. 1997; SPERT et al. 1998). They report both acute and chronic cell specific changes in receptor expression. Acute changes include decreases in some $GABA_A$ receptor mRNAs and increases in others (SPERT et al. 1998). Chronic changes include loss of receptors in the pyramidal cell layer presumably reflecting neuronal degeneration. There is also an overall increase in a number of receptor subunits ($\alpha 1$, $\alpha 2$, $\alpha 4$, $\alpha 5$, $\beta 1$, $\beta 2$, $\beta 3$, $\gamma 2$, δ) in the molecular layer of the dentate gyrus. Again, this may represent a protective response to enhanced excitability. There are clearly differences in the subunit regulation observed in this study in comparison to that observed, e.g. in the single cell study of BROOKS-KAYAL et al. discussed above (BROOKS-KAYAL et al. 1998), the reasons for which are not clear.

Systemic γ-hydroxybutyric acid administration leads to the development of absence-like seizures in rats. Since these seizures are thought to critically involve the thalamus, Banerjee and colleagues measured changes in $GABA_A$ receptor subunit mRNAs in this animal model (BANERJEE et al. 1998). They observed a transient increase in $\alpha 1$ and decrease in $\alpha 4$ mRNA in the thalamic relay nuclei after the seizure, which returned to normal levels after 24 h. Whether this is translated into changes in protein was not determined, and how this relates to the pathogenesis of absence epilepsy in man is unclear.

II. $GABA_A$ Receptors and Absence Epilepsy

As discussed above, absence epilepsy is thought to arise from the thalamocortical circuitry comprising neocortical neurons, thalamic relay neurons and neurons within the reticular nucleus of the thalamus. The thalamocortical cells excite GABAergic neurons of the reticular nucleus, which in turn leads to recurrent inhibitory post-synaptic potentials on the thalamic relay neurons. Subsequent excitation by the relay neurons feeds back onto the reticular neurons, and the cycle begins again (HUGUENARD et al. 1994). Pharmacological agents have been used to demonstrate the central role of $GABA_A$ receptors in this process, and parenthetically, the key role of this circuitry in absence epilepsy. However, there is a curious anomaly in the action of $GABA_A$ receptor drugs in the treatment of absence epilepsy. While benzodiazepines such as clonazepam are effective (MATTSON 1995), barbiturates (PENRY and SO 1981) and steroids such as ganaxolone (at least in animal models) (SNEAD 1998) are ineffective and may actually exacerbate seizures. A key difference between these agents, as discussed above, is their receptor subtype selectivity. Clonazepam has a degree of receptor subtype selectivity, potentiating only receptor subtypes comprising $\alpha1$, $\alpha2$, $\alpha3$ or $\alpha5$ (coassembled with β and $\gamma2$ subunits), but not receptors containing $\alpha4$ or $\alpha6$ (coassembled with a β and $\gamma2$), or receptors containing a δ or ε subunit. In contrast, both barbiturates and steroids show no significant subtype selectivity. Neurons of the reticular nucleus express $GABA_A$ receptors, and in-situ hybridisation experiments tend to suggest that a possible subunit combination would be $\alpha3\beta1/\beta3\gamma2$ (WISDEN et al. 1992; HUNTSMAN et al. 1996; KULTAS-ILINSKY et al. 1998), which is clonazepam sensitive. It has been suggested that clonazepam mediates its effect by facilitating the recurrent inhibition in the reticular nucleus, thereby decreasing the inhibitory output onto relay neurons (HUGUENARD and PRINCE 1994). One could thus speculate that barbiturates and steroids are also able to mediate this effect, but in addition potentiate the activities of other $GABA_A$ receptor subtypes which are insensitive to clonazepam (e.g. $\alpha4\beta\gamma2$ and $\alpha4\beta\delta$, both of which are thought to exist in the thalamus) (SUR et al. 1999) leading to enhanced inhibition within the thalamic circuit. There is evidence for functionally diverse $GABA_A$ receptors in the thalamus, with receptors in the reticular nucleus having slower decay times than those in the ventrobasal nuclei (ZHANG et al. 1997). Correlating these functional properties with the functional properties of individual receptor subtypes is a key step towards identification of the appropriate receptor subtype to target, in this case, for absence epilepsy. Since $\alpha3\beta2\gamma2$ receptors have a slower inactivation rate than, e.g. $\alpha1\beta2\gamma2$ receptors (GINGRICH et al. 1995), and $\alpha3$ is expressed in the reticular nucleus (HUNTSMAN et al. 1996; KULTAS-ILINSKY et al. 1998), and $\alpha3\beta\gamma2$ containing receptors are sensitive to clonazepam, one could speculate that this subtype is such a target. However further studies, including approaches utilising the combination of both electrophysiology and single cell PCR (see BROOKS-KAYAL et al. 1998) would be useful in further refining such a hypothesis.

An interesting recent insight has come from the use of knockout mice. While these are discussed in detail elsewhere (see Chap.) it is of interest to note the phenotype of the $GABA_A$ receptor $\beta 3$ knockout mouse in the context of absence epilepsy. In the rodent $\beta 3$ is present in the reticular nucleus and essentially absent in the relay nuclei (WISDEN et al. 1992). In $\beta 3$ knockout mice the $GABA_A$ mediated inhibition in the reticular formation is ablated, while it remains essentially normal in the relay neurons (HUNTSMAN et al. 1999). Furthermore, the oscillatory synchrony of activity in this nucleus is greatly intensified. Hypersynchrony is symptomatic of absence epilepsy. This leads one to consider $GABA_A$ $\beta 3$ containing receptors as a possible therapeutic target for absence epilepsy.

III. Alterations in GABA Levels and $GABA_A$ Receptors in Human Epilepsy

Recent studies in man have either been in vivo using (a) positron emission tomography (PET) scanning with isotopically labelled ligands or (b) proton magnetic resonance spectroscopy, or have used postmortem or surgical tissue for autoradiographic or immunocytochemical studies.

PET scanning studies have usually employed [^{11}C]-flumazenil to assess alterations in the expression of GABA-benzodiazepine receptors in patients with epilepsy. A series of studies (SAVIC et al. 1988, 1990, 1996) has provided evidence that in focal (partial) epilepsies there is a reduction in the binding of flumazenil that commonly extends beyond the focal pathology or EEG ictal focus. This loss of flumazenil binding is less extensive than the reduction in glucose metabolism detected by [^{18}F]-fluorodeoxyglucose, but it provides a similarly reliable indication of the lateralisation of the focus in temporal lobe epilepsy. (HENRY et al. 1993). In contrast some patients with focal cortical dysgenesis show focal enhancement of flumazenil binding (RICHARDSON et al. 1996). In generalised seizures there is evidence for a slight increase in cortical, thalamic and cerebellar flumazenil binding (PREVETT et al. 1995; KOEPP et al. 1997).

Protein spectroscopy of the occipital lobe shows that GABA levels are reduced in patients with poor seizure control (PETROFF et al. 1996a). GABA content is markedly enhanced by vigabatrin and modestly increased by gabapentin (PETROFF et al. 1996b,c; NOVOTNY et al. 1999).

F. GABAergic Agents in Status Epilepticus

There have been many suggestions that status epilepticus is related to a failure of GABAergic inhibition occurring as a consequence of seizure activity. Recent experimental studies include the demonstration by KAPUR and MACDONALD (1997) that there is a functional change in $GABA_A$ receptors in the rat hippocampus during the course of status epilepticus such that the ED_{50}

for suppression of seizure activity by diazepam changes 10-fold between 10 min and 45 min after seizure onset.

Any of the anticonvulsants that act by enhancing GABA-mediated inhibition is potentially a treatment for status epilepticus. Barbiturates have been widely used, and chlormethiazole can also be effective. Tiagabine and vigabatrin have been shown to be effective in experimental models (HALONEN et al. 1995, 1996). Benzodiazepines, principally diazepam and lorazepam, are however the most widely used GABA-related agents and in many centres are regarded as the first line of treatment. Their efficacy has recently been confirmed in a major controlled trial (TREIMAN et al. 1998).

G. Conclusions: Future Prospects for Anti-Epileptic Drugs Acting on GABAergic Transmission

Recent developments concerning the selective regional expression of $GABA_A$ subunits and their altered expression and function in some epilepsy syndromes have given rise to the view that improved therapy can be achieved by identifying drugs that are highly selective for the particular subunit combinations that participate in seizure generation. Using cell systems expressing specific subunit combinations it is possible to screen novel benzodiazepines, or compounds acting at the BZ or other sites to identify drugs that will be selective for specific epilepsy syndromes and also may show reduced sedative or myorelaxant side effects.

References

Abdul-Ghani AS, Norris PJ, Smith CCT, Bradford HF (1981) Effects of γ-acetylenic GABA and γ-vinyl GABA on synaptosomal release and uptake of GABA. Biochem Pharmacol 30: 1203–1209

Amano K, Hamada K, Yagi K, Seino M (1998) Antiepileptic effects of topiramate on amygdaloid kindling in rats. Epilepsy Res 31:123–128

Ashton D, Fransen J, Heeres J, Clinke GHC, Janssen PAJ (1992) In-vivo studies on the mechanism of action of the broad spectrum anticonvulsant loreclezole. Epilepsy Res 11:27–36

Banerjee PK, Tillakaratne NJ, Brailowsky S, Olsen RW, Tobin AJ, Snead OC III (1998) Alterations in $GABA_A$ receptor alpha 1 and alpha 4 subunit mRNA levels in thalamic relay nuclei following absence-like seizures in rats. Exp Neurol 154:213–23

Beekman M, Ungard JT, Gasior M, Carter RB, Dijkstra D, Goldberg SR, Witkin JM (1998) Reversal of behavioral effects of pentylenetetrazol by the neuroactive steroid ganaxolone. J Pharmacol Exp Ther Mar 284:868–77

Belelli D, Lambert JJ, Peters JA, Wafford KA, Whiting PJ (1997) The interaction of the general anaesthetic etomidate with the γ-aminobutyric acid type A receptor is influenced by a single amino acid. Proc Natl Acad Sci USA 94:11031–11036

Benke D, Honer M, Michel C, Mohler H (1996) $GABA_A$ receptor subtypes differentiated by their gamma-subunit variants: prevalence, pharmacology and subunit architecture. Neuropharmacology 35:1413–23

Borden LA, Murali Dhar TG, Smith KE, Weinshank RL, Branchek TA, Gluchowski C (1994) Tiagabine, SK&F 89976-A, CI-966, and NNC-711 are selective for the cloned GABA transporter GAT-1. Europ J Pharmacol 269:219–224
Blümcke I, Beck H, Nitsch R, et al. (1996) Preservation of calretinin-immunoreactive neurons in the hippocampus of epilepsy patients with Ammon's horn sclerosis. J Neuropathol Exp Neurol 55:329–341
Brooks-Kayal AR, Shumate MD, Jin H, Rikhter TY, Coulter DA (1998) Selective changes in single cell GABA(A) receptor subunit expression and function in temporal lobe epilepsy. Nat Med 4:1166–72
Buhl EG, Otis IS, Mody I (1996) Zinc-induced collapse of augmented inhibition by GABA in a temporal lobe epilepsy model. Science 271:369–373
Carter RB, Wood PL, Wieland S, Hawkinson JE, Belelli D, Lambert JJ, White HS, Wolf HH, Mirsadeghi S, Tahir SH, Bolger MB, Lan NC, Gee KW (1997) Characterization of the anticonvulsant properties of ganaxolone (CCD 1042; 3-alpha-hydroxy-3-beta-methyl-5-alpha-pregnan-20-one), a selective, high-affinity, steroid modulator of the gamma-aminobutyric acid(A) receptor. J Pharmacol Exp Ther 280:1284–95
Chadwick D (1999) Safety and efficacy of vigabatrin and carbamazepine in newly diagnosed epilepsy: a multicentre randomised double-blind study. Lancet 354:13–19
Coenen AML, Blezer EHM, Van Luijtelaar ELJM (1995) Effects of the GABA-uptake inhibitor tiagabine on electroencephalogram, spike-wave discharges and behaviour of rats. Epilepsy Res 21:89–94
Coulter DA (1999) Chronic epileptogenic cellular alterations in the limbic system after status epilepticus. Epilepsia 40 [Suppl 1]:S23–S33
Cross AJ, Stirling JM, Robinson TN, Bowen DM, Francis PT, Green AR (1989) The modulation by chlormethiazole of the GABAA-receptor complex in rat brain. Br J Pharmacol 98:284–290
Dalby NO, Thomsen C, Fink-Jensen A, Lundbeck J, Sokilde B, Man C-M, Sorensen PO, Meldrum B. (1997) Anticonvulsant properties of two GABA uptake inhibitors NNC 05–2045 and NNC-05–2090, not acting preferentially on GAT1. Epilepsy Res 28:63–72
DeBiasi S, Vitellaro-Zuccarello L, Brecha NC (1998) Immunoreactivity for the GABA-transporter-1 and GABA-transporter-3 is restricted to astrocytes in the rat thalamus. A light and electron-microscopic immunolocalization. Neurosci 83:815–828
DeLorey TM, Handforth A, Anagnostaras S, Homanics GE, Minassian BA, Asatourian A, Fanselow F, Delgado-Escueta A, Ellison G, Olsen RW (1998) Mice lacking the $\beta 3$ subunit of the $GABA_A$ receptor have the epilepsy phenotype and many of the behavioural characteristics of Angelman syndrome. J Neurosci 18:8505–8514
During MJ, Ryder KM, Spencer DD (1995) Hippocampal GABA transporter function in temporal-lobe epilepsy. Nature 376:174–177
Eckstein-Ludwig U, Fei J, Schwarz W (1999) Inhibition of uptake, steady state currents, and transient charge movements generated by the neuronal GABA transporter by various anticonvulsant drugs. Brit J Pharmacol. 128:92–102
Evans RH (1979) Potentiation of the effects of GABA by pentobarbitone. Brain Res 171:113–120
Gingrich KJ, Roberts WA, Kass RS (1995) Dependence of the GABAA receptor gating kinetics on the alpha-subunit isoform: implications for structure-function relations and synaptic transmission. J Physiol (Lond) 489:529–543
Green AR (1998) Clomethiazole (Zendra) in acute ischemic stroke: basic pharmacology and biochemistry and clinical efficacy. Pharmacol Ther 80:123–147
Green AR, Misra A, Hewitt KE, Snape MF, Cross AJ (1998). An investigation of the possible interaction of clomethiazole with glutamate and ion channel sites as an explanation of its neuroprotective activity. Pharmacol Toxicol 83:90–94
Hadingham KL, Garrett EM, Wafford KA, Bain C, Heavens RP, Sirinathsinghji DJ, Whiting PJ (1996) Cloning of cDNAs encoding the human γ-aminobutyric acid

type A receptor $\alpha 6$ subunit and characterisation of the pharmacology of $\alpha 6$ containing receptors. Mol Pharmacol 49:253–259
Hales TG, Lambert JJ (1992) Modulation of $GABA_A$ and glycine receptors by chlormethiazole. Eur J Pharmacol 210:239–246
Halonen T, Miettinen R, Toppinen A, Tuunanen J, Kotti T, Riekkinen PJ (1995) Vigabatrin protects against kainic acid induced neuronal damage in the rat hippocampus. Neurosci Lett 195:13–16
Halonen T, Nissinen J, Jansen JA, Pitkanen A. (1996) Tiagabine prevents seizures, neuronal damage and memory impairment in experimental status epilepticus. Eur J Pharmacol 299:69–81
Harvey PK, Higenbottam TW, Loh L (1975) Chlormethiazole in treatment of status epilepticus. Br Med J 2:603–605
Henry TR, Frey KA, Sackellares JC, Gilman S, Koeppe RA, Brunberg JA, Ross DA, Berent S, Young AB, Kuhl DE (1993) In vivo cerebral metabolism and central benzodiazepine-receptor binding in temporal lobe epilepsy. Neurology 43:1998–2006
Hosford DA, Wang Y (1997) Utility of the lethargic (lh/lh) mouse model of absence seizures in predicting the effects of lamotrigine, vigabatrin, tiagabine, gabapentin and topiramate against human absence seizures. Epilepsia 38:408–414
Houser CR, Esclapez M, Fritschy JM, Möhler H (1995) Decreased expression of the $\alpha 5$ subunit of the $GABA_A$ receptor in a model of temporal lobe epilepsy. Soc Neurosci Abstr 21:1475
Huguenard JR, Prince DA Kapur J, Lothman EW, DeLorenzo RJ (1994) Intrathalamic rhythmicity studied in vitro: nominal T-current modulation causes robust antioscillatory effects. J Neurosci 14:5485–5502
Huguenard JR, Prince DA (1994) Clonazepam suppresses $GABA_B$-mediated inhibition in thalamic relay neurons through effects in nucleus reticularis. J Neurophysiol 71:2576–2581
Huntsman MM, Leggio MG, Jones EG (1996) Nucleus-specific expression of GABA(A) receptor subunit mRNAs in monkey thalamus. J Neurosci 16: 3571–3589
Huntsman MM, Porcello DM, Homanics GE, DeLorey TM, Huguenard JR (1999) Reciprocal inhibitory connections and network synchrony in the mammalian thalamus. Science 283:541–543
Kälviäinen R, Halonen T, Pitkänen A, Riekkinen PJ (1993) Amino acid levels in the cerebrospinal fluid of newly diagnosed epileptic patients: effect of vigabatrin and carbamazepine monotherapies. J Neurochem 60:1244–1250
Kälviäinen R, Brodie M, Duncan J, Chadwick D, Edwards D, Lyby K (1998) A double blind, placebo-controlled trial of tiagabine given three times daily as add-on therapy for refractory partial seizures. Epilepsy Res 30:31–40
Kälviäinen R, Nousiäinen I, Mantyjarvi M, Riekkinen PPJ (1999) Initial vigabatrin monotherapy is associated with increased risk of visual field constriction; a comparative follow-up study with patients on initial carbamazepine monotherapy and healthy controls. Epilepsia 39 [Suppl 6]:72
Kapur J, Lothman EW, DeLorenzo RJ (1994) Loss of $GABA_A$ receptors during partial status epilepticus. Neurology 44:2407–2408
Kapur J, Coulter DA (1995) Experimental status epilepticus alters gamma-aminobutyric acid type A receptor function in CA1 pyramidal neurons. Ann Neurol 38:893–900
Kapur J, Macdonald RL (1997) Rapid-seizure-induced reduction of benzodiazepine and Zn^{2+} sensitivity of hippocampal dentate granule cell $GABA_A$ receptors. J Neurosci 17:7532–7540
Kash SF, Johnson RS, Tecott LH, Noebels JL, Mayfield RD, Hanahan D, Baekkeskov S (1997) Epilepsy in mice deficient in the 65-kDa isoform of glutamic acid decarboxylase. Proc Nat Acad Sci USA 94:14060–14065
Koepp MJ, Richardson MP, Brooks DJ, Cunningham VJ, Duncan JS (1997) Central benzodiazepine/gamma-aminobutyric acid A receptors in idiopathic generalized

epilepsy: an [^{11}C]-flumazenil positron emission tomography study. Epilepsia 38:1089–1097
Kokaia M, Pratt GD, Elmer E, Bengzon J, Fritschy JM, Kokaia Z, Lindvall O, Möhler H (1994) Biphasic differential changes of $GABA_A$ receptor subunit mRNA levels in dentate gyrus granule cells following recurrent kindling-induced seizures. Brain Res Mol Brain Res 23:323–332
Kultas-Ilinsky K, Leontiev V, Whiting PJ (1998) Expression of 10 GABA(A) receptor subunit messenger RNAs in the motor-related thalamic nuclei and basal ganglia of Macaca mulatta studied with in situ hybridization histochemistry. Neuroscience 85:179–204
Luddens H, Pritchett DB, Kohler M, Killisch I, Keinanen K, Monyer H, Sprengel R, Seeburg PH (1990) Cerebellar $GABA_A$ receptor selective for a behavioral alcohol antagonist. Nature 346:648–651
Macdonald RL, Meldrum BS (1995) Principles of antiepileptic drug action. In: Levy RH, Mattson RH, Meldrum BS (eds) Antiepileptic drugs, 4th edn. Raven Press, New York, pp 61–77
Marson, AG, Kadir ZA, Hutton JL, Chadwick DW (1997) The new antiepileptic drugs: a systematic review of their efficacy and tolerability. Epilepsia 38:859–880
Martin PJ, Millac PA (1994) Status epilepticus: management and outcome of 107 episodes. Seizure 3:107–113
Mathers DA, Barker JL (1980) (-)Pentobarbital opens ion channels of long duration in cultured mouse spinal neurons. Science 209:507–509
Mattson RH (1995) General principles: selection of antiepileptic drug therapy. In: Levy RH, Mattson RH, Meldrum BS (eds) Antiepileptic drugs, 4th edn. Raven Press, New York, pp 123–135
Meldrum BS, Horton R (1978) Blockade of epileptic responses in the photosensitive baboon, Papio papio, by two irreversible inhibitors of GABA-transaminase, γ-acetylenic GABA (4-amino-hex-5-ynoic acid and γ-vinyl GABA (4-amino-hex-5-enoic acid). Psychopharmacol 69:47–50
Metcalf BW (1979) Inhibitors of GABA metabolism. Biochem Pharmacol 28: 1712–1715
Mienville J-M (1998) Persistent depolarizing action of GABA in rat Cajal-Retzius cells. J Physiol 512:809–817
Miller P, Kovar I (1983) Chlormethiazole in the treatment of neonatal status epilepticus. Postgrad Med J 59:801–802
Minassian BA, DeLorey TM, Olsen RW, Philippart M, Zhang Q, Bronstein Y, Guerrini R, Van Ness P, Livet MO, Delgado-Escueta AV (1998) Angelman syndrome: correlations between epilepsy phenotypes and genotypes. Ann Neurol 43:485–493
Minelli A, Brecha NC, Karschin C, DeBiasi S, Conti F (1995) GAT-1, a high affinity GABA plasma membrane transporter, is localized to neurons and astroglia in the cerebral cortex. J Neurosci 15:7734–7746
Minelli A, DeBiasi S, Brecha NC, Zucharello LV, Conti F (1996) GAT-3, a high-affinity GABA plasma membrane transporter, is localized to astrocytic processes, and it is not confined to the vicinity of GABAergic synapses in the cerebral cortex. J Neurosci 16:6255–6264
Morimoto K, Sato H, Yamamoto Y, Watanabe T, Suwaki H (1997) Antiepileptic effects of tiagabine, a selective GABA uptake inhibitor, in the rat kindling model of temporal lobe epilepsy. Epilepsia 38:966–974
Nakamura J, Tamura S, Kanda T, Ishii A, Ishihara K, Serikawa T, Yamada J, Sasa M (1994) Inhibition by topiramate of seizures in spontaneously epileptic rats and DBA/2 mice. Eur J Pharmacol 254:83–89
Nelson N (1998) The family of Na^+/Cl^- neurotransmitter transporters. J Neurochem 71:1785–1803
Novotny EJ, Hyder F, Shevell M, Rothman DL (1999) GABA changes with vigabatrin in the developing human brain. Epilepsia 40:462–466

Nusser Z, Hajos N, Somogyi P, Mody I (1998) Increased number of synaptic GABA(A) receptors underlies potentiation at hippocampal inhibitory synapses. Nature 395:172–7
Penry JK, So E (1981) Refractoriness of absence seizures and phenobarbital. Neurology 31:158
Petroff OAC, Rothman DL, Behar KL, Lamoureux D, Mattson RH (1996c) The effect of gabapentin on brain gamma-aminobutyric acid in patients with epilepsy. Ann Neurol 39:95–99
Petroff OAC, Rothman DL, Behar KL, Mattson RH (1996b) Human brain GABA levels rise after initiation of vigabatrin therapy but fail to rise further with increasing dose. Neurology 46:1459–1463
Petroff OAC, Rothman DL, Behar KL, Mattson RH (1996a) Low brain GABA level is associated with poor seizure control. Ann Neurol 40:908–911
Poulter MO, Brown LA, Tynan S, Willick G, William R, McIntyre DC (1999) Differential expression of $\alpha1$, $\alpha2$, $\alpha3$ and $\alpha5$ GABAA receptor subunits in seizure-prone and seizure-resistant rat models of temporal lobe epilepsy. J Neurosci 19:4654–4661
Prevett MC, Lammertsma AA, Brooks DJ, Bartenstein PA, Patsalos PN, Fish DR, Duncan JS (1995) Benzodiazepine-$GABA_A$ receptors in idiopathic generalized epilepsy measured with [^{11}C]-flumazenil and positron emission tomography. Epilepsia 36:113–121
Pritchett DB, Sontheimer H, Shivers BD, Ymer S, Kettenmann H, Schofield PR, Seeburg PH (1989) Importance of a novel GABA A receptor subunit for benzodiazepine pharmacology. Nature 338:582-585
Quirk K, Gillard NP, Ragan CI, Whiting PJ, McKernan RM (1994) Model of subunit composition of GABA-A receptor subtypes expressed in rat cerebellum with respect to their α and γ/δ subunits. J Biol Chem 269:16020–16028
Qume M, Fowler LJ (1997) Effect of chronic treatment with the GABA transaminase inhibitors gamma-vinyl GABA and ethanolamine-*O*-sulphate on the in vitro GABA release from rat hippocampus. Br J Pharmacol 122:539–545
Rentmeester T, Janssen A, Hulsman J, Scholtes F, van der Kleij B, Overweg J, Meijer J, de Beukelaar F (1991) A double-blind, placebo-controlled evaluation of the efficacy and safety of loreclezole as add-on therapy in patients with uncontrolled partial seizures. Epilepsy Res 9:59–64
Rice A, Rafiq A, Shapiro SM, Jakoi ER, Coulter DA, DeLorenzo RJ (1996) Long-lasting reduction of inhibitory function and gamma-aminobutyric acid type A receptor subunit mRNA expression in a model of temporal lobe epilepsy. Proc Natl Acad Sci USA 93:9665–9669
Richardson MP, Koepp MJ, Brooks DJ, Fish DR, Duncan JS (1996) Benzodiazepine receptors in focal epilepsy with cortical dysgenesis: an 11-C-flumazenil PET study. Ann Neurol 40:188–198
Richens A, Chadwick DW, Duncan JS, Dam M, Gram L, Mikkelsen M, Morrow J, Mengel H, Shu V, McKelvy JF, Pierce MW (1995) Adjunctive treatment of partial seizures with tiagabine: a placebo-controlled trial. Epilepsy Res 21:37–42
Rivera C, Voipio J, Payne JA, Ruusuvuori E, Lahtinen H, Lamsa K, Pirvola U, Saarma M, Kaila K (1999) The K/Cl co-transporter KCC2 renders GABA hyperpolarizing during neuronal maturation. Nature 397:251–255
Roepstorff A, Lambert JDC (1992) Comparison of the effect of the GABA uptake blockers, tiagabine and nipecotic acid, on inhibitory synaptic efficacy in hippocampal CA1 neurones. Neurosci Lett 146:131–134
Rogers CJ, Twyman RE, Macdonald RL (1994) Benzodiazepine and beta-carboline regulation of single $GABA_A$ receptor channels of mouse spinal neurones in culture. J Physiol (Lond) 475:69–82
Savic I, Roland P, Pearson A, Paulic S, Sedwell G, Widen L (1988) In vivo demonstration of reduced benzodiazepine receptor binding in human epileptic foci. Lancet 2:864–866

Savic I, Widen L, Thorell JO, Blomqvist G, Ericson K, Roland P (1990) Cortical benzodiazepine receptor binding in patients with generalized and partial epilepsy. Epilepsia 31:724–730
Savic I, Svanborg E, Thorell JO (1996) Cortical benzodiazepine receptor changes are related to frequency of partial seizures: a positron emission tomography study epilepsia 37:236–244
Saxena NC, Macdonald RL (1996) Properties of putative cerebellar gamma aminobutyric acid A receptor isoforms. Mol Pharmacol 49:567–579
Schwarzer C, Tsunashima K, Wanzenbock C, Fuchs K, Sieghart W, Sperk G (1997) GABA(A) receptor subunits in the rat hippocampus. 2. Altered distribution in kainic acid-induced temporal lobe epilepsy. Neuroscience 80:1001–1017
Shank RP, Gardocki JF, Vaught JL, Davis CB, Schupsky JJ, Raffa RB, Dodgson SJ, Nortey SO, Maryanoff BE (1994) Topiramate: preclinical evaluation of structurally novel anticonvulsant. Epilepsia 35:450–460
Shank RP (1995) Preclinical profile of topiramate, a novel anticonvulsant. Adv AED Therapy 1:6
Slany A, Zezula J, Fuchs K, Sieghart W (1995) Allosteric modulation of [^{3}H]flunitrazepam binding to recombinant $GABA_A$ receptors. Eur J Pharmacol 291:99–105
Smith SE, Parvez NS, Chapman AG, Meldrum BS (1995) The γ-aminobutyric acid uptake inhibitor, tiagabine, is anticonvulsant in two animal models of reflex epilepsy. Eur J Pharmacol 273:259–265
Snead OC III (1998) Ganaxolone, a selective, high-affinity steroid modulator of the gamma-aminobutyric acid-A receptor, exacerbates seizures in animal models of absence. Ann Neurol 44:688–91
Sperk G, Schwarzer C, Tsunashima K, Fuchs K, Sieghart W (1997) $GABA_A$ receptor subunits in the hippocampus. 1. Immunocytochemical distribution of 13 subunits. Neuroscience 80:987–1000
Spert G, Scharzer C, Tsunashima K, Kandlhofer S (1998) Expression of $GABA_A$ receptor subunits in the hippocampus of the rat after kainic acid-induced lesions. Epilepsy Res 32:129–139
Sur C, Farrar S, Kerby J, Whiting PJ, Atack J, McKernan RM (1999) Preferential co-assembly of α4 and δ subunits of the GABA-A receptor in rat thalamus. Mol Pharmacol 56:110–115
Taverna S, Sancini G, Mantegazza M, Franceschetti S, Avanzini G (1999) Inhibition of transient and persistent Na^+ current fractions by the new anticonvulsant topiramate. J Pharmacol Exp Ther 288:960–968
Thompson SA, Whiting PJ, Wafford KA (1996) Alpha subunits influence the action of pentobarbital on recombinant GABA-A receptors. Br J Pharmacol 117:521–527
Thompson SM, Gähwiler BH (1992) Effects of the GABA uptake inhibitor tiagabine on inhibitory synaptic potentials in rat hippocampal slice cultures. J Neurophysiol 67:1698–1701
Titulaer MNG, Kamphuis W, Pool CW, Heerikhuize JJ van, Lopes da Silva FH (1994) Kindling induces time dependent and regional specific changes in the [^{3}H] muscimol binding in the rat hippocampus: a quantitative autoradiographic study. Neuroscience 59:817–826
Titulaer MN, Ghijsen WE, Kamphuis W, De Rijk TC, Lopes da Silva FH (1995) Opposite changes in $GABA_A$ receptor function in the CA1-3 area and fascia dentata of kindled rat hippocampus. J Neurochem 64:2615–2621
Treiman DM, Meyers PD, Walton NY, Collins JF, Colling C, Rowan J, Handforth A, Faught E, et al (1998) A comparison of four treatments for generalized convulsive status epilepticus. New Engl J Med 339:792-798
Tsunashima K, Schwarzer C, Kirchmair E, Sieghart W, Sperk G (1997) GABA(A) receptor subunits in the rat hippocampus. 3. Altered messenger RNA expression in kainic acid-induced epilepsy. Neuroscience 80:1019–1032
Vigevano F, Cilio MR (1997) Vigabatrin versus ACTH as first-line treatment for infantile spasms: a randomized, prospective study. Epilepsia 38:1270–1274

Wafford KA, Bain CJ, Quirk K, McKernan RM, Wingrove PB, Whiting PJ, Kemp J (1994) A novel allosteric site on the GABA-A receptor β subunit. Neuron 12:775–782

Wafford KA, Thompson SA, Sikela J, Wilcox AS, Whiting PJ (1996) Functional characterisation of human GABA-A receptors containing the α4 subunit. Mol Pharmacol 50:670–678

Walton NY, Treiman DM (1988) Response of status epilepticus induced by lithium and pilocarpine to treatment with diazepam. Exp Neurol 101:267–75

Walker MC, Sander JWAS (1996) Topiramate: a new epileptic drug for refractory epilepsy. Seizure 5:199–203

Waymire KG, Mahuren JD, Jaje JM, Guilarte TR, Coburn SP, MacGregor GR (1995) Mice lacking tissue non-specific alkaline phosphatase die from seizures due to defective metabolism of vitamin B-6. Nature Genet 11:45–51

White HS, Brown SD, Woodhead JH, Skeen GA, Wolf HH (1997) Topiramate enhances GABA-mediated chloride flux and GABA-evoked chloride currents in murine brain neurons and increases seizure threshold. Epilepsy Res 28:167–179

Whiting PJ, McKernan RM, Wafford KA (1995) Structure and function of vertebrate $GABA_A$ receptor subtypes. In: Bradley RJ, Harris RA (eds) Intl Rev Neurobiol, 38. Academic Press, pp 95–138

Whiting PJ, McAllister G, Vasilatis D, Bonnert T, Heavens RP, Smith DW, Hewson L, O'Donnell R, Rigby M, Sirinathsinghji DJS, Marshall G, Thompson SA, Wafford KA (1997) Neuronal restricted RNA splicing regulates the expression of a novel $GABA_A$ receptor subunit conferring atypical functional properties. J Neurosci 17:5027–5037

Wild JM, Martinez C, Reinshagen G, Harding GFA (1999) Characteristics of a unique visual field defect attributed to vigabatrin. Epilepsia 40:1784–1794

Williamson A, Telfeian AE, Spencer DD (1995) Prolonged GABA responses in dentate granule cells in slices isolated from patients with temporal lobe sclerosis. J Neurophysiol 74:378–387

Wingrove PB, Wafford KA, Bain C, Whiting PJ (1994) The modulatory action of loreclezole at the γ-aminobutyric acid type A receptor is determined by a single amino acid in the β2 and β3 subunit. Proc Natl Acad Sci USA 91:4569–4573

Wingrove PB, Thompson SA, Wafford KA, Whiting PJ (1997) Key amino acids in the gamma subunit of the gamma-aminobutyric acid A receptor that determine ligand binding and modulation at the benzodiazepine site. Molec Pharmacol 52:874-881

Wisden W, Laurie DJ, Monyer HM, Seeburg PH (1992) The distribution of 13 $GABA_A$ receptor subunit mRNAs in rat brain. I. Telencephalon, diencephalon, mesencephalon. J Neurosci 12:1040–1062

Wisden, W, Herb A, Weiland H, Keinanen K, Luddens H, Seeburg PH (1991) Cloning, pharmacological characteristics and expression pattern of rat $GABA_A$ receptor α4 subunit. FEBS Lett 289:227–230

Zhang SJ, Huguenard JR, Prince DA (1997) $GABA_A$ receptor-mediated Cl^- currents in rat thalamic reticular and relay neurons. J Neurophysiol 78:2280–2286

Zezula J, Slany A, Sieghart W (1996) Interaction of allosteric ligands with $GABA_A$ receptors containing one, two, or three different subunits. Eur J Pharmacol 301:207–214

Zhong Y, Simmonds MA (1997) Interactions between loreclezole, chlormethiazole and pentobarbitone at GABA(A) receptors: functional and binding studies. Br J Pharmacol 121:1392–1396

CHAPTER 7

Heterologous Regulation of $GABA_A$ Receptors: Protein Phosphorylation

T.G. SMART, P. THOMAS, N.J. BRANDON and S.J. MOSS

A. Introduction

Heterologous regulation of ligand-gated ion channels has the potential for acute and chronic modulation of ion channel activity. This has important consequences for the control of neuronal excitability particularly when this involves the type A γ-aminobutyric acid ($GABA_A$) receptor. GABA, a neurotransmitter widely known to initiate the majority of inhibitory synaptic neurotransmission in the central nervous system (CNS), activates these receptors. There are numerous ways of regulating $GABA_A$ receptors and under normal physiological conditions these receptors will inevitably be subjected to a variety of inter- and intracellular homeostatic mechanisms with the purpose of regulating not just receptor function, but also assembly and cell surface number and location. One such ubiquitous and diverse mechanism for regulating $GABA_A$ receptors involves protein phosphorylation (MOSS and SMART 1996; SMART 1997). This type of regulation involves the short- or long-term covalent modification of receptor/ ion channel structure by the transfer of a charged phosphate group(s) from adenosine triphosphate to specific serine, threonine or tyrosine residues. This structural modification can lead to alterations in receptor function at the level of ligand-activated ion channel gating and also regulate mechanisms affecting receptor turnover and assembly.

Phosphorylation is a process catalysed by numerous enzymes classified as protein kinases. These are further sub-classified into serine/threonine second messenger-dependent protein kinases, including, cAMP-dependent protein kinase (PKA), cGMP-dependent protein kinase (PKG) and the family of kinases denoted as protein kinase C (PKC) exhibiting various dependencies on Ca^{2+} and phospholipid for activation (SCOTT and SODERLING 1992; FRANCIS and CORBIN 1994; TANAKA and NISHIZUKA 1994). Another major class of protein kinases in addition to serine/threonine kinases, is formed by tyrosine kinases which can be sub-classified into receptor and non-receptor tyrosine kinase families (XU et al. 1997; VAN DER GEER et al. 1994); the latter includes the prototypic member, Src, which specifically phosphorylates tyrosine residues.

The duration for which a particular protein remains phosphorylated is under dynamic control and is a function of the activity of protein kinases and phosphoprotein phosphatases whose function is to cleave phosphate groups from proteins (NAIRN and SHENOLIKAR 1992; MUMBY and WALTER 1993). Interestingly, the expression levels of many of these kinases and phosphatases is highest in the central nervous system which would suggest an important role(s) in neuronal function (WALAAS and GREENGARD 1991; LEVITAN 1994). This chapter discusses recent developments concerning the phosphorylation and dephosphorylation of $GABA_A$ receptors by protein kinases and phosphatases respectively, and the consequences for receptor regulation.

B. Physiological Role of $GABA_A$ Receptors

Activation of $GABA_A$ receptors in neurones results in the rapid flux of predominantly Cl^- ions through an integral ion channel. At a typical inhibitory synapse, the rapid presynaptic release of GABA and consequent postsynaptic $GABA_A$ receptor activation leads to the graded production of inhibitory postsynaptic potentials (IPSPs) in native neurones. The level of released GABA (typically 500 μmol/l – 1 mmol/l) is predicted to saturate postsynaptic $GABA_A$ receptors. For the majority of neurones in the CNS, the spontaneous release of GABA produces an incessant low-grade bombardment of postsynaptic neurones resulting in almost continuous spontaneous or miniature IPSP activity. In embryonic or immature neurones, quite often GABA activates a depolarisation of the membrane frequently resulting in the generation of action potential firing. In contrast, in postnatal, adult neurones, GABA has a predominantly hyperpolarising action leading to a cessation of action potential firing. The hyperpolarisation *per se is* not necessary to inhibit action potential firing since the underlying membrane Cl^- conductance increase is sufficient to *shunt* all excitatory synaptic currents and currents underlying action potential activity even without any change in the membrane potential. It is the abrupt cessation of action potential firing following the stimulus-evoked release of GABA that led to the classification of this molecule as a fast inhibitory neurotransmitter in the CNS (KAILA 1994; MACDONALD and OLSEN 1994; SMART 1998, for review).

C. Molecular Structure of $GABA_A$ Receptors

I. $GABA_A$ Receptor Subunit Families

$GABA_A$ receptors are widely distributed throughout the CNS and are the main sites of action for a variety of clinically relevant therapeutic agents, including the benzodiazepines, barbiturates and selected general anaesthetics in addition to non-therapeutic ethanol, neurosteroids and a range of cations (SIEGHART 1995). Cloning studies have revealed that $GABA_A$ receptors are members of a ligand-gated ion channel superfamily that comprises the following members: nicotinic acetylcholine (nAChR), glycine and 5HT-3 serotonergic receptors. This ion channel family exhibits many conserved structural features including a large glycosylated N-terminal extracellular domain with presumed disulphide bridge(s), 4 transmembrane domains (TM1–4) and a major intracellular domain between TM3 and TM4 (BARNARD et al. 1987; UNWIN 1993). Native $GABA_A$ receptors, like all members of this family, are believed to be pentameric in structure and formed from individual subunits selected from the following discrete families of vertebrate and chick species and classified according to their amino acid homologies: α(1–6), β(1–4), γ(1–4), δ(1), ε(1) and π(1) (RABOW et al. 1995; DAVIES et al. 1997; HEDBLOM and KIRKNESS 1997). Whilst α, β and γ subunit families appear quite frequently in the CNS and possess multiple members, $GABA_A$ receptors containing the single δ or ε subunits are thought to represent less frequent receptor isoforms. The π subunit has, so far, only been located in peripheral tissues where its function and presumed subunit partners are unknown (HEDBLOM and KIRKNESS 1997).

There are also an additional three homologous subunits, classified as ρ1–3, which are expressed principally but not exclusively in the retina. These subunits differ from the preceding families since they form bicuculline-insensitive receptors and exhibit minimal desensitisation after GABA exposure. Despite their molecular similarity to the $GABA_A$ receptor subunits, the distinct pharmacological profile and their inability to be co-expressed with $GABA_A$ receptor subunits has led to the designation of a separate class, the $GABA_C$ receptors. (CUTTING et al. 1991; BORMANN and FEIGENSPAN 1995; SHINGAI et al. 1996).

II. Domain Structures and Alternative Splicing

Analysis of the different domains of individual $GABA_A$ receptor subunits indicates that the greatest areas of structural diversity are to be found within the large intracellular domains between TM3 and TM4 (SIEGHART 1995; MACDONALD and OLSEN 1994). This diversity is increased by the ability of mRNAs for the α6, β2, β4, γ2 and ρ1 subunits to be alternatively spliced yielding two discrete proteins usually denoted as 'short' and 'long' forms (WHITING et al. 1990; BATESON et al. 1991; KOFUJI et al. 1991; HARVEY et al. 1994; KORPI et al. 1994; MCKINLEY et al. 1995). For the majority of these subunits, the structural diversity generated by the splicing events occurs principally within the large intracellular domain between TM3 and TM4, the exceptions being α6 and ρ1 subunits where splicing affects the extracellular N-terminal domains.

For the γ2 subunit alternative splicing results in the insertion of 8 amino acids within the large intracellular loop between TM3 and TM4 (WHITING et al. 1990; KOFUJI et al. 1991). The inserted sequence contains a serine residue forming part of a consensus site for phosphorylation by a number of protein kinases, including PKC. Similarly, alternative splicing of the chicken or human β2 subunit, also within the TM3/TM4 loop, results in the insertion of 17 and 38 amino acids into the long forms of the β2 subunit, respectively (HARVEY et al. 1994; MCKINLEY et al. 1995). Both these insertions contain consensus sites for phosphorylation which in the case of the human insertion encodes a strong consensus for PKA phosphorylation (MCKINLEY et al. 1995).

III. Subunit Heterogeneity and Co-Assembly

Using *in situ* hybridisation and immunohistochemistry to structures within the central nervous system, considerable temporal and spatial $GABA_A$ receptor subunit heterogeneity has been revealed (LAURIE et al. 1992; WISDEN et al. 1992; FRITSCHY et al. 1992; POULTER et al. 1992). There are distinctive expression profiles for a number of receptor subunits with α, β and γ subunits featuring throughout most areas of the brain. Of interest is the discrete localisation of the $\alpha6$ subunit to cerebellar granule cells and the close association with the development of δ subunit expression in these cells contrasts with the widespread expression of $\beta2/3$ subunits. These various expression profiles all support the notion of $GABA_A$ receptor heterogeneity throughout the central nervous system.

Heterologous expression of $GABA_A$ receptor cDNAs has been used to explore the properties of recombinant $GABA_A$ receptor subunits, deduce which subunits can co-assemble and determine the minimum subunit requirement for functional GABA-gated Cl^- channels. Generally, with the exception of the $\beta1$ and $\beta3$ subunits, single subunit expression of $\alpha1$, $\beta2$ and $\gamma2L$ does not result in the formation of functional ion channels. Instead these proteins are retained intracellularly within the endoplasmic reticulum (CONNOLLY et al. 1996a). Co-expression of α and β subunits produces robust GABA-gated currents which are modulated by barbiturates, inhibited by bicuculline, picrotoxin and Zn^{2+}, but are not enhanced by benzodiazepines (LEVITAN et al. 1988; PRITCHETT et al. 1989; SIGEL et al. 1990; MACDONALD and OLSEN 1994). However, the combinations $\alpha1\gamma2L$ or $\beta2\gamma2L$ fail to result in cell surface functional ion channels following their retention in the endoplasmic reticulum (CONNOLLY et al. 1996a,b). The inclusion of the γ subunit into receptors containing α and β subunits to form $\alpha1\beta2\gamma2L$, confers a sensitivity to the benzodiazepines and relative insensitivity to inhibition by Zn^{2+} (PRITCHETT et al. 1989; DRAGUHN et al. 1990; SMART et al. 1991). $GABA_A$ receptors can also be expressed as α, β and δ or ε subunits, resulting in the loss of benzodiazepine sensitivity. Overall, recombinant studies suggest that the majority of native neuronal $GABA_A$ receptors will contain a selection of α, β, and $\gamma2$ subunits.

D. Consensus Sites for Protein Phosphorylation

Elucidating where protein phosphorylation occurs on numerous proteins has allowed a number of consensus sites to be identified representing the minimal sequence requirement for phosphorylation by particular protein kinases (KENNELLY and KREBS 1991; PEARSON and KEMP 1991). The consensus sites are

Table 1. Consensus sequences for selected serine/threonine and tyrosine protein kinases

Kinase	Consensus sequence
PKA	RRX **S/T**>>RXX **S/T**>RX **S/T**
PKC	R/K $X_{(1-3)}$ **S/T** $X_{(1-3)}$ R/K>>**S/T** $X_{(1-3)}$ R/K>R/K $X_{(1-3)}$ **S/T**
PKG	R/KR/KX **S/T**>>R/KXX **S/T**>R/KX **S/T**
CaM KII	RXX **S/T**
Casein kinase 1	pS $X_{(1-3)}$ **S/T**>D/E$X_{(1-3)}$ **S/T**
Casein kinase 2	**S/T** $X_{(1-3)}$ D/E/pS
vSRC	E/DEEI**Y**G/EEF
Insulin receptor	XEEE**Y**MMMM

Consensus sites are indicated for selected serine/threonine kinases based on evidence accrued from numerous studies on protein kinase substrates (obtained from Kennelly and Krebs 1991; Pearson and Kemp 1991). For the tyrosine kinases, preferred peptide substrates are shown based on observations derived from peptide studies only (taken from Songyang et al. 1995). The identity of the phosphoacceptor group (**S**, **T** or **Y**) is underlined and in bold. X is a recognition neutral site and can be any amino acid. pS represents phosphoserine.

characterised by short amino acid sequences, surrounding the site(s) of phosphorylation, containing the minimum combination of amino acids required for substrate recognition (Table 1). These might include charged residues or residues with large hydrophobic side chains. Consensus site classification is, however, relatively imprecise since most protein kinases display a broad substrate specificity allowing only broad consensus site boundaries to be classified. It is usually a truism that the presence of a consensus site within a protein does not guarantee that phosphorylation will occur; neither does it categorically identify the kinase responsible if phosphorylation does indeed occur. Thus, consensus sites at best serve only as a guide for likely phosphorylation and probable involvement of kinases. Definitive evidence requires experimentation. An additional layer of complexity is that the GABA$_A$ receptor tertiary structure has not yet been resolved. Therefore predictions of membrane topology are largely based on hydropathy profiles derived from primary amino acid sequences. Since protein kinases and protein phosphatases are almost exclusively intracellular molecules, the accurate prediction of which residues are likely kinase substrates depends on accurately defining the intracellular domains of receptor subunits and their tertiary structure.

E. Identifying Phosphorylation Sites Within GABA$_A$ Receptor Subunits

I. Phosphorylation of Neuronal GABA$_A$ Receptors

Neuronal GABA$_A$ receptors, purified using benzodiazepine affinity columns, can be phosphorylated by a number of different protein kinases. PKA and PKC both appear to phosphorylate subunits deduced to be "β-type" from their relative molecular masses (53–57 kDa) observed following SDS-PAGE (KIRKNESS et al. 1989; BROWNING et al. 1990; TEHRANI and BARNES 1994). Moreover, polyclonal antisera directed against the large intracellular domain of the $\beta 1$ subunit blocked phosphorylation by both PKA and PKC (BROWNING et al. 1993). A receptor-associated kinase, which is not stimulated by either phorbol esters or cyclic nucleotides, can phosphorylate an "α-type" subunit (again deduced from a molecular mass of 51 kDa) (SWEETNAM et al. 1988; BUREAU and LASCHET 1995). Purified $GABA_A$ receptors are also substrates for the non-receptor tyrosine kinase, vSrc, which phosphorylates both "β-" and "γ-type" subunits (VALENZUELA et al. 1995). However, these experiments are hampered by the heterogeneous nature of affinity-purified receptor preparations and the low abundance of $GABA_A$ receptors in the brain causing the precise identity of the subunits phosphorylated in these studies to remain unclear.

II. Consensus Phosphorylation Sites in the Large Intracellular Domains

Examination of the major intracellular domains of $GABA_A$ receptor subunits reveals a number of consensus sites for both serine/threonine and tyrosine protein kinases (Fig. 1). Not all the receptor subunits contain these sites, though the receptor β subunit family seems best endowed with consensus sites for PKA, PKG, PKC and tyrosine kinases. Furthermore, $\gamma 2$ subunits contain consensus sites for PKC and tyrosine kinases, and the γ2L subunit contains an

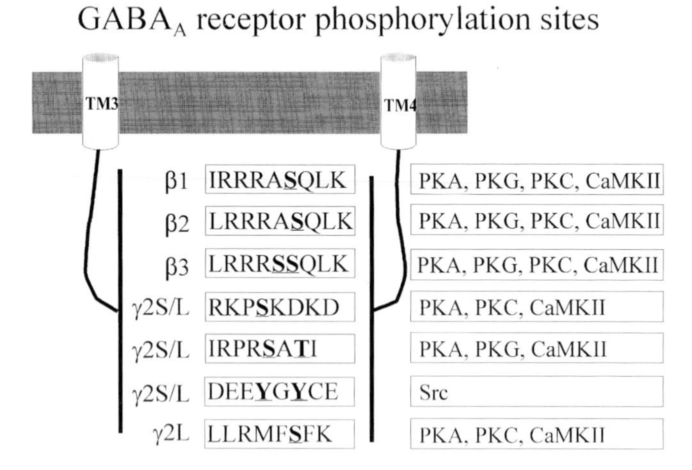

Fig. 1. Schematic diagram of the phosphorylation consensus sequences of the large intracellular domain of $GABA_A$ receptors. Sequences for only $\beta 1$–3 and $\gamma 2$ subunits are illustrated and enlarged between transmembrane domains (TM) 3 and 4. The phosphorylated residue(s) is shown *bold and underlined*. The *right panel* indicates the protein kinases capable of phosphorylating these residues

additional site for phosphorylation by PKC. For the α subunit family, only the $\alpha6$ subunit encodes a strong consensus site for phosphorylation by a number of kinases, including PKA. To date, the $\rho1$ subunit possesses a number of consensus phosphorylation sites particularly for PKC (CUTTING et al. 1991).

III. Phosphorylation of Recombinant $GABA_A$ Receptors

1. Use of Fusion Proteins

To obviate the problems associated with identifying phosphorylation sites within purified neuronal $GABA_A$ receptor subunits, the large intracellular domains of β and γ subunits have been expressed as soluble glutathione-*S*-transferase (GST) fusion proteins in *E. Coli*, allowing purification under native conditions (SMITH and JOHNSON 1988). By using site-directed mutagenesis, the murine $\beta1$ subunit intracellular domain was clearly demonstrated to be phosphorylated by PKA, PKC, PKG and CaMKII on Serine (S) 409 (MOSS et al. 1992a; MCDONALD and MOSS 1994). This conserved residue (S410 for the $\beta2$ subunit) is also phosphorylated by the same spectrum of these kinases in both the $\beta2$ and $\beta3$ subunits (MCDONALD et al. 1998). Additional serines, S383 in $\beta1$ and S384 in $\beta3$ subunits, can also be phosphorylated by CaMKII (MCDONALD and MOSS 1994; MCDONALD et al. 1998). This analytical approach also demonstrated that the $\beta1$ subunit fusion protein is a substrate for vSrc; however, the site(s) of phosphorylation were not identified (VALENZUELA et al. 1995).

Phosphorylation of both forms of the $\gamma2$ subunit has been analysed using similar methodologies (WHITING et al. 1990; KOFUJI et al. 1991). Within the 8 amino acid insert differentiating $\gamma2S$ from $\gamma2L$, is a high affinity substrate site (S343) for both PKC and CaM KII (WHITING et al. 1990; MOSS et al. 1992a; MACHU et al. 1993; MCDONALD and MOSS 1994). In comparison, both $\gamma2S$ and $\gamma2L$ are phosphorylated by PKC on S327 and by CaM KII on S348 and Threonine (T) 350 (MOSS et al. 1992a; MCDONALD and MOSS 1994). The $\gamma2L$ intracellular domain can also be phosphorylated by vSrc, but the phosphorylated residue(s) are unidentified (VALENZUELA et al. 1995). In contrast to β and γ subunit fusion proteins, there appears to be no significant phosphorylation of any α subunits by PKA, PKC, PKG, CaMKII or vSrc.

2. Use of Receptor Subunits

The studies with fusion proteins clearly indicated that β and γ subunits are major targets for protein kinases; however, these fusion proteins represent only a small fragment of the receptor protein subunit and thus phosphorylation of complete whole receptor subunits is necessary to validate the identification of substrate sites.

Typically, cDNAs encoding for various $GABA_A$ receptor subunits are used to transfect a secondary cell line, e.g. human embryonic kidney cells (HEK), which are then exposed to ^{32}P-orthophosphoric acid in the presence of kinase activators. Receptor subunits are then purified by selective antisera and subjected to phosphopeptide mapping and ultimately phosphoamino acid analysis. This procedure is performed on wild-type subunits and then essentially iterated on selected mutant subunits removing the postulated serine/threonine or tyrosine residues believed to be substrates for respective protein kinases. The phosphopeptide maps and phosphoamino acid analyses determine the precise location and number of the kinase substrate sites on individual $GABA_A$ receptor subunits.

In accordance with the previous work on fusion proteins, murine $GABA_A$ receptors composed of either $\alpha1\beta1$ or $\alpha1\beta1\gamma2S$ subunits expressed in HEK cells are phosphorylated by PKA on S409 of the $\beta1$ subunit (MOSS et al. 1992b). Using similar receptor constructs, the $\beta3$ subunit is phosphorylated on two adjacent residues S408 and S409, but surprisingly, the $\beta2$ subunit was not phosphorylated at the conserved position S410 by PKA. Protein kinase C, which has a similar substrate selectivity to PKA, also phosphorylated $\beta1$ on S409 and $\beta3$ subunits at S408 and S409 and, curiously, $\beta2$ subunits on S410 in $\alpha1\beta x$ and $\alpha1\beta x\gamma2$ subunits (where x = 1–3) (MCDONALD et al. 1998).

Apart from serine/threonine kinases, the $\beta1$ subunit can also be phosphorylated by vSrc on tyrosines (Y) 385 and Y387. The same kinase can also phosphorylate the $\gamma2L$ subunit, when co-expressed with $\alpha1\beta1$, on residues Y365 and Y367. $GABA_A$ receptors can also be tyrosine phosphorylated *in situ* in rat dorsal horn neurones as demonstrated by immunoprecipitating $\beta2/\beta3$ subunits and western blotting with phosphotyrosine antibodies (WAN et al. 1997a). Overall there is a close correlation between the phosphorylation of fusion proteins and their receptor subunit counterparts.

F. $GABA_A$ Receptor Phosphorylation: Consequences for Ion Channel Function

The demonstration that $GABA_A$ receptor subunits can be phosphorylated at defined residues does not indicate the likely physiological function of this process. Since these receptors incorporate integral ion channels, much attention has been devoted to assessing the effect of phosphorylation on native and recombinant ion channel function using electrophysiological methods of analysis.

I. cAMP-Dependent Protein Kinase

1. Native Neurones

PKA-induced phosphorylation of $GABA_A$ receptors has been reported to have a full spectrum of effects ranging from broad potentiation of receptor function to overall inhibition. For native neuronal preparations, PKA activation increased $GABA_A$ receptor desensitisation in cortical neurones (TEHRANI et al. 1989; but cf. TICKU and MEHTA 1990) and reduced GABA-activated currents in cultured neurones (HARRISON and LAMBERT 1989; PORTER et al. 1990;

Moss et al. 1992b; Robello et al. 1993). Furthermore, ^{36}Cl flux was reduced in synaptoneurosomes or microsacs after activating PKA with cAMP or by directly using the catalytic subunit of PKA (Heuschneider and Schwartz 1989; Schwartz et al. 1991; Leidenheimer et al. 1991).

In contrast to the general inhibitory effects of PKA, enhancements of $GABA_A$ receptor mediated currents have also been reported using a variety of G-protein coupled receptors to activate PKA with concomitant effects on $GABA_A$ receptor function. Using rat retinal neurones and cerebellar Purkinje cells, vasoactive intestinal peptide, VIP (Veruki and Yeh 1992, 1994; Wang et al. 1997) and noradrenaline (Waterhouse et al. 1982; Cheun and Yeh 1992; Parfitt et al. 1990; Llano and Gerschenfeld 1993) both enhanced GABA-activated responses and these effects may be mediated by PKA. Interestingly, in rabbit retina, VIP caused an inhibition of GABA-activated currents probably by a mechanism that is independent of PKA (Gillette and Dacheux 1995, 1996). A similar potentiation of GABA-activated currents to that produced by noradrenaline in Purkinje neurones can be achieved by using membrane-permeable 8-Br-cAMP. This potentiation was blocked by a specific PKA inhibitor peptide, PKIP (Kano and Konnerth 1992). More direct effects of PKA were reported using intracellular dialysis with the catalytic subunit of PKA. In rat retina, neurones exposed to internal PKA displayed enhanced GABA-activated responses, (Feigenspan and Bormann 1994a). Moreover, application of dopamine, histamine, adenosine, VIP, somatostatin and Leu or Met-enkephalins, all enhanced $GABA_A$ receptor function and were assumed to be activating adenylate cyclase (Feigenspan and Bormann 1994a). A direct potentiation of GABA-activated currents has also been observed in hippocampal dentate granule neurones (Kapur and Macdonald 1996).

Further evidence that PKA can differentially modulate native $GABA_A$ receptor function has now been obtained at rat hippocampal synapses. In pyramidal neurones, PKA activation reduced the amplitude of GABA-mediated inhibitory postsynaptic currents (IPSCs) whereas in granule cells in the dentate gyrus, PKA was ineffective (Poisbeau et al. 1999) (Fig. 2). These results may be explained by expression of native $GABA_A$ receptors with differing β subunit complement (see below).

2. Recombinant Receptors

The variable effects of PKA on native neuronal $GABA_A$ receptor function may result from heterogeneity amongst $GABA_A$ receptors differentially expressed in different cell types, from differences in the methods used to activate the kinases, or from using different animal species of receptor subunits. The precise elucidation of the effects of PKA on $GABA_A$ receptor regulation required the use of a simpler cell system allowing electrophysiological and biochemical measurements to be made in the same cell background expressing either defined or a limited number of receptor subunits.

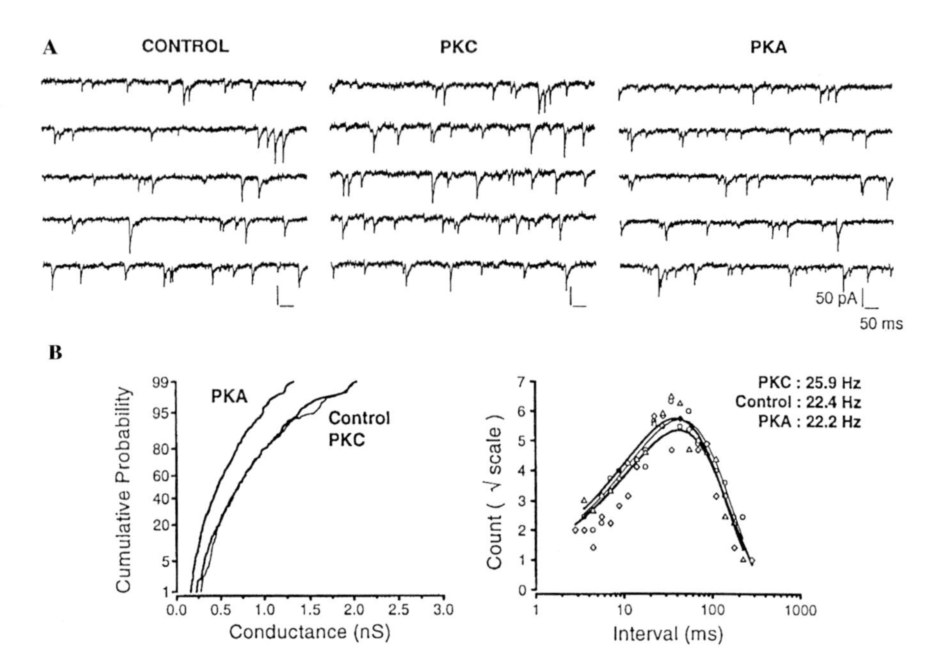

Fig. 2A,B. Modulation of GABA-mediated IPSCs in hippocampal neurones by PKA and PKC. **A** Whole-cell recording of mIPSCs with normal pipette electrolyte, or one containing 6 μg/ml PKC or 6 μg/ml PKA. Note the reduced mIPSC amplitudes in the PKA exposed neurone and little apparent effect after PKC treatment. **B** *Left panel*, cumulative probability distributions of mIPSCs peak conductances in control cells, and those internally dialysed with PKA and PKC. The reduced mIPSC amplitudes by PKA are manifest by a lateral, leftward shift in the distribution; *right panel*, inter-event intervals were log binned and plotted against the square root of their occurrence (count). The frequency of events is unaffected by PKC (*diamonds*) or PKA (*triangles*). The *lines* are exponential probability density functions indicating the random occurrence of the IPSCs. The mean frequencies for the 3 conditions are indicated. Taken from POISBEAU et al. (1999) with permission

Early studies in HEK cells revealed that $GABA_A$ receptors composed of either $\alpha1\beta1$ or $\alpha1\beta1\gamma2S$ subunits were functionally inhibited by activation of PKA and this inhibition was prevented by mutating S409 to alanine (A) (Moss et al. 1992b). The desensitisation rate for GABA-activated currents on $\alpha1\beta1$ heteromers was slowed by cAMP or by co-expressing the catalytic subunit of PKA, $C\alpha$. This effect was also prevented by the S409A mutation in the $\beta1$ subunit. An additional effect of PKA has been reported following transfection of $\alpha1\beta1\gamma2S$ cDNAs into cell lines with high, intermediate or low levels of catalytically active PKA. The largest GABA-activated membrane currents were recorded from cells with high PKA activity suggesting that chronic exposure to PKA was enhancing $GABA_A$ receptor function. This effect was not observed with $\alpha1\beta1$ receptors and, although it is difficult to compare GABA-activated currents between cells due to varying transfection efficiencies, the correlation between PKA activity and current amplitude was also not observed when expressing the $\beta1$(S409A) mutant with $\alpha1$ and $\gamma2S$ subunits (ANGELOTTI et al. 1993).

These early studies, however, offered no clear explanation as to why PKA regulation of $GABA_A$ receptor function in neurones should be so variable. To examine this aspect further the role of the other two β subunits in PKA regulation of receptor function was studied using patch clamp recording. A differential effect of PKA on $GABA_A$ receptor function was not expected following an exchange of the β subunits in the receptor complex given the similarity in the PKA consensus sequences for the β subunits. However, GABA-activated currents recorded from $\alpha1\beta2$ or $\alpha1\beta2\gamma2S$ $GABA_A$ receptors in HEK cells were insensitive to PKA activity. This result contradicted earlier work on $\beta2$ fusion proteins demonstrating that the intracellular loop could be phosphorylated (MCDONALD and MOSS 1997) but was in accordance with later work clearly indicating that the $\beta2$ subunit was not a substrate for PKA (MCDONALD et al. 1998). Whole-cell recording from HEK cells expressing $\alpha1\beta3$, $\alpha1\beta3\gamma2S$ or $\beta3$ homomers, revealed that activation of PKA following intracellular dialysis of cAMP resulted in a potentiation of ligand-gated currents (MCDONALD et al. 1998). This potentiation was abolished by mutating the only sites for PKA phosphorylation in the $\beta3$ subunit, S408 and S409 to alanines (Fig. 3).

Interestingly, these two serines are not functionally equivalent following PKA phosphorylation. Expressing $\beta3$(S408A) leaving only S409 available for PKA phosphorylation resulted in declining GABA-activated currents following dialysis with cAMP. This result concurs with previous data obtained with the $\beta1$ subunit which can only be phosphorylated on S409 resulting in inhibition of GABA-gated currents (MOSS et al. 1992b). The corresponding mutant, $\beta3$(S409A), leaving only S408 to be phosphorylated, was insensitive to modulation following PKA activation. Thus, although phosphorylation at S408 appears to be functionally silent, it is necessary to act in concert with S409 phosphorylation in the $\beta3$ subunit to observe a potentiation of GABA -activated currents (MCDONALD et al. 1998).

Thus the phosphorylation profile of the $\beta3$ subunit could be converted to that of the $\beta1$ subunit by simply mutating S408 to alanine. The interconversion of the post-phosphorylation functional behaviour of $\beta3$ subunit-containing receptors was further investigated by mutating alanine 408 in the $\beta1$ subunit to serine, reproducing the substrate sites normally found in the $\beta3$ subunit. Expressing $\alpha1\beta1$(A408S)$\gamma2S$ receptors in HEK cells resulted in basal phosphorylation of both S408 and S409. Intracellular dialysis of cAMP now caused a potentiation of GABA-activated currents rather than the inhibition associated with phosphorylation at S409 alone in the $\beta1$ subunit. Similar to the $\beta3$ subunit, if S409 was mutated to alanine leaving only A408S in the mutant $\beta1$ subunit, phosphorylation had no effect on GABA-activated currents. Thus, studies on both $\beta1$ and $\beta3$ subunits indicate that phosphorylation of S408 and

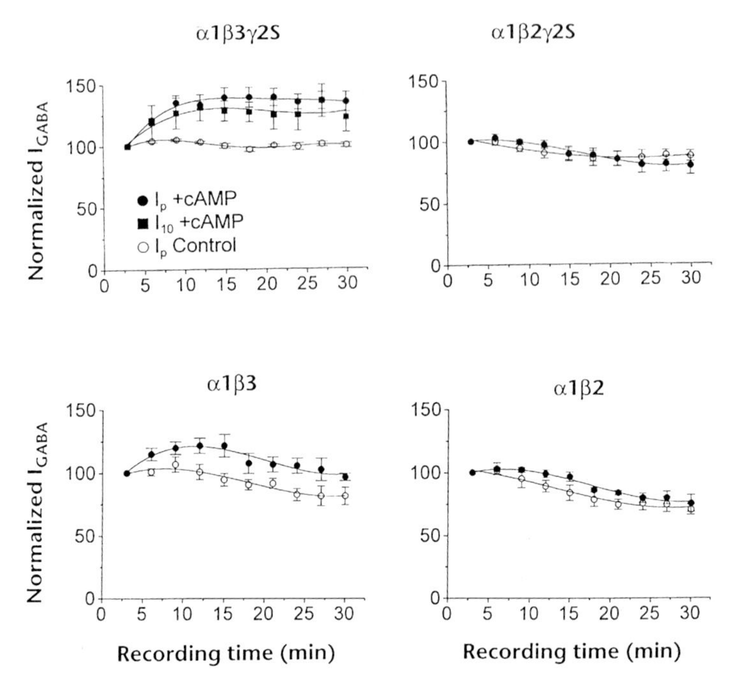

Fig. 3. Regulation of recombinant $GABA_A$ receptors by PKA. Membrane currents were activated by 10 μmol/l GABA applied rapidly to HEK cells expressing $\alpha1\beta3\gamma2S$, $\alpha1\beta2\gamma2S$, $\alpha1\beta3$ or $\alpha1\beta2$ $GABA_A$ receptor subunits at a holding potential of –40 mV. Currents were recorded at various times after formation of whole-cell recording mode defined as t = 0. Cells were either exposed to a control pipette solution (*open symbols*) or one containing 300 μmol/l cAMP (*closed symbols*) to activate PKA. GABA-activated currents were normalised to the response recorded at t = 3 in each cell (= 100%). Each point represents the mean +/– s.e.m. I_p and I_{10} represent the peak current and current after 10 s following GABA application. Note the enhanced responses in the $\beta3$ subunit containing cells and the lack of effect of cAMP in the $\beta2$ subunit expressing cells

S409 is required to potentiate receptor function, while receptor inhibition requires phosphorylation of S409 alone (McDONALD et al. 1998).

These results with the recombinant receptors now offer a plausible explanation for the wide variety of regulatory effects observed when PKA phosphorylates native $GABA_A$ receptors. The phosphorylation of distinct β subunit isoforms potentiates, inhibits or has no effect on $GABA_A$ receptor function, allowing greater fidelity in the control of synaptic inhibition. In the CNS, β subunits do display different spatial and temporal expression patterns (LAURIE et al. 1992). Furthermore, individual neurones could express different β subunit isoforms, either exclusively, or, if mixed populations are present, these isoforms could be targeted to specific synapses, particularly since in Madin Darby canine kidney (MDCK) cells β subunits are important for the subcellular localisation of $GABA_A$ receptors (CONNOLLY et al. 1996b). With this in mind, PKA-induced phosphorylation could differentially regulate the function of $GABA_A$ receptors at particular synapses, even within the same neurones, providing a sculpted inhibitory response rather than a blanket up- or down-regulation of neuronal excitability.

II. cGMP-Dependent Protein Kinase

cGMP-dependent protein kinase can phosphorylate $GABA_A$ receptors on β subunits using similar residues to those phosphorylated by PKA (McDONALD and MOSS 1997). There are relatively few studies that have directly determined the effect of PKG on $GABA_A$ receptor function. Nitric oxide (NO) can inhibit $GABA_A$ receptor function in the retina (WEXLER et al. 1998), cerebral cortex and cerebellar granule cells (ZARRI et al. 1994; ROBELLO et al. 1996), possibly by reducing single GABA channel open probability (ROBELLO et al. 1998). A major role for NO is the activation of guanylate cyclase causing accumulation of cGMP and consequent activation of PKG. Inhibitors of PKG prevented all or some of the actions of NO on these preparations suggesting a role for PKG-induced phosphorylation causing inhibition of $GABA_A$ receptor function.

Nitric oxide also inhibited GABA-activated currents on recombinant $\alpha1\beta2\gamma2S$ receptors but had no effect on $\alpha1\beta2$ constructs, unless activated by high GABA concentrations when potentiation was observed (FUKAMI et al. 1998). These authors concluded that NO acted directly on the $GABA_A$ receptor and was dependent upon the presence of the $\gamma2S$ subunit since the membrane permeant cGMP analogue, 8-Br-cGMP, was inactive. In contrast, cGMP increased GABA-activated currents on $\alpha1\beta2\gamma2L$ constructs expressed in oocytes (LEIDENHEIMER 1996). This effect was prevented by PKG inhibitor peptide but mutation of S410, a site that was phosphorylated by PKG in large intracellular loop fusion proteins, failed to prevent the action of PKG (LEIDENHEIMER 1996). Thus, although phosphorylation appeared to affect the GABA-activated current, it may not involve phosphorylation of the receptor $\beta2$ subunit *per se*. Moreover, as for PKA, it is unclear whether the $\beta2$ subunit is actually phosphorylated by PKG at S410.

III. Ca^{2+}/Phospholipid Dependent Protein Kinase

Early experiments using *Xenopus* oocytes injected with either rat or chick brain mRNA were used to assess whether expressed $GABA_A$ receptors were modulated by PKC (SIGEL and BAUR 1988; MORAN and DASCAL 1989). Activation of PKC using phorbol esters resulted in reduced GABA-activated whole-cell currents suggesting that phosphorylation was acting in an inhibitory manner. Subsequent studies using the heterologous expression of receptor cDNAs demonstrated that phorbol ester-induced PKC activity can inhibit the function of a range of receptors constructed from: $\alpha1,3,5$, $\beta1$–2 and $\gamma2$ subunits (SIGEL et al. 1991; LEIDENHEIMER et al. 1992, 1993). The specificity of phorbol ester action was also examined in $\alpha1\beta1\gamma2L$ subunit-containing receptors, where PKC inhibitory peptide (PKCI) blocked the effect of PKC (LEIDENHEIMER et al. 1992). The role of specific phosphorylation sites for PKC within the predicted large intracellular domains of individual subunits has been examined using site specific mutagenesis.

Presently, the inhibitory action of PKC has been studied using $GABA_A$ receptors composed of $\alpha1\beta x$ and $\alpha1\beta x\gamma2S/L$ (where x = 1 or 2). Functional studies of selected receptor subunit mutations revealed that multiple phosphorylation sites are involved, including S409 in the $\beta1$ subunit, S410 in the $\beta2$ subunit, S327 in both the $\gamma2S$ and $\gamma2L$ subunits, and S343 exclusively within the $\gamma2L$ subunit (KELLENBERGER et al. 1992; KRISHEK et al. 1994). Analyses of GABA concentration response curves demonstrated that PKC phosphorylation caused a non-competitive depression in these curves with usually greater inhibitions observed at high GABA concentrations particularly noticeable for receptors incorporating the $\gamma2L$ subunit. Systematic mutation of these serine residues revealed that phosphorylation at any of the sites on the $\beta1$ or $\gamma2$ subunits is sufficient to underwrite the negative modulation of receptor function, with phosphorylation at S343 within the 8 extra amino acids within the $\gamma2L$ subunit producing the largest inhibitory effect. These phosphorylation sites were therefore suggested to be functionally non-equivalent (KRISHEK et al. 1994). In contrast to the reports of down-regulation of receptor function, studies employing intracellular dialysis of trypsin-cleaved rat brain PKC, leading to constitutive activation, have observed potentiation of responses to GABA recorded from $\alpha1\beta1\gamma2L$ subunit $GABA_A$ receptors (LIN et al. 1994). This enhancement was blocked by the PKCI peptide, and also by mutating either S409 ($\beta1$ subunit), S327 ($\gamma2S$ or $\gamma2L$ subunits) or S343 ($\gamma2L$ subunit) to alanines (LIN et al. 1994, 1996). Whether the different results obtained with PKC regulation of $GABA_A$ receptors reflects the different expression systems used is unclear. What is more important is the method chosen to activate PKC. Most studies employ phorbol esters to activate endogenous PKC and rely on inactive congeners or mutant receptor subunits as controls. Intracellular dialysis with activated PKC will enable this kinase to access and phosphorylate many proteins that normally would be inaccessible through compartmentalisation and this may consequently affect receptor function. Nevertheless, the mutant subunits should also control for this unless PKC is having another, as yet unidentified, effect on receptor function (SMART 1997).

Apart from regulating receptor function, PKC-induced phosphorylation may also affect the ability of other modulators that bind to discrete sites on the receptor protein, to affect $GABA_A$ receptor function. Serine 343 in the $\gamma2L$ subunit has been suggested to affect potentiation of receptor function by ethanol (WAFFORD et al. 1991; WAFFORD and WHITING 1992; cf. SIGEL et al. 1993). However, potentiation by ethanol and other alcohols can also be achieved when PKC is inhibited (MARSZALEC et al. 1994). Moreover, in sensory ganglionic neurones, GABA-activated responses were unaffected by ethanol under conditions where S343 should be phosphorylated (ZHAI et al. 1998). Finally, the creation of a transgenic mouse containing only the γ2S subunit isoform thus lacking S343, did not affect the ethanol sensitivity of GABA-activated responses compared to wild-type mice (HOMANICS et al. 1999). This result suggested that phosphorylation at S343 is not pre-requisite for ethanol modulation of the $GABA_A$ receptor. In comparison, potentiation of GABA-gated responses on $\alpha1\beta2\gamma2L$ subunit-containing receptors by 3α, 21-dihydroxy-5α pregnan-20-one (THDOC) is enhanced by prior exposure of cells to phorbol esters, suggesting PKC phosphorylation can affect neurosteroid regulation of receptor function (LEIDENHEIMER and CHAPELL 1997). Furthermore, benzodiazepine and barbiturate-induced potentiation of GABA-activated responses was also enhanced following activation of PKC (LEIDENHEIMER et al. 1993).

An examination of the effects of PKC activation on native neuronal $GABA_A$ receptors has suggested a largely inhibitory role. Using cerebellar microsacs GABA-induced chloride flux was selectively inhibited by PKC activators (LEIDENHEIMER et al. 1992), but PKC does not appear to modulate receptor desensitisation in spinal cord microsacs (TICKU and MEHTA 1990). Utilising complete cells, GABA-activated responses in sympathetic neurones are inhibited by phorbol ester treatment but not by the inactive α-phorbols (KRISHEK et al. 1994). Similar results were obtained from rabbit retinal bipolar neurones with the PKC inhibitors staurosporine and calphostin C blocking the inhibition (GILLETTE and DACHEUX 1996).

Regulation of $GABA_A$ receptor function can also be achieved by activation of G-protein coupled receptor families that are known to activate PKC. For example activating neurokinin receptors in bullfrog primary sensory neurones inhibited GABA-activated currents in a manner dependent upon Pertussis toxin-insensitive G-proteins (YAMADA and AKASU 1996). PKC inhibitor peptide blocked this effect and the PKC activator, *sn*–1,2-dioctanoylglycerol (DOG; a diacylglycerol analogue) reproduced the inhibition of the GABA response. Regulation of $GABA_A$ receptor function by PKC may also be relevant at inhibitory synapses. WEINER et al. (1994) demonstrated that a PKC inhibitor peptide enhanced IPSPs in hippocampal brain slices and concurs with many recombinant receptor studies demonstrating a reduction in GABA-activated responses following phosphorylation by PKC. However, recently, in adult hippocampal slices, constitutively-active PKC had no effect on IPSCs in pyramidal neurones but potentiated IPSCs in granule neurones (POISBEAU et al. 1999).

The ρ subunits forming the $GABA_C$ receptors can also be regulated following PKC activation. GABA-activated responses recorded from neuronal $GABA_C$ receptors in rat retinal bipolar cells were inhibited by intracellular phorbol esters, an effect prevented by the PKC inhibitor tamoxifen or by alkaline phosphatase (FEIGENSPAN and BORMANN 1994b). Recombinant $\rho1$ subunits expressed in *Xenopus* oocytes were also modulated by PKC causing inhibition of GABA-gated currents (KUSAMA et al. 1995). Inspection of the intracellular loops of $\rho1$ and $\rho2$ subunits revealed six and one potential phosphorylation consensus sequences for PKC respectively (KUSAMA et al. 1998); however, replacing those residues thought to be phosphorylated by

PKC did not affect the inhibition of GABA-gated currents by PKC, suggesting that these sites, and possibly direct phosphorylation of the $GABA_C$ receptor, is not involved in the modulation by PKC. Interestingly, a fusion protein formed from the intracellular loop of the $\rho 1$ subunit is not a substrate for PKC, PKA, PKG and CaMKII. (S.J. Moss and J. Hanley, unpublished observations). Thus PKC regulation of $GABA_C$ receptors may proceed via phosphorylation of an intermediary protein possibly affecting cell surface expression and/or ion channel function.

IV. Ca^{2+}/Calmodulin-Dependent Protein Kinase II and Ca^{2+}-Dependent Phosphatases

Intracellular Ca^{2+} homeostasis appears to have a prominent impact in the regulation of GABA-activated currents (AKAIKE 1990). It is not clear whether the various actions of Ca^{2+}, including both potentiation and inhibition of GABA-gated currents, is dependent upon phosphorylation involving Ca^{2+} dependent kinases such as PKC or CaMKII. The activity of PKC is relatively well documented but studies on CaMKII and $GABA_A$ receptors are quite scarce. In rat dorsal horn neurones the catalytic subunit of CaMKII potentiated GABA-activated currents and IPSP amplitudes with a reduction in $GABA_A$ receptor desensitisation (WANG et al. 1995). Interestingly, calyculin-A, an inhibitor of protein phosphatases 1 and 2A, also potentiated the response to GABA (WANG et al. 1995) suggesting that the $GABA_A$ receptor was probably subject to basal phosphorylation and that this could regulate receptor function.

The involvement of Ca^{2+} in $GABA_A$ receptor function has also been observed in isolated hippocampal neurones. Exposure of these cells to glutamate or *N*-methyl-D-aspartate (NMDA) reduced the GABA-activated response and this effect was abolished by removing extracellular Ca^{2+} (STELZER and SHI 1995; CHEN and WONG 1995). This suggested that Ca^{2+} influx via the NMDA receptor was regulating $GABA_A$ receptor function. Subsequent studies suggested that Ca^{2+} was activating a Ca^{2+}/calmodulin-dependent phosphatase, calcineurin (phosphatase 2B), and that the down-regulation occurred via dephosphorylation of the $GABA_A$ receptor (STELZER and SHI 1995; CHEN and WONG 1995; ROBELLO et al. 1997) (Fig. 4). The protein kinase thought to be basally phosphorylating these $GABA_A$ receptors is currently unknown. Moreover, it is unclear whether these receptors are actually basally phosphorylated in the absence of any biochemical studies. A recent study has indicated that inhibition of calcineurin reduced desensitisation of GABA-gated responses in hippocampal neurones (MARTINA et al. 1996).

Modulation of $GABA_A$ receptors by CaMKII may also be physiologically relevant. In cerebellar Purkinje neurones, activation of the excitatory climbing fibre pathway produced a potentiation of postsynaptic GABA responses and IPSC amplitudes, a phenomenon known as rebound potentiation (KANO et al. 1992). The potentiation was dependent upon a postsynaptic increase in Ca^{2+} influx (KANO et al. 1996; HASHIMOTO et al. 1996) and could be blocked by

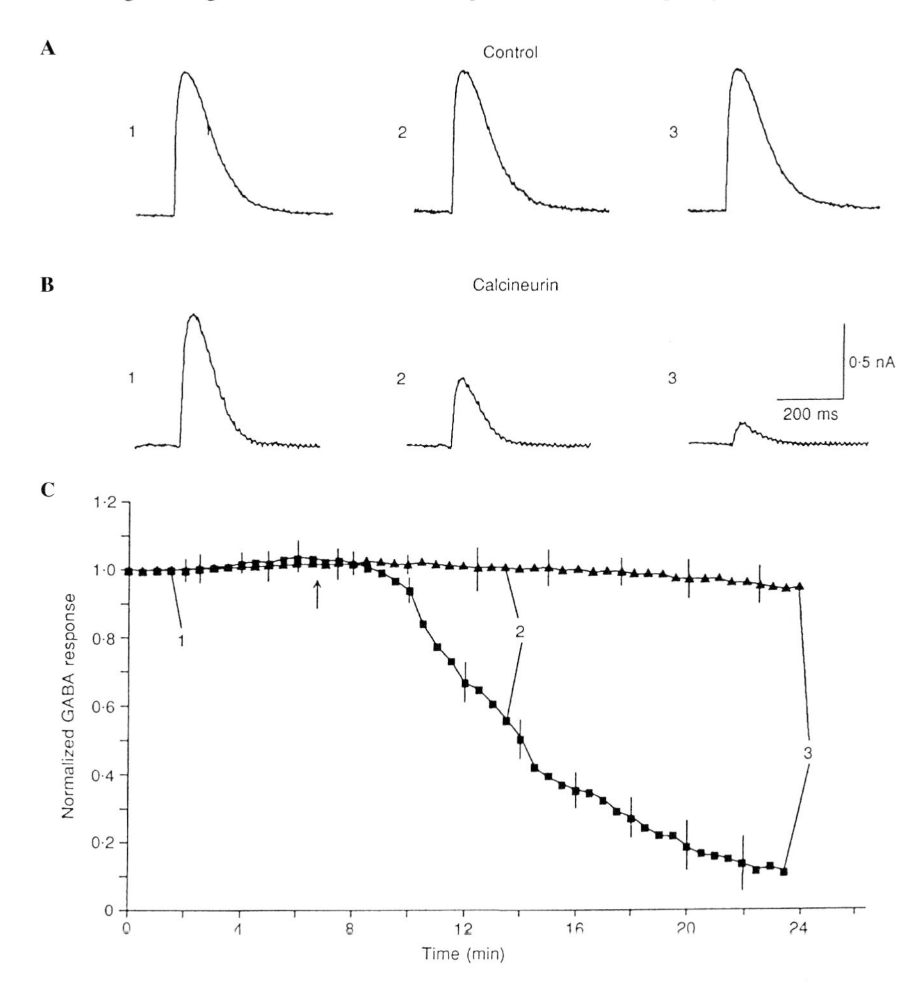

Fig. 4A–C. Suppression of GABA-activated responses by calcineurin: **A** 100 μmol/l GABA-activated currents in dissociated hippocampal neurones in control Krebs; **B** following intracellular application of 0.15 μmol/l calcineurin; **C** time course plot revealing the down-regulation of GABA-activated responses in 5–7 cells exposed to control Krebs (■) or following perfusion with calcineurin (▲). The *numbers* refer to time points when the current records in **A** and **B** were obtained. Data taken with permission from CHEN and WONG (1995)

inhibitors of CaMKII (e.g. KN62) (KANO et al. 1996). It is yet to be established that the cerebellar Purkinje cell $GABA_A$ receptors are subject to direct phosphorylation by CaMKII. Indeed, although CaMKII has been shown to phosphorylate fusion proteins of the large intracellular loops of $GABA_A$ receptor subunits (McDONALD and MOSS 1994) there are no studies detailing phosphorylation of $GABA_A$ receptors in neurones or in heterologous expression systems by this kinase.

V. Tyrosine Kinases

Receptor and non-receptor tyrosine kinases represent a large group of enzymes capable of phosphorylating a variety of proteins. The $GABA_A$ receptor was noted as a potential substrate for tyrosine kinases with consensus sequences identified on $\beta 1$ and $\gamma 2$ subunits (Moss et al. 1995). Biochemical studies revealed that $GABA_A$ receptors composed of $\alpha 1$, $\beta 1$ and $\gamma 2L$ subunits, coexpressed in HEK cells with the constitutively-active tyrosine kinase, vSrc, were phosphorylated on residues Y365 and Y367 in the $\gamma 2L$ subunit. Furthermore, tyrosine phosphorylation was also observed on residues Y370 and Y372 of the $\beta 1$ subunit and this could be increased by mutating Y365 and Y367 to phenylalanines in the $\gamma 2L$ subunit, suggesting that the preferential substrates for tyrosine kinases were located on the $\gamma 2L$ subunit (Moss et al. 1995). Functional studies, also in HEK cells, demonstrated that GABA-activated currents were potentiated by tyrosine phosphorylation on Y365 and Y367 of the $\gamma 2L$ subunit, and by expressing the mutant $\gamma 2L$ subunit incorporating Y365F and Y367F, this potentiating effect was prevented. Interestingly, the phosphorylated tyrosines in $\beta 1$ subunits appeared not to have any functional effect on GABA-gated currents (Moss et al. 1995). The results from recombinant receptors were also reproduced in native neuronal $GABA_A$ receptors of sympathetic ganglia. The tyrosine kinase, Src, potentiated GABA-gated currents indicating the involvement of tyrosine phosphorylation. GABA-activated currents recorded from these cells could also be inhibited by intracellular application of genistein, suggesting these receptors might be basally-phosphorylated. In accordance with the concept of basal phosphorylation the tyrosine phosphatase inhibitor, sodium vanadate, potentiated GABA-activated responses (Moss et al. 1995).

In broad agreement with the previous results on recombinant $GABA_A$ receptors, VALENZUELA et al. (1995) used tyrosine kinase inhibitors such as genistein, and observed a reduction in the amplitude of GABA-activated responses recorded from *Xenopus* oocytes expressing $\alpha 1\beta 1\gamma 2L$ and $\alpha 1\beta 1$ $GABA_A$ receptor constructs. These authors concluded that the prevention of phosphorylation of tyrosine residues in the $\beta 1$ and $\gamma 2L$ subunits was the cause of the inhibitory effects of the tyrosine kinase inhibitors.

Whether phosphorylation of the β subunit has any functional effect has received further attention from WAN et al. (1997a). Intracellular dialysis with $pp60^{c\text{-}Src}$ into cultured spinal dorsal horn neurones caused a progressive increase in GABA-gated currents, an effect prevented by pretreatment with the inhibitor genistein. Immunoprecipitation of $\beta 2/3$ receptor subunits after c-Src dialysis, followed by Western blotting of the neuronal homogenates with a phosphotyrosine antibody revealed that $\beta 1$ subunits were tyrosine phosphorylated. Pretreatment of the cells with genistein reduced the level of tyrosine phosphorylation. Recombinant $\alpha 1\beta 2$ receptors expressed in HEK cells were also sensitive to externally-applied genistein which inhibited the responses to GABA. Although no phosphorylation was detected for the $\gamma 2$

subunit this might have reflected the low levels of isolated $\gamma 2$ subunit from Western blotting. In contrast, both VALENZUELA et al. (1995) and Moss et al. (1995) observed phosphorylation on the $\gamma 2$ subunit after pre-incubation with $pp60^{c\text{-}Src}$ on purified bovine brain $GABA_A$ receptors (VALENZUELA et al. 1995) or recombinant $GABA_A$ receptors (Moss et al. 1995; VALENZUELA et al. 1995).

Additional functional assays utilising GABA mediated Cl^- flux from brain microsacs also observed an inhibition by the tyrosine kinase inhibitors, genistein and the typhostins, B-42 and B-44 (VALENZUELA et al. 1995). Single channel recording from rat sympathetic neurones indicated that tyrosine phosphorylation increased the mean open time and the probability of GABA ion channel opening (Moss et al. 1995).

1. $GABA_A$ Receptor: Response Rundown and Washout

Whole-cell recording of ligand-gated membrane currents opened a new vista on the properties of receptors and their associated ion channels; however, this mode of recording also revealed the propensity for many ligand-activated currents to undergo a reduction in amplitude with the duration of the recording. This phenomenon, often referred to as 'rundown' or 'washout', can be inconvenient but also indicated that soluble second messengers may be important for the maintenance of the response and possibly of the underlying membrane bound receptors. Phosphorylation of receptors by unidentified kinases or conditions conducive to phosphorylation have been implicated in preventing GABA response rundown in mammalian neurones (STELZER et al. 1988; GYENES et al. 1988, 1994; CHEN et al. 1990). The study by Moss et al. (1995) also concluded that tyrosine phosphorylation may be a means of potentiating or maintaining $GABA_A$ receptor function. Recently, the down-regulation of $200\,\mu mol/l$ GABA-gated currents recorded from HEK cells transfected with $\alpha 3\beta 2\gamma 2$ subunits was measured in cells exposed to low levels of ATP and relatively high levels of buffered Ca^{2+}. These conditions caused a reduction in the maximum currents induced by GABA and a smaller GABA EC_{50} (HUANG and DHILLON 1998). This down-regulation could also be induced by inhibiting tyrosine kinases with genistein or lavendustin-A. This phenomenon was completely prevented or attenuated by lowering resting Ca^{2+} levels and increasing intracellular ATP or by inhibiting tyrosine phosphatase with vanadate (HUANG and DHILLON 1998). Interestingly, inhibiting the activity of calcineurin also prevented rundown. Calcineurin would be activated by increased intracellular Ca^{2+} (Ca^{2+}/calmodulin-dependent) implying that serine/threonine phosphorylation is important in maintaining the GABA response. Although stimulation of PKA or PKC failed to affect the degree of response rundown, presumably the importance of ATP is due to this molecule being a substrate for protein tyrosine kinases.

This study suggested that phosphorylation by a tyrosine kinase clearly maintains the function of $GABA_A$ receptors and although the site of phosphorylation has not been resolved, the ability of calcineurin to induce response rundown indicates that another site is also involved in the maintenance of receptor activity. The importance of tyrosine phosphorylation and ATP for maintaining responses to GABA has also been noted in neurones forming the diagonal band of Broca in the forebrain (JASSAR et al. 1997).

A complicating factor in the regulation of $GABA_A$ receptors by ATP involves the possibility that this molecule may *directly* affect receptor function. Using rat nucleus tractus solitarii neurones, SHIRASAKI et al. (1992) observed that GABA-activated currents were reduced in the absence of intracellular ATP. However, the involvement of phosphorylation was questioned since intracellular application of alkaline phosphatase did not affect GABA responses, and inhibition of phosphatases using okadaic acid similarly was ineffective. The GABA EC_{50} concentration was increased by removing intracellular ATP with a competitive style lateral displacement of the GABA concentration response curve. The authors concluded that ATP may directly regulate the activity of the $GABA_A$ receptor (SHIRASAKI et al. 1992). This does not completely discount a role for phosphorylation in receptor regulation particularly since the effect, if any, of protein kinases was not studied.

Many of the agents used to modulate the activity of protein kinases often have secondary non-specific actions on ion channel function that can lead to confusion when interpreting data and can occasionally result in the false identification of the involvement of phosphorylation (LEIDENHEIMER et al. 1990; WHITE et al. 1992; LAMBERT and HARRISON 1990). Recent evidence suggests that tyrosine kinase inhibitors must now also be treated with caution (DUNNE et al. 1998). Extracellular application of genistein or the inactive control compound, daidzein, to $\alpha1\beta1\gamma2S$ $GABA_A$ receptors expressed in *Xenopus* oocytes resulted in a non-competitive depression of the GABA concentration response curve. Non-specificity was suspected when these compounds similarly inhibited mutant receptors devoid of tyrosine phosphorylation sites in the $\gamma2S$ subunit following their conversion to phenylalanines (Y365F, Y367F). Interestingly, using alternative tyrosine kinase inhibitors, such as typhostin A25, which avoid the genistein-susceptible ATP binding site by targeting the substrate binding site, also resulted in the inhibition of responses to GABA on wild-type $\alpha1\beta1\gamma2S$ and tyrosine mutant receptors (DUNNE et al. 1998). This study concluded that intracellular application (cf. Moss et al. 1995) is the most specific method for using these inhibitors, and that mutated receptors should be used as controls to ensure that the effects observed are solely due to phosphorylation.

G. Regulation of $GABA_A$ Receptor Cell Surface Expression

Phosphorylation may also be involved in regulating the cell surface expression of receptors in addition to affecting ion channel function directly. For example, insulin enables the translocation of $GABA_A$ receptors ($\alpha1\beta2\gamma2$) from intracellular compartments to the surface membrane of HEK cells (WAN et al. 1997b). This effect appeared to be dependent upon the $\beta2$ subunit and was blocked by genistein implicating the insulin receptor tyrosine kinase in this process. Interestingly, in hippocampal neurones, insulin also increased the levels of cell membrane $\beta2/\beta3$ subunits, assessed using antibodies to these subunits. This increase involved the up-regulation of functional $GABA_A$ receptors since postsynaptic sensitivity to GABA increased and the amplitude of mIPSCs were potentiated by 30% after insulin application (WAN et al. 1997b). It remains unclear whether direct tyrosine phosphorylation of the receptor is necessary for translocation or if intermediary proteins are required.

In addition to tyrosine kinases, expressing $GABA_A$ receptors in cells exhibiting chronic activation of PKA has been reported to enhance the assembly of $GABA_A$ receptors (ANGELOTTI et al. 1993). Three cell lines were selected with different levels of constitutive PKA activity denoted as: RAB 10, L929 and Cα12, which possess PKA activities of 5, 100 and 500 kinase units/mg protein respectively. $GABA_A$ receptors composed of $\alpha1\beta1\gamma2S$ subunits were transiently expressed in these cell lines and for Cα12 cells, the whole-cell currents activated by GABA were much larger compared to currents in L929 and RAB 10 cells. Similar experiments with receptors composed of $\alpha1\beta1$ subunits revealed no enhancement of GABA-activated currents suggesting this effect was specific for $\gamma2$ subunit-containing receptors. The potentiation was blocked by expressing the mutated $\beta1$ subunit, $\beta1$(S409A) which is devoid of the only PKA phosphorylation site in receptors composed of $\alpha1\beta1\gamma2S$ subunits (MOSS et al. 1992b). This result implicated the involvement of S409 in $\beta1$ subunits in the potentiation of GABA responses, but why it is not apparent in $\alpha1\beta1$ receptors is unknown. The GABA EC_{50} was unaffected by the cell line chosen for expression; however, accurate comparison of GABA-induced current amplitudes measured in different cell lines is complicated by variations in transfection efficiencies.

The prospect of regulating $GABA_A$ receptor subunit expression by intracellular cAMP levels was suggested following the observation that the adenylate cyclase activator, forskolin, increased the expression of the $GABA_A$ receptor $\alpha1$ subunit while reducing the level of $\alpha6$ subunit expression (THOMPSON et al. 1996). It is presently unclear whether PKA phosphorylation is involved, perhaps by directly phosphorylating receptor subunits (although $\alpha1$ and $\alpha6$ subunits are not obvious substrates for PKA), or whether this process affects surface expression by regulating factors controlling receptor subunit DNA transcription.

The ability of protein kinases to affect cell-surface receptor expression has received support from three studies investigating the effects of PKC on $GABA_A$ and $GABA_C$ receptors. Direct effects of PKC phosphorylation on $GABA_A$ receptor ion channel function have been studied in detail (MOSS and SMART 1996; SMART 1997); however, the mechanisms by which PKC activation reduces GABA-induced currents may be more complex. Recording from

HEK cells expressing $\alpha 1\beta 2\gamma 2L$ subunits at 31 °C, activation of PKC using phorbol esters reduced the GABA response and this effect was not prevented by mutating all the available sites for phosphorylation in the $GABA_A$ receptor $\beta 2$ and $\gamma 2L$ subunits ($\beta 2$(S410A); $\gamma 2L$(S327A, S343A) (CONNOLLY et al. 1999). Confocal microscopic analysis coupled with epitope tagging of the receptor subunits revealed that PKC at 31 °C was enabling the effective internalisation of the receptor and thereby causing a reduction in GABA-activated current. Further confocal analysis indicated that both $\alpha 1\beta 2$ and $\alpha 1\beta 2\gamma 2$ receptors could endocytose constitutively and this appeared to be unaffected by PKC. Thus the intracellular accumulation of $\alpha 1\beta 2\gamma 2$ receptors after activating PKC suggests that this kinase is hindering the recycling of the receptor back to the cell membrane. The ability of PKC to promote internal accumulation suggests that either PKC is phosphorylating the receptor subunits at one or more sites distinct from those previously reported (KRISHEK et al. 1994), although this seems unlikely, or PKC is possibly phosphorylating intermediary or accessory proteins that regulate the cell surface stability of these receptors. These proteins have not yet been identified. The earlier studies of PKC modulation of $GABA_A$ receptor function at temperatures less than 20 °C, particularly in *Xenopus* oocytes (SIGEL and BAUR 1988; KELLENBERGER et al. 1992; KRISHEK et al. 1994) would not have resolved receptor internalisation since this process would be expected to be largely inoperative at such low temperatures. However, recent evidence by CHAPELL et al. (1998) indicates that $GABA_A$ receptor ($\alpha 1\beta 2\gamma 2L$ or $\alpha 1\beta 2$) internalisation can indeed occur in oocytes at ambient temperatures. Phorbol ester induced reduction in the GABA-activated responses was not prevented by mutation of the known PKC phosphorylation site in the $\beta 2$ subunit (S410A). Moreover, by using green fluorescent protein fusions to the C-terminal domain of the $\alpha 1$ subunit, a clear reduction in fluorescence was observed in accordance with PKC-induced receptor internalisation. Thus it appears that PKC-induced reductions in GABA responses may be mediated by direct phosphorylation of the $GABA_A$ receptor protein and also by down-regulation possibly involving intermediary proteins.

PKC activation may also cause the internalisation of $GABA_C$ receptors expressed in HEK or COS-7 cells. Even including ATP in the patch pipette electrolyte did not prevent the down-regulation of GABA-activated currents which was alleviated by KN-62, an inhibitor of CaMKII, or by staurosporine which will also inhibit PKC (FILIPPOVA et al. 1999). Curiously, recordings were quite stable in the absence of internal ATP. Intracellular dialysis with the catalytic subunit of PKC reduced GABA responses and these responses could be transiently enhanced by alkaline phosphatase. Interestingly, mutation of three consensus sites for phosphorylation in the $\rho 1$ subunit did not affect the time-dependent decrease in GABA-activated current which may involve the actin cytoskeleton. The reduction in current amplitude was markedly accentuated by raising the temperature to 32 °C, indicative of an internalisation process, whereas, as expected at the lower temperature of 16 °C, no down-regulation was observed. Membrane capacitance was also reduced concomitant with the reduction in GABA-activated responses; however, the expression of Kv1.4 potassium channels showed no down-regulation suggesting that this process was not simply due to a non-specific loss of cell membrane (FILIPPOVA et al. 1999).

In conclusion, receptor internalisation, possibly not involving a direct phosphorylation of the receptor protein, may be an additional mechanism to regulate the function of GABA receptors simply by controlling the number expressed on the cell surface. This would be expected to have clear implications at active inhibitory synapses.

H. Conclusion

It has become apparent from many studies that phosphorylation of $GABA_A$ and $GABA_C$ receptors can have an important role to play in their regulation. Considerable attention has been targeted on the direct control of channel function by phosphorylation but it is now becoming clear that phosphorylation can effect numerous other important aspects of receptor regulation including: assembly, synaptic targeting, anchoring, and also receptor turnover. Some inroads into the elucidation of potential anchoring molecules have been made recently using the yeast two-hybrid system for resolving interacting molecules. For $GABA_A$ receptors, a novel protein termed GABARAP has been identified and putatively designated as a molecule that may allow the $GABA_A$ receptor to associate with, or anchor to, the cell cytoskeleton. GABARAP appears to interact with the large intracellular domain of the $\gamma 2$ subunit (WANG et al. 1999). In addition, the glycine receptor anchoring molecule, gephryin, is important for the clustering of $GABA_A$ receptors. Examining cortical neurones obtained from animals lacking the $\gamma 2$ subunit also revealed a parallel loss of gephyrin. In addition, inhibiting gephryin expression using antisense oligonucleotides also resulted in a loss of $GABA_A$ receptor clusters involving $\alpha 2$ and $\gamma 2$ subunits (ESSRICH et al. 1998). However, it is unclear whether the $\gamma 2$ subunit can directly interact with gephryin or whether an intermediary protein is required. Moreover, the involvement of phosphorylation, if any, in this process has not been addressed. The production of transgenic mice devoid of selected well-characterised phosphorylation sites will provide insight into the potential importance of these sites for receptor anchoring molecules. Another anchoring molecule, the microtubule-associated protein MAP-1B, interacts with the $\rho 1$ subunit of the $GABA_C$ receptor in preference to homomeric $\beta 3$ subunit $GABA_A$ receptors (HANLEY et al. 1999). In addition, MAP-1B and $\rho 1$ appeared to colocalise on postsynaptic sites of bipolar cell axons in the retina suggesting a physiological role for this interaction (HANLEY et al. 1999).

A second area of interest involving phosphorylation concerns those molecules necessary for kinases and phosphatases to anchor onto, or near, receptor subunits to enable their engagement and subsequent phosphorylation/dephosphorylation of the receptor protein. These anchoring molecules could compartmentalise the subcellular distribution of kinases and phosphatases. The regulatory and functional role that kinase and phosphatase anchoring molecules could have on the phosphorylation of $GABA_A$ receptor remains undetermined until they have been unequivocally identified. However, it is presumed that $GABA_A$ receptors will contain receptors for activated C kinase (RACK) (MOCHLY-ROSEN et al. 1995; MOCHLY-ROSEN and GORDON 1998) and also be receptive to A-kinase binding proteins (AKAPs) (DELL'ACQUA and SCOTT 1997), simply due to previous demonstrations that PKA and PKC can directly phosphorylate and regulate the function of $GABA_A$ receptors. These molecules may also be relevant to the regulation of $GABA_A$ receptors by parallel activation of G-protein coupled receptor families linked to numerous second messenger transduction pathways. The identification of these anchoring molecules will clearly enable several critical pieces of the intracellular jigsaw to be put in place regarding the regulation of this important receptor class that underlies inhibitory synaptic transmission in the CNS.

Acknowledgements. We gratefully acknowledge support by the Medical Research Council and The Wellcome Trust.

References

Akaike N (1990) GABA-gated Cl^- currents and their regulation by intracellular free Ca^{2+}. In: Chloride channels and carriers in nerve muscle and glial cells. Ed. FJ Alvarez-Leefmans, JM Russell. Plenum Press, New York, pp. 261–273

Angelotti TP, Uhler MD, Macdonald RL (1993) Enhancement of recombinant γ-aminobutyric acid type A receptor currents by chronic activation of cAMP-dependent protein kinase. Mol Pharmacol 44:1202–1210

Barnard EA, Darlison MG, Seeburg PH (1987) Molecular biology of the $GABA_A$ receptor: the receptor channel superfamily. Trends in Neurosci 10:502–509

Bateson AN, Lasham A, Darlison MG (1991) γ-Aminobutyric $acid_A$ receptor heterogeneity is increased by alternative splicing of a novel β subunit gene transcript. J Neurochem 56:1437–1440

Bormann J, Feigenspan A (1995) $GABA_C$ receptors. Trends in Neurosc. 18:515–519

Browning MD, Bureau M, Dudek EM, Olsen RW (1990) Protein kinase C and cAMP-dependent protein kinase phosphorylate the beta subunit of the purified γ-aminobutyric $acid_A$ receptor. Proc Natl Acad Sci USA 87:1315–1318

Browning MD, Endo S, Smith GB, Dudek E.M, Olsen RW (1993) Phosphorylation of the $GABA_A$ receptor by cAMP-dependent protein kinase and by protein kinase C: analysis of the substrate domain. Neurochem Res 1:95–100

Bureau MH, Laschet JJ (1995) Endogenous phosphorylation of distinct γ-aminobutyric acid type A receptor polypeptides by Ser/Thr and Tyr kinase activities associated with the purified receptor. J Biol Chem 270:26482–26487

Chapell R, Bueno OF, Alvarez-Hernandez X, Robinson LC, Leidenheimer NJ (1998) Activation of protein kinase C induces γ-aminobutyric acid type A receptor internalisation in *Xenopus* oocytes. J Biol Chem 273:32595–32601

Chen QX, Stelzer A, Kay AR, Wong R (1990) $GABA_A$ receptor function is regulated by phosphorylation in acutely dissociated guinea-pig hippocampal neurones. J Physiol 420:207–221

Chen QX, Wong RKS (1995) Suppression of $GABA_A$ receptor responses by NMDA application in hippocampal neurones acutely isolated from the adult guinea pig. J Physiol 482.2:353–362
Cheun JE, Yeh HH (1992) Modulation of $GABA_A$ receptor-activated current by norepinephrine in cerebellar Purkinje cells. Neurosci 51:951–960
Connolly CN, Krishek BJ, McDonald BJ, Smart TG, Moss SJ (1996a) Assembly and cell surface expression of heteromeric and homomeric γ-aminobutyric acid type A receptors. J Biol Chem 271:89–96
Connolly CN, Wooltorton JR, Smart TG, Moss SJ (1996b) Subcellular localization of γ-aminobutyric acid type A receptors is determined by receptor β subunits. Proc Natl Acad Sci USA 93:9899–9904
Connolly CN Kittler JT, Thomas P, Uren JM, Brandon NJ, Smart TG, Moss SJ (1999) Cell surface stability of γ-aminobutyric acid type A receptors. Dependence on protein kinase C activity and subunit composition. J Biol Chem 274:36565–36572
Cutting GR, Luo L, O'Hara BF, Kasch LM, Montrose-Rafizadeh C, Donovan DM, Shimada S, Antonarakis SE, Guggino WB, Uhl GR, Kazazian HH (1991) Cloning of the γ-aminobutyric acid (GABA) ρ1 cDNA: a GABA receptor subunit highly expressed in the retina. Proc Natl Acad Sci USA 88:2673–2677
Davies PA, Hanna MC, Hales TG, Kirkness EF (1997) Insensitivity to anaesthetic agents conferred by a class of $GABA_A$ receptor subunit. Nature 385:820–823
Dell'Acqua ML, Scott JD (1997) Protein kinase A anchoring. J Biol Chem 272:12881–12884
Draguhn A, Verdorn TA, Ewert M, Seeburg PH, Sakmann B (1990) Functional and molecular distinction between recombinant rat $GABA_A$ receptor subtypes by Zn^{2+}. Neuron 5:781–788
Dunne EL, Moss SJ, Smart TG (1998) Inhibition of $GABA_A$ receptor function by tyrosine kinase inhibitors and their inactive analogues. Mol Cell Neurosci 12:300–310
Essrich C, Lorez M, Benson JA, Fritschy J-M, Luscher B (1998) Postsynaptic clustering of major $GABA_A$ receptor subtypes requires the γ2 subunit and gephyrin. Nature Neurosci. 1:563–571
Feigenspan A, Bormann J (1994a) Facilitation of GABAergic signaling in the retina by receptors stimulating adenylate cyclase. Proc Natl Acad Sci USA 91:10893–10897
Feigenspan A, Bormann J (1994b) Modulation of $GABA_C$ receptors in rat retinal bipolar cells by protein kinase C. J Physiol 481:325–330
Filippova N, Dudley R Weiss DS (1999) Evidence for phosphorylation-dependent internalization of recombinant human ρ1 $GABA_C$ receptors. J Physiol 518:385–399
Francis SH, Corbin JD (1994) Structure and function of cyclic nucleotide-dependent protein kinases. Ann Rev Physiol 56:237–272
Fritschy JM, Benke D, Mertens S, Oertel WH, Bachi T, Mohler H (1992) Five subtypes of type A γ-aminobutyric acid receptors identified in neurons by double and triple immunofluorescence staining with subunit-specific antibodies. Proc Natl Acad Sci USA 89:6726–6730
Fukami S, Uchida I, Mashimo T, Takenoshita M, Yoshiya I (1998) Gamma subunit dependent modulation by nitric oxide (NO) in recombinant $GABA_A$ receptor. Neuroreport 9:1089–1093
Gillette MA, Dacheux R (1995) GABA- and glycine-activated currents in the rod bipolar cell of the rabbit retina. J Neurophysiol 74:856–875
Gillette MA, Dacheux RF (1996) Protein kinase modulation of $GABA_A$ currents in rabbit retinal rod bipolar cells. J Neurophysiol 76:3070–3086
Gyenes M, Farrant M, Farb DH (1988) "Run-down" of γ-aminobutyric $acid_A$ receptor function during whole-cell recording: a possible role for phosphorylation. Mol Pharmacol 34:719–723
Gyenes M, Wang Q, Gibbs TT, Farb DH (1994) Phosphorylation factors control neurotransmitter and neuromodulator actions at the γ-aminobutyric acid type A receptor. Mol Pharmacol 46:542–549

Harrison NL, Lambert NA (1989) Modification of $GABA_A$ receptor function by an analog of cyclic AMP. Neurosci Lett 105:137–142
Hashimoto T, Ishii T, Ohmori H (1996) Release of Ca^{2+} is the crucial step for the potentiation of IPSCs in the cultured cerebellar Purkinje cells of the rat. J Physiol 497.3:611–627
Hanley JG, Koulen P, Bedford F, Gordon-Weeks PR, Moss SJ (1999) The protein MAP-1B links $GABA_C$ receptors to the cytoskeleton at retinal synapses. Nature 397: 66–69
Harvey RJ, Chinchetru MA, Darlison MG (1994) Alternative splicing of a 51-nucleotide exon that encodes a putative protein kinase C phosphorylation site generates two forms of the chicken γ-aminobutyric $acid_A$ receptor $\beta 2$ subunit. J Neurochem 62:10–16
Hedblom E, Kirkness EF (1997) A novel class of $GABA_A$ receptor subunit in tissues of the reproductive system. J Biol Chem 272:15346–15350
Heuschneider G, Schwartz RD (1989) cAMP and forskolin decrease γ-aminobutyric acid-gated chloride flux in rat brain synaptoneurosomes. Proc Natl Acad Sci USA 86:2938–2942
Homanics GE, Harrison NL, Quinlan JJ, Krasowski MD, Rick CEM, de Blas AL, Mehta AK, Kist F, Mihalek RM, Aul JJ, Firestone LL (1999) Normal electrophysiological and behavioural responses to ethanol in mice lacking the long splice variant of the $\gamma 2$ subunit of the γ-aminobutyrate type A receptor. Neuropharmacol 38:253–265
Huang R-Q, Dillon GH (1998) Maintenance of recombinant type A γ-aminobutyric acid receptor function: Role of protein tyrosine phosphorylation and calcineurin. J Pharm Exp Ther 286:243–255
Jassar BS, Ostashewski PM, Jhamandas JH (1997) $GABA_A$ receptor modulation by protein tyrosine kinase in the rat diagonal band of Broca. Brain Res 775:127–133
Kaila K (1994) Ionic basis of $GABA_A$ receptor channel function in the nervous system. Prog Neurobiol 42:489–537
Kano M, Konnerth A (1992) Potentiation of GABA-mediated currents by cAMP-dependent protein kinase. Neuroreport, 3:563–566
Kano M, Rexhausen U, Dreessen J, Konnerth A (1992) Synaptic excitation produces a long-lasting rebound potentiation of inhibitory synaptic signals in cerebellar Purkinje cells. Nature, 356:601–604
Kano M, Kano M, Fukunaga K, Konnerth A (1996) Ca^{2+}-induced rebound potentiation of γ-aminobutyric acid-mediated currents requires activation of Ca2+/calmodulin-dependent kinase II. Proc Natl Acad Sci USA 93:13351–13356
Kapur J, Macdonald RL (1996) Cyclic-AMP-dependent protein kinase enhances hippocampal dentate granule cell $GABA_A$ receptor currents. J Neurophysiol 76: 2626–2634
Kellenberger S, Malherbe P, Sigel E (1992) Function of the $\alpha 1\beta 2\gamma 2S$ γ-aminobutyric acid type A receptor is modulated by protein kinase C via multiple phosphorylation sites. J Biol Chem 267:25660–25663
Kennelly PJ, Krebs EG (1991) Consensus sequences as substrate specificity determinants for protein kinases and protein phosphatases. J Biol Chem 266:15555–15558
Kirkness EF, Bovenkerk CF, Ueda T, Turner AJ (1989) Phosphorylation of γ-aminobutyrate (GABA)/benzodiazepine receptors by cyclic AMP-dependent protein kinase. Biochem J 259:613–616
Kofuji P, Wang JB, Moss SJ, Huganir RL, Burt DR (1991) Generation of two forms of the γ-aminobutyric $acid_A$ receptor $\gamma 2$-subunit in mice by alternative splicing. J Neurochem 56:713–715
Korpi ER, Kuner T, Kristo P, Kohler M, Herb A, Luddens H, Seeburg PH (1994) Small N-terminal deletion by splicing in cerebellar $\alpha 6$ subunit abolishes $GABA_A$ receptor function. J Neurochem 63:1167–1170
Krishek BJ, Xie X, Blackstone C, Huganir RL, Moss SJ, Smart TG (1994) Regulation of $GABA_A$ receptor function by protein kinase C phosphorylation. Neuron 5:1081–1095

Kusama T, Sakurai M, Kizawa Y, Uhl GR, Murakami H (1995) GABA $\rho 1$ receptor: inhibition by protein kinase C activators. Eur J Pharmacol 291:431–434
Kusama T, Hatama K, Sakurai M, Kizawa Y, Uhl GR, Murakami H (1998) Consensus phosphorylation sites of human $GABA_C/GABA_\rho$ receptors are not critical for inhibition by protein kinase C activation. Neurosci Lett 255:17–20
Laurie DJ, Wisden W, Seeburg PH (1992) The distribution of thirteen $GABA_A$ receptor subunit mRNAs in the rat brain. III. Embryonic and postnatal development. J Neurosci 12:4151–4172
Lambert NA, Harrison NL (1990) Analogs of cyclic AMP decrease γ-aminobutyric $acid_A$ receptor-mediated chloride current in cultured rat hippocampal neurons via an extracellular site. J Pharm Exp Ther 255:90–94
Leidenheimer NJ, Browning MD, Dunwiddie TV, Hahner LDx Harris RA (1990) Phosphorylation-independent effects of second messenger system modulators on γ-aminobutyric $acid_A$ receptor complex function. Mol Pharmacol 38:823–828
Leidenheimer NJ, Machu TK, Endo S, Olsen RW, Harris RA, Browning MD (1991) Cyclic AMP-dependent protein kinase decreases γ-aminobutyric $acid_A$ receptor-mediated $^{36}Cl^-$ uptake by brain microsacs. J Neurochem, 57:722–725
Leidenheimer NJ, McQuilkin SJ, Hahner LD, Whiting P, Harris RA (1992) Activation of protein kinase C selectively inhibits the γ-aminobutyric $acid_A$ receptor: role of desensitization. Mol Pharmacol, 41:1116–1123
Leidenheimer NJ, Whiting P, Harris RA (1993) Activation of calcium-phospholipid-dependent protein kinase enhances benzodiazepine and barbiturate potentiation of the $GABA_A$ receptor. J Neurochem, 60:1972–1975
Leidenheimer NJ (1996) Effect of PKG activation on recombinant $GABA_A$ receptors. Mol Brain Res 42:131–134
Leidenheimer NJ, Chapell R (1997) Effects of PKC activation and receptor desensitisation on neurosteroid modulation of $GABA_A$ receptors. Mol Brain Res 52:173–181
Levitan IB (1994) Modulation of ion channels by protein phosphorylation and dephosphorylation. Ann Rev Physiol 56:193–212
Levitan ES, Schofield PR, Burt DR, Rhee LM, Wisden W, Kohler M, Fujita N, Rodriguez HF, Stephenson A, Darlison MG, Barnard EA, Seeburg PH (1988) Structural and functional basis for $GABA_A$ receptor heterogeneity. Nature, 335:76–79
Lin Y-F, Browning MD, Dudek EM, Macdonald RL (1994) Protein kinase C enhances recombinant bovine $\alpha 1\beta 1\gamma 2L$ $GABA_A$ receptor whole-cell currents expressed in L929 fibroblasts. Neuron 3:1421–1431
Lin Y-F, Angelotti TP, Dudek EM, Browning MD, Macdonald RL (1996) Enhancement of recombinant $\alpha 1\beta 1\gamma 2L$ γ-aminobutyric $acid_A$ receptor whole-cell currents by protein kinase C is mediated through phosphorylation of both $\beta 1$ and $\gamma 2L$ subunits. Mol Pharmacol 50:185–195
Llano I, Gerschenfeld HM (1993) β-adrenergic enhancement of inhibitory synaptic activity in rat cerebellar stellate and Purkinje cells. J Physiol 468:210–224
Macdonald RL, Olsen RW (1994) $GABA_A$ receptor channels. Ann Rev Neurosci 17:569–602
Machu TK, Firestone JA, Browning MD (1993) Ca^{2+}/calmodulin-dependent protein kinase II and protein kinase C phosphorylate a synthetic peptide corresponding to a sequence that is specific for the $\gamma 2L$ subunit of the $GABA_A$ receptor. J Neurochem, 61:375–377
Marszalec W, Kurata Y, Hamilton BJ, Carter DB, Narahashi T (1994) Selective effects of alcohols on γ-aminobutyric acid A receptor subunits expressed in human embryonic kidney cells. J Pharm Exp Ther 269:157–163
Martina M, Mozrzymas JW, Boddeke HWGM, Cherubini E (1996) The calcineurin inhibitor cyclosporin A-cyclophilin A complex reduces desensitisation of $GABA_A$-mediated responses in acutely dissociated rat hippocampal neurons. Neurosci Lett 215:95–98
McDonald BJ, Moss SJ (1994) Differential phosphorylation of intracellular domains of γ-aminobutyric acid type A receptor subunits by calcium/calmodulin type 2-depen-

dent protein kinase and cGMP-dependent protein kinase. J Biol Chem 269:18111–18117
McDonald BJ, Moss SJ (1997) Conserved phosphorylation of the intracellular domains of $GABA_A$ receptor $\beta 2$ and $\beta 3$ subunits by cAMP-dependent protein kinase, cGMP-dependent protein kinase, protein kinase C and Ca2+/calmodulin type II-dependent protein kinase. Neuropharmacology 36:1377–1385
McDonald BJ, Amato A, Moss SJ, Smart TG (1998) Adjacent phosphorylation sites on $GABA_A$ receptor β subunits determine regulation by cAMP-dependent protein kinase. Nature Neurosci 1:23–28
McKinley DD, Lennon DJ, Carter DB (1995) Cloning, sequence analysis and expression of two forms of mRNA coding for the human $\beta 2$ subunit of the $GABA_A$ receptor. Mol Brain Res 28:175–179
Mochly-Rosen D, Smith BL, Chen CH, Disatnik MH, Ron D (1995) Interaction of protein kinase C with RACK1, a receptor for activated C-kinase: a role in beta protein kinase C mediated signal transduction. Biochem Soc Trans 23:596–600
Mochly-Rosen D, Gordon AS (1998) Anchoring proteins for protein kinase C: a means for isozyme selectivity. FASEB J 12:35–42
Moran O, Dascal N (1989) Protein kinase C modulates neurotransmitter responses in *Xenopus* oocytes injected with rat brain mRNA. Mol Brain Res 5:193–202
Moss SJ, Doherty CA, Huganir RL (1992a) Identification of the cAMP-dependent protein kinase and protein kinase C phosphorylation sites within the major intracellular domains of the $\beta 1$, $\gamma 2S$ and $\gamma 2L$ subunits of the γ-aminobutyric acid type A receptor. J Biol Chem 267:14470–14476
Moss SJ, Smart TG, Blackstone CD, Huganir RL (1992b) Functional modulation of $GABA_A$ receptors by cAMP-dependent protein phosphorylation. Science 257:661–665
Moss SJ, Gorrie GH, Amato A, Smart TG (1995) Modulation of $GABA_A$ receptors by tyrosine phosphorylation. Nature 377:344–348
Moss SJ, Smart TG (1996) Modulation of amino acid-gated ion channels by protein phosphorylation. Int J Neurobiol 39:1–52
Mumby MC, Walter G (1993) Protein serine/threonine phosphatases: Structure, regulation, and functions in cell growth. Physiol Rev 73:673–699
Nairn AC, Shenolikar S (1992) The role of protein phosphatases in synaptic transmission, plasticity and neuronal development. Curr Opin in Neurobiol 2:296–301
Parfitt KD, Hoffer BJ, Bickford-Wimer PC (1990) Potentiation of gamma-aminobutyric acid-mediated inhibition by isoproterenol in the cerebellar cortex: receptor specificity. Neuropharmacol 29:909–916
Pearson RC, Kemp BE (1991) Protein kinase phosphorylation site sequences and consensus specificity motifs: tabulations. Meth Enzymol 200:62–81
Poisbeau P, Cheney MC, Browning MD, Mody I (1999) Modulation of synaptic $GABA_A$ receptor function by PKA and PKC in adult hippocampal neurons. J Neurosci 19:674–683
Porter NM, Twyman RE, Uhler MD, Macdonald RL (1990) Cyclic AMP-dependent protein kinase decreases $GABA_A$ receptor current in mouse spinal neurons. Neuron 5:789–796
Poulter MO, Barker JL, O'Carroll AM, Lolait SJ, Mahan LC (1992) Differential and transient expression of $GABA_A$ receptor alpha-subunit mRNAs in the developing rat CNS. J Neurosci 12:2888–2900
Pritchett DB, Sontheimer H, Shivers BD, Ymer S, Kettenmann H, Schofield PR, Seeburg PH (1989) Importance of a novel $GABA_A$ receptor subunit for benzodiazepine pharmacology. Nature, 338:582–585
Rabow LE, Russek SJ, Farb DH (1995) From ion currents to genomic analysis: Recent advances in $GABA_A$ receptor research. Synapse 21:189–274
Robello M, Amico C, Cupello A (1993) Regulation of $GABA_A$ receptor in cerebellar granule cells in culture: differential involvement of kinase activities. Neuroscience 53:131–138

Robello M, Amico C, Bucossi G, Cupello A, Rapallino MV, Thellung S (1996) Nitric oxide and $GABA_A$ receptor function in the rat cerebral cortex and cerebellar granule cells. Neurosci 74:99–105

Robello M, Amico C, Cupello A (1997) A dual mechanism for impairment of $GABA_A$ receptor activity by NMDA receptor activation in rat cerebellum granule cells. Eur J Biophys 25:181–187

Robello M, Amico C, Cupello A (1998) Cerebellar granule cell $GABA_A$ receptors studied at the single channel level: Modulation by protein kinase G. Biochem Biophys Res Commun 253:768–773

Schwartz RD, Heuschneider G, Edgar PP, Cohn JA (1991) cAMP analogs inhibit γ-aminobutyric acid-gated chloride flux and activate protein kinase A in brain synaptoneurosomes. Mol Pharmacol, 39:370–375

Scott JD, Soderling TR (1992) Serine/threonine protein kinases. Curr Opin in Neurobiol 22:289–295

Shingai R, Yanagi K, Fukushima T, Sakata K, Ogurusu T (1996) Functional expression of GABA $\rho3$ receptors in *Xenopus* oocytes. Neurosci Res 26:387–390

Shirasaki T, Aibara K, Akaike N (1992) Direct modulation of $GABA_A$ receptor by intracellular ATP in dissociated nucleus tractus solitarii neurones of rat. J Physiol 449:551–572

Sieghart W (1995) Structure and pharmacology of γ-aminobutyric $acid_A$ receptor subtypes. Pharmacol Rev 47:182–224

Sigel E, Baur R (1988) Activation of protein kinase C differentially modulates neuronal Na^+, Ca^{2+},and γ-aminobutyrate type A channels. Proc Natl Acad Sci USA 85:6192–6196

Sigel E, Baur R, Trube G, Mohler H, Malherbe P (1990) The effect of subunit composition of rat brain $GABA_A$ receptors on channel function. Neuron 5(5):703–711

Sigel E, Baur R, Malherbe P (1991) Activation of protein kinase C results in down-modulation of different recombinant $GABA_A$-channels. FEBS Lett 291:150–152

Sigel E, Baur R, Malherbe P (1993) Recombinant $GABA_A$ receptor function and ethanol. FEBS Lett 324:140–142

Smart TG (1997) Regulation of excitatory and inhibitory neurotransmitter-gated ion channels by protein phosphorylation. Curr Opin in Neurosci 7:358–367

Smart TG (1998) Electrophysiology of $GABA_A$ receptors. In Amino Acid neurotransmission, Ed. FA Stephenson, AJ Turner, Portland Press, London, pp: 37–63

Smart TG, Moss SJ, Xie X, Huganir RL (1991) $GABA_A$ receptors are differentially sensitive to zinc: dependence on subunit composition. Br J Pharmacol, 103:1837–1839

Smith DB, Johnson KS (1988) Single-step purification of polypeptides expressed in *Escherichia coli* as fusions with glutathione *S*-transferase. Gene 67:31–40

Songyang Z, Carraway KL, Eck MJ, Harrison SC, Feldman RA, Mohammadi M, Schlessinger J, Hubbard SR, Smith DP, Eng C, Lorenzo MJ, Ponder BAJ, Mayer BJ, Cantley LC (1995) Catalytic specificity of protein-tyrosine kinases is critical for selective signalling. Nature 373:536–539

Stelzer A, Kay AR, Wong RKS (1988) $GABA_A$-receptor function in hippocampal cells is maintained by phosphorylation factors. Science 241:339–341

Stelzer A, Shi H (1995) Impairment of $GABA_A$ receptor function by *N*-methyl-D-aspartate-mediated calcium influx in isolated CA1 pyramidal cells. Neurosci 62:813–828

Sweetnam PM, Lloyd J, Gallombardo P, Malison RT, Gallager DW, Tallman JF, Nestler EJ (1988) Phosphorylation of the $GABA_A$/benzodiazepine receptor-subunit by a receptor-associated protein kinase. J Neurochem 51:1274–1284

Tanaka C, Nishizuka Y (1994) The protein kinase C family for neuronal signaling. Ann Rev Neurosci 17:551–567

Tehrani MH, Hablitz JJ, Barnes EM Jr (1989) cAMP increases the rate of $GABA_A$ receptor desensitization in chick cortical neurons. Synapse 4:126–131

Tehrani MH, Barnes EM Jr (1994) $GABA_A$ receptors in mouse cortical homogenates are phosphorylated by endogenous protein kinase A. Mol Brain Res 24:55–64

Thompson CL, Pollard S, Stephenson FA (1996) Bidirectional regulation of $GABA_A$ receptor $\alpha1$ and $\alpha6$ subunit expression by a cyclic AMP-mediated signalling mechanism in cerebellar granule cells in primary culture. J Neurochem 67:434–437

Ticku MK, Mehta AK (1990) γ-Aminobutyric $acid_A$ receptor desensitization in mice spinal cord cultured neurons: lack of involvement of protein kinases A and C. Mol Pharmacol, 38:719–724

Unwin N (1993) Neurotransmitter action: opening of ligand gated ion channels. Cell, Suppl 72:31–41

Van der Geer P, Hunter T, Lindberg RA (1994) Receptor protein-tyrosine kinases and their signal transduction pathways. Ann Rev Cell Biol 10:251–337

Valenzuela CF, Machu TK, McKernan RM, Whiting P, Van Renterghem BB, McManaman JL, Brozowski SJ, Smith GB, Olsen RW, Harris RA (1995) Tyrosine phosphorylation of $GABA_A$ receptors. Mol Brain Res 31:165–172

Veruki ML, Yeh HH (1992) Vasoactive intestinal polypeptide modulates $GABA_A$ receptor function in bipolar cells and ganglion cells of the rat retina. J Neurophysiol 67:791–797

Veruki ML, Yeh HH (1994) Vasoactive intestinal polypeptide modulates $GABA_A$ receptor function through activation of cyclic AMP. Vis Neurosci 11:899–908

Wafford KA, Burnett DM, Leidenheimer NJ, Burt DR, Wang JB, Kofuji P, Dunwiddie TV, Harris RA, Sikela JM (1991) Ethanol sensitivity of the $GABA_A$ receptor expressed in *Xenopus* oocytes requires 8 amino acids contained in the $\gamma2L$ subunit. Neuron 7:27–33

Wafford KA, Whiting PJ (1992) Ethanol potentiation of $GABA_A$ receptors requires phosphorylation of the alternatively spliced variant of the $\gamma2$ subunit. FEBS Lett 313:113–117

Walaas SI, Greengard P (1991) Protein phosphorylation and neuronal function. Pharmacol Rev 43:300–334

Wan Q, Man HY, Braunton J, Wang W, Salter MW, Becker L, Wang YT (1997a) Modulation of $GABA_A$ receptor function by tyrosine phosphorylation of β subunits. J Neurosci 17:5062–5069

Wan Q, Xiong ZG, Man HY, Ackerley CA, Braunton J, Lu WY, Becker LE, MacDonald JF, Wang YT (1997b) Recruitment of functional $GABA_A$ receptors to postsynaptic domains by insulin. Nature 388:686–690

Wang H-L, Li A, Wu T (1997) Vasoactive intestinal polypeptide enhances the GABAergic synaptic transmission in cultured hippocampal neurons. Brain Res 746:294–300

Wang RA, Cheng G, Kolaj M, Randic M (1995) α-subunit of calcium/calmodulin-dependent protein kinase II enhances γ-aminobutyric acid and inhibitory synaptic responses of rat neurons in vitro. J Neurophysiol 73:2099–2106

Wang H, Bedford FK, Brandon NJ, Moss SJ, Olsen RW (1999) $GABA_A$-receptor-associated protein links $GABA_A$ receptors and the cytoskeleton. Nature 397:69–72

Waterhouse BD, Moises HC, Yeh HH, Woodward DJ (1982) Norepinephrine enhancement of inhibitory synaptic mechanisms in cerebellum and cerebral cortex: mediation by beta adrenergic receptors. J Pharm Exp Ther 221:495–506

Weiner JL, Zhang L, Carlen PL (1994) Potentiation of $GABA_A$-mediated synaptic current by ethanol in hippocampal CA1 neurons: possible role of protein kinase C. J Pharm Exp Ther 268:1388–1395

Wexler EM, Stanton PK, Nawy S (1998) Nitric oxide depresses $GABA_A$ receptor function via coactivation of cGMP-dependent kinase and phosphodiesterase. J Neurosci 18:2342–2349

White G, Li C, Ishac E (1992) 19-Dideoxyforskolin does not mimic all cAMP and protein kinase A independent effects of forskolin on GABA activated ion currents in adult rat sensory neurons. Brain Res 586:157–161

Whiting P, McKernan RM, Iversen LL (1990) Another mechanism for creating diversity in γ-aminobutyrate type A receptors: RNA splicing directs expression of two forms of $\gamma2$ subunit, one of which contains a protein kinase C phosphorylation site. Proc Natl Acad Sci USA 87:9966–9970

Wisden W, Laurie DJ, Monyer H, Seeburg PH (1992) The distribution of 13 $GABA_A$ receptor subunit mRNAs in the rat brain. I. Telencephalon, diencephalon, mesencephalon. J Neurosci 12:1040–1062

Xu W, Harrison SC, Eck MJ (1997) Three-dimensional structure of the tyrosine kinase c-Src. Nature 385:595–602

Yamada K, Akasu T (1996) Substance P suppresses $GABA_A$ receptor function via protein kinase C in primary sensory neurones of bullfrogs. J Physiol 496.2:439–449

Zarri I, Bucossi G, Cupello A, Rapallino MV, Robello M (1994) Modulation by nitric oxide of rat brain $GABA_A$ receptors. Neurosci Lett 180:239–242

Zhai J, Stewart RR, Friedberg MW, Li C (1998) Phosphorylation of the $GABA_A$ receptor γ2L subunit in rat sensory neurons may not be necessary for ethanol sensitivity. Brain Res 805:116–122

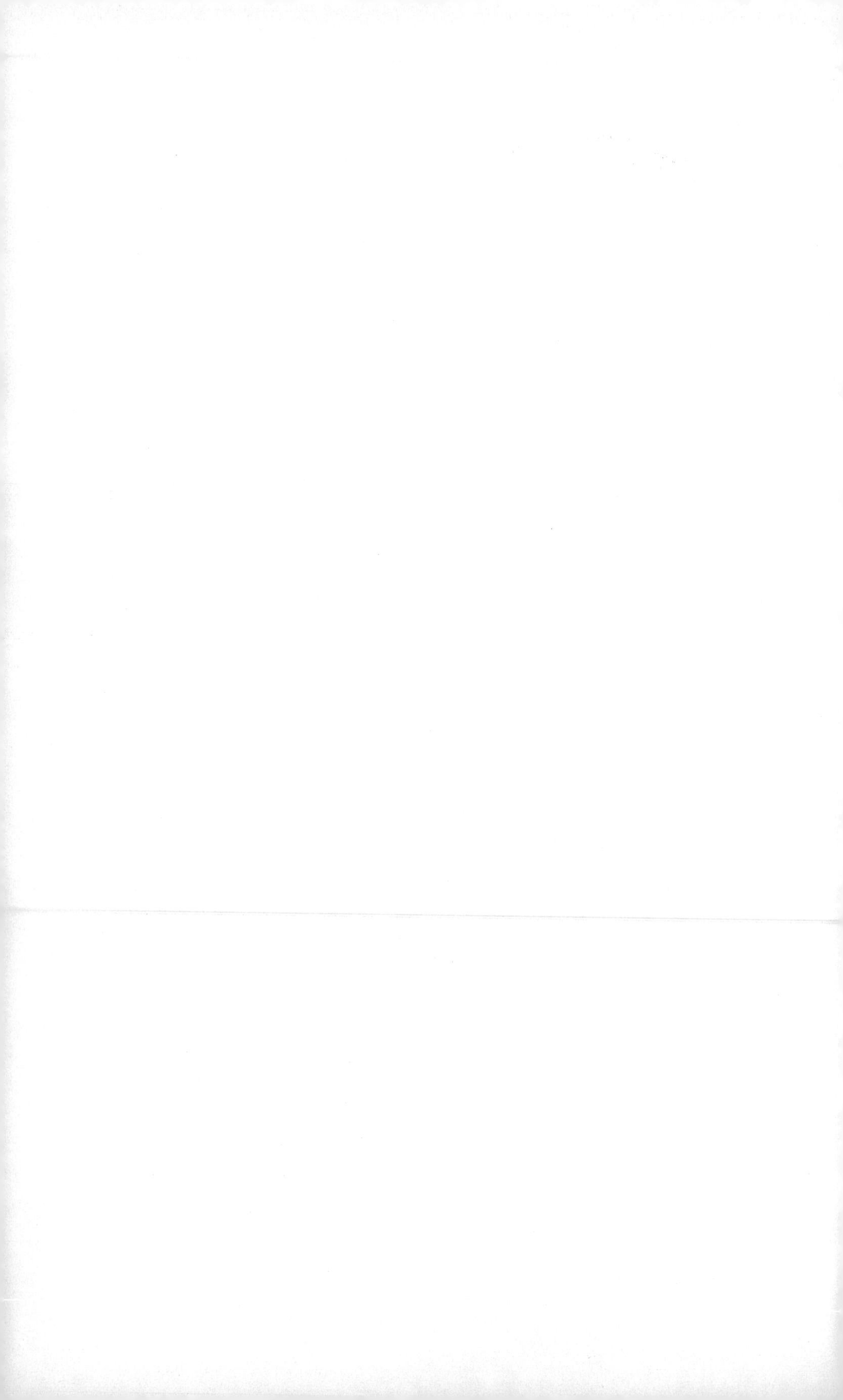

CHAPTER 8

Tolerance and Dependence to Ligands of the Benzodiazepine Recognition Sites Expressed by $GABA_A$ Receptors

E. COSTA, J. AUTA and A. GUIDOTTI

A. A Mechanistic Hypothesis on the Tolerance and Dependence to the Ligands of *Benzodiazepine Recognition Sites* (BZ-RS) Expressed by $GABA_A$ Receptors

Synaptic junctions, including γ-aminobutyric acid (GABA)-gated Cl^- channels ($GABA_A$ receptors), are expressed in almost every brain neuron. In the neocortex, they are expressed in apical dendrites, somata, and initial axon segments of pyramidal neurons in which $GABA_A$ receptors play an important role in synchronizing rhythmic columnary activity and other firing patterns that subserve integrative processes of cortical functions. The intrinsic activity of GABA-gating at $GABA_A$ receptors depends on the structure of the subunits assembled in these pentameric channels (for a review see MACDONALD and OLSEN 1994; COSTA and GUIDOTTI 1996; COSTA 1998), and on the expression of recognition sites for endogenous molecules (endozepines and neurosteroids), which modulate GABA-gated Cl^--current intensity (COSTA and GUIDOTTI 1991; GUIDOTTI and COSTA 1998; MATSUMOTO et al. 1999).

Anxiolytic ligands of BZ-RS bind with various affinities to specific sites expressed by certain $GABA_A$ receptor subtypes that include an α (α_1 or α_2 or α_3 or α_4 or α_5) and a γ_2 or γ_3 subunit and thereby allosterically amplify GABA-gated Cl^--current intensities (for review see COSTA and GUIDOTTI 1996; BARNARD et al. 1998). According to Henry La Chatelier's principle, when a system at equilibrium is perturbed it will shift in a direction that minimizes the perturbation (COLQUHOUN 1999). When anxiolytic ligands of BZ-RS are abused or prescribed for protracted time schedules they trigger tolerance, which is associated with compensatory structural changes in $GABA_A$ receptors directed to minimize functional consequences of the persistent amplification of GABA-gated Cl^--current intensities induced by BZ-RS occupancy (GALLAGER and PRIMUS 1993; KLEIN and HARRIS 1996; MILLER and GREENBLATT 1996; IMPAGNATIELLO et al. 1996; LONGONE et al. 1996; PESOLD et al. 1997). When a long-lasting treatment with anxiolytic drugs is abruptly terminated a syndrome emerges, which in part reflects the inadequacy of the $GABA_A$ receptor structural modifications induced by the BZ-RS ligands to maintain an acceptable function of neuronal circuits after the ligand is cleared from tissues. This withdrawal syndrome is usually taken as an evidence for drug dependence (WOOD et al. 1992).

The onset of tolerance to each behavioral response elicited by BZ-RS ligands occurs after a well-defined latency. For instance, during a protracted treatment with anxiolytic full agonists of BZ-RS, sedation is the first response to develop tolerance; this is followed by tolerance to amnesia, then anticonvulsant activity tolerance ensues, and ultimately, anxiolytic action develops tolerance (NUTT 1990). Since tolerance and the associated compensatory structural change of $GABA_A$ receptors minimize the consequences of long-term occupancy of BZ-RS by exogenous ligands, one might postulate that the different time course for the onset of tolerance to the various action of anxiolytic BZ-RS ligands might reflect an intrinsic difference in the transcription activation of the 17 genes that encode the various $GABA_A$ receptor subunits. Hence, the assessment of the changes in $GABA_A$ receptor subunit expression during the $GABA_A$ receptor adaptation is an important clue that helps increase the understanding of the molecular mechanisms that are operative in BZ-RS ligand tolerance.

B. Tools to study changes in $GABA_A$ receptor subunit assembly

Unfortunately, there are no appropriate methods to analyze the stoichiometry and degree of isomerism in the subunit assembly of various $GABA_A$ receptor subtypes (COSTA 1998). Moreover, we are not yet able to decipher the molecular language of a presumed code regulating the order in which subunits must assemble to form various $GABA_A$ receptor subtypes (COSTA 1998). Although with the use of immunochemistry and immunohistochemistry, we are able to assess neuronal colocalization of various $GABA_A$ receptor subunits, the accuracy of such assessment is limited by the specific antibody affinity for each subunit – the degree of this affinity often prevents detection of subunits expressed in relatively low amounts (CARUNCHO and COSTA 1994; FRITSCHY and MOHLER 1995). Even though we are able to detect the expression of two or three subunits in a neuron, we never know which of these subunits is repeated so as to construct the pentameric subunit assembly that is characteristic of various $GABA_A$ receptor subtypes. In this regard, it is appropriate to note that the $GABA_A$ receptor classification presented by the International Union of Pharmacology (Pharmacological Review, vol. 50, no. 2; BARNARD et al. 1998) has used a three-subunit coding system to define the structure of $GABA_A$ receptor subtypes.

The clustering of an α, β, or γ subunit gene in chromosomes 4, 5, and 15 suggests that similar neuronal colocalization of α, β, or γ subunit may subserve an important aspect of brain function. Estimation of the physical distance, using in-situ hybridization to cells in interphase and gene localization using hybridization in cells in metaphase, demonstrates the existence of β-α-α-γ gene clusters in cytogenetic bands of chromosomes 4 (p12) and 5 (q34). Remarkably, phylogenetic-tree analysis predicts the existence of a β-α-γ ancestral gene cluster in which internal duplication of ancestral α was followed by cluster duplication (RUSSEK 1999). Although the three-subunit coding proposed by the International Union of Pharmacology (BARNARD et al. 1998) might have a genetic justification, it contains an inherent ambiguity that may require revision when we can improve our methodology to determine how the stoichiometry and isomerization relates to the coding of subunit sequences in $GABA_A$ receptor subunit assembly. A methodology to distinguish the intracellular immunostaining of populations of $GABA_A$ receptors, which belong to receptors that are either being disbanded or synthesized, also remains unclear – a distinction between these two populations of neuronal $GABA_A$ receptor assemblies would be important to examine adaptive structural changes of $GABA_A$ receptors associated with tolerance to anxiolytic full agonists of BZ-RS (for details, consult FRITSCHY and MOHLER 1995).

Another concern is our present inability to determine contiguity among various subunits assembled to form native $GABA_A$ receptors; in fact, such understanding is essential to distinguish whether the pocket for the high affinity binding of BZ-RS ligands is suitable to express the allosteric modulation of GABA-gated current intensity when appropriate ligands are bound. For such ligand binding, not only is the presence of an α (α_1, α_2, α_3, α_4, or α_5) and a γ_2 or a γ_3 subunit required, but it is also necessary that the $\alpha\ \gamma_2$ or $\alpha\ \gamma_3$ subunits are contiguous. TRETTER et al. (1997) developed a method to determine ratios of dimeric complexes operative in the subunit assembly of multimeric proteins. They concluded that during transfections of cDNAs encoding, for the different subunits to be expressed in recombinant $GABA_A$ receptors subtypes each such cDNA sequence will express only 50% of receptors with a subunit configuration with $\alpha_X\ \gamma_2$ contiguity but the rest of the receptor configurations that are expressed will lack such a subunit contiguity.

C. Limitations in Interpreting Studies of $GABA_A$ Receptor Chimerae With and Without Single Amino Acid Mutations

PRITCHETT and SEEBURG (1991) showed that transiently expressed recombinant $GABA_A$ receptor subtypes transfected with cDNAs encoding for α_1, β_2, and γ_2, or α_3, β_2, and γ_2 subunits, include BZ-RS that differ by more than tenfold in their affinity for a specific ligand (for instance, when the ligand CL218872 is used as a $[^3H]$flumazenil displacer). To study mechanistically the characteristics of these above mentioned differences in BZ-RS ligand affinity, PRITCHETT and SEEBURG (1991) have pioneered the transfection of specifically constructed chimeric cDNAs of mutated α subunits together with cDNAs of native β_2 and

γ_2 subunits in $GABA_A$ receptors and, using these chimeric receptors, have also studied the relationship between the function and structure of these chimeric recombinant $GABA_A$ receptors. These authors cautiously stated that any interpretation derived from this experimental approach must be based on two assumptions: (1) the ligands used must bind competitively because of steric overlaps and thereby preclude simultaneous occupation of their respective high affinity binding site by other ligands; and (2) there must be a direct interaction between amino acids identified in chimeric subunits by mutation analysis and compounds showing altered affinity. Both assumptions apply to specific high affinity ligands and, even in this case, one must be mindful that the substitution of even a single amino acid in a given chimera subunit sequence might alter protein structures at a distant site, for instance, destroying a salt bridge and thereby causing the appearance of false-positive results. Although chimeric studies of receptors are attractive and fashionable, they may be particularly problematic in their interpretation when these studies are directed at the definition of the structure of binding sites for low affinity ligands, such as those operative in mediating barbiturate- and ethanol-induced modifications of $GABA_A$ receptor responses associated with tolerance to these two drugs of abuse.

D. Characterization of BZ-RS Ligands Endowed with Anxiolytic and Anticonvulsant Actions

There is considerable interest in the availability of an effective $GABA_A$ receptor-based anxiolytic drug that would not share the major problems that are affecting the therapeutic use of anxiolytic drugs now on the market. At this point, this practical problem has been provisionally resolved by the use of antagonists of catecholamine, dopamine, and serotonin receptors (PRICE et al. 1995), which presumably have anxiolytic action because they modulate GABAergic interneurons; however, with these antagonists there are also problems concerning therapeutic specificity and side effects. Treatment of anxiety and panic disorders has become a healing art where medications are selected by "ex-adjuvantibus" criteria; clearly there is a great need for a rationale in a treatment selection. In the case of BZ-RS ligands, it is likely that many problems derive from the close proximity among BZ doses used to treat anxiety and panic, or some convulsive disorders, and BZ doses that elicit unwanted side effects (see Fig. 1 for an example using alprazolam in the rat model). A similar relationship is operative for most of the anxiolytic ligands of BZ-RS currently on the market. This situation has created the belief that every anxiolytic BZ must have similar safety problems. However, these problems might be shared only by BZs endowed with full positive-allosteric modulatory activity of $GABA_A$ receptors but not by BZs endowed with partial positive-allosteric modulatory activity (HAEFELY 1994; COSTA and GUIDOTTI 1996). To

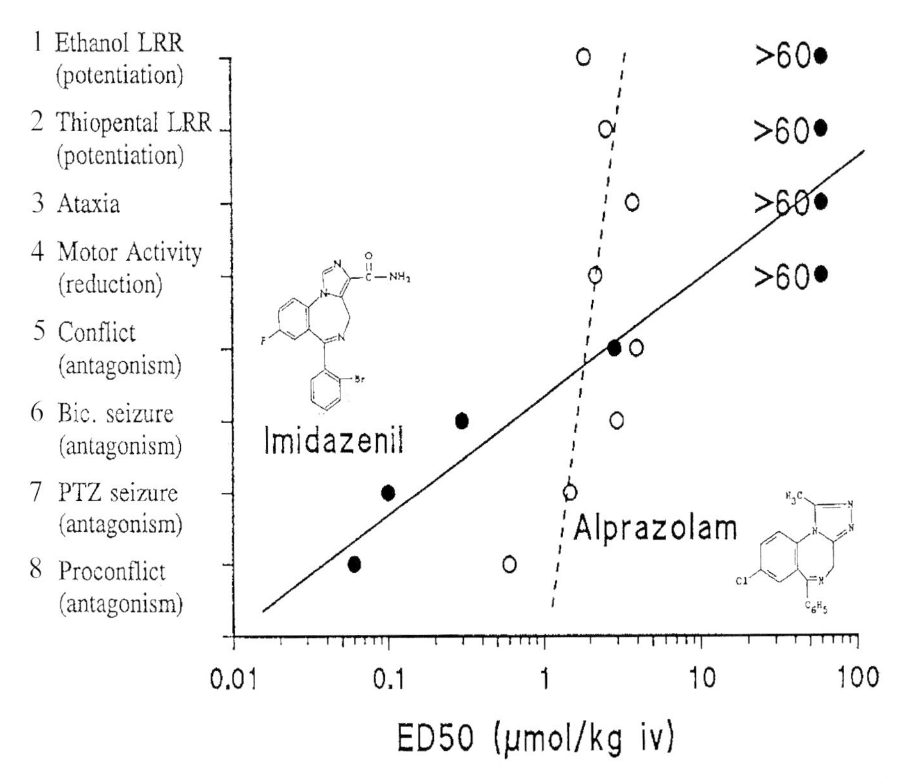

Fig. 1. Pharmacological profiles if imidazenil (a partial positive-allosteric modulator) and alprazolam (a full positive-allosteric modulator). Abscissa: ED_{50} for imidazenil and alprazolam. *Ordinate*: behavioral tests predicting side effects (1, 2, 3) or clinically useful (4) sedative, (5) anxiolytic, (6, 7) anticonvulsant, and (8) antipanic activity. For details, see THOMPSON et al. (1994)

assess this point, in Fig. 1 we have contrasted the dose-dependent action of alprazolam – a full positive allosteric modulator (Tables 1 and 2) – with that of imidazenil – a partial positive allosteric modulator (Tables 1 and 2). Imidazenil displays a high affinity for BZ-RS and a low clearance rate (in rats $T_{1/2}$ is 90 min; in the monkey $T_{1/2}$, longer than 6 h). Figure 1 shows that imidazenil tends to elicit side effects when given in doses that are at least two orders of magnitude greater than those that elicit anxiolytic, antipanic, and anticonvulsant action. Since imidazenil is a partial agonist in at least the eight subtypes of $GABA_A$ receptors in which it was tested (COSTA and GUIDOTTI 1996), it never maximizes the intensity of GABA-gated Cl^--currents (Table 1) and is virtually devoid of tolerance and dependence liability (Table 2).

These considerations motivated the following classification of BZ-RS ligands based on their overall intrinsic activity in terms of their amplification modes of GABA-gated Cl^--current intensities:

Table 1. Examples of maximal intrinsic efficacy of positive – full, partial, and selective – allosteric modulators (10^{-5} mol/l) on the GABA ED_{50} at various recombinant $GABA_A$ receptor subtypes. (From COSTA and GUIDOTTI 1996 – consult this reference for a complete list of the recombinant receptors tested)

Recombinant receptor Subunit Composition	GABA ED_{50} (μmol/l)	Full modulator Diazepam[a]	Partial modulator Imidazenil[a]	Selective modulator Zolpidem[a]
$\alpha_1\ \beta_1\ \gamma_2$	4.5	150	80	230
$\alpha_2\ \beta_1\ \gamma_2$	7.5	280	60	210
$\alpha_3\ \beta_1\ \gamma_2$	15.0	400	140	280
$\alpha_5\ \beta_1\ \gamma_2$	2.4	100	45	15
$\alpha_3\ \beta_2\ \gamma_2$	4.5	125	60	5

[a] Amplification as a percent of current intensity elicited by the GABA ED_{50} for each recombinant $GABA_A$ receptor.

1. *Full positive-allosteric modulators* that maximize GABA-gated Cl^--current intensities at several $GABA_A$ receptor subtypes (see Table 1); these compounds have also been termed *full-agonists* (see Table 2).
2. *Partial positive-allosteric modulators* that partially amplify GABA-gated Cl^- channel-current intensities at several $GABA_A$ receptor subtypes (Table 1) – these compounds have also been termed *partial agonists* (see Table 2).
3. *Selective-positive-allosteric modulators* of GABA-gated Cl^--current intensities at some selected $GABA_A$ receptor subtypes (see Table 1) – these compounds have been also termed *selective agonists* (Table 2).
4. High affinity ligands of BZ-RS that are devoid of intrinsic activity on GABA-gated Cl^--current intensities, but antagonize the pharmacologically-induced positive- or negative-modulation of GABA-gated Cl^--current intensities. These compounds have been termed *antagonists* (Table 2).

The probability of finding other partial allosteric modulators in various chemical classes of BZ-RS ligands is theoretically high considering the various chemical classes of drugs endowed with high affinity binding to BZ-RS (see Table 2). This high probability is supported by stereochemical considerations inherent in the mechanisms of allosteric modulation at $GABA_A$ receptors. In fact, there are two topographically and stereochemically distinct sites mediating the action of allosteric modulators acting on $GABA_A$ receptors. The BZ-RS is located on a $GABA_A$ receptor regulatory pocket. The binding of high affinity positive allosteric modulator ligands to this pocket brings about a rapidly reversible allosteric transition of the pentameric conformation of the $GABA_A$ receptor protein. This transition modifies the intrinsic activity of GABA in gating Cl^--channels. Hence, the allosteric modification of $GABA_A$ receptors caused by the binding of BZ-RS ligands includes a number of possible intermediary constraints operative in this transition that allows for a high degree of variability in the overall response. Thus, cooperative interactions at

Table 2. Ligands of BZ-RS and their pharmacological profile (Adapted from COSTA and GUIDOTTI 1996; BARNARD et al. 1998)

Class	Chemical	Allosteric modulation of GABA gated Cl^- current intensities (intrinsic efficacy, IE)	Tolerance liability (intensity)	Flumazenil-precipitated withdrawal (intensity)
Full positive allosteric modulators or full agonists	Classical 1,4 BZs (alprazolam, clonazepam, diazepam, flurazepam, flunitrazepam, lorazepam, midazolam, triazolam)	↑↑↑ Maximal IE at many $GABA_A$ receptor subtypes	+++	+++
Selective positive allosteric modulators or selective agonists	Triazolopyridazines (CL-218872), pyrazolopyridines (CGS20625, ICI 190622), Thienopyrimidines, (NNC14-0590), Imidazoquinazolines (NNC14-0185, NNC14-0189), imidazopyridines (zolpidem), β-carbolines (abecamil), 1,4 BZ (2- oxoquazepam)	↑↑ Maximal IE at some $GABA_A$ receptor subtypes	++	++
Partial positive allosteric modulators or *partial agonists*	Imidazobenzodiazepine carboxamides (imidazenil), imidazobenzodiazepinones (bretazenil, FG8205), benzoquinolizones (Ro19-8022)	↑ Low IE at many $GABA_A$ receptor subtypes	+/–	–
Antagonists Devoid of intrinsic modulatory activity	1,4 BZ (flumazenil, ZG63, Ro147437), *β*-carbolines (ZK93426), pyrazoloquinolinones (CGS 8216)	None	–	–

both sites of the allosteric complex (CHANGEUX and EDELSTEIN 1998) may participate in the amplification of GABA-gated Cl^- current intensity.

The $GABA_A$ receptor Cl^- channel opens into three different opening states with mean durations of 0.5 ms, 2.6 ms, and 7.6 ms. (MACDONALD and OLSEN 1994). The average opening time duration increases with the increase of GABA concentrations. The amplification of GABA-gating by BZ-RS ligands differs mechanistically from the increase in current intensity elicited by increase of GABA concentrations. In fact clinically effective concentrations of BZs increase the frequency of both channel openings and bursts, but the average channel opening time and burst duration remain unchanged. MACDONALD and OLSEN (1994) suggested that positive-allosteric ligands of BZ-RS can induce the channel openings (O_1) of $GABA_A$ receptors by selectively increasing the affinity of only one (C_1) of the two GABA biding sites located on $GABA_A$ receptors (see Scheme 1). This could explain why opening time or

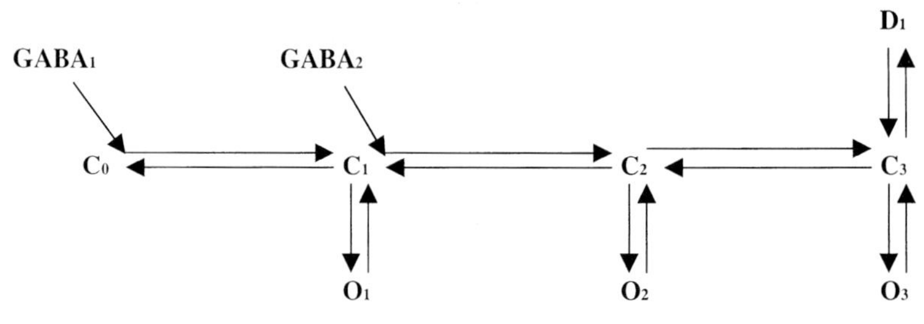

Scheme 1.

burst duration fail to change. An alternative explanation could be that BZs also reduce the rate of channel desensitization (in Scheme 1, C_3–D_1 transition).

It has been suggested (see MACDONALD and OLSEN 1994) that full positive-allosteric modulators that bind to BZ-RS maximize GABA-gating amplification because they increase the opening probabilities of monoligated $GABA_A$ receptors and facilitate C_1–C_2 transition, thereby enhancing Cl^--current intensity by accelerating C_2–O_2 transition rates and decreasing the rates of channel desensitization (C_3–D_1 transition). In contrast, using partial positive-allosteric modulators at concentrations that cause anxiolytic and anticonvulsant actions in the absence of side effects (see Fig. 1), it is likely that the opening probability of monoligated $GABA_A$ receptors is increased and C_2–O_2 transition rates are modestly increased (see Scheme 1) while C_3–D_1 transition frequency remains unchanged. Thus, partial agonists increase the frequency of channel openings and bursts by a smaller extent than that of full agonists even if applied in a range of doses that are about 2–3 order of magnitude greater than the doses that elicit anticonvulsant and anxiolytic activity; however, the above mechanistic hypothesis to explain the difference between partial and full agonist requires further testing.

It may be suggested that a selective agonist may preferentially bind to the BZ-RS pocket expressed by a specific $GABA_A$ receptor subtype, increasing the affinity of the C_1 GABA binding site of this $GABA_A$ receptor subtype. Perhaps this subtle difference between the selective and partial agonists was not focused consistently in early reports and thereby created some confusion regarding the specific mode of action of these two different classes of compounds. In fact, a few years ago, abecarnil was considered a partial allosteric modulator but indeed its pharmacological profile strongly suggest that it is a selective-allosteric modulator (see Table 2).

E. Can the Subunit Expression Modification Associated with BZ Tolerance Explain the Decreased Intrinsic Activity of Full Positive-Allosteric Modulators at $GABA_A$ Receptors?

Most GABA-based anxiolytic drugs currently in clinical use are full-allosteric modulators of BZ-RS and therefore possess high tolerance liability, which limits their protracted clinical use (Table 2). Although modifications of pharmacokinetic processes due to enzyme induction can theoretically account for drug tolerance liability, direct lines of investigation indicate that tolerance to BZ-RS ligands used therapeutically is never associated with an increase in the degradation rate (HIGH and FEELY 1988; AUTA et al. 1994; MILLER and GREENBLATT 1996).

The hypothesis that $GABA_A$ receptor subunit assembly changes during or following tolerance development was initially suggested by the experiments of Gallager and coworkers (HENINGER et al. 1990; PRIMUS and GALLAGER 1992; GALLAGER and PRIMUS 1993). These authors also reported that in rats receiving long-term diazepam treatment, the decrease of its anticonvulsant efficacy could be temporally related to a reduced sensitivity of $GABA_A$ receptors expressed in cortical and dorsal raphe nucleus neurons, leading to a reduction in GABA-mediated neurotransmission in these brain structures (GALLAGER et al. 1984). Today, several lines of evidence suggest that during the development of tolerance to the sedative, amnestic, anticonvulsant, and anxiolytic actions of full-allosteric modulators of $GABA_A$ receptors, the following adaptive changes may be operative: *i) changes in $GABA_A$ receptor subunit assembly, or ii) phosphorylation-dependent uncoupling of $GABA_A$ receptor allosteric modulation.*

I. Changes in GABA Receptor Subunit Assembly

1. Studies on Ligand Binding to BZ-RS

A plausible mechanism for the onset of sedative and anticonvulsant tolerance after a long-term treatment with full-agonist BZ-RS ligands could be a decrease in affinity and/or expression of $GABA_A$ receptors endowed with BZ-RS. However, in most of the studies conducted in vitro or in vivo in BZ-tolerant rats, either there were no changes in BZ-RS expression and affinity (BRAESTRUP et al. 1979; GALLAGER et al. 1984; FARB et al. 1984; STEPHENS and SCHNEIDER 1985; IMPAGNATIELLO et al. 1996; LONGONE et al. 1996; PRIMUS et al. 1996; KLEIN and HARRIS 1996), or the changes in affinity and expression of BZ-RS were modest (ROSENBERG and CHIU 1981; CRAWLEY et al. 1982; TIETZ et al. 1986; MILLER et al. 1988). Collectively, these results would appear to rule out that either change in $GABA_A$ receptor affinity for and/or the expression density of $[^3H]$BZ binding sites play a major role for the tolerance to the sedative, antiepileptic, and anxiolytic tolerance to full-agonist BZ-RS ligands.

2. Changes in $GABA_A$ Receptor Subunit mRNA Expression

Several laboratories have examined whether there are changes in expression level of mRNA encoding for specific $GABA_A$ receptor subunits in cortex, hippocampus, and other brain structures of rats receiving a protracted treatment with full-agonist ligands of BZ-RS. Northern blots of α, β, and γ subunits from mRNA extracted from brains of rats receiving saline or diazepam via minipump infusion for three weeks (HENINGER et al. 1990; PRIMUS and GALLAGER 1992) or an equipotent dose of lorazepam infused in mice for 4 weeks (KANG and MILLER 1991) showed a decrease in α_1 and γ_2 but not in β_1 subunit mRNA expression in cortex but similar changes were not detected in hippocampus or cerebellum. Analogous changes in α, β, and γ subunit expression were observed in cortex and hippocampus of rats treated for 4 weeks with 100–150 mg/kg per day of flurazepam orally (ZHAO et al. 1994; TIETZ et al. 1993). This administration schedule of flurazepam decreases GABA-mediated feed-forward and recurrent inhibition in the hippocampus CA1 pyramidal cell region (XIE and TIETZ 1991; ZHENG et al. 1993), GABA-mediated inhibitory postsynaptic potentials (ZHENG et al. 1993), and produces tolerance to sedative and anticonvulsant actions of this BZ (ROSENBERG et al. 1991). In these rats after 4 weeks of treatment, using Northern blot and in situ hybridization to monitor expression of mRNAs encoding for α_1, α_5, and γ_2 subunits, there was a localized decrease of α_1 mRNA expression in the CA1 hippocampal region and in layers II, III, and IV of the cortex. The same authors (ZHAO et al. 1994; TIETZ et al. 1994) reported a transient reduction in cortical and hippocampal expression of α_5 mRNA from 4 h to 2 weeks, which returned to basal value after 4 weeks. A similar transient decrease of α_5 subunit mRNA in rat brain during the first 2 weeks of daily treatment with flurazepam (40 mg/kg i/p.) was reported by O'DONOVAN et al. (1992).

Long-term treatment with ligands acting as full-agonist BZ-RS elicits a rapidly occurring tolerance to sedation that appears to be associated to a decrease in the mRNA encoding for α_5 subunits, while the onset of anticonvulsant tolerance occurs later and is associated with a decrease in the cortical and hippocampal expression of mRNA encoding for α_1 and γ_2 subunits of $GABA_A$ receptor. However, due to diversity in the intrinsic activity of various BZ-RS ligands or in the treatment duration and doses, and some uncertainty in the degree and type of tolerance observed and, most important, because of the use of nonquantitative methodology to determine the expression of mRNAs encoding for different $GABA_A$ receptor subunits, it has been difficult to correlate the degree of tolerance with the extent and quality of possible $GABA_A$ receptor subtype modifications. In fact, in these studies one can only correlate tolerance with changes in the steady state of mRNA expression but it is difficult to determine whether the expression of the translation products or the $GABA_A$ receptor subtypes have changed.

When the expression of ten $GABA_A$ receptor subunit mRNAs was measured with quantitative RT-PCR technology in discrete brain areas of rats exhibiting a well-defined degree of anticonvulsant and amnestic tolerance following 14 days treatment with increasing doses of diazepam (up to 60 mg/kg per day), IMPAGNATIELLO et al. (1995) and LONGONE et al. (1996) found a large decrease (40%–50%) in expression of α_1 and γ_2 (short and long variant) and an increase by approximately 30% in α_5 subunit in frontoparietal motor cortex, but no changes in subunit mRNAs expression were detected in the adjacent frontoparietal somatosensory cortex. Also, there was a decrease in α_1 subunit mRNA (20%) in the hippocampus without changes in α_5 and γ_2, α_2, α_3, α_4, and β_2, or γ_1, γ_2, and δ subunit transcripts. Importantly, in these studies it was also demonstrated that in the same group of rats, tolerance to the anticonvulsant or amnestic actions of diazepam, and the increase in the expression of mRNA subunits virtually return to basal values 72 h after the BZ treatment is terminated.

Taken together, these studies demonstrate two important facets of $GABA_A$ receptor regulation, which are presumably associated with tolerance to BZs:

1. Changes in $GABA_A$ receptor subunit mRNA expression are brain area-specific and these differences are highly significant.
2. Changes in $GABA_A$ receptor subunit mRNAs do not occur as a consequence of generalized nonspecific BZ action on DNA transcription rate or stability.

It is currently considered that these changes may be part of an adaptive response to a persistent and extensive upregulation of GABA-gated Cl^--current intensities caused by the BZ treatment. It is important to note that, after a long-term treatment with full-allosteric modulators of BZ-RS, the lack of changes in $GABA_A$ receptor subunit mRNA expression in the sensory cortex suggests that the modifications of mRNA expression by BZ might be selectively targeted to the function of specific cortical areas; however, the mechanism of such specificity is not well understood. In fact, the expression density of the various subunits of native $GABA_A$ receptors in the sensory cortex is very similar to that of the adjacent motor cortex. Indirect support that such selectivity in mRNA regulation may relate to specific changes in GABAergic function present in one cortical area but not in another, comes from the observation that:

1. Lesions of thalamic afferents to the cortex, which decrease pyramidal neuron columnary activity in visual cortex, change the expression pattern of $GABA_A$ receptor subunits showing a selective and fairly localized decrease in α_1 and γ_2 subunits and an increase in α_5 subunit mRNA expression in layer IV of the visual cortex without any change in other layers (MOHLER et al. 1995).
2. In monkeys, visual deprivation induced by unilateral intraocular injection of tetrodotoxin resulted in downregulation of α_1, β_2, and γ_2 subunits in layer IV of the primary visual cortex that is reversible when tetrodotoxin is cleared (HENDRY et al. 1994).

Thus, a selective regional inhibition of columnary firing by BZs might explain the differences found between the somatosensory and motor cortex following treatment with the full agonist, diazepam; however, further studies are needed to elucidate the nature of such a selective action of BZs.

3. Changes in $GABA_A$ Receptor Subunit Expression

Since changes in neuronal expression of mRNA encoding for $GABA_A$ receptor subunits may not reflect changes in the levels of various $GABA_A$ receptor subtypes and since it is impossible from measurements of mRNA to infer which receptor subtypes are modified, using immunohistochemistry with specific $GABA_A$ receptor subunit antibodies and gold immunolabeling, PESOLD et al. (1997) quantified whether the expression density of these subunits is also altered in areas where the expression of mRNAs encoding these subunits is changed. In the same experimental conditions used by IMPAGNATIELLO et al. (1995) and LONGONE et al. (1996) to induce and evaluate tolerance to diazepam, PESOLD et al. (1997) reported a selective decrease of α_1 (37%) subunit in layers II and V of the frontoparietal motor cortex and a concomitant increase in expression of α_5 subunit (150%) with only minor and virtually insignificant changes in the expression of $GABA_A$ receptor subunits in the frontoparietal somatosensory cortex.

Thus, it is possible to propose that a long-term exposure to full-allosteric modulator ligands of BZ-RS changes the expression of proteins that are assembled in $GABA_A$ receptors and, very likely, the $GABA_A$ receptor subtypes. For instance, it can be inferred that $GABA_A$ receptors including α_1 and γ_2 subunits may be decreased and receptor subtypes including the α_5 subunit may be increased. Full-allosteric modulators have been shown to require γ_2 subunit to express a maximal intrinsic modulatory activity; in turn, amplification of GABA action at $GABA_A$ receptors including α_1 subunits is greater than that of receptors which include the α_5 subunit (COSTA and GUIDOTTI 1996). Since $GABA_A$ receptors assembled with α_5, α_6, and δ subunits have low sensitivity to full-allosteric modulators of $GABA_A$ receptors (BARNARD et al. 1998), in the future it would be important to establish whether δ subunits also change in selective brain areas of rats tolerant to diazepam or other full-agonist BZ-RS ligands.

We believe that the frontoparietal motor cortex may not be the only cortical area in which changes in $GABA_A$ receptor structure occur following long-term administration of full-allosteric modulators of BZ-RS. It should be noted that, just as protection against convulsions is only one of many pharmacological properties of diazepam, changes in expression of $GABA_A$ receptor subunit in pyramidal apical dendrites, and/or neuronal somata, and initial axon segments of pyramidal neurons in the frontoparietal motor cortex may be only one of many cortical areas in which expression of $GABA_A$ receptors subtypes is changing to compensate for persistent amplification of GABA-gated Cl^--current intensities elicited by long-term treatment with BZs. However, it is necessary to explore why the amplification of GABA-gated Cl^--current intensities elicited by BZ has such selectivity for certain cortical areas. For instance, BZs may also produce their anxiolytic effects by amplifying GABA-gated Cl^--current intensities in $GABA_A$ receptors expressed in selective limbic structures of the Papez circuit (amygdaloid nuclei) that have been implicated in modulation of emotions (PRATT et al. 1998). They may impair cognitive function by acting on GABAergic circuits in hippocampus and limbic cortex, whereas their ataxic action may be due to a functional modification of $GABA_A$ receptor expression in striatum, cerebellum, or spinal cord. These areas of the CNS may became the target of $GABA_A$ receptors assembly modifications during temporally specific phases of sedative, ataxic, amnestic, anticonvulsant, and anxiolytic tolerance development that occurs in long-term exposure to full-agonist ligands of BZ-RS. Therefore, it would be important to conduct pertinent studies of subunit expression in discrete brain regions that are believed to be operative in the expression of specific actions of BZs to detect whether tolerance to these actions temporally coincides with changes of specific $GABA_A$ receptor subunit expression, and presumably $GABA_A$ receptor subtype. Indeed, electron microscopic studies coupled with neurophysiological recording in slices of the above-mentioned structures at various times during the development of tolerance are needed to make a more precise correlation and resolve some of the many questions that are still pending.

II. $GABA_A$ Receptor Subunit Allosteric Uncoupling

One consistent feature of a protracted treatment with full-agonist ligands of BZ-RS is the uncoupling of GABA and BZ–RS interactions in the absence of changes in expression density of a specific BZ-RS. For example, GALLAGER et al. (1984) observed a 50% decrease in GABA-dependent increase of [^{3}H]flunitrazepam binding in brain synaptic membrane preparations from rats that became tolerant to diazepam's sedative, amnestic, and anticonvulsant actions. Similar uncoupling has been reproduced in primary neuronal cultures of chick (FRIEDMAN et al. 1996), mouse (HU and TICKU 1994), and cultures of cells transfected with various $GABA_A$ receptor subunits (PRIMUS et al. 1996; KLEIN and HARRIS 1996), and then exposed for extended time periods to full-agonist ligands of BZ-RS. In neuronal cultures, the uncoupling of GABA and BZ recognition sites appears to require several hours of BZ exposure (18–60h) and the magnitude of the uncoupling is proportional to the ligand intrinsic efficacy at the specific $GABA_A$ receptor expressed (PRIMUS et al. 1996; FRIEDMAN et al. 1996). In contrast, in recombinant $GABA_A$receptors expressed in cells stably transfected with various combinations of cDNAs encoding for $GABA_A$ receptor subunits, exposure to BZs for a few hours uncouples (GABA)–(BZ-RS) binding interactions in the absence of appreciable changes in $GABA_A$ receptor subunit expressed (PRIMUS et al. 1996; KLEIN and HARRIS 1996). In these recombinant receptors, the rate of uncoupling depends both on $GABA_A$ receptor subtypes and on the intrinsic activity (GABA-shift) of the specific BZ-RS ligand tested. The extent of allosteric uncoupling is greater for a full-allosteric modulator than for a partial-allosteric modulator, or that for a selective allosteric modulator, and depends on the $GABA_A$ receptor subtype (PRIMUS et al. 1996). Since in the experiments with recombinant receptors the uncoupling elicited by BZ-RS ligands occurs in the absence of GABA, one might infer that induction of uncoupling is dependent on the intrinsic efficacy but is independent from the affinity of these BZ-RS ligands.

To explain the uncoupling that occurs in the absence of changes in $GABA_A$ receptor subunit expression, two mechanisms have been considered: (a) phosphorylation of receptor proteins; and (b) receptor internalization or recycling.

Some lines of evidence indicate that phosphorylation of PKA or PKC consensi expressed by γ_2 or β_2 subunits affects $GABA_A$ receptor function (for a review see KLEIN and HARRIS 1996). Possibly, the $GABA_A$ receptor subunit conformational changes resulting from BZ-RS ligand intrinsic activity may play some role in the modulation of γ_2 or β_2 subunit phosphorylation. The changes of $GABA_A$ receptor phosphorylation following a protracted administration of full agonists of BZ-RS ligands might also favor receptor internalization. In recombinant receptors, including α_1, β_2, and γ_2 $GABA_A$ receptor subunits, a protracted exposure to diazepam facilitates internalization. In fact, such exposure increases the cytosolic content of $GABA_A$ receptor subunits, whereas the subunit expression in cell membranes is decreased (TEHRANI and BARNES 1993). However, the documentation to support a phosphorylation dependent (GABA)-(BZ-RS) uncoupling following a protracted $GABA_A$ receptor occupancy by full agonist BZ-RS ligands remains incomplete.

F. Are Changes in $GABA_A$ Receptor Subunit Assembly Relevant to BZ Dependence?

The appearance of tolerance and consequent need for dose escalation to maintain specific therapeutic effects of anxiolytic BZ is a consistent liability factor for long-term therapy with anxiolytic 1–4 BZ derivatives with full-agonist activity.

In experimental animals and in human subjects exposed to long-term treatment with high doses of diazepam, alprazolam, lorazepam, and flurazepam, a withdrawal syndrome may occur after an abrupt termination of the treatment (Wood et al. 1992). In rats, this syndrome is characterized by tremors, wet dog shakes, piloerection, anxiety, and myoclonic jerks; it begins at 2–5 days of latency and peaks at about 8–11 days, a time when tolerance has already disappeared (Ryan and Boisse 1983). It is essential to keep in mind that tolerance to various pharmacological actions of full-agonist ligands of BZ-RS is relatively short-lived and, in fact, disappears 48–72 h after treatment discontinuation (Gent et al. 1985; Zhao et al. 1994; Impagnatiello et al. 1996; Longone et al. 1996).

Most attempts at understanding the molecular mechanisms underlying the dependence to full-agonist BZ-RS ligands have been directed to study changes in the subunit assembly of $GABA_A$ receptors and their functional consequences in the regulation of GABAergic tone. However, the modification of $GABA_A$ receptors subunit assembly and the changes in GABAergic function associated with tolerance disappear before the onset of a withdrawal syndrome (Gent et al. 1985; Longone et al. 1996). Frequently, the changes in $GABA_A$ receptor subunit expression have disappeared when the susceptibility to withdrawal syndromes is maximal (Ryan and Boisse 1983; Zhao et al. 1994; Cash et al. 1997).

Steppuhn and Turski (1993) were the first to report that glutamate receptor antagonists reduced the severity of the withdrawal syndrome elicited by an abrupt discontinuation of a long-term treatment with full-agonist BZ-RS ligands.

More recent studies in mice have confirmed the results obtained in rats (Tsuda et al. 1998; Dunworth and Stephens 1998; Koff et al. 1997). Some reports suggest that the expression of the withdrawal syndrome caused by an abrupt discontinuation of a long-term treatment with diazepam is temporally associated with an increase in AMPA receptor subunit expression in neocortex and hippocampus but not in cerebellum (Guidotti et al. 1997), and may even be associated with an increase in NMDA receptor subunit expression (Tsuda et al. 1998).

A possible explanatory hypothesis is that the protracted and maximal amplification of GABA-gated Cl^--current intensities elicited by diazepam leads to a compensatory enhancement of glutamate receptor expression, which tends to minimize the functional consequences of the aforementioned GABAergic tone imbalance by improving the equilibrium between glutamatergic and GABAergic tone. The normalization of the compensatory increase of glutamatergic tone is slower than that of BZ-elicited changes in $GABA_A$ receptor subunit assembly. The reason for this difference may reside in the intrinsic time-constant properties of $GABA_A$, AMPA, and NMDA receptor subunits turnover rates.

G. Development of Tolerance and Dependence Liability After Long-Term Treatment with Selective-Positive-Allosteric Modulators of $GABA_A$ Receptors

Zolpidem and abecarnil are the best-studied selective-positive-allosteric modulators of $GABA_A$ receptors. Other selective allosteric modulators, which are listed in Table 2, are poorly characterized and will not be discussed here.

I. Zolpidem

This imidazopyridine is a ligand for BZ-RS with potent hypnotic/sedative properties but a weak anticonvulsant action. The difference between the pharmacological profile of zolpidem and typical full-allosteric modulators, such as diazepam, very likely resides in their specific binding affinity to various $GABA_A$ receptor subtypes and in the various clearance rates for the two BZ-RS ligands (Arbilla et al. 1985). Diazepam (a full-allosteric modulator) binds to the BZ-RS expressed by any $GABA_A$ receptor, including a γ_2 subunit contiguous with an α_1, α_2, α_3, and α_5 subunit. In contrast, zolpidem binds with high affinity and expresses high intrinsic efficacy by preferentially binding to any $GABA_A$ receptor that includes an α_1 subunit contiguous to a γ_2 subunit. High concentrations of zolpidem (up to 10^{-5} mol/l) may also amplify GABA-gated Cl^- currents in $GABA_A$ receptor subunit expressing an α_3-subunit contiguous to a γ_2 subunit (see Table 1 and Costa and Guidotti 1996, Barnard et al. 1998).

When zolpidem is given to rats for long time periods in doses equivalent to those used in humans to elicit sedation and to facilitate sleep induction, it failed to produce tolerance; however, when given in higher doses (i.e., those required for anticonvulsant activity), it produced tolerance (Arbilla et al. 1985; Evans et al. 1990). No zolpidem dependence is reported in human subjects when the drug is prescribed as a short-acting hypnotic, even on a protracted time schedule.

II. Abecarnil

The pharmacological profile of abecarnil was initially defined as anxioselective, because its anxiolytic and anticonvulsant properties appeared after doses that were smaller than those required to produce sedation (Stephens et al. 1990). Abecarnil binds preferentially and with high affinity and intrinsic activity to α_1- and α_3-containing $GABA_A$ receptors and acts as a partial agonist in receptors expressing α_5 subunits (Stephens et al. 1991; Knoflach et al. 1993; Pribilla et al. 1993). In rodents, a long-term administration of anxiolytic and anticonvulsant doses of abecarnil produced anticonvulsant tolerance and an abrupt discontinuation of this treatment elicited signs of withdrawal (Loscher et al. 1996). Moreover, when abecarnil was given to human subjects for 3 weeks in sedative doses, anxiolytic tolerance and, after its abrupt discontinuation, mild to moderate withdrawal syndromes were observed (BALLENGER et al. 1991). Despite the limitations associated with the presence of a withdrawal syndrome in humans, it is fair to add that abecarnil has lower anticonvulsant tolerance and dependence liability in rodents than full-positive-allosteric ligands of BZ-RS (SERRA et al. 1994; NATOLINO et al. 1996; HOLT et al. 1996).

Little is known about whether long-term administration of selective positive-allosteric ligands of BZ-RS in doses that cause tolerance changes the expression of $GABA_A$ receptor subunit mRNAs; however, in a recent study HOLT et al. (1996) demonstrated that in rats injected subcutaneously once a day for 7 or 14 days with 6 mg/kg of abecarnil in sesame oil, there was a significant decrease of neocortical β_2 and γ_2 subunit mRNAs without changes in expression of mRNA encoding for α subunits. An equivalent dose of diazepam for 14 days (15 mg/kg) decreased the cortical expression of α_1, β_2, and γ_2 subunit mRNAs and increased that of α_4, γ_3, and α_5 subunit mRNAs.

In conclusion, although abecarnil or zolpidem (two selective BZ-RS ligands, which are slightly different in their $GABA_A$ receptor subtype selectivity) are not devoid of tolerance and dependence liability when used in a dose range one to two orders of magnitude greater than their respective anxiolytic or sedative doses, both are less potent than full agonists of BZ-RS in eliciting tolerance to their respective sedative or anticonvulsant activities.

H. Lack of Tolerance or Dependence Following Long Term Treatment with Partial-Positive-Allosteric BZ-RS Ligands

An ideal partial positive-allosteric modulator of $GABA_A$ receptors should possess a high affinity and low intrinsic activity for most $GABA_A$ receptor subtypes, it should not produce metabolites endowed with full allosteric modulatory activities, and it should have a good bioavailability and a relatively long half-life (COSTA and GUIDOTTI 1996). Table 2 lists ligands for BZ-RS with a documented putative partial-allosteric modulatory profile. In this list, imidazenil is the compound that complies most closely to the above-mentioned criteria defining an ideal partial agonist (COSTA and GUIDOTTI 1996).

Bretazenil is a partial-allosteric modulator in vitro, but in vivo it has encountered a limited use because of its fast metabolic rates leading to the formation of a metabolite with sedative activity and tolerance liability in humans and rats (AUTA et al. 1994; 1995; BUSTO el al. 1994).

Imidazenil, unlike bretazenil, is slowly metabolized (a half-life of 90 min in rats, and 6 h or more in monkeys) (AUTA et al. 1994, 1995), and in doses 60-fold greater than those that antagonize the sedative and ataxic action of diazepam fails to cause accumulation of metabolites that act as full-agonist ligands of BZ-RS (GIUSTI et al. 1993; COSTA and GUIDOTTI 1996).

An important aspect of the imidazenil pharmacological profile is its ability to elicit a mild amnesia and a possible weak tolerance to the anxiolytic and anticonvulsant action only at doses about two order of magnitude greater than those that elicit an anxiolytic (Paronis et al. 1997; Giusti et al. 1993) and anticonvulsant action (Giusti et al. 1993) in rats and monkeys. This sequence of events establishes a remarkably distinctive difference between the pharmacological profiles of partial, full or selective-positive-allosteric modulators of GABA-gated Cl^--current intensities as prospective drugs to treat convulsive state or anxiety disorders. In fact, with selective modulators, the tolerance liability appears with doses that are very close to those that cause anxiolytic and anticonvulsant actions, whereas in the case of partial positive-modulators, sedation, amnesia, tolerance, and dependence liability are virtually undetectable for a wide range of doses significantly (two to three orders of magnitude) above the doses to be used therapeutically. While it is not known whether selective agonists cause anxiolytic and anticonvulsant action in animals tolerant to diazepam, the partial agonist, imidazenil, can cause anticonvulsant action in rats tolerant to the anticonvulsant action of diazepam (Impagnatiello et al. 1996). However, imidazenil cannot be used as a sedative or to induce sleep, and zolpidem must be preferred as a sleep inducer.

I. Imidazenil is Devoid of Tolerance and Dependence Liability in Rodents

Imidazenil has an affinity for BZ recognition sites that is ten times higher than that of diazepam, and in addition, when it is administered in a single dose (one-tenth that of diazepam), it can antagonize the sedative and ataxic actions of diazepam, and these same doses antagonize bicuculline-induced seizures for a period that lasts longer than that of diazepam (Giusti et al. 1993; Auta et al. 1994). This property suggests that imidazenil can be an attractive drug to be tested in the treatment of convulsive states. When imidazenil is administered to rats for 21 days, 3 times daily, in doses progressively increasing from 2.5 μmol/kg to 7.5 μmol/kg, it does not induce anticonvulsant or anxiolytic tolerance (Auta et al. 1994). These doses of imidazenil are equipotent as anticonvulsant to 17.6–58.2 μmol/kg of diazepam (Auta et al. 1994) but, unlike imidazenil, these doses of diazepam cause anticonvulsant tolerance and, after abrupt discontinuation, elicit withdrawal symptoms. In other experiments, with an equipotent treatment schedule of diazepam and imidazenil, diazepam tolerance occurred after a few (5–7) days, whereas tolerance to imidazenil failed to occur even after 130 days of administration (Zanotti et al. 1996). Similarly, the repeated administration of imidazenil (0.1 mg/kg i.p.) to mice (3 times daily for 30 days) failed to induce tolerance (Ghiani et al. 1994).

This difference in tolerance liability of imidazenil and diazepam cannot be attributed to an imidazenil failure to occupy the $GABA_A$ receptor population that is occupied by diazepam, but may be due to a different amplification degree of GABA-gated current intensity which is greater for diazepam than imidazenil. In a series of structurally different recombinant $GABA_A$ receptors, imidazenil elicited a consistently modest degree of GABA-gated Cl^-

current amplification intensity, which was always much lower than that caused by diazepam (Table 1, and also see GIUSTI et al. 1993; COSTA and GUIDOTTI 1996).

Because the anticonvulsant action of imidazenil persists unabated in rats that are tolerant to diazepam, one might surmise that the modification of $GABA_A$ receptor assembly triggered by long term treatment with diazepam is still susceptible to the slight amplification of GABA-gated Cl^--currents intensity of imidazenil, and probably such modest amplification is sufficient to antagonize bicuculline convulsion. To support such hypotheses, it became important to determine whether the persistent occupancy of BZ-RS by imidazenil can also modify, in a manner different from diazepam, the $GABA_A$ receptor subunit composition in selected brain areas. Remarkably, imidazenil fails to change the expression of 10 mRNAs encoding for the corresponding $GABA_A$ receptor subunits when given 3 times daily for 14 days in daily total doses ranging between 7.5 and 30 μmol/kg, doses that are at least 25 to 100 times greater than the imidazenil ED_{50} to inhibit bicuculline-induced convulsions and five times greater than equipotent doses of diazepam (IMPAGNATIELLO et al. 1996).

Moreover, imidazenil does not elicit signs of dependence either after the abrupt discontinuation of a protracted treatment or after flumazenil doses that precipitated a withdrawal syndrome in rats receiving diazepam (AUTA et al. 1994).

II. Imidazenil is Devoid of Tolerance and Dependence Liability in Monkeys

In Patas monkeys working on a complex behavioral task of repeated acquisition (learning) and performance components, alprazolam (1 μmol/kg orally) decreases the response rate and increases the percent errors in acquisition while having little or no effect on performance (Fig. 2). Imidazenil, in oral doses as small as 0.025 μmol /kg, failed per se to alter acquisition or performance, but when given 1 h before alprazolam, attenuated the disruptive effects elicited by this drug on the acquisition component (THOMPSON et al. 1995; AUTA et al. 1995). However, as shown in Fig. 2, imidazenil in a dose of 12.5 μmol/kg, which is a dose 500 times greater than the minimal active dose that inhibits the cognitive deficit elicited by alprazolam (THOMPSON et al. 1994), causes only a modest disruption of acquisition, but remarkably antagonizes the large cognitive deficit elicited by alprazolam. Modest disruptive effects of were elicited by the first oral dose (12.5 μmol/kg), of imidazenil but these disruptive effects virtually vanished when the same dose was repeated the following day (see Fig. 2). Remarkably, this dose of imidazenil repeated for 7 days continues to antagonize the cognitive deficits elicited by a single injection of alprazolam suggesting a virtual lack of tolerance (see Fig. 2). Importantly, after 10 days of treatment with 12.5 μmol/kg of imidazenil, no overt signs of withdrawal were observed after abrupt discontinuation. These data suggest that

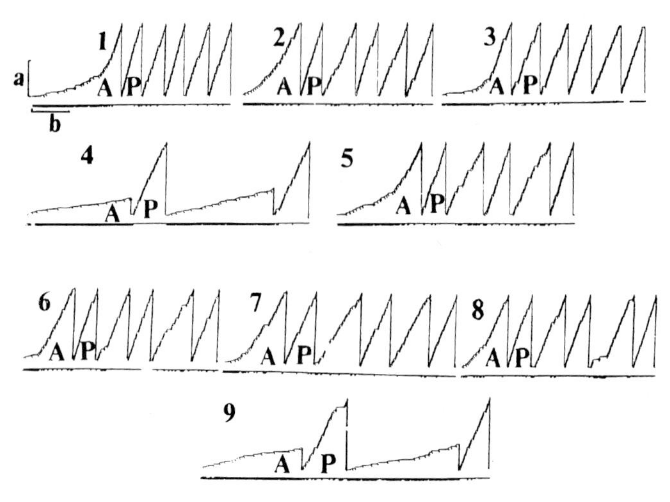

Fig. 2. Cumulative records from a Patas monkey showing the pattern of responding under a multiple schedule with acquisition (*A*) and performance (*P*) components. Each record represents a complete session (60 reinforcements), except for alprazolam alone which shows the first two acquisition and performance components respectively during a 2-h session. The response pen stepped upward with each correct response (scale *a* – *ordinate* – represents 100 correct responses) and was deflected downward upon completion of the four-response chain. Errors are indicated by the event pen (below each record). The scale *b* – *abscissa* – represents a 5 min responding. The respective cumulative recordings represent sessions preceded by administration of: (*1*) vehicle 60 min presession; (*2*) imidazenil 0.25 μmol/kg 60 min presession; (*3*) imidazenil 0.5 μmol/kg; (*4*) alprazolam 1 μmol/kg 60 min presession; (*5*) imidazenil 1.25 μmol/kg per day 1; (*6*) imidazenil 1.25 μmol/kg per day 2; (*7*) imidazenil 1.25 μmol/kg per day 3, 60 min before alprazolam 1 μmol/kg; (*8*) imidazenil 1.25 μmol/kg per day 7 60 min before alprazolam 1 μmol/kg; (*9*) alprazolam 1 μmol/kg 60 min presession administered on day 2 following discontinuation of imidazenil. In this experiment 1.25 μmol/kg of imidazenil was administered orally once a day from day 1 to day 14 (For details, see Auta et al. 2000)

imidazenil is a prototype of a new generation of anxiolytic and anticonvulsant imidazo-1-4-benzodiazepines that has minimal disruptive effects on learning and memory in doses 500 times greater than a minimal pharmacologically active dose. Imidazenil is virtually devoid of tolerance and dependence liability and other unwanted side effects exhibited by full-agonist BZ-RS ligands when tested in rodents and primates. This suggests that it is not the BZ-RS occupancy per se but the intrinsic efficacy of the ligand that determines consequent changes in subunit assemblies of $GABA_A$ receptors that might be responsible for the tolerance and dependence liabilities of full-agonist 1-4-benzodiazepines that have $GABA_A$ receptor-based anxiolytic actions.

In conclusion, imidazenil could become the first anxiolytic drug partially devoid of a consistent sedative action even in doses that are 500 times greater than the minimal active dose. With regard to its anticonvulsant activity, it is pertinent to mention that its potency is not altered even when the intrinsic activity of GABA at $GABA_A$ receptor is altered as it might be in the case of human pathology associated with convulsive states.

Acknowledgment. This work was supported by National Institute of Healt, USA Grant 56500 to F.C., Grant 49486 to A.G. and Grant 56890 to A.G.

References

Arbilla S, Depoortere H, George P, Langer SZ (1985) Pharmacological profile of the imidazopyridine zolpidem at benzodiazepine receptors and electrocorticogram in rats. Naunyn-Schmiedeberg's Arch Pharmacol 330:2248–2251
Auta J, Fraust WB, Lambert P, Guidotti A, Costa E, Moerschbaccher JM (1995) Comparison of the effects of full and partial allosteric modulators of $GABA_A$ receptors on complex behavioral processes in monkeys. Behav Pharmacol 6:323–332
Auta J, Giusti P, Guidotti A, Costa E (1994) Imidazenil, a partial positive allosteric modulator of $GABA_A$ receptor, exhibits low tolerance and dependence liabilities in rats. J Pharmacol Exp Ther 270:1262–1269
Auta J, Guidotti A, Costa E (2000) Imidazenil prevention of Alprazolam-induced aquisition deficit in patas monkeys is devoid of tolerance. Proc Natl Acad Sci USA 97:2314–2319
Ballenger JC, McDonald S, Noyes R, Rickels K, Sussman N, Woods S, Patin J, Singer J (1991) The first double-blind, placebo-controlled trial of a partial benzodiazepine agonist abecarnil (ZK 112-119) in generalized anxiety disorder. Psychopharmacol Bull 27:171–179
Barnard EA, Skolnick P, Olsen RW, Mohler H, Sieghart W, Biggio G, Braestrup C, Bateson AN, Langer SZ (1998) International Union of Pharmacology. XV. Subtypes of γ-aminobutyric $acid_A$ receptors: classification on the basis of subunit structure and receptor function. Pharmacol Rev 50:291–313
Braestrup C, Nielsen M, Squires RF (1979) No changes in rat benzodiazepine receptors after withdrawal from continuous treatment with lorazepam and diazepam. Life Sci 24:347–350
Busto U, Kaplan HL, Zawertailo L, Sellers EM (1994) Pharmacologic effects and abuse liability of bretazenil, diazepam, and alprazolam in humans. Clin Pharmacol Ther 55:451–463
Caruncho, HJ, Costa E (1994) Double-immunolabeling analysis of $GABA_A$ receptor subunits in label-fracture replicas of cultured rat cerebellar granule cells. Receptor Channels 2:143–153
Cash DJ, Serfozo P, Allan AM (1997) Desensitization of γ-aminobutyric $acid_A$ receptors in rat is increased by chronic treatment with chlordiazepoxide: a molecular mechanism of dependence. J Pharmacol Exp Ther 283:704–711
Changeux JP, Edelstein SJ (1998) Allosteric receptors after 30 years. Neuron 21:959–980
Colquhoun D (1999) GABA and the single oocyte: relating binding to gating. Nature Neurosci 2:201–202
Costa E, Guidotti A (1991) Diazepam binding inhibitor (DBI): a peptide with multiple biological actions. Life Sci 49:235–344
Costa E, Guidotti A (1996) Benzodiazepines on trial: a research strategy for their rehabilitation. Trends Pharm Sci 17:192–200
Costa E (1998) From $GABA_A$ receptor diversity emerges a unified vision of GABAergic inhibition. Ann Rev Pharmacol Toxicol 38:321–350
Crawley JN, Marangos PJ, Stivers J, Goodwin FK (1982) Chronic clonazepam administration induces benzodiazepine receptor subsensitivity. Neuropharmacology 21:85–89
Dunworth SJ, Stephens DN (1998) Sensitization to repeated withdrawal in mice treated chronically with diazepam is blocked by NMDA receptor antagonist. Psychopharm 136:308–310

Evans SM, Funderburk FR, Griffiths RR (1990) Zolpidem and triazolam in humans: behavioral and subjective effects and abuse liability. J Pharmacol Exp Ther 255:1246–1255

Farb DH, Borden LA, Chan CY, Czajkowski CM, Gibbs TT, Schiller GD (1984) Modulation of neuronal function through benzodiazepine receptors: biochemical and electrophysiological studies of neurons in primary monolayer cell culture. Ann NY Acad Sci 435:1–31

Friedman LK, Gibbs TT, Farb DH (1996) γ-Aminobutyric acid$_A$ receptor regulation: heterologous uncoupling of modulatory site interactions induced by chronic steroid, barbiturate, benzodiazepine, or GABA treatment in culture. Brain Res 707:100–109

Fritschy JM, Mohler H (1995) $GABA_A$ receptor heterogeneity in the adult rat brain: differential regional and cellular distribution of seven major subunits. J Comp Neurol 359:154–194

Gallager DW, Primus RJ (1993) Benzodiazepine tolerance and dependence. $GABA_A$ receptor complex locus of change. Biochem Soc Symp 59:1135–1141

Gallager DW, Lakoski JM, Gonsalves SF, Rauch SL (1984) Chronic benzodiazepine treatment decreases postsynaptic GABA sensitivity. Nature, 308:74–77

Gent JP, Feely MP, Haigh JR (1985) Differences between tolerance characteristics of two anticonvulsant benzodiazepines. Life Sci 37:849–856

Ghiani CA, Serra M, Motzo C, Giusti P, Cuccheddu T, Porceddu, Biggio G (1994) Chronic administration of an anticonvulsant dose of imidazenil fails to induce tolerance of $GABA_A$ receptor function in mice. Eur J Pharm 254:299–302

Giusti P, Ducic I, Puia G, Arban R, Walser A, Guidotti A, Costa E (1993) Imidazenil, a new partial positive allosteric modulator of γ-aminobutyric acid$_A$ (GABA) action at $GABA_A$ receptors. J Pharmacol Exp Ther 266:1018–1028

Guidotti A, Costa E (1998) Can the antidysphoric and anxiolytic profiles of SSRIs be related to their ability to increase brain 3a, 5a, tetrahydroprogesterone (ALLO) availability? Biol Psychiatry 44:865–873

Guidotti A. Impagnatiello F, Longone P, Costa E (1997) Relevance of AMPA receptor subunit changes in the mechanisms of dependence following protracted diazepam treatment. Neurosci Abs 551:5, 1398

Haefely W (1994) Allosteric modulation of the $GABA_A$ receptor channel: a mechanism for interaction with a multitude of central nervous system functions. In: Mohler H, DaPrada M (eds) The challenge of neuropharmacology. Editiones Roche, Basel, Switzerland, pp 15–39

Haigh JRM, Feely M (1988) Tolerance to the anticonvulsant effect of benzodiazepines. Trends Pharm Sci 9:361–366

Hendry SH, Huntsman MM, Vinuela A, Mohler H, DeBlas AL, Jones EG (1994) $GABA_A$ receptor subunit immunoreactivity in primate visual cortex: distribution in macaques and humans and regulation by visual input in adulthood. J Neurosci 14:2383–2401

Heninger C, Saito N, Tallman JF, Garrett KM, Vitek MP, Duman RS, Gallager DW (1990) Effects of continuous diazepam administration on $GABA_A$ receptor subunit mRNA in rat brain. J Mol Neurosci 2:100–107

Holt RA, Bateson AN, Martin IL (1996) Chronic treatment with diazepam or abecarnil differentially affects the expression of the $GABA_A$ receptor subunit mRNAs in the rat cortex. Neuropharmacology 35:1457–1463

Hu J, Ticku MK (1994) Chronic benzodiazepine agonist treatment produces functional uncoupling of the γ-aminobutyric acid-benzodiazepine receptor ionophore complex in cortical neurons. Mol Pharmacol 45:618–625

Impagnatiello F, Pesold C, Longone P, Caruncho H, Fritschy JM, Costa E, Guidotti A (1996) Modifications of γ-aminobutyric acid$_A$ receptor subunit expression in rat neocortex during tolerance to diazepam. Mol Pharm 49:822–831

Kang I, Miller LG (1991) Decreased $GABA_A$ receptor subunit mRNA concentrations following chronic lorazepam administration. Br J Pharmacol 103:1285–1287

Klein RL, Harris RA (1996) Regulation of $GABA_A$ receptor structure and function by chronic drug treatment in vivo and stably transfected cells. Jpn J Pharmacol 70:1–15
Knoflach F, Drescher U, Scheurer L, Malherbe P, Mohler H (1993) Full and partial agonism display by benzodiazepine receptor ligands at recombinant γ-aminobutyric $acid_a$ receptor subtypes. J Pharmacol Exp Ther 266:385–391
Koff JM, Pritchard GA, Greenblatt DJ, Miller LG (1997) The NMDA receptor competitive antagonist CPP modulates benzodiazepine tolerance and discontinuation. Pharmacology 55:217–227
Longone P, Impagnatiello F, Guidotti A, Costa E (1996) Reversible modification of $GABA_A$ receptor subunit mRNA expression during tolerance to diazepam-induced cognition dysfunction. Neuropharmacology 35:1467–1473
Loscher W, Rundfeldt C, Honack D, Ebert U (1996) Long-term studies on anticonvulsant tolerance and withdrawal characteristics of benzodiazepine receptor ligands in different seizure models in mice. I. Comparison of diazepam, clonazepam, clobazam, and abecarnil. J Pharmacol Exp Ther 279:561–572
MacDonald RL, Olsen RW (1994) $GABA_A$ receptor channels. Annu Rev Neurosci 17:569–602
Matsumoto K, Uzunova V, Pinna G, Taki K, Uzunov DP, Watanabe H, Mienville JM, Guidotti A, Costa E (1999) Permissive role of brain allopregnanolone content in the regulation of pentobarbital-induced righting reflex loss. Neuropharmacology (in press, 1999)
Miller LG, Greenblatt DJ, Barnhill JG, Shader RI (1988) Chronic benzodiazepine administration. I. Tolerance is associated with benzodiazepine receptor downregulation and decreased γ-aminobutyric $acid_A$ receptor function. J Pharmacol Exp Ther 246:170–176
Miller LG, Greenblatt DJ (1966) Benzodiazepine discontinuation syndrome: clinical and experimental aspects . In: Schuster CR, Kuhar MJ (eds) Handbook of Experimental Pharmacology, 118. Springer, Berlin Heidelberg New York, pp263–267
Mohler H, Benke D, Benson J, Luscher B, Fritschy JM (1995) $GABA_A$ receptor subtypes in vivo: cellular localization, pharmacology and regulation. In Biggio G, Sanna E, Serra M, Costa E (eds) GABA receptors and anxiety: from neurobiology to treatment. Raven Press, New York, pp41–56
Natolino F, Zanotti A, Contarino A, Lipartiti M, Giusti P (1996) Abecarnil, a beta-carboline derivative, does not exhibit anticonvulsant tolerance or withdrawal effects in mice. Naunyn-Schmiedeberg's Arch Pharmacol 354:612–617
Nutt D (1990) Pharmacological mechanisms of benzodiazepine withdrawal. J Psychiatry Res 24:105–110
O'Donovan MC, Buckland PR, Spurlock G, McGuffin P (1992) Bi-directional changes in the levels of messenger RNAs encoding γ-aminobutyric $acid_A$ receptor subunits after flurazepam treatment. Eur J Pharm Mol Pharm 226:335–341
Paronis CA, Costa E, Guidotti A, Bergman J (1997) The effects of bretazenil and imidazenil on schedule-controlled behavior in monkeys. Neurosci Abs 551:2, 1397
Pesold C, Caruncho H, Impagnatiello F, Berg MJ, Fritschy JM, Guidotti A, Costa E (1997) Tolerance to diazepam and changes in $GABA_A$ receptor subunit expression in rat neocortical areas. Neuroscience 79:477–487
Pratt JA, Brett RR, Laurie DJ (1998) Benzodiazepine dependence: from neural circuits to gene expression. Pharmacol Biochem Behav 59:925–934
Pribilla L, Neuhaus R, Huba R, Hillman M, Turner JD, Stephens DN, Schneider HH (1993) Abecarnil is a full agonist at some and a partial agonist at other recombinant $GABA_A$ receptor subtypes. In Stephens DN (ed) Anxiolytic β-carbolines from molecular biology to the clinic. Springer, Berlin Heidelberg New York, pp50–61
Price LH, Goddard AW, Barr L, Goodman WK (1995) Pharmacological challenges in anxiety disorders. In Bloom F, Kupfer D (eds) Psychopharmacology: the fourth generation of progress. Raven Press, New York, pp1311–1323
Primus RJ, Yu J, Xu J, Hartnett C, Meyyappan M, Kostas C, Ramabhadran TV, Gallager DW (1996) Allosteric uncoupling after chronic benzodiazepine exposure of

recombinant γ-aminobutyric $acid_A$ receptors expressed in Sf9 cells: ligand efficacy and subtype selectivity. J Pharmacol Exp Ther 276:882–890
Primus RJ, Gallager DW (1992) $GABA_A$ receptor subunit mRNA levels are differentially influenced by chronic FG 7142 and diazepam exposure. Eur J Pharmacol 226:21–28
Pritchett DB, Seeburg PH (1991) γ-Aminobutyric $acid_A$ receptor point mutation increases the affinity of compounds for the benzodiazepine site. Proc Natl Acad Sci 88:1421–1425
Rosenberg HC, Chiu TH (1981) Tolerance during chronic benzodiazepine treatment is associated with decreased receptor binding. Eur J Pharmacol 70:453–460
Rosenberg HC, Tietz EI, Chiu TH (1991) Differential tolerance to the antipentylenetetrazol activity of benzodiazepines in flurazepam treated rats. Pharm Biochem Behav 39:711–716
Russek SJ (1999) Evolution of $GABA_A$ receptor diversity in human genome. Gene 227:213–222
Ryan GP, Boisse NR (1983) Experimental induction of benzodiazepine tolerance and dependence. J Pharmacol Exp Ther 226:100–107
Serra M, Ghiani CA, Motzo C, Porceddu ML, Biggio G (1994) Long-term treatment with abecarnil fails to induce tolerance in mice. Eur J Pharmacol 259:1–6
Stephens DN, Schneider HH (1985) Tolerance to the benzodiazepine diazepam in an animal model of anxiolytic activity. Psychopharmacology 87:322–327
Stephens DN, Schneider HH, Kehr W, Andrews JS, Rettig KJ, Turski L, Schmiechen R, Turer JD, Jensen LH, Petersen EN, Honore T, Bondo-Hansen J (1990) Abecarnil, a metabolically stable, anxioselective β-carboline acting at benzodiazepine receptors. J Pharmacol Exp Ther 253:334–343
Steppuhn KG, Turski L (1993) Diazepam dependence prevented by glutamate antagonist. Proc Natl Acad Sci USA 90:6889–6893
Tehrani MH, Barnes EM Jr (1993) Identification of $GABA_A$/benzodiazepine receptors on clathrin-coated vesicles from rat brain. J Neurochem 60:1755–1761
Thompson DM, Auta J, Guidotti A, Costa E (1995) Imidazenil, a new anxiolytic and anticonvulsant drug attenuates a benzodiazepine-induced cognition deficit in monkeys. J Pharmacol Exp Ther 273:1307–1312
Tietz EI, Huang X, Weng X, Rosenberg HC, Chiu TH (1993) Expression of α1, β5 and β2 $GABA_A$ receptor subunit measured in situ in rat hippocampus and cortex following chronic flurazepam administration. J Mol Neurosci 4:277–292
Tietz EI, Rosenberg HC, Chiu TH (1986) Autoradiographic localization of benzodiazepine receptor downregulation. J Pharmacol Exp Ther 236:284–292
Tretter V, Ehya N, Fuchs K, Sieghart W (1997) Stoichiometry and assembly of recombinant $GABA_A$ receptor subtype. J Neurosci 17:2728–2737
Tsuda M, Suzuki T, Misawa M (1998) Region-specific changes in [3H]dizolcipine binding in diazepam-withdrawn rats. Neurosci Lett 24:113–115
Wood JH, Katz LA, Winger G (1992) Benzodiazepines: use, abuse, and consequences. Pharmacol Rev 44:151–347
Xie XH, Tietz EI (1991) Chronic benzodiazepine treatment of rats induces reduction of paired-pulse inhibition in CA1 region of in vitro hippocampus. Brain Res 561:69–76
Zanotti A, Mariot R, Contarino A, Lipartiti M, Giusti P (1996) Lack of anticonvulsant tolerance and benzodiazepine receptor down-regulation with imidazenil in rats. Br J Pharmacol 117:647–652
Zhao T, Chiu TH, Rosenberg HC (1994) Reduced expression of γ-aminobutyric acid type A/benzodiazepine receptor γ2 and α5 subunit mRNAs in brain regions of flurazepam-treated rats. J Pharm Exp Ther 45:657–663
Zheng X, Xie XH, Tietz EI (1993) Fast monosynaptic inhibition is reduced in CA1 region of hippocampus after chronic flurazepam treatment. Soc Neurosci Abst 19:1141

CHAPTER 9

$GABA_A$ Receptors and Disease

H. Y. KIM and R.W. OLSEN

1. Introduction

Studies of human diseases of the nervous system have demonstrated that many of the disorders result from disruption of normal developmental processes which promote organization and maturation of neuronal circuitry. Such disorders include many different forms of inherited childhood epilepsies in which genetic factors contribute to the abnormal development. Adult disorders can be influenced by genetic factors as well. Linkage analysis and family- or population-based association studies are useful in finding genes that are responsible for simple disorders, but not for complex human disorders which involve multiple genes. In addition to genetic mutations and variations, many disorders in adult are influenced by environmental insults. Based on the dual function of GABA as the main inhibitory neurotransmitter in adult and as a developmental factor during embryogenesis and early postnatal life, one can envision the perturbing consequences of aberrant GABA actions on neurophysiology. Recent advances in gene knock-out technology have yielded a wealth of evidence suggesting that, indeed, mutations within the GABA signaling pathway can result in a wide array of neurodevelopmental abnormalities, as well as abnormalities in non-neuronal structures such as the palate (CULIAT et al. 1995; WAYMIRE et al. 1995; ASADA et al. 1997; KASH et al. 1997).

GABA exerts its actions via a chloride channel, the $GABA_A$ receptors, and the G-protein coupled $GABA_B$ receptors. Binding of GABA to the $GABA_A$ receptors can cause either inhibition or excitation of neurons depending on the type of neurons and their microenvironment, local circuits, or perhaps, types of $GABA_A$ receptors present in the particular neuron (CHERUBINI et al. 1991). $GABA_A$ responses are chloride currents, which are generally inhibitory. These can be depolarizing if cellular chloride transport is weak, as happens especially early in life (RIVERA et al. 1999). Although less is known about the $GABA_B$ receptors, binding of GABA to the $GABA_B$ receptors appears to cause only inhibition, and subsequent inhibition of action potential generation, or inhibition of transmitter release from nerve endings.

There are at least 19 separate $GABA_A$ receptor subunits (α [1–6], β [1–4], γ [1–3], δ, ρ [1–3], ε, and π) identified to date (BARNARD et al. 1998) and these subunits have overlapping but distinct expression patterns both in time and space (LÜDDENS and WISDEN 1991). At least three different subunit polypeptides form a pentameric $GABA_A$ receptor complex (MACDONALD and OLSEN 1994). Expression studies of recombinant $GABA_A$ receptors in *Xenopus* oocytes and mammalian cell lines have shown that receptors comprised of different subunit combinations display distinct electrophysiological and pharmacological profiles (LUDDENS et al. 1995; SIEGHART 1995; MCKERNAN and WHITING 1996). In development, the expression pattern of $GABA_A$ receptor subunits differs from that in the adult (reviewed in KIM et al. 1996). Generally, the most prominent subunits in the developing brain are expressed less in adults and the less abundant subunits in the developing brain become more prominent in adults. Therefore, the molecular biology of $GABA_A$ receptors indicates that the brain can possess myriad $GABA_A$ receptors with varied function. Further, mutations which affect types (therefore function) and distribution of the $GABA_A$ receptors can alter cell-cell and cell-environment interactions. In this review we consider the role of GABA and $GABA_A$ receptors in diseases that arise from both developmental aberrations and environmental insults later in life.

B. Diseases of Development and $GABA_A$ Receptors

GABA and $GABA_A$ receptors are present as early as embryonic day 14 in rodent embryos, and this has led researchers to suggest that GABA may play a role in modulating brain development via acting through the $GABA_A$ receptors (COYLE and ENNA 1976; LAUDER et al. 1986; LAURIE et al. 1992; MA et al. 1993). A series of studies using cultured primary neurons, brain slices, as well as intact animals further suggested that GABA promotes neuronal migration, cytodifferentiation and synaptogenesis (HANSEN et al. 1984; MEIER et al. 1984, 1985; SPOERRI 1987; WOLFF et al. 1987; BARBIN et al. 1993; KIM et al. 1993; BEHAR et al. 1994; MITCHELL and REDBURN 1996), partly by increasing intracellular calcium concentration in immature neurons (CONNOR et al. 1987; YUSTE and KATZ 1991; LEINEKUGEL et al. 1995; OBRIETAN and VAN DEN POL 1996; OWENS et al. 1996; CHERUBINI et al. 1998). Stronger evidence, however, comes from studies using genetically engineered mouse models which allow a more direct assessment of relationship between genotype and phenotype in an intact animal. Knockout gene targeting strategies have the potential for revealing physiologic and pathophysiologic roles of a given gene product with certain caveats. Many null mutations are embryonically or perinatally lethal. In null mutants, loss of a gene product in all cells, throughout life, is a drastic situation, more severe than many human genetic diseases, which may result from altered rather than loss-of-function mutations. Other knockouts may show little or no phenotypes. In addition, whether or not a compensatory gene activity occurs in response to absence of a gene certainly introduces questions, making it difficult to establish a direct relationship between a gene function and the mutant phenotype. Nevertheless, one advantage of gene targeting is that it can produce a large number of animals carrying an identical mutation that can be analyzed. Extensive analysis of these mice will reveal clues not only to the role of the targeted gene in the phenotype, but also to the compensatory events which will provide information about interacting genetic pathways as well as provide additional targets for therapies for specific disorders. Finally, variable expression of phenotypic characteristics in mouse strains of differing genetic background can lead to the discovery of genes whose products modify the expression of the targeted gene; these in turn, can lead to identification of phenotype pathways.

To date, at least four (α_6, β_3, γ_2, and δ) different GABA$_A$ receptor subunit genes have been disrupted in mice (GUNTHER et al. 1995; HOMANICS et al. 1997a,b; JONES et al. 1997; OLSEN et al. 1997b). Each one of these mutant mice displays a wide variety of overlapping yet distinct set of physiological and behavioral deficits allowing the dissection of the role of individual GABA$_A$ receptor subunit genes.

Among the aforementioned GABA$_A$ receptor knock-out experiments, the β_3 subunit gene disruption produced mice with a particularly interesting array of phenotypic characteristics, presumably due to lack of the β_3 subunit function during embryonic development (HOMANICS et al. 1997a). Mice with the GABA$_A$ receptor β_3 subunit disruption exhibit electroencephalographic abnormalities, seizures, learning and memory deficits, poor motor skills on a repetitive task, hyperactivity, and a disturbed rest-activity cycle (DELOREY et al. 1998). These same abnormal behaviors are associated with Angelman syndrome (AS), a neurodevelopmental genetic disorder characterized by severe mental retardation and epilepsy. This is first direct evidence associating a GABA$_A$ receptor subunit gene with an inherited human disorder. Further, the β_3 knockout mice exhibited increased oscillatory synchrony in the thalamic reticular nucleus, which suggests that the GABA$_A$ receptor mediated inhibition is critical for normal (non-seizure) modulation of neuronal rhythms (HUNTSMAN et al. 1999).

In addition to mice with targeted mutations, there are several genetic rodent models of epilepsy in which alterations in the GABA$_A$ receptor genes are thought to influence the animal's seizure threshold. These include the tottering mice, E1 mice, genetically epilepsy-prone rats (GEPR), genetic absence epilepsy rats of Strasbourg (GAERS), and seizure-susceptible gerbils. The tottering mouse, a model of absence and myoclonic seizures, has impaired GABA$_A$ receptor function (TEHRANI and BARNES 1995) and increased levels of GABA$_A$ receptor α_2 and β_2 mRNAs (TEHRANI et al. 1997), although the mutation is in the α_{1A} subunit of voltage-sensitive calcium channel (FLETCHER et al. 1996). Decreased hippocampal GABA uptake is believed to be involved in seizure activities of the E1 mouse, a model of temporal lobe epilepsy (JANJUA et al. 1991). Impaired GABA$_A$ receptor function may be responsible for convulsive seizures in GEPRs, a model of generalized motor seizures (EVANS et al. 1994). GAERS show decreased GABA$_A$ receptor binding in hippocampus and altered $GABA_B$ receptor function (SNEAD et al. 1992). Genetically seizure-susceptible gerbils exhibit decreased expression of $GABA_A$ receptors in their substantia nigra and midbrain regions (OLSEN et al. 1986). Our current understanding of whether, and, if so, how these $GABA_A$ receptor alterations contribute to seizure susceptibility is limited. Since we do know that the $GABA_A$ receptors are involved in early CNS development, careful analysis of mutant embryos of $GABA_A$ receptor expression and function may allow identification of perturbed steps that might lead to the abnormal phenotypes.

In human studies, $GABA_A$ receptor subunit genes serve as major candidates for inherited epilepsy disorders. For example, SANDER et al. (1997) performed a linkage analysis between subtypes of idiopathic generalized epilepsy and the $GABA_A$ receptor α_5, β_3, γ_3 gene cluster on chromosome 15, and found a possible linkage between families of juvenile myoclonic epilepsy (JME) patients and the $GABA_A$ receptor gene cluster. On the contrary, a different study ruled out an association between a separate group of JME patients and dinucleotide repeat allelic variants of the α_5 or β_3 gene (GUIPPONI et al. 1997). Although the results are somewhat inconclusive with the families tested, one cannot exclude a better linkage or association in different affected families since epilepsy disorders are both heterogeneous and also may arise from multiple mutations.

C. Diseases of Adult and $GABA_A$ Receptors

The power of genetic engineering now allows the identification of individual $GABA_A$ receptor genes that can cause neurological or psychiatric disease, and further, provides a mechanism to distinguish the developmental defects from the defects which may arise from the absence of a particular gene product in the adult brain (DYMECKI 1996; MARTH 1996). These techniques are crucial in understanding the role of $GABA_A$ receptor genes, since most subunit genes display highly complex expression patterns, suggesting functional diversity not only among different subunit genes but also of the same gene depending on time and space. Studies using such conditional gene targeting approaches are still in their infancy.

We summarize here the results obtained from conventional biochemical, pharmacological, behavioral, and anatomical methods that suggest a role of $GABA_A$ receptor function in human and animal disorders. In many cases it is unclear whether the altered $GABA_A$ receptors cause a particular disorder, or they merely represent a plastic change in response to some physiological trauma. Regardless of whether the changes are causal or consequential, understanding the mechanism of such changes in the disease cascade will provide the basis for developing valuable therapies for the particular disorder.

I. $GABA_A$ Receptor Function in Adult Epilepsy

Perhaps the main human disorder associated with defects in the $GABA_A$ receptor system is epilepsy (OLSEN and AVOLI 1997). In many models of epilepsy, both genetic and experimental, GABA$_A$ receptor function is altered in some fashion – either enhanced or attenuated. The alterations in the GABA$_A$ receptor function may represent the brain's way of adjusting to a trauma brought on by the seizures themselves. Ironically, in some cases, these changes can contribute to further problems. One classical example of such case is temporal lobe epilepsy (TLE) in human, one of the most prevalent seizure disorders in adults. TLE is characterized by development of spontaneous seizures after a brain injury. The process of seizure development includes extensive loss of hilar principal cells and GABAergic interneurons as well as synaptic reorganization of the dentate gyrus (reviewed in HOUSER 1992; OBENAUS et al. 1993).

In an experimental model of TLE, rats are induced to have status epilepticus with a pilocarpine injection. These rats develop increased seizure susceptibility long after – 2 or more weeks – their initial status epilepticus episode. Using this model, BROOKS-KAYAL et al. (1998) demonstrated that the dentate granule cells from epileptic rats had alterations in GABA$_A$ receptor function including increased zinc sensitivity and decreased zolpidem enhancement, as well as changes in the mRNA levels of several GABA$_A$ receptor subunit genes. The authors further demonstrated that these alterations precede the onset of epilepsy, suggesting a causal relationship between the altered GABA$_A$ receptors, apparently of a new subunit composition, and the subsequent development of epilepsy. Also in the pilocarpine-induced status epilepticus model, HOUSER et al. (1995) used in situ hybridization to demonstrate a reduced level of GABA$_A$ receptor α_5 subunit mRNA in CA1 at the time that spontaneous seizures developed. Similarly, in the kainic acid-induced TLE model, the initial status epilepticus leads to massive changes in the GABA$_A$ receptor composition in cells throughout hippocampus (FRIEDMAN et al. 1994; SCHWARZER et al. 1997; TSUNASHIMA et al. 1997). Some of these changes in GABA$_A$ receptor subunit expression may be transient, and suggest that alterations in other genetic pathways are responsible for kindling-induced increase of seizure susceptibility (KOKAIA et al. 1994). In human TLE tissue, a decrease in the number of GABA transporters and the subsequent decrease in glutamate-induced GABA release were observed (DURING et al. 1995).

Likewise, human studies using PET scanning or surgical samples of patients have reported altered GABA$_A$ receptor binding and function (SAVIC et al. 1988; MCDONALD et al. 1991; JOHNSON et al. 1992; OLSEN et al. 1992; HENRY et al. 1993; GIBBS et al. 1996; SHUMATE et al. 1998). The results from human studies can, however, be variable, perhaps due to heterogeneity of epilepsies among patients, normal population variability, difficulties in working with surgical or post-mortem tissue samples, and difficulties in making precise anatomical comparisons between patients. Generally speaking, patients with focal epilepsy have reduced GABA$_A$ receptor binding, although the reduction may be a result of cell loss in the damaged brain (OLSEN et al. 1992). However, functional changes of GABA$_A$ receptors seen in isolated neurons from TLE patients resemble changes seen in corresponding cells of epileptic rats (Gibbs et al. 1996; Shumate et al. 1998).

$GABA_A$ receptor alterations were observed in neocortex from TLE surgical samples from patients with brain pathologies of differing severity. In the first set, neurosteroid enhancement of [^{3}H]flunitrazepam binding was increased in samples from patients with hippocampal sclerosis but not in samples from TLE patients with tumors, or normal autopsy cases (Van Ness et al. 1995). In the second set, neurosteroid enhancement of [^{35}S]TBPS binding and diazepam-insensitive binding of [^{3}H]RO15–4513 were increased in samples from patients with severe sclerosis and sprouting in the dentate gyrus, but not in samples from patients with less or no sprouting (Olsen et al. 1995). Increased neurosteroid modulation and [^{3}H]RO15–4513 binding suggests that $GABA_A$ receptor subunit composition may have been altered in the brain of these patients, perhaps nature's way to compensate for the increased activity during seizures. It remains possible that the receptor changes contribute to the epileptogenesis. Interestingly, increased level of the diazepam-insensitive receptors – containing the α_4 subunit – and increased steroid modulation of $GABA_A$ receptors were observed in the chronic intermittent ethanol (CIE) treated rat withdrawal kindling model (Mahmoudi et al. 1997; Kang et al. 1998). The altered $GABA_A$ receptor function was accompanied by increased seizure susceptibility (Kang et al. 1996). A similar increase in the α_4 subunit was observed in animals withdrawn from chronic ethanol (Devaud et al. 1995a, 1997), or withdrawn from chronic neurosteroid administration (Smith et al. 1998a, 1998b), and in the rat kindling model of epilepsy (Clark et al. 1994).

Kindling involves intermittent exposure to a subconvulsant dose of a convulsant chemical or electrical stimulus, which after sufficient number of repeats increases seizure susceptibility (Goddard et al. 1969; Lewin et al. 1989). Other $GABA_A$ receptor subunit changes were reported in both CA1 (Kamphuis et al. 1995) and dentate gyrus (Nusser et al. 1998) of kindled rats. On the other hand, the α_4 subunit decreased and α_1 increased in the thalamus of rats undergoing experimental absence seizures (Banerjee et al. 1998a) along with decreased steroid modulation of binding (Banerjee et al. 1998b). The possible relevance of these subunits to epilepsy is reviewed in Olsen et al. (1999). Removal of chronic, intracortical GABA infusion into rats or monkeys leads to localized seizures at the infusion site, the "GABA withdrawal syndrome" (Brailowsky et al. 1988). Salazar et al. (1994) suggested that the seizures may stem from reduced rate of GABA synthesis in the infused cortex, although GABA levels did not correlate with the seizure time duration, suggesting other more persistent alterations. Finally, intrahippocampal infusion of antisense oligodeoxynucleotides for the γ_2 subunit in adult rat results in limbic status epilepticus and neurodegeneration (Karle et al. 1998), suggesting that impairment of the $GABA_A$ receptor function can directly cause epilepsy.

II. $GABA_A$ Receptor Function in Anxiety

There are several different anxiety disorders in human – panic disorder, generalized anxiety disorder, post-traumatic stress disorder, social phobia, simple phobias, and obsessive compulsive disorder – and benzodiazepines are effective in treating most of these anxiety disorders with the exception of phobic anxieties. Given that the benzodiazepines work via a subset of $GABA_A$ receptors (COSTA 1998), much work in psychopharmacology has concentrated on studies to understand the mechanism of how GABA/benzodiazepines receptor activation might alleviate anxiety. Here, we describe some of the recent studies that illustrate relationship between specific $GABA_A$/benzodiazepines receptor function and specific human anxiety disorders. Many of these findings come from studies using animal models, developed to mimic human anxiety disorders. The hope is to use the animal models to study the various neurobiological mechanisms of anxiety as well as test for better anxiolytic drugs.

Among some of the well-validated and often used tests of anxiety is the social interaction test, which probably best reflects the generalized anxiety disorder in human (reviewed in FILE 1995). It measures the time spent in social interaction when rats are placed in an unfamiliar or brightly lit environment. Administration of anxiolytic drugs allows rats to feel at ease, and therefore spend more time in such an environment. Using the social interaction test, SANDERS and SHEKHAR (1995b) demonstrated that muscimol-induced $GABA_A$ receptor activation in the rat central nucleus of the amygdala results in anxiolytic effects but not in basolateral amygdala. Injection of a benzodiazepine (chlordiazepoxide) into basolateral amygdala, however, resulted in anxiolytic effects (SANDERS and SHEKHAR 1995a), suggesting that different types of $GABA_A$ receptors regulate different neural pathways leading to the development of anxiety (COSTA 1998). Furthermore, it predicts that regional variations in the $GABA_A$/benzodiazepine receptor complex will contribute to different states of anxiety. Further, clinical anxiety may be accompanied by, and possibly result from, experiences that change $GABA_A$ receptors. In support of this argument, conflict behavior training and shock treatments, applied to rats in animal conflict paradigms, also used to screen drugs for anxiolytic effects, result in selective changes in the $GABA_A$ α_1 or α_2 subunits, respectively (ZHANG et al. 1998).

Conditions that induce acute stress in rats or panic attacks in humans result in elevated levels of neuroactive steroids (PURDY et al. 1991; BARBACCIA et al. 1997). In the pseudopregnant rat model, $GABA_A$ receptor modulating neurosteroid 3α-hydroxy-5α-pregnan-20-one (allopregnanolone) withdrawal results in increased anxiety, as measured by the elevated plus maze protocol, and the increased anxiety is accompanied by an increase in $GABA_A$ receptor α_4 subunit mRNA and protein levels (SMITH et al. 1998a,b). Allopregnanolone, a metabolite of progesterone, and other neuroactive steroids enhance GABA-induced chloride currents by allosteric modulation of ligand binding to the

$GABA_A$ receptor (Majewska et al. 1986; Gee et al. 1988; Turner et al. 1989; Nguyen et al. 1995; Olsen and Sapp 1995). Interestingly, increased α_4 subunit is often accompanied by increased steroid modulation of $GABA_A$ receptor binding and increased hyperexcitability (Olsen et al. 1999). These observations perhaps imply a reciprocal relationship between the α_4 subunit and steroid enhancement of the $GABA_A$ receptors, consistent with the observation that the α_4 subunit is often associated with the δ subunit which appears to reduce steroid sensitivity (Zhu et al. 1996). Therefore, assuming that the endogenous neurosteroids help maintain brain activity, one might imagine a compensatory increase in the neurosteroid enhancement when the increased level of α_4 subunit produces less functional $GABA_A$ receptors. Conversely, when a stressful situation leads to elevated levels of neuroactive steroids, thus leading to excessive GABAergic activity, the body may respond by producing modified $GABA_A$ receptors that are less sensitive to neurosteroid-enhanced inhibition (Concas et al. 1998).

In human panic disorder patients, the saccadic eye movement paradigm is often used to test effects of benzodiazepines (Roy-Byrne et al. 1990). Using this test, Roy-Byrne et al. (1990) demonstrated that patients with panic disorder are less sensitive than controls to diazepam, implicating genetic differences in the GABA/benzodiazepine receptor function in panic disorder patients.

In addition to data from experimental model systems, recent development in sophisticated imaging techniques has led to the documentation of altered benzodiazepine receptor function in the brains of panic disorder patients. ^{123}I-iomazenil single photon emission tomography (SPECT) study showed increased benzodiazepine binding in the prefrontal cortex of patients with panic disorder (Kuikka et al. 1995). On the contrary, decreased benzodiazepine binding was observed in the frontal, occipital and temporal cortices of a different set of panic disorder patients when compared to epileptic patients (Schlegel et al. 1994). Likewise, using PET, Malizia et al. (1998) observed reduction of benzodiazepine binding in the brain of panic disorder patients, with the most dramatic reduction in the right orbitofrontal cortex and the right insula, regions which are thought to be essential in the mediation of anxiety.

III. $GABA_A$ Receptor Function in Alcoholism

A growing body of evidence suggests that ethanol mediates some of its effects via the $GABA_A$ receptor complex (reviewed in Faingold et al. 1998; Harris et al. 1998). Electrophysiological studies using isolated neuronal cultures, recombinant $GABA_A$ receptor expression in mammalian cell lines, and brain slices demonstrate that ethanol enhances GABA-induced hyperpolarizing effect of the $GABA_A$ receptors (Narahashi et al. 1991; Snead et al. 1992; Frye et al. 1996; Soldo et al 1998). Pharmacological and behavioral analyses show that GABA$_A$ receptors can modulate ethanol intake behavior in rats (HODGE et al. 1995). Studies comparing alcohol-preferring rats vs alcohol-non-preferring rats show that GABA$_A$ receptors can influence ethanol dependence and withdrawal when associated with certain genetic conditions (ALLAN and HARRIS 1991; NOWAK et al. 1998). The mechanism which underlies the ethanol potentiation is less clear, although subunit composition (WAFFORD et al. 1991; CRISWELL et al. 1995; CREWS et al. 1996), protein phosphorylation of receptor subunit polypeptides (SAMSON and HARRIS 1992; WAFFORD and WHITING 1992; LIN et al. 1993; WAFFORD et al. 1993; HARRIS et al. 1995, 1998; KLEIN and HARRIS 1996), and modulation by neuroactive steroids are suggested as possible means to regulate ethanol sensitivity of the GABA$_A$ receptors (DEVAUD et al. 1995a; MEHTA and TICKU 1998). For example, some investigators believe that phosphorylation of the γ_{2L} subunit is important (SAMSON and HARRIS 1992; WAFFORD and WHITING 1992; WAFFORD et al. 1993) in GABA-induced potentiation by ethanol, but some studies clearly do not agree. First, recombinant receptors in *Xenopus* oocytes do not show differential response to ethanol when γ_{2S} subunit is replaced by γ_{2L} subunit (SIGEL et al. 1993). Second, ethanol enhances GABA-induced currents in immature cerebellar Purkinje cells which express γ_{2S} and not γ_{2L} (SAPP and YEH 1998). Similarly, GABA$_A$ receptor-activated currents in primary cultures of rat dorsal root ganglion (DRG) cells are insensitive to ethanol, despite the expression of phosphorylated γ_{2L} subunit in the cells (ZHAI et al. 1998). Finally, γ_{2L} knock-out mice are not affected in any assay of ethanol sensitivity (HOMANICS et al. 1998).

Chronic ethanol treatment in cultured cells and animals can result in changes (either increase or decrease) in levels of specific GABA$_A$ receptor subunit mRNAs and proteins (MHATRE and TICKU 1992, 1994; MHATRE et al. 1993; DEVAUD and MORROW 1994; DEVAUD et al. 1995b; HIROUCHI et al. 1993; KLEIN et al. 1995; KLEIN and HARRIS 1996; TABAKOFF and HOFFMAN 1996). Likewise, in humans, long term alcohol consumption results in variations in GABA$_A$ receptor function, possibly resulting from altered subunit composition (DODD 1994; LEWOHL et al. 1996). For example, increased levels of the GABA$_A$ receptor β_3 and the α_1 subunit mRNAs were observed in alcoholic postmortem frontal cortical sections (MITSUYAMA et al. 1998). Similarly, THOMAS et al. (1998) found an increased level of the GABA$_A$ receptor α_1 subunit mRNA in the motor cortex, as well as increased levels of the α_3, and the β_3 subunit mRNAs in the frontal cortex of the postmortem tissues of alcoholics who also had cirrhosis of the liver. The postmortem tissues of alcoholics who did not have cirrhosis of the liver showed no change in the corresponding GABA$_A$ receptor mRNA levels. In a separate study, LEWOHL et al. (1997) reported an increased level of α_1 mRNA expression in postmortem cortical tissues of both groups of alcoholics, with and without cirrhosis of liver, when compared to non-alcoholic control groups. Such alteration, or adaptation, can produce GABA$_A$ receptors with decreased GABA-mediated Cl^- flux, that are less responsive to many of its agonists such as ethanol itself, pentobarbital and flunitrazepam, yet more responsive to benzodiazepine inverse agonists

(ALLAN and HARRIS 1987; MORROW et al. 1988; BUCK and HARRIS 1990; MHATRE and TICKU 1989, 1992; MHATRE et al. 1993). Consequently, these changes following chronic exposure to ethanol are proposed, at least in part, to underlie physical dependence and withdrawal, and aspects of alcoholism. Similarly, rats treated with chronic intermittent ethanol (CIE) show reduction of $GABA_A$ receptor function, accompanied by increased seizure susceptibility (KANG et al. 1996), increased levels of α_4 subunit (MAHMOUDI et al. 1997), and altered $GABA_A$ receptor pharmacology in hippocampal slices, consistent with altered subunit composition (KANG et al. 1998).

The types and degrees of $GABA_A$ receptor adaptation in response to chronic ethanol exposure appear to depend on the animal's genetic variation as well as gender (ALLAN and HARRIS 1991; DEVAUD et al. 1995b, 1998). More direct evidence demonstrating the importance of $GABA_A$ receptors in the development of alcoholism comes from two recent studies, both using human subjects, which found an intriguing causal association between risk for alcoholism and CA dinucleotide repeats of two $GABA_A$ receptor subunit genes, α_3 and β_3 (PARSIAN and CLONINGER 1997; NOBLE et al. 1998). The mechanism of how these $GABA_A$ receptor subunit allelic variants contribute to the development of alcoholism is unknown. It is also worth noting that when the individual possesses the 'alcoholic' allele of the $GABA_A$ receptor β_3 gene, in addition to the D2 dopamine receptor (DRD2) Al allele, the risk for developing alcoholism increases (NOBLE et al. 1998). Since alcoholism encompasses a wide range of physiological and emotional changes, one should perhaps expect multiple genes to modulate genetic pathways that can lead to the development of alcoholism.

D. Conclusion

Alterations in $GABA_A$ receptor function have been implicated in several pathological conditions other than the ones mentioned in this chapter. Mutations in the $GABA_A$ receptor β_3 subunit produce cleft palate and sleep disorder in mice (CULIAT et al. 1995; HOMANICS et al. 1997a). Recently, HUNTSMAN et al. (1998) demonstrated that the ratio of $GABA_A$ receptor γ_{2L} and γ_{2S} mRNAs was altered – reduction of γ_{2L} and increase of γ_{2S} – in prefrontal cortex of schizophrenics. A similar change was seen in primary cultured neurons exposed to chronic barbiturates (TYNDALE et al. 1997) and in the CIE rat model of alcohol dependence (R.F. Tyndale, D.W. Sapp and R.W. Olsen, unpublished results), and may represent aberrant plasticity in $GABA_A$ receptors. Further, modulation of hormone secretion in pancreas may involve specific $GABA_A$ receptor subtypes (RORSMAN et al. 1989; BORBONI et al. 1994; YANG et al. 1994; VON BLANKENFELD et al. 1995). In an experimental model of Huntington's disease (HD), administration of quinolinic acid in rats results in a lesion in the striatum (NICHOLSON et al. 1995), which is accompanied by increases and alterations in the $GABA_A$ receptor subtypes in the substantia nigra (NICHOLSON et al. 1996), analogous to increased GABA binding seen in striatal output regions in HD patients (ENNA et al. 1976). Finally, a ^{123}I-iomazenil SPECT study found a correlation between severity of motor impairment in Parkinson's disease and decrease in ^{123}I-iomazenil uptake (KAWABATA and TACHIBANA 1997). Again, one needs to determine that the reduction is specific for GABA$_A$ receptors and not due to cell loss which can be substantial in neurodegenerative diseases.

Given the ubiquitous nature of GABA$_A$ receptors and their interactions with a wide variety of clinically relevant drugs, it is very tempting to associate changes in GABA$_A$ receptors with various aspects of human illnesses. The challenge would be not only to determine which changes are causal for a particular illness, but to understand clearly what certain changes mean and to use this knowledge to develop new and improved therapeutics for that particular illness. This implies that we still have to elucidate the role of an ever-increasing number of subunits and the mechanisms responsible for modulating these subunits. The continued development of new technologies, especially in gene targeting, provides much hope. We now have abilities to remodel the mouse genome by using site-specific recombination systems such as Cre-loxP (reviewed in MARTH 1996) and FLP/FRT (reviewed in DYMECKI 1996). One can manipulate multiple genes – perhaps in a particular genetic pathway – by using both systems either sequentially or simultaneously in the same cell. This type of approach is crucial in understanding disorders that involve defects in more than one gene. Furthermore, the site-specific recombination systems can be exploited in combination with an inducible system (MANSUY et al. 1998) to achieve even higher specificity and complexity. The result can be a creation of a multipotential mouse in which genes can be turned on or off at will. As many illnesses are complex and involve multiple pathways, the key to our understanding of gene function and human phenotype may lie in successful use of conditional genetics.

Acknowledgement. Supported by NIH grants NS28772 and AA07680.

References

Allan AM, Harris RA (1987) Acute and chronic ethanol treatments alter GABA-receptor operated chloride channels. Pharmacol Biochem Behav 27:665–670
Allan AM, Harris RA (1991) Neurochemical studies of genetic differences in alcohol action. In: Crabbe JC Jr, Harris RA (eds) The genetic basis of alcohol and drug actions. Plenum Press, New York, pp 105–152
Asada H, Kawamura Y, Maruyama K, Kume H, Ding RG, Kanbara N, Kuzume H, Sanbo M, Yagi T, Obata K (1997) Cleft palate and decreased brain γ-aminobutyric acid in mice lacking the 67-kDa isoform of glutamic acid decarboxylase. Proc Natl Acad Sci USA 94:6496–6499
Banerjee PK, Tillakaratne NJ, Brailowsky S, Olsen RW, Tobin AJ, Snead OC (1998a) Alterations in GABA$_A$ receptor α_1 and α_4 subunit mRNA levels in thalamic relay nuclei following absence-like seizures in rats. Exp Neurol 154:213–223
Banerjee PK, Olsen RW, Tillakaratne NJ, Brailowsky S, Tobin AJ, Snead OC (1998b) Absence seizures decrease steroid modulation of t-[^{35}S]butylbicyclophoshorothionate binding in thalamic relay nuclei. J Pharmacol Exp Ther 287:766–772

Barbaccia ML, Roscetti G, Trabucchi M, Purdy RH, Mostallino MC, Concas A, Biggio G (1997) The effects of inhibitors of GABAergic transmission and stress on brain and plasma allopregnanolone concentrations. Brit J Pharmacol 120:1582–1588
Barbin G, Pollard H, Gaiarsa JL, Ben-Ari Y (1993) Involvement of $GABA_A$ receptors in the outgrowth of cultured hippocampal neurons. Neurosci Lett 152:150–154
Barnard EA, Skolnick P, Olsen RW, Möhler H, Sieghart W, Biggio G, Braestrup C, Bateson AN, Langer SZ (1998) International Union of Pharmacology. XV. Subtypes of γ-aminobutyric $acid_A$ receptors: classification on the basis of subunit structure and receptor function. Pharmacol Rev 50:291–313
Behar TN, Schaffner AE, Colton CA, Somogyi R, Olah Z, Lehel E, Barker JL (1994) GABA induced chemokinesis and NGF-induced chemotaxis of embryonic spinal cord neurons. J Neurosci 14:29–38
Borboni P, Porzio O, Fusco A, Sesti G, Lauro R, Marlier LN (1994) Molecular and cellular characterization of the $GABA_A$ receptor in the rat pancreas. Mol Cell Endocrinol 103:157–163
Brailowsky S, Kunimoto M, Menini C, Silva-Barrat S, Riche D, Naquet TR (1988) The GABA-withdrawal syndrome: A new model of focal epileptogenesis. Brain Res 442:175–179
Brooks-Kayal AR, Shumate MD, Jin H, Rikhter TY, Coulter DA (1998) Selective changes in single cell $GABA_A$ receptor subunit expression and function in temporal lobe epilepsy. Nat Med 4:1166–1172
Buck KJ, Harris RA (1990) Benzodiazepine agonist and inverse agonist actions on $GABA_A$ receptor-operated chloride channels. II. Chronic effects of ethanol. J Pharmacol Exp Ther 253:713–719
Cherubini E, Gaiarsa J, Ben-Ari Y (1991) GABA: an excitatory transmitter in early postnatal life. Trends Neurosci 14:515–519
Cherubini E, Martina M, Sciancalepore M, Strata F (1998) GABA excites immature CA3 pyramidal cells through bicuculline-sensitive and -insensitive chloride-dependent receptors. Perspect Devel Neurobiol 5:289–304
Clark M, Massenburg GS, Weiss SR, Post RM (1994) Analysis of the hippocampal $GABA_A$ receptor system in kindled rats by autoradiographic and in situ hybridization techniques: contingent tolerance to carbamazepine. Brain Res Mol Brain Res 26:309–319
Concas A, Mostallino MC, Porcu P, Follesa P, Barbaccia ML, Trabucchi M, Purdy RH, Grisenti P, Biggio G (1998) Role of brain allopregnanolone in the plasticity of γ-aminobutyric acid $type_A$ receptor in rat brain during pregnancy and after delivery. Proc Natl Acad Sci USA 95:13284–13289
Connor JA, Tseng HY, Hockberger PE (1987) Depolarization- and transmitter-induced changes in intracellular Ca^{2+} of rat cerebellar granule cells in explant cultures. J Neurosci 7:1384–1400
Costa E (1998) From $GABA_A$ receptor diversity emerges a unified vision of GABAergic inhibition. Ann Rev Pharmacol Toxicol 38:321–350
Coyle JT, Enna SJ (1976) Neurochemical aspects of the ontogenesis of GABAergic neurons in the rat brain. Brain Res 111:119–133
Crews FT, Morrow AL, Criswell H, Breeese G (1996) Effects of ethanol on ion channels. Int Rev Neurobiol 39:283–367
Criswell HE, Simson PE, Knapp DJ, Devaud LL, McCown TJ, Duncan GE, Morrow AL, Breese GR (1995) Effect of zolpidem on γ-aminobutyric acid (GABA)-induced inhibition predicts the interaction of ethanol with GABA on individual neurons in several rat brain regions. J Pharmacol Exp Ther 273:526–536
Culiat CT, Stubbs LJ, Woychik RP, Russell LB, Johnson DK, Rinchik EM (1995) Deficiency of the β_3 subunit of the type A γ-aminobutyric acid receptor causes cleft palate in mice. Nat Genet 11:344–346
DeLorey TM, Handforth A, Homanics GE, Minassian BA, Asatourian A, Anagnostaras SG, Fanselow MS, Ellison G, Delgado-Escueta AV, Olsen RW (1998) Mice lacking the $\beta3$ subunit of the $GABA_A$ receptor have the epilepsy phenotype and

many of the behavioral characteristics of Angelman syndrome. J Neurosci 18:8505–8514
Devaud LL, Morrow AL (1994) Effects of chronic ethanol administration on [^{3}H]zolpidem binding in rat brain. Eur J Pharmacol 267:243–247
Devaud LL, Purdy RH, Morrow AL (1995a) The neurosteroid, 3 α-hydroxy-5 α-pregnan-20-one, protects against bicuculline-induced seizures during ethanol withdrawal in rats. Alcohol Clin Exp Res 19:350–355
Devaud LL, Smith FD, Grayson DR, Morrow AL (1995b) Chronic ethanol consumption differentially alters the expression of γ-aminobutyric acid A receptor subunit mRNAs in rat cerebral cortex: competitive, quantitative reverse transcriptase-polymerase chain reaction analysis. Mol Pharmacol 48:861–868
Devaud LL, Fritschy JM, Sieghart W, Morrow AL (1997) Bidirectional alterations of GABA$_A$ receptor subunit peptide levels in rat cortex during chronic ethanol consumption and withdrawal. J Neurochem 69:126–130
Devaud LL, Fritschy JM, Morrow AL (1998) Influence of gender on chronic ethanol-induced alterations in GABA$_A$ receptor in rats. Brain Res 796:222–230
Dodd PR (1994) GABA$_A$ receptors in damaged cerebral cortex areas in human chronic alcoholics. Alcohol Alcohol Suppl 2:187–191
During MJ, Ryder KM, Spencer DD (1995) Hippocampal GABA transporter function in temporal-lobe epilepsy. Nature 376:174–177
Dymecki SM (1996) FLP recombinase promotes site-specific DNA recombination in embryonic stem cells and transgenic mice. Proc Natl Acad Sci USA 93:6191–6196
Enna SJ, Bennett JP, Bylund DB, Snyder SH, Bird ED, Iversen LL (1976) Alterations of brain neurotransmitter receptor binding in Huntington's chorea. Brain Res 116:531–537
Evans MS, Viola-McCabe KE, Caspary DM, Faingold CL (1994) Loss of synaptic inhibition during repetitive stimulation in genetically epilepsy-prone rats (GEPR). Epilepsy Res 18:97–105
Faingold CL, N'Goueno P, Riaz A (1998) Ethanol and neurotransmitter interactions – from molecular to integrative effects. Prog Neurobiol 55:509–535
File SE (1995) Animal models of different anxiety states. In: Biggio G, Sanna E, Costa E (eds) GABA$_A$ receptors and anxiety: from neurobiology to treatment. Raven Press, New York, pp 93–113
Fletcher CF, Lutz CM, O'Sullivan TN, Shaughnessy JD Jr, Hawkes R, Frankel WN, Copeland NG, Jenkins NA (1996) Absence epilepsy in tottering mutant mice is associated with calcium channel defects. Cell 87:607–617
Friedman LK, Pellegrini-Giampietro DE, Sperber EF, Bennett MV, Moshe SL, Zukin RS (1994) Kainate-induced status epilepticus alters glutamate and GABA$_A$ receptor gene expression in adult rat hippocampus: an in situ hybridization study. J Neurosci 14:2697–2707
Frye GD, Fincher AS, Grover CA, Jayaprabhu S (1996) Lanthanum and zinc sensitivity of GABA$_A$ activated currents in adult medial septum/diagonal band neurons from ethanol dependent rats. Brain Res 720:101–110
Gee KW, Bolger MB, Brinton RE, Coirini H, McEwen BS (1988) Steroid modulation of the chloride ionophore in rat brain: structure-activity requirements, regional dependence and mechanism of action. J Pharmacol Exp Ther 246:803–812
Gibbs JW, Zhang YF, Kao CQ, Holloway KL, Oh KS, Coulter DA (1996) Characterization of GABA$_A$ receptor function in human temporal cortical neurons. J Neurophysiol 75:1458–1471
Goddard GV, McIntyre DC, Leech CK (1969) A permanent change in brain function resulting from daily electrical stimulation. Exp Neurol 25:295–330
Guipponi M, Thomas P, Girard-Reydet C, Feingold J, Baldy-Moulinier M, Malafosse A (1997) Lack of association between juvenile myoclonic epilepsy and GABRA5 and GABRB3 genes. Am J Med Genet 74:150–153
Gunther U, Benson J, Benke D, Fritschy JM, Reyes G, Knoflach F, Crestani F, Aguzzi A, Arigoni M, Lang Y, Buethmann H, Möhler H, Luscher B (1995)

Benzodiazepine-insensitive mice generated by targeted disruption of the γ_2 subunit gene of γ-aminobutyric acid type A receptors. Proc Natl Acad Sci USA 92:7749–7753

Hansen GH, Meier E, Schousboe A (1984) GABA influences the ultrastructure composition of cerebellar granule cells during development in culture. Int J Dev Neurosci 2:247–257

Harris RA, McQuilkin SJ, Paylor R, Abeliovich A, Tonegawa S, Wehner JM (1995) Mutant mice lacking the γ isoform of protein kinase C show decreased behavioral actions of ethanol and altered function of γ-aminobutyrate type A receptors Proc Natl Acad Sci USA 92:3658–3662

Harris RA, Valenzuela CF, Brozowski S, Chuang L, Hadingham K, Whiting PJ (1998) Adaptation of γ-aminobutyric acid type A receptors to alcohol exposure: studies with stably transfected cells. J Pharm Exp Ther 284:180–188

Henry TR, Frey KA, Sackellares JC, Gilman S, Koeppe RA, Brunberg JA, Ross DA, Berent S, Young AB, Kuhl DE (1993) In vivo cerebral metabolism and central benzodiazepine-receptor binding in temporal lobe epilepsy. Neurology 43:1998–2006

Hirouchi M, Hashimoto T, Kuriyama K (1993) Alteration of $GABA_A$ receptor α_1 subunit mRNA in mouse brain following continuous ethanol inhalation. Eur J Pharmacol 247:127–130

Hodge CW, Chappelle AM, Samson HH (1995) GABAergic transmission in the nucleus accumbens is involved in the termination of ethanol self-administration in rats. Alcohol Clin Exp Res 19:1486–1493

Homanics GE, DeLorey TM, Firestone LL, Quinlan JJ, Handforth A, Harrison NL, Krasowski MD, Rick CEM, Korpi ER, Makela R, Brilliant MH, Hagiwara N, Ferguson C, Snyder K, Olsen RW (1997a) Mice devoid of γ-aminobutyrate type A receptor β_3 subunit have epilepsy, cleft palate, and hypersensitive behavior. Proc Natl Acad Sci USA 94:4143–4148

Homanics GE, Ferguson C, Quinlan JJ, Daggett J, Snyder K, Lagenaur C, Mi ZP, Wang XH, Grayson DR, Firestone LL (1997b) Gene knockout of the α_6 subunit of the γ-aminobutyric acid type A receptor: lack of effect on responses to ethanol, pentobarbital, and general anesthetics. Mol Pharmacol 51:588–596

Homanics GE, Harrison NL, Quinlan JJ, Krasowski MD, Rick CEM, DeBlas AL, Mehta AK, Mihalek RM, Aul JJ, Firestone LL (1998) Mice lacking the long splice variant of the γ_2 subunit of the $GABA_A$ receptor demonstrate increased anxiety and enhanced behavioral responses to benzodiazepine receptor agonists, but not to ethanol. Neuropharmacology 38:253–265

Houser CR (1992) Morphological changes in the dentate gyrus in human temporal lobe epilepsy. Epilepsy Res Suppl 7:223–234

Houser CR, Esclapez M, Fritschy JM, Möhler H (1995) Decreased expression of the α_5 subunit of the $GABA_A$ receptor in a model of temporal lobe epilepsy. Abstr Soc Neurosci 21:1475, no.576.3

Huntsman MM, Tran B-T, Potkin SG, Bunney WE, Jones EG (1998) Altered ratios of alternatively spliced long and short γ_2 subunit mRNAs of the γ-amino butyrate type A receptor in prefrontal cortex of schizophrenics. Proc Natl Acad Sci USA 95:15066–15071

Huntsman MM, Porcello DM, Homanics GE, DeLorey TM, Huguenard JR (1999) Reciprocal inhibitory connections and network synchrony in the mammalian thalamus. Science 283:541–543

Janjua NA, Mori A, Hiramatsu M (1991) γ-Aminobutyric acid uptake is decreased in the hippocampus in a genetic model of human temporal lobe epilepsy. Epilepsy Res 8:71–74

Johnson EW, de Lanerolle NC, Kim JH, Sundaresan S, Spencer DD, Mattson RH, Zoghbi SS, Baldwin RM, Hoffer PB, Seibyl JP, Innis RB (1992) Central and peripheral benzodiazepine receptors: opposite changes in human epileptogenic tissue. Neurology 42:811–815

Jones A, Korpi ER, McKernan RM, Pelz R, Nusser Z, Makela R, Mellor JR, Pollard S, Bahn S, Stephenson FA, Randall AD, Sieghart W, Somogyi P, Smith AJH, Wisden W (1997) Ligand-gated ion channel subunit partnerships: GABA$_A$ receptor α_6 subunit gene inactivation inhibits δ subunit expression. J Neurosci 17:1350–1362

Kamphuis W, De Rijk TC, Lopes da Silva FH (1995) Expression of GABA$_A$ receptor subunit mRNAs in hippocampal pyramidal and granular neurons in the kindling model of epileptogenesis: an in situ hybridization study. Brain Res Mol Brain Res 31:33–47

Kang M, Spigelman I, Sapp DW, Olsen RW (1996) Persistent reduction of GABA$_A$ receptor-mediated inhibition in rat hippocampus after chronic intermittent ethanol treatment. Brain Res 709:221–228

Kang MH, Spigelman I, Olsen RW (1998) The sensitivity of GABA$_A$ receptors to allosteric modulatory drugs in rat hippocampus following chronic intermittent ethanol treatment. Alcohol Clin Exp Res 22:2165–2173

Karle J, Woldbye DPD, Elster L, Diemer NH, Bolwig TG, Olsen RW, Nielsen M (1998) Antisense oligonucleotide to GABA$_A$ Receptor γ_2 subunit induces limbic status epilepticus. J Neurosci Res 54:863–869

Kash SF, Johnson RS, Tecott LH, Noebels JL, Mayfield RD, Hanahan D, Baekkeskov S (1997) Epilepsy in mice deficient in the 65-kDa isoform of glutamic acid decarboxylase. Proc Natl Acad Sci USA 94:14060–14065

Kawabata K, Tachibana H (1997) Evaluation of benzodiazepine receptor in the cerebral cortex of Parkinson's disease using ^{123}I-iomazenil SPECT. Japanese J Clin Med 55:244–248

Kim HY, Sapp DW, Olsen RW, Tobin AJ (1993) GABA alters GABA$_A$ receptor mRNAs and increases ligand binding. J Neurochem 61:2334–2337

Kim HY, Olsen RW, Tobin AJ (1996) GABA and GABA$_A$ receptors: development and regulation. In: Shaw CA, (ed) Receptor dynamics in neural development. CRC Press, Boca Raton, FL, pp 59–72

Klein RL, Mascia MP, Whiting PJ, Harris RA (1995) GABA$_A$ receptor function and binding in stably transfected cells: chronic ethanol treatment. Alcohol Clin and Exp Res 19:1338–1344

Klein RL, Harris RA (1996) Regulation of GABA$_A$ receptor structure and function by chronic drug treatments in vivo and with stably transfected cells. Jpn J Pharmacol 70:1–15

Kokaia M, Pratt GD, Elmer E, Bengzon J, Fritschy JM, Kokaia Z, Lindvall O, Möhler H (1994) Biphasic differential changes of GABA$_A$ receptor subunit mRNA levels in dentate gyrus granule cells following recurrent kindling-induced seizures. Brain Res Mol Brain Res 23:323–332

Kuikka JT, Pitkanen A, Lepola U, Partanen K, Vainio P, Bergstrom KA, Wieler HJ, Kaiser KP, Mittelbach L, Koponen H, et al. (1995) Abnormal regional benzodiazepine receptor uptake in the prefrontal cortex in patients with panic disorder. Nuc Med Comm 16:273–280

Lauder JM, Han VKM, Henderson P, Verdoorn T, Towle AC (1986) Prenatal ontogeny of the GABAergic system in the rat brain: an immunocytochemical study. Neuroscience 19:465–493

Laurie DJ, Wisden W, Seeburg PH (1992) The distribution of thirteen GABA$_A$ receptor subunit mRNAs in the rat brain. III. Embryonic and postnatal development. J Neurosci 12: 4151–4172

Leinekugel X, Tseeb V, Ben-Ari Y, Bregestovski P (1995) Synaptic GABA$_A$ activation induces Ca^{2+} rise in pyramidal cells and interneurons from rat neonatal hippocampal slices. J Physiol 487:319–329

Lewin E, Peris J, Bleck V, Zahniser NR, Harris RA (1989) Chemical kindling decreases GABA-activated chloride channels of mouse brain. Eur J Pharmacol 160:101–106

Lewohl JM, Crane DI, Dodd PR (1996) Alcohol, alcoholic brain damage, and GABA$_A$ receptor isoform gene expression. Neurochem Int 29:677–684

Lewohl JM, Crane DI, Dodd PR (1997) A method for the quantitation of the α_1, α_2, and α_3 isoforms of the $GABA_A$ receptor in human brain using competitive PCR. Brain Res Brain Res Proto 1:347–356

Lin AM, Freund RK, Palmer MR (1993) Sensitization of gamma-aminobutyric acid-induced depressions of cerebellar Purkinje neurons to the potentiative effects of ethanol by beta adrenergic mechanisms in rat brain. J Pharm Exp Ther 265:426–432

Lüddens H, Wisden W (1991) Function and pharmacology of multiple $GABA_A$ receptor subunits. Trends Pharmacol Sci 12:49–51

Lüddens H, Korpi ER, Seeburg PH (1995) $GABA_A$/benzodiazepine receptor heterogeneity: neurophysiological implications. Neuropharmacology 34:245–254

MacDonald RL, Olsen RW (1994) $GABA_A$ receptor channels. Ann Rev Neurosci 17: 569–602

Ma W, Saunders PA, Somogyi R, Poulter MO, Barker JL (1993) Ontogeny of $GABA_A$ receptor subunit mRNAs in rat spinal cord and dorsal root ganglia. J Comp Neurol 338:337–359

Mahmoudi M, Kang MH, Tillakaratne N, Tobin AJ, Olsen RW (1997) Chronic intermittent ethanol treatment in rats increases $GABA_A$ receptor α_4-subunit expression: possible relevance to alcohol dependence. J Neurochem 68:2485–2492

Majewska MD, Harrison NL, Schwartz RD, Barker JL, Paul SM (1986) Steroid hormone metabolites are barbiturate-like modulators of the GABA receptor. Science 232:1004–1007

Malizia AL, Cunningham VJ, Bell CJ, Liddle PF, Jones T, Nutt DJ (1998) Decreased brain $GABA_A$-benzodiazepine receptor binding in panic disorder: preliminary results from a quantitative PET study. Arch Gen Psychiat 55:7150720

Mansuy IM, Winder DG, Moallem TM, Osman M, Mayford M, Hawkins RD, Kandel ER (1998) Inducible and reversible gene expression with the rtTA system for the study of memory. Neuron 21:257–265

Marth JD (1996) Recent advances in gene mutagenesis by site-directed recombination. J Clin Investigat 97:1999–2002

McDonald JW, Garofalo EA, Hood T, Sackellares JC, Gilman S, McKeever PE, Troncoso JC, Johnston MV (1991) Altered excitatory and inhibitory amino acid receptor binding in hippocampus of patients with temporal lobe epilepsy. Ann Neurol 29:529–541

McKernan RM, Whiting PJ (1996) Which $GABA_A$-receptor subtypes really occur in the brain? Trends Neurosci 19:139–143

Mehta AK, Ticku MK (1998) Chronic ethanol administration alters the modulatory effect of 5 α-pregnan-3α-ol-20-one on the binding characteristics of various radioligands of $GABA_A$ receptors. Brain Res 805:88–94

Meier E, Drejer J, Schousboe A (1984) GABA induces functionally active low-affinity GABA receptors on cultured cerebellar granule cells. J Neurochem 43:1737–1744

Meier E, Hansen GH, Schousboe A (1985) the trophic effect of GABA on cerebellar granule cells is mediated by GABA receptors. Int J Dev Neurosci 3:401–407

Mhatre MC, Ticku MK (1989) Chronic ethanol treatment selectively increases the binding of inverse agonists for benzodiazepine binding sites in cultured spinal cord neurons. J Pharmacol Exp Ther 251:164–168

Mhatre MC, Ticku MK (1992) Chronic ethanol alters γ-aminobutyric $acid_A$ receptor gene expression. Mol Pharm 42:415–422

Mhatre MC, Pena G, Sieghart W, Ticku MK (1993) Antibodies specific for $GABA_A$ receptor alpha subunits reveal that chronic alcohol treatment down-regulates alpha-subunit expression in rat brain regions. J Neurochem 61:1620–1625

Mhatre MC, Ticku MK (1994) Chronic ethanol treatment upregulates the GABA receptor β subunit expression. Mol Brain Res 23:246–252

Mitchell CK, Redburn DA (1996) GABA and GABA-A receptors are maximally expressed in association with cone synaptogenesis in neonatal rabbit retina. Brain Res Develop Brain Res 95:63–71

Mitsuyama H, Little KY, Sieghart W, Devaud LL, Morrow AL (1998) GABA$_A$ receptor α_1, α_4, and β_3 subunit mRNA and protein expression in the frontal cortex of human alcoholics. Alcohol Clin Exp Res 22:815–822
Morrow AL, Suzdak PD, Karanian JW, Paul SM (1988) Chronic ethanol administration alters γ-aminobutyric acid, pentobarbital and ethanol-mediated $^{36}Cl^-$ uptake in cerebral cortical synaptoneurosomes. J Pharmacol Exp Ther 246:158–164
Narahashi T, Arakawa O, Nakahiro M, Twombly DA (1991) Effects of alcohols on ion channels of cultured neurons. Ann NY Acad Sci 625:26–36
Nguyen Q, Sapp, DW, Van Ness PC, Olsen RW (1995) Modulation of GABA$_A$ receptor binding in human brain by neuroactive steroids: Species and brain regional differences. Synapse 19:77–87
Nicholson LF, Faull RL, Waldvogel HJ, Dragunow M (1995) GABA and GABA$_A$ receptor changes in the substantia nigra of the rat following quinolinic acid lesions in the striatum closely resemble Huntington's disease. Neuroscience 66:507–521
Nicholson LF, Waldvogel HJ, Faull RL (1996) GABA$_A$ receptor subtype changes in the substantia nigra of the rat following quinolinate lesions in the striatum: a correlative in situ hybridization and immunohistochemical study. Neuroscience 74:89–98
Noble EP, Zhang X, Ritchie T, Lawford BR, Grosser SC, Young R, Sparkes RS (1998) D_2 dopamine receptor and GABA$_A$ receptor β_3 subunit genes and alcoholism. Psychiat Res 81:133–147
Nowak KL, McBride WJ, Lumeng L, Li TK, Murphy JM (1998) Blocking GABA$_A$ receptors in the anterior ventral tegmental area attenuates ethanol intake of the alcohol-preferring P rat. Psychopharm 139:108–116
Nusser Z, Hajos N, Somogyi P, Mody I (1998) Increased number of synaptic GABA$_A$ receptors underlies potentiation at hippocampal inhibitory synapses. Nature 395:172–177
Obenaus A, Esclapez M, Houser CR (1993) Loss of glutamate decarboxylase mRNA-containing neurons in the rat dentate gyrus following pilocarpine-induced seizures. J Neurosci 13:4470–4485
Obrietan K, van den Pol AN (1996) Growth cone calcium elevation by GABA. J Comp Neurol 372:167–175
Olsen RW, Wamsley JK, Lee RJ, Lomax P (1986) Benzodiazepine/ barbiturate/GABA receptor-chloride ionophore complex in a genetic model for generalized epilepsy. Adv Neurol 44:365–378
Olsen RW, Bureau M, Houser CR, Delgado-Escueta AV, Richards JG, Mohler H (1992) GABA/benzodiazepine receptors in human focal epilepsy. Epilepsy Res Suppl 8:389–397
Olsen RW, Houser CR, Makela R, Delgado-Escueta AV (1995) Altered GABA$_A$ receptor subtype-specific and allosteric binding properties in neocortex resected from focal epilepsy tissue associated with severe hippocampal pathology and sprouting. Soc Neurosci Abstr 21:1517, no. 597.12
Olsen RW, Sapp DW (1995) Neuroactive steroid modulation of GABA$_A$ receptors. In: Biggio G, Sanna E, Serra M, Costa E (eds) GABA$_A$ receptors and anxiety: from neurobiology to treatment (Adv Biochem Psycopharm 48). Raven Press, New York, pp 57–74
Olsen RW, Avoli M (1997) GABA and epileptogenesis. Epilepsia 38:399–407
Olsen RW, DeLorey TM, Handforth A, Ferguson C, Mihalek RM, Homanics GE (1997) Epilepsy in mice lacking GABA$_A$ receptor delta (δ) subunits. Epilepsia 38 [Suppl 8]:123, no. E.03
Olsen RW, DeLorey TM, Gordey M, Kang MH (1999) Regulation of GABA receptor function and epilepsy. In: Delgado-Escueta AV, Wilson W, Olsen RW, Porter RJ (eds) Jasper's basic mechanisms of the epilepsies, vol III. Lippincott-Williams & Wilkins, New York, pp 499–510

Owens DF, Boyce LH, Davis MB, Kriegstein AR (1996) Excitatory GABA responses in embryonic and neonatal cortical slices demonstrated by gramicidin perforated-patch recordings and calcium imaging. J Neurosci 16:6414–6423
Parsian A, Cloninger CR (1997) Human $GABA_A$ receptor α_1 and α_3 subunits genes and alcoholism. Alcohol Clin Exp Res 21:430–433
Purdy RH, Morrow AL, Moore PH, Paul SM (1991) Stress-induced elevations of γ-aminobutyric acid type A receptor-active steroids in the rat brain. Proc Natl Acad Sci USA 88:4553–4557
Rivera C, Voipio J, Payne JA, Ruusuvuori E, Lahtinen H, Lamsa K, Pirvola U, Saarma M, Kaila K (1999) The K^+/Cl^- co-transporter KCC2 renders GABA hyperpolarizing during neuronal maturation. Nature 397:251–255
Rorsman P, Berggren PO, Bokvist K, Ericson H, Möhler H, Ostenson CG, Smith PA (1989) Glucose-inhibition of glucagon secretion involves activation of $GABA_A$-receptor chloride channels. Nature 341:233–236
Roy-Byrne PP, Cowley DS, Greenblatt DJ, Shader RI, Hommer D (1990) Reduced benzodiazepine sensitivity in panic disorder. Arch Gen Psychiat 47:534–538
Salazar P, Montiel T, Brailowsky S, Tapia R (1994) Decrease of glutamate decarboxylase activity after in vivo cortical infusion of γ-aminobutyric acid. Neurochem Int 24:363–368
Samson HH, Harris RA (1992) Neurobiology of alcohol abuse. Trends Pharmacol Sci 13:206–211
Sander T, Kretz R, Williamson MP, Elmslie FV, Rees M, Hildmann T, Bianchi A, Bauer G, Sailer U, Scaramelli A, Schmitz B, Gardiner RM, Janz D, Beck-Mannagetta G (1997) Linkage analysis between idiopathic generalized epilepsies and the $GABA_A$ receptor α_5, β_3 and γ_3 subunit gene cluster on chromosome 15. Acta Neurol Scand 96:1–7
Sanders SK, Shekhar A (1995a) Anxiolytic effects of chlordiazepoxide blocked by injection of GABA and benzodiazepine receptor antagonists in the region of the anterior basolateral amygdala of rats. Biol Psychiat 37:473–476
Sanders SK, Shekhar A (1995b) Regulation of anxiety by $GABA_A$ receptors in the rat amygdala. Pharm Biochem Behav 52:701–706
Sapp DW, Yeh HH (1998) Ethanol-$GABA_A$ receptor interactions: a comparison between cell lines and cerebellar Purkinje cells. J Pharmacol Exp Ther 284:768–776
Savic I, Persson A, Roland P, Pauli S, Sedvall G, Widen L (1988) In vivo demonstration of reduced benzodiazepine receptor binding in human epileptic foci. Lancet 2:863–866
Schlegel S, Steinert H, Bockisch A, Hahn K, Schloesser R, Benkert O (1994) Decreased benzodiazepine receptor binding in panic disorder measured by IOMAZENIL-SPECT. A preliminary report. Eur Arch Psychiat Clinic Neurosci 244:49–51
Schwarzer C, Tsunashima K, Wanzenbock C, Fuchs K, Sieghart W, Sperk G (1997) $GABA_A$ receptor subunits in the rat hippocampus II: altered distribution in kainic acid-induced temporal lobe epilepsy. Neuroscience 80:1001–1017
Shumate MD, Lin DD, Gibbs JW, Holloway KL, Coulter DA (1998) $GABA_A$ receptor function in epileptic human dentate granule cells: comparison to epileptic and control rat. Epilepsy Res 32:114–128
Sieghart W (1995) Structure and pharmacology of $GABA_A$ receptor subtypes. Pharmacol Rev 47:181–234
Sigel E, Baur R, Malherbe P (1993) Recombinant $GABA_A$ receptor function and ethanol. FEBS Lett 324:140–142
Smith SS, Gong QH, Hsu FC, Markowitz RS, Ffrench-Mullen JM, Li X (1998a) $GABA_A$ receptor α_4 subunit suppression prevents withdrawal properties of an endogenous steroid. Nature 392:926–930
Smith SS, Gong QH, Li X, Moran MH, Bitran D, Frye CA, Hsu FC (1998b) Withdrawal from 3α-OH-5α-pregnan-20-one using a pseudopregnancy model alters the kinetics of hippocampal $GABA_A$-gated current and increases the $GABA_A$

receptor α_4 subunit in association with increases anxiety. J Neurosci 18:5275–5284
Snead OC, Depaulis A, Banerjee PK, Hechler V, Vergnes M (1992) The GABA$_A$ receptor complex in experimental absence seizures in rat: an autoradiographic study. Neurosci Lett 140:9–12
Soldo BL, Proctor WR, Dunwiddie TV (1998) Ethanol selectively enhances the hyperpolarizing component of neocortical neuronal responses to locally applied GABA. Brain Res 800:187–197
Spoerri PE (1987) GABA-mediated developmental alterations in a neuronal cell line and in cultures of cerebral and retinal neurons. In: Redburn DA (ed) Neurotrophic activity of GABA during development. Alan R. Liss, New York, pp 189–220
Tabakoff B, Hoffman PL (1996) Alcohol addiction: an enigma among us. Neuron 16:909–912
Tehrani MH, Barnes EM Jr (1995) Reduced function of γ-aminobutyric acid A receptors in tottering mouse brain: role of cAMP-dependent protein kinase. Epilepsy Res 22:13–21
Tehrani MH, Baumgartner BJ, Liu SC, Barnes EM Jr (1997) Aberrant expression of GABA$_A$ receptor subunits in the tottering mouse: an animal model for absence seizures. Epilepsy Res 28:213–223
Thomas GJ, Harper CG, Dodd PR (1998) Expression of GABA$_A$ receptor isoform genes in the cerebral cortex of cirrhotic and alcoholic cases assessed by S1 nuclease protection assays. Neurochem Int 32:375–385
Tsunashima K, Schwarzer C, Kirchmair E, Sieghart W, Sperk G (1997) GABA$_A$ receptor subunits in the rat hippocampus III: altered messenger RNA expression in kainic acid-induced epilepsy. Neuroscience 80:1019–1032
Turner DM, Ransom RW, Yang JS, Olsen RW (1989) Steroid anesthetics and naturally occurring analogs modulate the γ-aminobutyric acid receptor complex at a site distinct from barbiturates. J Pharmacol Exp Ther 248:960–966
Tyndale RF, Bhave SV, Hoffmann E, Hoffmann PL, Tabakoff B, Tobin AJ, Olsen RW (1997) Pentobarbital decreases the γ-aminobutyric acidA receptor subunit $\gamma 2$ long/short mRNA ratio by a mechanism distinct from receptor occupation. J Pharmacol Exp Ther 283:350–357
Van Ness PC, Awad IA, Estes M, Nguyen Q, Olsen RW (1995) Neurosteroid modulation of benzodiazepine binding to neocortical GABA$_A$ receptors in human focal epilepsy varies with pathology. Soc Neurosci Abstr 21, 1516, 597.5
von Blankenfeld G, Turner J, Ahnert-Hilger G, John M, Enkvist MO, Stephenson F, Kettenmann H, Wiedenmann B (1995) Expression of functional GABA$_A$ receptors in neuroendocrine gastropancreatic cells. Pflugers Archiv Eur J Physiol 430:381–388
Wafford KA, Burnett DM, Leidenheimer NJ, Burt DR, Wang JB, Kofuji P, Dunwiddie TV, Harris RA, Sikela JM (1991) Ethanol sensitivity of the GABA$_A$ receptor expressed in *Xenopus* oocytes requires 8 amino acids contained in the γ_{2L} subunit. Neuron 7:27–33
Wafford KA, Whiting PJ (1992) Ethanol potentiation of GABA$_A$ receptors requires phosphorylation of the alternatively spliced variant of the γ_2. FEBS Lett 313:113–117
Wafford KA, Burnett D, Harris RA, Whiting PJ (1993) GABA$_A$ receptor subunit expression and sensitivity to ethanol. Alcohol Alcohol Suppl 2:327–330
Waymire KG, Mahuren JD, Jaje JM, Guilarte TR, Coburn SP, MacGregor GR (1995) Mice lacking tissue non-specific alkaline phosphatase die from seizures due to defective metabolism of vitamin B-6. Nature Genet 11:45–51
Wolff JR, Ferenc J, Kasa P (1987) Synaptic, metabolic and morphogenetic effects of GABA in the superior cervical ganglion of rats. In: Redburn DA (ed) Neurotrophic activity of GABA during development. Alan R. Liss, New York, pp 221–252
Yang W, Reyes AA, Lan NC (1994) Identification of the GABA$_A$ receptor subtype mRNA in human pancreatic tissue. FEBS Lett 346:257–262

Yuste R, Katz LC (1991) Control of postsynaptic Ca^{2+} influx in developing neocortex by excitatory and inhibitory neurotransmitters. Neuron 6:333–344
Zhai J, Stewart RR, Friedberg MW, Li C (1998) Phosphorylation of the $GABA_A$ receptor gamma γ_{2L} subunit in rat sensory neurons may not be necessary for ethanol sensitivity. Brain Res 805:116–122
Zhang L, Rubinow DR, Ma W, Marks JM, Feldman AN, Barker JL, Tathan TA (1998) GABA receptor subunit mRNA expression in brain of conflict, yoked control and control rats. Brain Res Mol Brain Res 58:16–26
Zhu WJ, Wang JF, Krueger KE, Vicini S (1996) δ Subunit inhibits neurosteroid modulation of $GABA_A$ receptors. J Neurosci 16:6648–6656

CHAPTER 10

$GABA_C$ Receptors*: Structure, Function and Pharmacology

J. BORMANN and A. FEIGENSPAN

A. Introduction

In the vertebrate central nervous system (CNS), γ-aminobutyric acid (GABA) is the most widely distributed neurotransmitter (SIVILOTTI and NISTRI 1991). Initially, GABA was found to activate bicuculline-sensitive Cl^- channels, but GABA-mediated activation of cation channels was discovered subsequently (see BORMANN 1988, for review). This lead to the notion of $GABA_A$ and $GABA_B$ receptors, which was introduced by HILL and BOWERY (1981). The $GABA_A$ receptor directly gates a Cl^- ionophore and has modulatory binding sites for benzodiazepines, barbiturates, neuosteroids and ethanol (MACDONALD and OLSON 1994; BORMANN 1988). By contrast, $GABA_B$ receptors couple to Ca^{2+} and K^+ channels via G-proteins and second-messenger systems (BORMANN 1988; BOWERY 1989; BETTLER et al. 1998). They are activated by baclofen and resistant to drugs that modulate $GABA_A$ receptors.

It now appears that GABA gates at least three classes of GABA receptors that are distinct both pharmacologically and structurally (see BORMANN and FEIGENSPAN 1995; JOHNSTON 1996; CHERUBINI and STRATA 1997 for review). Early studies by Johnston and colleagues indicated that the partially folded GABA analogue *cis*-4-aminocrotonic acid (CACA) selectively activates a third class of GABA receptor in the mammalian CNS (JOHNSTON et al. 1975). These receptors, which were tentatively designated $GABA_C$ (DREW et al. 1984), are insensitive to both bicuculline and baclofen.

Several lines of evidence now indicate that $GABA_C$ receptors are composed of ρ-subunits. When heterologously expressed, ρ-subunits form homo-oligomeric receptors with similar electrophysiological and pharmacological properties compared with $GABA_C$ receptors. Bicuculline-resistant $GABA_C$ responses and ρ-subunits have been colocalized in the same retinal neurons and studied at the molecular level. This review summarizes current knowledge on the structure, function and pharmacology of $GABA_C$ receptors.

---

* Since $GABA_C$-receptors are GABA-gated chloride channels they can be classified as $GABA_A$-receptors. The term $GABA_C$-receptor is therefore not recommended by the IUPHAR nomenclature committee (Pharmacol. Rev. 50, 291–313, 1998). In the present volume the term is used solely for the take of brevity (H.M.).

B. Structure of $GABA_C$ Receptors

I. Cloning of Vertebrate ρ-Subunits

The first member of the class of GABA-receptor ρ-subunits was cloned by Cutting and colleagues in an attempt to identify new proteins encoding chloride channels (CUTTING et al. 1991). The highly conserved transmembrane regions M2–M3 of $GABA_A$ and glycine receptor subunits were utilized for PCR amplification of human DNA sequences and finally isolating and cloning the ρ1-cDNA from a retinal cDNA library. The mature protein predicted from this sequence shares only 30%–38% similarity with other GABA receptor subunits. The human ρ2-subunit is 74% identical to ρ1, the highest degree of amino acid divergence residing in the large intracellular loop between M3 and M4 (CUTTING et al. 1992). By analogy to the nicotinic acetylcholine (UNWIN 1995) and $GABA_A$ receptors (NAYEEM et al 1994) the ρ-subunits assemble into a pentameric receptor channel with a central pore for Cl^- ions.

ρ-Subunits have been cloned from a variety of vertebrate species. Three ρ-subunits are known in the rat: ρ1 (ENZ et al. 1995; ZHANG et al. 1995; WEGELIUS et al. 1996), ρ2 (ENZ et al. 1995; ZHANG et al. 1995; OGURUSU et al. 1995) and ρ3 (OGURUSU and SHINGAI 1996). They display 88%–99% similarity, at the protein level, to the human counterparts. Other species include chick (ρ1–2) (ALBRECHT and DARLISON 1995), mouse (ρ1–3) (GREKA et al. 1998) and perch (ρ1–3) (QIAN et al. 1997, 1998).

II. Subunit Composition of $GABA_C$ Receptors

A prominent feature of ρ-subunits is their ability to form functional homooligomeric GABA receptor Cl^- channels (CUTTING et al. 1991; WANG et al. 1994; OGURUSU and SHINGAI 1996). This contrasts with the situation of $GABA_A$ receptors, where a combination of different subunits, typically $\alpha\beta\gamma$, is needed for the receptors to express the full range of physiological and pharmacological functions (SIGEL et al. 1990). Also, there is evidence from coexpression studies that ρ-subunits neither coassemble with $GABA_A$ receptor α-, β- or γ-subunits, nor with the glycine receptor β-subunit (SHIMADA et al. 1992; KUSAMA et al. 1993a). However, the ρ-subunits are capable to interact amongst themselves to form functional GABA receptors (ENZ and CUTTING 1998), e.g. $\rho1\rho2$ heterooligomers (ZHANG et al. 1995; ENZ and CUTTING 1999). The expression pattern in the brain of homo- and heterooligomeric $GABA_C$ receptors should depend on the distribution of ρ-transcripts.

C. Neuronal Localization

Whereas the existence of $GABA_C$ receptors has first been shown outside the retina (JOHNSTON et al. 1975; see BORMANN and FEIGENSPAN 1995; JOHNSTON 1996 for review), recent work has identified the vertebrate retina as the richest source for GABA$_C$ receptors. Bicuculline- and baclofen-insensitive GABA$_C$ have been identified on rod bipolar cells in the rat retina (FEIGENSPAN et al. 1993; FEIGENSPAN and BORMANN 1994a). GABA$_C$ receptors in non-mammalian retinae have been detected on rod-driven horizontal cells of white perch (QIAN and DOWLING 1993, 1994), on hybrid bass bipolar cells (QIAN and DOWLING 1995), and on cone-driven horizontal cells in catfish retinae (DONG and WERBLIN 1994). In the tiger salamander retina, GABA$_C$ receptors have been localized to bipolar cell terminals (LUKASIEWICZ and WERBLIN 1994; LUKASIEWICZ et al. 1994).

Combining reverse transcriptase-polymerase chain reaction (RT-PCR) with in situ hybridization has demonstrated a differential distribution of $\rho1$-, $\rho2$- and $\rho3$-subunits in the retina and brain of the rat (ENZ et al. 1995; WEGELIUS et al. 1998) and chick (ALBRECHT and DARLISON 1995; ALBRECHT et al. 1997). Whereas $\rho1$ was restricted to this tissue, $\rho2$ was detected in all brain regions, although with highest level of expression in the retina. In situ hybridization of retinal sections revealed that $\rho1$ and $\rho2$ transcripts are present in the inner nuclear layer, and by studying isolated retinal cells, both ρ-subunits could be localized to rod bipolar cells (ENZ et al. 1995). The ρ_2 isoform could also be detected in ganglion cells (YEH et al. 1996); however, the ganglion cells tested did not display bicuculline-resistant responses to GABA. $\rho2$-Transcripts were found in most other CNS structures, notably in the hippocampus, spinal cord, cerebellum and the thalamus/basal ganglia (ENZ et al. 1995; WEGELIUS et al. 1998). Expression of $\rho3$ is strong in the adult hippocampus (WEGELIUS et al. 1998), but may also be present in other areas (BOUE-GRABOT et al. 1998). It is very likely that the GABA$_C$ receptor-like responses observed in various parts of the brain were due to the presence of $\rho2$-homooligomeric or $\rho2\rho3$-heterooligomeric GABA$_C$ receptors.

The immunocytochemical localization of ρ-subunits was studied after raising polyclonal antibodies against the N-terminus of the rat $\rho1$ isoform (ENZ et al. 1996). This region is different from that of the known GABA$_A$ receptor (e.g. $\alpha1$–3, $\beta1$–3, $\gamma2$, δ) or glycine receptor ($\alpha1$, β) subunits, and antibodies against $\rho1$ do not recognize GABA$_A$ or glycine receptor subunits. Since the N-terminal region of $\rho1$ is very similar to the $\rho2$ (82%) and $\rho3$ (78%) isoforms, the polyclonal antibody labels all three ρ-subunits. In vertical retinal sections, strong punctate immunoreactivity was found throughout the inner plexiform layer, at the axon terminals of different types of bipolar cells. The dendrites of rod bipolar cells were also labeled by the antibody (ENZ et al. 1996). A comparable staining pattern was demonstrated for mammalian (rat, cat, mouse, rabbit, monkey), goldfish and chick retinas (ENZ et al. 1996; KOULEN et al. 1997; WÄSSLE et al. 1998). Interestingly, the rat antibody did not label horizontal cells in the fish retina, although GABA$_C$ responses have originally been described for this cell type in the white perch retina (QIAN and DOWLING 1993). It is possible, however, that the perch ρ-subunits (QIAN et al. 1997, 1998) are not recognized by the rat antiserum. The use of an antibody specific for the $\rho1$-subunit on rat cerebellum revealed the presence of this subunit in the soma and dendrites of Purkinje neurons (BOUE-GRABOT et al. 1998).

D. Functional Properties of $GABA_C$ Receptors

I. Identification of $GABA_C$ Receptors

Retinal bicuculline-insensitive $GABA_C$ receptors were first observed by Miledi and colleagues after expressing mRNA from bovine retina in *Xenopus* oocytes (POLENZANI et al. 1991). The ρ-subunits, that were originally cloned from a human retinal cDNA library (CUTTING et al. 1991, 1992), form homooligomeric channels with characteristic $GABA_C$ receptor pharmacology, when expressed in *Xenopus* oocytes (CUTTING et al. 1991; SHIMADA et al. 1992; KUSAMA et al. 1993a,b; WANG et al. 1994; SHINGAI et al. 1996; QIAN et al. 1997, 1998). In the retina, native $GABA_C$ receptors have been described in horizontal cells of the white perch and catfish retina (QIAN and DOWLING 1993, 1994; DONG and WERBLIN 1994) as well as in bipolar cells (Fig. 1A,B) of various vertebrate species (FEIGENSPAN et al. 1993; LUKASIEWICZ et al. 1994; QIAN and DOWLING 1995; LUKASIEWICZ and WONG 1997; QIAN et al. 1997). Recently, a $GABA_C$ receptor-mediated Cl^- current has been identified in porcine cones where it may participate in feedback from horizontal cells to cones (PICAUD et al. 1998).

II. GABA Affinity and Ion Selectivity

The $GABA_C$ receptor is more sensitive to GABA than the $GABA_A$ receptor: the concentration of GABA producing half-maximal response (EC_{50}) at $GABA_C$ receptors is 1–4 μmol/l (POLENZANI et al. 1991; WOODWARD et al. 1992b; QIAN and DOWLING 1993, 1994; FEIGENSPAN and BORMANN 1994a; WANG et al. 1994). Since the GABA response of rat retinal bipolar cells is mediated by both $GABA_A$ and $GABA_C$ receptors, the affinity of both receptor subtypes for GABA could be directly compared (FEIGENSPAN and BORMANN 1994a). The concentration-response curve recorded in the presence of bicuculline displayed an average EC_{50} value of 4.2 μmol/l and a Hill coefficient (n) of n = 1.3 (Fig. 1C). In contrast, the $GABA_A$ receptor-mediated portion of the bipolar cell GABA response revealed an average EC_{50} value of 27.1 μmol/l and a Hill coefficient of n = 2.0. Thus, the $GABA_A$ and $GABA_C$ receptors of bipolar cells exhibit a sevenfold difference in binding affinity for GABA, but a similar degree of cooperativity for agonist binding.

A series of site-directed mutations have been constructed in the human GABA $\rho 1$ receptor subunit to determine domains conferring affinity and activation properties of $GABA_C$ receptor channels (AMIN and WEISS 1994; KUSAMA et al. 1994). Five amino acids located in the N-terminal region of the $\rho 1$-subunit are important for GABA-mediated activation (AMIN and WEISS 1994). These five mutations could be grouped into two domains which corre-

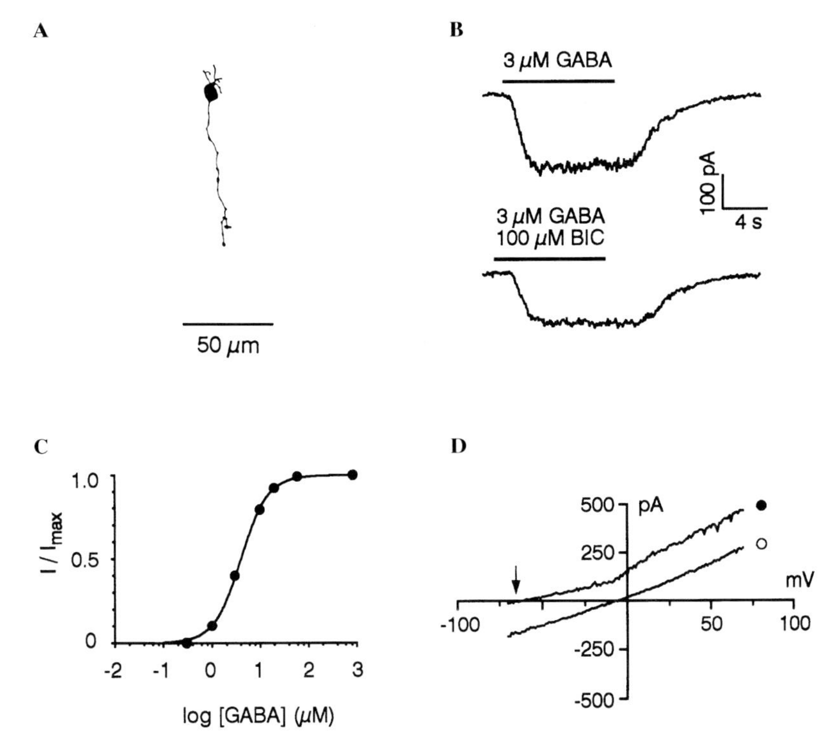

Fig. 1A–D. GABA$_C$ receptors in retinal bipolar cells. **A** Camera lucida drawing of a bipolar cell that was injected with Lucifer Yellow in a retinal slice culture. **B** Identification of GABA$_C$ receptors in bipolar cells. The total GABA response (*top trace*) was only partially blocked by the GABA$_A$ receptor antagonist bicuculline (BIC), isolating a residual current which was mediated by GABA$_C$ receptors (*lower trace*). **C** Activation properties of GABA$_C$ receptor channels. Peak amplitudes of bicuculline-insensitive GABA-induced currents (I) were normalized relative to the current obtained at a saturating GABA concentration of 1 mmol/l (I_{max}). The ratio I/I_{max} was plotted vs GABA concentration. The dose-response curve indicates an EC_{50} value of 4.0 μmol/l and a Hill coefficient (n) of 1.5. **D** Chloride selectivity of GABA$_C$ receptor channels. Whole-cell current-voltage relations were obtained by ramping the command potential from –70 mV to +70 mV. With equal extra- and intracellular Cl^- concentrations, the reversal potential of the bicuculline-insensitive GABA response was close to 0 mV. Upon partial replacement of internal Cl^- by equal amounts of the impermeable anion gluconate, the reversal potential shifted to the left (–59 ± 4 mV)

spond to the GABA-binding regions found on β2-subunits. However, only three residues correspond directly to the analogous position in β2, which is likely to account for the different activation and gating properties of GABA$_C$ receptors. The affinity of homomeric ρ1 GABA$_C$ receptors is significantly diminished when a position in the conserved N-terminal cysteine loop is changed (Kusama et al. 1994). Likewise, the Hill coefficient is increased by a mutation in the extracellular loop between transmembrane regions 2 and 3.

The conducting element of $GABA_C$ receptors is an integral membrane ionophore, similar to other ligand-gated ion channels. When the transmembrane Cl^- gradient changes, the reversal potential for $GABA_C$ receptor-mediated responses is altered as predicted by the Nernst equation (Fig. 1D), indicating that $GABA_C$ receptors are Cl^--selective pores (FEIGENSPAN et al. 1993; QIAN and DOWLING 1993; DONG et al. 1994; WANG et al. 1994).

III. Single Channel Characteristics

The single-channel conductance of retinal $GABA_C$ receptors has been studied in outside-out patches taken from the cell bodies of cultured (neonatal) or isolated (adult) bipolar cells (FEIGENSPAN et al. 1993; FEIGENSPAN and BORMANN 1994a). When $GABA_A$ receptors were blocked by bicuculline, GABA induced single-channel inward currents at negative holding potentials (Fig. 2A). The conductances which were obtained from the slope of the linear current-voltage relations (Fig. 2B), were 7.4 pS for cultured bipolar cells and 7.9 pS for isolated bipolar cells.

In the absence of bicuculline, GABA induced single-channel currents with two amplitudes of 0.5 pA and 2 pA (FEIGENSPAN and BORMANN 1994a). The 0.5 pA current could not be blocked by bicuculline, and thus corresponds to channel openings mediated by $GABA_C$ receptors. The 2 pA events, corresponding to a conductance of ~30 pS, were no longer visible in the presence of bicuculline, indicating that they were mediated by $GABA_A$ receptors. Furthermore, the value of 30 pS is in good agreement with conductance measurements from retinal amacrine cells known to exclusively express the $GABA_A$ receptor subtype (FEIGENSPAN et al. 1993; FEIGENSPAN and BORMANN 1994a), and other CNS neurons (for review see BORMANN 1988; SIVILOTTI and NISTRI 1991; MACDONALD and OLSEN 1994). When the gating properties of $GABA_A$ and $GABA_C$ receptors were examined, $GABA_A$ receptor channels of cultured amacrine cells revealed a mean open time of 25 ms, whereas $GABA_C$ receptors showed a sixfold longer mean open time of 150 ms.

IV. Pore Size

Work on the IPSPs in spinal synapses has shown that postsynaptic Cl^- channels are permeable to a variety of small inorganic and organic anions (COOMBS et al. 1955). These results were interpreted that ion channels act as molecular sieves and discriminate between different ions according to their hydrated size. An important question was whether or not the small conductance of $GABA_C$ receptors was due to a smaller open channel diameter when compared to $GABA_A$ receptors (FEIGENSPAN and BORMANN 1994a). $GABA_C$ receptor channels conduct other small anions up to the size of acetate (Fig. 2C), suggesting a pore diameter of 5.1 Å (Fig. 2D), comparable to the values of 4.9 Å and 5.6 Å obtained for native $GABA_A$ receptors in cultured amacrine cells (FEIGENSPAN and BORMANN 1998) and cultured spinal neurons, respec-

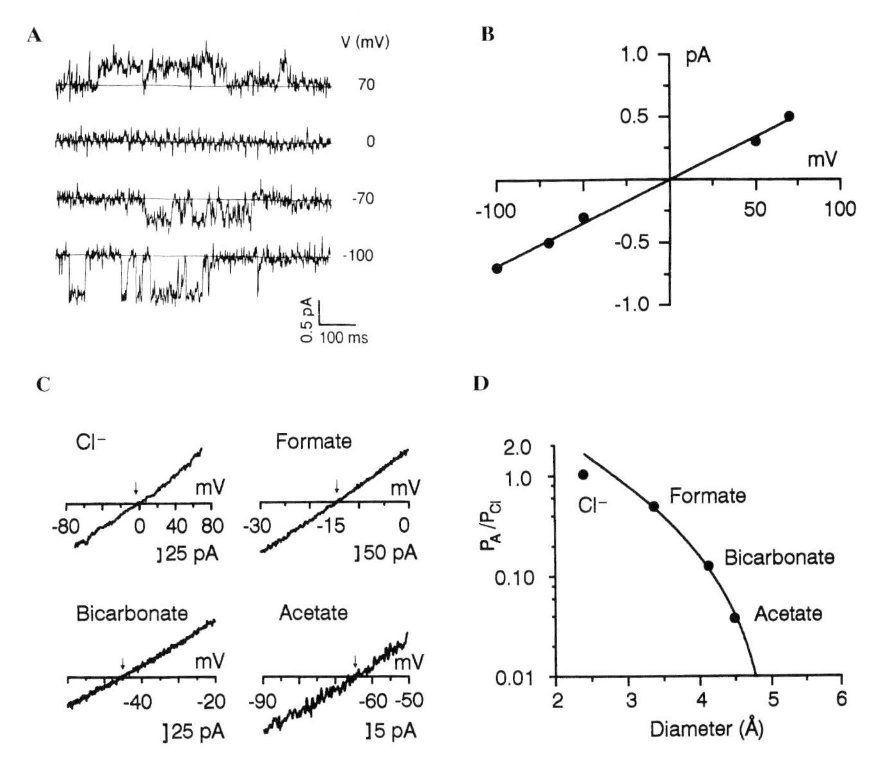

Fig. 2A–D. Conductance and pore size of GABA$_C$ receptors. **A** Outside-out patch recordings obtained from a bipolar cell. GABA-induced single-channel currents were recorded at the holding potentials indicated. **B** The slope of the linear current-voltage relation reveals a single-channel conductance of 7 pS. **C** Reversal potential measurements of GABA$_C$ receptor-mediated currents in bipolar cells upon partial replacement of internal Cl^- by various monovalent anions. The reversal potentials determined by ramping the command voltage are 2 mV (Cl^-), –15 mV (formate), –46 mV (bicarbonate), and –65 mV (acetate). **D** Permeability of the various test anions (P_A) relative to Cl^- permeability (P_{Cl}) derived from the biionic reversal potential measurements shown in **C**. Data points were fitted with a model which assumes that the anions are spherical ions, and that the permeability depends upon the ionic diameter and frictional forces within the channel. The pore diameter estimated from the fit is 5.1 Å

tively (Bormann et al. 1987). Thus, GABA$_A$ and GABA$_C$ receptors do not differ significantly in their open channel diameter.

V. Desensitization

Desensitization of ionotropic receptors is most likely a mechanism which allows these receptors to operate in the physiological concentration range of the endogenous ligand (DeVries and Schwartz 1999). Interestingly, GABA$_A$ and GABA$_C$ receptor-mediated Cl^--currents differ markedly in their time courses. During prolonged application of agonist, GABA$_A$ responses are transient, reaching a peak and then desensitizing to a lower steady current. In contrast, binding of GABA to $GABA_C$ receptors generates a sustained current, showing very little if any desensitization in the maintained presence of the agonist (for comparison of the decay times see BORMANN and FEIGENSPAN 1995). This is in line with the fast desensitizing responses of $GABA_A$ receptors expressed in *Xenopus* oocytes, and the sustained currents recorded from homooligomeric $\rho 1$ receptors in the same expression system (AMIN and WEISS 1994). Recently, structural motifs that confer agonist-induced desensitization were identified by expressing chimeras constructed from $\rho 1$- and $\beta 2$-subunits in *Xenopus* oocytes (LU and HUANG 1998). Regions in both the amino- and carboxyterminal domains of the $\beta 2$-subunit are important determinants for the desensitization properties of the receptor.

E. Pharmacology

I. $GABA_C$ Agonists

It has been suggested that folded analogues of GABA may interact selectively with bicuculline- and baclofen-insensitive GABA receptors (JOHNSTON et al. 1975; DREW et al. 1984). Converting the single covalent bond between carbon atoms C2 and C3 of the GABA molecule into a double bond fixes these atoms in a plane thereby generating two isomers: *cis*- and *trans*-4-aminocrotonic acid (CACA and TACA). The chemical structures of these conformationally restricted GABA analogues are shown in Fig. 3A in both fully extended and folded conformations.

The most potent $GABA_C$ receptor agonists besides GABA are muscimol (WOODWARD et al. 1993; QIAN and DOWLING 1993, 1994; KUSAMA et al. 1993a,b; DONG et al. 1994; WANG et al. 1994) and TACA (FEIGENSPAN et al. 1993; KUSAMA et al. 1993a,b; WOODWARD et al. 1993; DONG et al. 1994; LUKASIEWICZ et al. 1994). Comparison of their chemical structures indicates that these agonists are effective in the fully extended GABA conformation. However, CACA and TACA have been used to reveal differences in the agonist binding profiles of $GABA_A$ and $GABA_C$ receptors (JOHNSTON et al. 1975). When applied to retinal bipolar cells, both the *cis*- and the *trans*-enantiomer induced inward currents (FEIGENSPAN et al. 1993). The folded compound CACA elicited small but consistent responses (Fig. 3B). The blocking effect of bicuculline on CACA-evoked responses was significantly less than its effect on GABA-induced currents, indicating a preference of CACA for $GABA_C$ receptors. The extended compound TACA was almost equipotent at both GABA receptor subtypes (Fig. 3C). TACA-induced whole-cell currents were comparable in amplitude to currents evoked by equal concentrations of GABA and could be blocked by bicuculline to a similar extent. Another pair of *cis*-and *trans*-enantiomers with differential activity at $GABA_A$ and $GABA_C$ receptors has been described for $\rho 1$- and $\rho 2$-subunits expressed in *Xenopus* oocytes (KUSAMA et al. 1993a,b). $GABA_C$ receptors are selectively activated by *cis*-2-

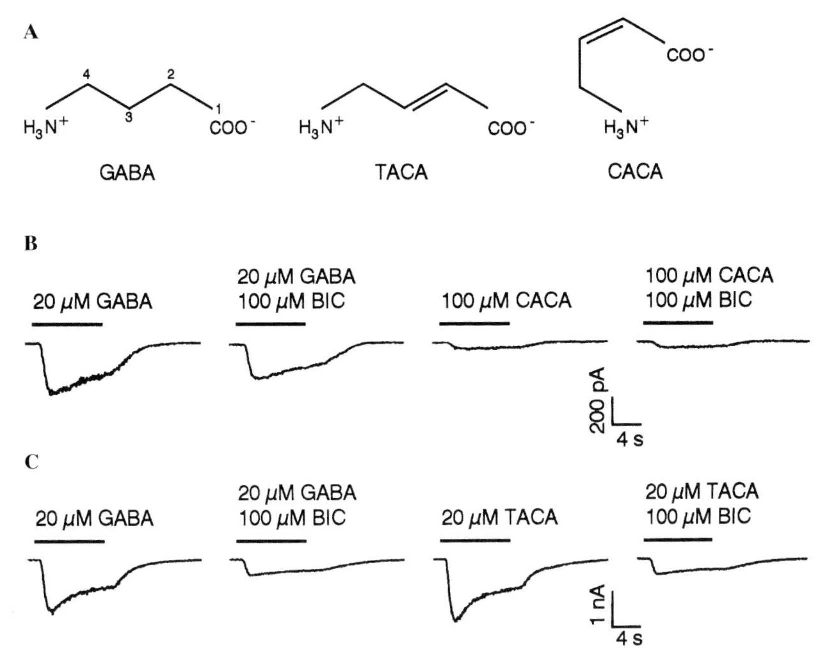

Fig. 3A–C. Agonist selectivity of GABA$_C$ receptors. **A** Chemical structures of three GABA$_C$ receptor agonists: GABA, TACA (*trans*-4-aminocrotonic acid), and CACA (*cis*-4-aminocrotonic acid). The conformationally restricted GABA analogues are shown in fully extended and folded conformations. **B** GABA-induced whole-cell currents recorded from a bipolar cell at −70 mV membrane potential. The bicuculline (BIC)-insensitive GABA$_C$ receptor-mediated response is shown in the *second trace*. CACA evoked only ~10% of the peak-current amplitude obtained with GABA (*third trace*). Bicuculline reduced the CACA response by 9%, compared with the 40% reduction seen in this cell with GABA (*fourth trace*), indicating a preference of CACA for GABA$_C$ receptors. **C** TACA produced currents that were ~30% larger than the currents evoked by GABA. The percentage inhibition by bicuculline was similar for the GABA and TACA response

aminomethyl-cyclopropane carboxylic acid (CAMP), while this compound is inert at GABA$_A$ receptors.

In a recent study, various C2, C3, C4 and N-substituted GABA and TACA analogues were examined for activity at GABA$_C$ receptors (Chebib et al. 1997). *trans*-4-Amino-2-fluorobut-2-enoic acid was found to be a potent agonist at homomeric ρ1 receptors expressed in *Xenopus* oocytes. In addition, the sulphinic acid analogue of GABA, homohypotaurine, is a potent partial agonist at GABA$_C$ receptors. In general, GABA$_C$ receptor agonists lose their potency when methyl or halo groups are substituted at the C3, C4 and N positions of the GABA and TACA molecules, whereas substitution at the C2 position is tolerated. Thus, the binding site of GABA$_C$ receptors for agonists or

Table 1. Pharmacological comparison of $GABA_A$ and $GABA_C$ receptors (FEIGENSPAN and BORMANN 1998)

Drug	Concentration (μmol/l)	I/I_C	
		$GABA_A$	$GABA_C$
Flunitrazepam	1	2.20 ± 0.73 (9)	0.98 ± 0.10 (6)
Zolpidem	1	3.31 ± 0.95 (23)	1.07 ± 0.07 (6)
CL-218,872	1	1.55 ± 0.41 (33)	0.99 ± 0.28 (9)
Pentobarbital	50	3.47 ± 1.35 (4)	1.05 ± 0.13 (5)
Alphaxalone	1	1.62 ± 0.66 (10)	0.92 ± 0.13 (5)
Picrotoxinin	10	0.48 ± 0.10 (9)	0.98 ± 0.15 (11)
	100	0.04 ± 0.04 (3)	0.81 ± 0.24 (8)
Strychnine	5	0.48 ± 0.07 (4)	1.01 ± 0.18 (4)
Zn^{2+}	50	0.99 ± 0.03 (3)	0.91 ± 0.08 (11)
SR-95531	10	0 (5)	0.86 ± 0.03 (6)
	100	0 (4)	0.48 ± 0.03 (7)
γ-HCH	10	0.60 ± 0.13 (11)	0.61 ± 0.14 (5)
	100	n.d.	0.39 ± 0.03 (6)
α-HCH	10	1.03 ± 0.09 (11)	1.11 ± 0.05 (6)
δ-HCH	10	1.65 ± 0.69 (5)	0.92 ± 0.12 (6)
Dieldrin	10	1.02 ± 0.09 (15)	1.06 ± 0.12 (6)

I/I_C indicates ratio of GABA-induced current in the presence of drug (I) relative to control current (I_C) ± SEM for *n* experiments.
n.d., not determined.

competitive antagonists might be smaller than that of $GABA_A$ and $GABA_B$ receptors (CHEBIB et al. 1997).

$GABA_C$ receptors do not respond to potent $GABA_A$ receptor modulators such as benzodiazepines, barbiturates and neurosteroids (POLENZANI et al. 1991; SHIMADA et al. 1992; FEIGENSPAN et al. 1993, 1994a; QIAN and DOWLING 1993, 1994; LUKASIEWICZ et al. 1994; DONG et al. 1994; WANG et al. 1994) (Table 1). The $GABA_B$ receptor agonist baclofen (POLENZANI et al. 1991; FEIGENSPAN et al. 1993; QIAN and DOWLING 1993, 1994) and antagonists such as phaclofen, saclofen and CGP-35348 (FEIGENSPAN et al. 1993; QIAN and DOWLING 1993; WOODWARD et al. 1993) are also inactive at $GABA_C$ receptors (Fig. 4).

II. $GABA_C$ Antagonists

The Cl^- channel blocker picrotoxinin has been shown to block $GABA_C$ receptor-mediated currents in fish, amphibians and ferrets and in oocytes expressing retinal poly(A^+) RNA (QIAN and DOWLING 1993; WOODWARD et al. 1993; LUKASIEWICZ et al. 1994; LUKASIEWICZ and WONG 1997). IC_{50} values for picrotoxinin measured in recombinant $GABA_C$ receptors are 48 μmol/l and 4.7 μmol/l for $\rho 1$ and $\rho 2$, respectively (WANG et al. 1995a). The highest affinity of picrotoxinin for homomeric subunits (IC_{50} = 0.68 μmol/l) has been demon-

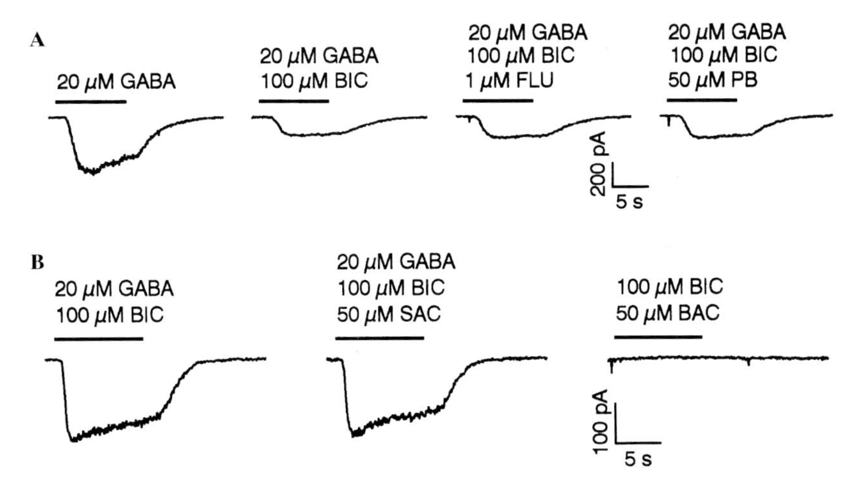

Fig. 4A,B. Pharmacology of GABA$_C$ receptors. **A** Effect of GABA$_A$ modulatory drugs. GABA-induced whole-cell currents were recorded from a bipolar cell at –70 mV holding potential. The bicuculline-insensitive GABA$_C$ response (*second trace*) was not affected by flunitrazepam (FLU) or pentobarbital (PB). **B** Effect of GABA$_B$ drugs. The GABA$_C$ component is not changed by the GABA$_B$ receptor antagonist 2-hydroxysaclofen (SAC), and the GABA$_B$ agonist baclofen (BAC) did not induce any measurable response

strated for rat ρ3-subunits expressed in *Xenopus* oocytes (Shingai et al. 1996). This value is very similar to the affinity for picrotoxinin of native GABA$_C$ receptors expressed in catfish cone horizontal cells (IC_{50} = 0.64 μmol/l; Dong and Werblin 1996).

In contrast, the rat bipolar cell GABA$_C$ receptor is rather insensitive to picrotoxinin (Feigenspan et al. 1993, Pan and Lipton 1995). The picrotoxinin insensitivity of rat retinal GABA$_C$ receptors is likely due to the ρ2-subunit (Zhang et al. 1995), which is expressed in rat bipolar cells (Enz et al. 1995, 1996). Site-directed mutagenesis has demonstrated that two different amino acid residues in transmembrane segment 2 of human and rat ρ-subunits confer picrotoxinin resistance (Enz and Bormann 1995; Wang et al. 1995a; Zhang et al. 1995). In addition, by substituting proline at position 309 with residues found at analogous position in the highly picrotoxinin-sensitive glycine α and GABA$_A$ receptor subunits, the competitive component of picrotoxinin inhibition was abolished (Wang et al. 1995a). The effect of picrotoxinin on GABA$_C$ receptors has been studied in cultured as well as in acutely isolated rat bipolar cells (Feigenspan and Bormann 1993, 1994a). At 100 μmol/l concentration, picrotoxinin reduced the peak GABA$_C$ response by only ~20% (Fig. 5), whereas GABA$_A$ receptors of retinal amacrine cells were completely blocked by the same concentration (Table 1).

The pyridazinyl-GABA derivative SR-95531 (gabazine) has been described as a selective and competitive GABA$_A$ receptor antagonist

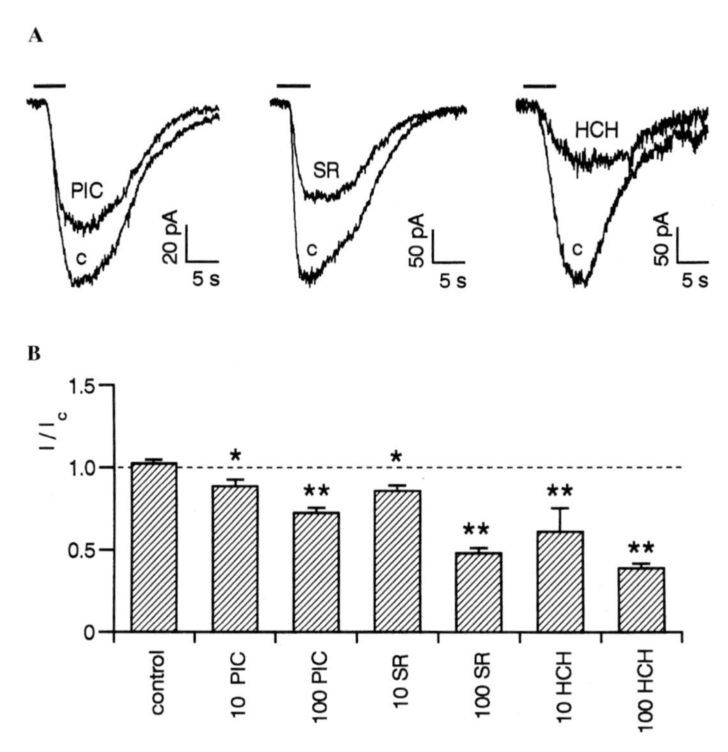

Fig. 5A,B. $GABA_C$ receptor antagonists. **A** Bicuculline-insensitive $GABA_C$ receptor-mediated control responses (*marked c*) were reduced by picrotoxinin (PIC, 100 μmol/l), SR-95531 (SR, 100 μmol/l), and γ-hexachlorocyclohexane (HCH, 100 μmol/l). **B** Summary of antagonistic drug effects at GABAC receptors. Each *bar* represents the average ratio I/I_C of the currents measured in the presence and absence of the drug tested. Numbers indicate drug concentrations in μmol/l. The *error bars* indicate SEM for groups of five cells. The no-effect level ($I/I_C = 1$) is indicated by the *dashed line*. *Asterisks* represent statistical differences from control ($*p \leq 0.05$, $**p \leq 0.01$; Student's *t*-test)

(MIENVILLE and VICINI 1987). When applied to bicuculline-insensitive GABA receptors expressed by bovine retinal poly(A^+) RNA, SR-95531 acted as a competitive inhibitor, although 240 times less potent than at $GABA_A$ receptors (WOODWARD et al. 1993). In rat retinal bipolar cells, SR-95531 exhibited moderate antagonistic activity at $GABA_C$ receptors (FEIGENSPAN and BORMANN 1994a) (Table 1), but it had no effect on the bicuculline-insensitive response of horizontal cells of the white perch retina (QIAN and DOWLING 1994). In addition, the glycine receptor antagonist strychnine which also inhibits $GABA_A$ receptors of hippocampal and retinal neurons (SHIRASAKI et al. 1991; FEIGENSPAN et al. 1993) has no effect on the $GABA_C$ receptor-mediated response (Table 1).

The partially folded GABA analogues isoguvacine, THIP, piperidine-4-sulfonic acid, isonipecotic acid, 3-aminopropyl sulfonic acid and Z-3-

amidinothiopropenoic acid (ZAPA) show slight antagonistic effects at $GABA_C$ receptors or no effect at all (CUTTING et al. 1992; WOODWARD et al. 1993; QIAN and DOWLING 1994). The extended GABA analogue, imidazole-4-acetic acid, is a strong antagonist at $GABA_C$ receptors (KUSAMA et al. 1993a; QIAN and DOWLING 1994). The $GABA_B$-selective agonist 3-aminopropyl-(methyl)phosphinic acid (3-APMPA) has been shown to bind to retinal $GABA_C$ receptors with low micromolar potency (WOODWARD et al. 1993). Recently, a hybrid of isoguvacine and 3-APMPA has been designed, which retains its affinity for $GABA_C$ receptors but interacts only weakly with $GABA_A$ or $GABA_B$ receptors (MURATA et al. 1996; RAGOZZINO et al. 1996). This compound, (1,2,5,6-tetrahydropyridine-4-yl)methylphosphinic acid (TPMPA), acts as a selective antagonist of human $GABA_C$ receptors expressed in *Xenopus* oocytes.

The effects of hexachlorocyclohexanes (HCH) on bicuculline-sensitive and -insensitive GABA receptors expressed in *Xenopus* oocytes have been described by Woodward and coworkers (WOODWARD et al. 1992a). In this expression system, the γ-enantiomer (lindane) was a potent inhibitor of both the $GABA_A$- and $GABA_C$-like currents. When γ-HCH was applied to isolated bipolar cells in the presence of bicuculline, the $GABA_C$ receptor-mediated current was reduced (Fig. 5). Likewise, γ-HCH inhibited the $GABA_A$ response in the same sample of bipolar cells. The isomers α- and δ-HCH as well as dieldrin had no effect on retinal $GABA_C$ receptors (FEIGENSPAN and BORMANN 1994a, 1998) (Table 1).

F. Modulation of $GABA_C$ Receptors

I. Extracellular Modulation

Zinc is widely distributed throughout the vertebrate central nervous system (HAUG 1967; FREDERICKSON 1989) and most likely acts as an endogenous neuromodulator at pre- and postsynaptic ion channels (ASSAF and CHUNG 1984; XIE and SMART 1991). It has been shown to modulate the function of both $GABA_A$ and $GABA_B$ receptors (WESTBROOK and MAYER 1987; LEGENDRE and WESTBROOK 1990; XIE and SMART 1991; HARRISON and GIBBONS 1994). In photoreceptors of the vertebrate retina, zinc is colocalized with glutamatergic synaptic vesicles, where it may act as a diffusible molecular switch (WU et al. 1993).

$GABA_C$ receptors are present in regions of the retina with high concentrations of the divalent metal ion Zn^{2+}. Native $GABA_C$ receptors present on bipolar and horizontal cells from the retina of cold-blooded vertebrates can be down-modulated by extracellular application of Zn^{2+} (DONG and WERBLIN 1995, 1996; QIAN and DOWLING 1995; QIAN et al. 1997), which acts as a mixed antagonist. Zn^{2+} binds to $GABA_C$ receptors with a half-inhibition concentration of 8.2 μmol/l (DONG and WERBLIN 1995). $GABA_C$ receptors expressed in *Xenopus* oocytes after injection of mRNA for $\rho 1$- or $\rho 2$-subunits are also inhibited by Zn^{2+}, and Zn^{2+} inhibition of GABA $\rho 1$ receptors displays both competitive and noncompetitive components (Calvo et al. 1994; Wang et al. 1994; Chang et al. 1995). In either system the effect of Zn^{2+} is independent of voltage (Wang et al. 1995b; Dong and Werblin 1996). The binding site for Zn^{2+} is located on the surface of the receptor molecule, as indicated by the effect of extracellular pH on Zn^{2+} inhibition (Wang et al. 1995b). Site-directed mutagenesis has revealed a single histidine residue (His 156) in the extracellular domain of $\rho 1$ critical for Zn^{2+}-sensitivity (Wang et al. 1995b). The divalent metal ions Ni^{2+} and Cd^{2+} also down-modulate $GABA_C$ receptor function with the order of potency $Zn^{2+}>Ni^{2+}>Cd^{2+}>Co^{2+}$ (Calvo et al. 1994; Kaneda et al. 1997). The same His 156 residue is also involved in inhibition of $GABA_C$ receptors by Ni^{2+} and Cd^{2+} (Wang et al. 1995b). In contrast to the potent inhibitory effects described above, $GABA_C$ receptors of the rat retina were only slightly blocked by extracellular Zn^{2+} (Feigenspan and Bormann 1998) (Table 1).

Recently, a positive modulation by extracellular Ca^{2+} of the $GABA_C$ response of retinal horizontal cells has been shown (Kaneda et al. 1997). Thus, the extracellular domain of the $GABA_C$ receptor is likely to have two functionally distinct binding sites mediating facilitation (Ca^{2+}) and inhibition (Zn^{2+}, Ni^{2+}, Cd^{2+} Co^{2+}).

Functional $GABA_C$ receptors which are formed in HEK 293 cells by transiently expressing the rat $\rho 1$-subunit, can be modulated by extracellular protons (Wegelius et al. 1996). A decrease in pH from 7.4 to 6.4 leads to an inhibition of $GABA_C$ receptor currents, whereas an increase in pH results in up-regulation of the GABA response. A regulatory binding site for protons on the $\rho 1$ subunit has been described for the inhibitory effect of Zn^{2+} (Wang et al. 1995b).

II. Intracellular Modulation by Protein Kinases

The ρ-subunits are composed of four membrane-spanning regions and a cytoplasmic loop between the third and fourth transmembrane domain (Cutting et al. 1991). The intracellular loop contains consensus sequence sites for phosphorylation by protein kinase C (PKC), suggesting a role for PKC in the modulation of $GABA_C$ receptor function. The presence of PKC has previously been demonstrated in rod bipolar cells, which stain selectively with an antibody directed against the α-isoenzyme of PKC (Greferath et al. 1990; Karschin and Wässle 1990).

An intracellular regulatory pathway has been identified in cultured retinal bipolar cells, which involves activation of PKC and results in the down-modulation of $GABA_C$ receptors (Feigenspan and Bormann 1994c). The effect of the phorbol ester PMA clearly indicates that the down-regulation of $GABA_C$ receptor function requires activation of PKC. The signaling pathway upstream of PKC involves the phospholipase C-mediated hydrolysis of phosphoinositol (PI) thereby producing diacylglycerol (DAG), the physiological activator of PKC (Fig. 6). As a consequence, PKC phosphorylation reduces the current through GABA$_C$ receptors, and thereby the inhibitory action of GABA. PKC-mediated inhibition of $\rho 1$ receptor function has been observed with the $\rho 1$-subunit expressed in *Xenopus* oocytes (KUSAMA et al. 1995). However, recent evidence indicates that consensus sequence sites in both $\rho 1$ and $\rho 2$ are not critical for inhibition by PKC of GABA$_C$ receptor function (KUSAMA et al. 1998).

Retinal bipolar cells receive glutamatergic input from photoreceptors, with kainate receptors present in the membrane of hyperpolarizing bipolar cells (DEVRIES and SCHWARTZ 1999) and metabotropic glutamate receptor subtype 6 (mGluR6) on the dendrites of depolarizing bipolar cells (NOMURA et al. 1994). The effect of *trans*-ACPD and L-AP4 on the GABA$_C$ response of rat retinal bipolar cells was studied (FEIGENSPAN and BORMANN 1994c). Both compounds act as ligands at metabotropic glutamate receptors (SCHOEPP and CONN 1993). Run-down of the GABA-induced current was enhanced in the presence of *trans*-ACPD and L-AP4 (Fig. 6A), suggesting that both metabotropic glutamate receptor agonists couple to PKC activation, and subsequently down-regulate GABA$_C$ receptor function. Neither glutamate agonist elicited an inward current, thus ruling out the possibility of modulating cyclic nucleotide-gated channels (NAWY and JAHR 1990, 1991; DE LA VILLA et al. 1995). The specific agonist at metabotropic glutamate receptors mGluR1 and mGluR5, *trans*-azetidine-2,4-dicarboxylic acid, and the mGluR agonist quisqualic acid decreased the GABA$_C$ receptor-mediated current in a rat retinal slice preparation (EULER and WÄSSLE 1998). In addition, extracellular application of serotonin also accelerated the run-down of the bicuculline-insensitive GABA response (Fig. 6A) (FEIGENSPAN and BORMANN 1994b). This effect appeared to be mediated by the 5-HT$_2$ receptor subtype, as it was mimicked by the more specific agonist α-methyl serotonin (Fig. 6A). Figure 6B shows the current model proposed for the modulation of retinal GABA$_C$ receptors by PKC.

Protein kinase A (PKA), which modulates GABA$_A$ receptor function in the retina and elsewhere in the CNS (KANO and KONNERTH 1992; VERUKI and YEH 1992; FEIGENSPAN and BORMANN 1994b), had no effect on GABA$_C$ receptors of rat retinal bipolar cells (Fig. 6A). However, in acutely isolated cone horizontal cells of the catfish retina, dopamine selectively reduced the GABA$_C$ receptor current (DONG and WERBLIN 1994). This effect is most likely mediated by D_1 dopamine receptors coupled to adenylyl cyclase, since it can be mimicked by the D_1 selective agonist SKF-38393 and forskolin. In bipolar cells of the tiger salamander retina, extracellular application of dopamine relieved the GABA$_C$ receptor-mediated inhibition of calcium entry and thus transmitter release (WELLIS and WERBLIN 1995). This effect is also likely due to binding of dopamine to D_1 receptors and subsequent activation of the cAMP second messenger pathway.

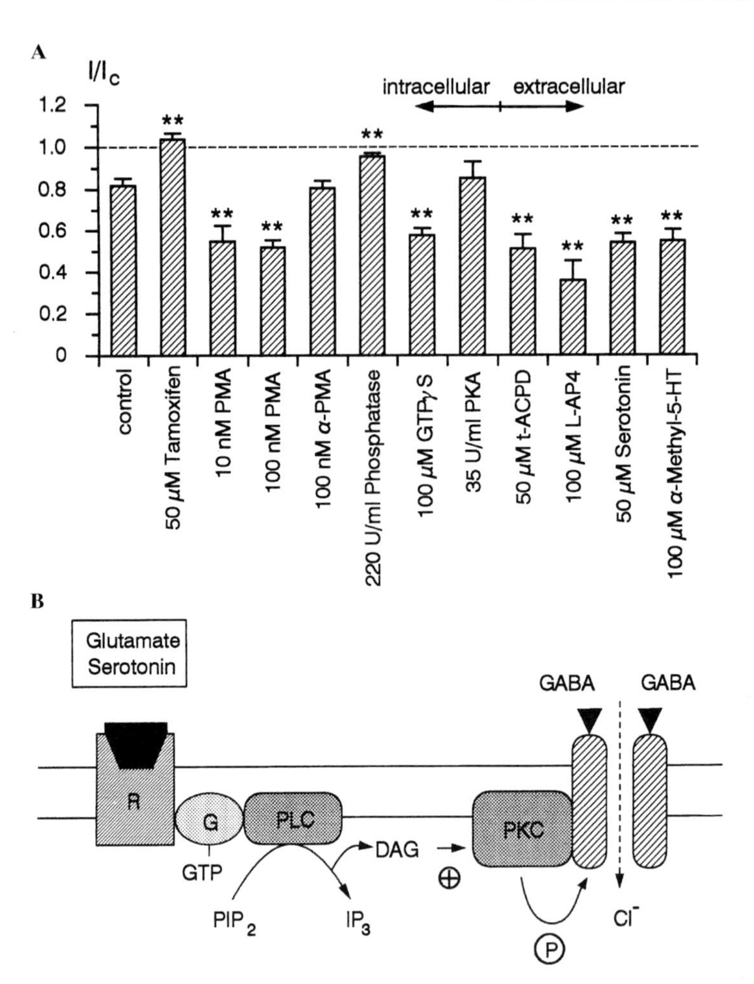

Fig. 6A,B. Modulation of $GABA_C$ receptors by protein kinase C (PKC). **A** Time-dependence of $GABA_C$ receptor-mediated whole-cell currents. The *bars* indicate I/I_C, the ratio of current measured after 20 min to the current measured at 1 min. *Error bars* denote SEM for ten control cells and five cells otherwise. In control experiments, the bicuculline-insensitive GABA response showed run-down of typically 18% ($I/I_C = 0.82$) after 20 min of recording. Various drugs were tested for their ability to modify the control response after intra- or extracellular application, as indicated. Extracellular drugs were applied for 30 s between consecutive GABA pulse, intracellular drugs were included in the pipette solution. *Asterisks* indicate statistical differences ($p \leq 0.01$) from control, as calculated with Student's *t*-test. Abbreviations: PMA, phorbol 12-myristate, 13-acetate; α-PMA, 4α-phorbol 12-myristate, 13-acetate; PKA, catalytic subunit of cAMP-dependent protein kinase; *t*-ACPD, *trans*-(+)-1-amino-1,3-cyclopentane dicarboxylate; L-AP4, 2-amino-4-phosphonobutyric acid. **B** Model illustrating the sequence of events which may lead to a reduction of $GABA_C$ receptor-mediated currents (*dashed arrow*). The *box* shows agonists that stimulate phospholipase C (PLC) activity following binding of glutamate and serotonin to metabotropic glutamate and $5\text{-}HT_2$ receptors, respectively. Abbreviations: R, receptor; G, G-protein; PIP_2, phosphatidylinositol 4,5-bisphosphate; DAG, diacylglycerol; IP_3, inositol 1,4,5-trisphosphate; P, phosphate group; +, activation

G. Physiological Function of $GABA_C$ Receptors

Bicuculline-baclofen-insensitive $GABA_C$ responses have been described in various parts of the vertebrate brain, including spinal cord (JOHNSTON et al. 1975), optic tectum (NISTRI and SIVILOTTI 1985; SIVILOTTI and NISTRI 1989), cerebellum (DREW et al. 1984; DREW and JOHNSTON 1992) and hippocampus (STRATA and CHERUBINI 1994; MARTINA et al. 1995). However, specific physiological functions of $GABA_C$ receptors in those brain areas are still elusive. It is conceivable though that in the developing hippocampus, $GABA_C$ receptors could be important for shaping the range of inhibitory synaptic functions required for the establishment of various forms of learning and memory.

More specific ideas have emerged as to the physiological role(s) of $GABA_C$ receptors in the vertebrate retina. Although $GABA_C$ receptors are common on horizontal and bipolar cells in lower vertebrates (QIAN and DOWLING 1993, 1994, 1995; DONG et al. 1994; LUKASIEWICZ et al. 1994), $GABA_C$ receptors in the mammalian retina are localized on bipolar cells, where they coexist with $GABA_A$ receptors (FEIGENSPAN et al. 1993; FEIGENSPAN and BORMANN 1994a; PAN and LIPTON 1995). GABA receptors on bipolar cell terminals have been shown to down-regulate voltage-dependent calcium channels, thereby decreasing presynaptic transmitter release (TACHIBANA et al. 1993; LUKASIEWICZ et al. 1994; MATTHEWS et al. 1994; PAN and LIPTON 1995; WELLIS and WERBLIN 1995). Since GABAergic amacrine cells make synapses onto bipolar cell terminals (MARC et al. 1978; YAZULLA et al. 1987; CHUN and WÄSSLE 1989; POURCHO and OWCZARZAK 1989), synaptically released GABA is likely to modulate transmitter release from bipolar cells. $GABA_C$ receptor-mediated inhibition of the excitatory synaptic transmission between bipolar and ganglion cells has been described in salamander retina (LUKASIEWIECZ and WERBLIN 1994; WELLIS and WERBLIN 1995). Interestingly, $GABA_C$ receptors appear to contribute to the control of acetylcholine (ACh) release in the rabbit retina (MASSEY et al. 1997). When $GABA_A$ receptors were completely blocked by saturating concentrations of SR-95531, picrotoxin caused a further increase in ACh release indicating a contribution of $GABA_C$ receptors. The inhibition responsible for directional selectivity, however, is exclusively mediated by $GABA_A$ receptors (MASSEY et al. 1997).

The high affinity of $GABA_C$ receptors for GABA and their sustained response properties make them ideally suited to fine tune bipolar cell output. GABA feedback inhibition from amacrine to bipolar cells is likely to control bipolar cell output. Thus, low GABA concentrations insufficient to activate type A receptors could nevertheless affect bipolar cell output via $GABA_C$ receptors. Since $GABA_A$ receptors may activate more rapidly than $GABA_C$ receptors (PAN and LIPTON 1995), the ratio of $GABA_A$ and $GABA_C$ receptors at bipolar cell terminals is likely to determine the kinetics of GABAergic inhibition. Comparing GABAergic synaptic responses of bipolar cell terminals and ganglion cells in the salamander retina, LUKASIEWICZ and SHIELDS (1998) could demonstrate that the temporal properties of the synaptic responses are determined by the combination of $GABA_A$ and $GABA_C$ receptors. Rod and cone bipolar cells in the rat retina display a differential pattern of $GABA_A$ vs $GABA_C$ receptors (EULER and WÄSSLE 1998). While 70% of the total GABA-induced current in rod bipolar cells was mediated by $GABA_C$ receptors, this fraction was only 20% in cone bipolar cells. In addition, the $GABA_C$-receptor-mediated fraction of the GABA response appears to differ between morphological types of cone bipolar cells (EULER and WÄSSLE 1998).

Further evidence for a $GABA_C$ receptor-mediated modulation of bipolar cell output has been obtained in the amphibian retina (ZHANG and SLAUGHTER 1995). When $GABA_A$ receptors were blocked by bicuculline or SR-95531, and $GABA_B$ receptors were saturated with baclofen, GABA preferentially reduced ON light responses in amacrine and ganglion cells, presumably through a presynaptic mechanism that inhibited bipolar cell output. Additionally, ZHANG and SLAUGHTER (1995) found that although the peak $GABA_A$ receptor-mediated current is about five times greater than the $GABA_C$ receptor-mediated current, the desensitized A-type current was less than that produced by the C-type receptor. Thus, the $GABA_C$ receptor may generate a small but sustained current, well suited to provide tonic inhibition to second- and third-order neurons. In contrast, rapidly desensitizing $GABA_A$ receptor currents may mediate more powerful but transient inhibition.

As pointed out by LUKASIEWICZ (1996), $GABA_A$ and $GABA_C$ receptors show an interesting pattern of distribution within the retina. Amacrine and ganglion cells which are spike-generating neurons only express $GABA_A$ receptors. $GABA_C$ receptor-mediated currents have been identified in horizontal and bipolar cells and in cone photoreceptors, all slow potential neurons that do not fire action potentials. Transmitter release from these neurons is continuous and graded with membrane potential. This may enable non-desensitizing, high-affinity $GABA_C$ receptors to respond to low synaptic levels of GABA and thereby precisely regulate membrane potential and transmitter release. The presence of $GABA_C$ receptors may expand the capacity of these cell types to respond to a broad range of synaptic GABA concentrations.

→

Fig. 7. Comparison of $GABA_A$, $GABA_B$ and $GABA_C$ receptors. The $GABA_A$ receptor (top) is a Cl^- pore with 4.9 Å diameter and modulatory sites for benzodiazepines, barbiturates and general anesthetics. The action of GABA is blocked by bicuculline and picrotoxinin. The $GABA_A$ responses of retinal amacrine cells are up-regulated upon intracellular phosphorylation of the receptor by protein kinase A. Each $GABA_A$ receptor-subunit consists of four membrane-spanning domains (*insert, top*). Five such subunits assemble into a pentameric structure (*insert, bottom*). The internal Cl^- pore is lined by amphiphilic transmembrane segments M2. The $GABA_B$ receptor (*middle*) is a member of the seven – transmembrane protein family and coupled to effect or systems (K^+ or Ca^{2+} channels) via G-proteins. The $GABA_C$ (*bottom*) receptor is a Cl^- pore (5.1 Å diameter) and is activated selectively by CACA. The action of GABA is blocked by TPMPA and picrotoxinin. The $GABA_A$ antagonist bicuculline and $GABA_A$ modulatory drugs have no effect. The $GABA_C$ responses are down-regulated upon intracellular activation of protein kinase C

GABA$_A$

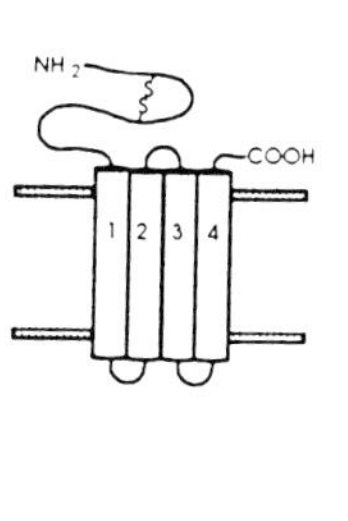

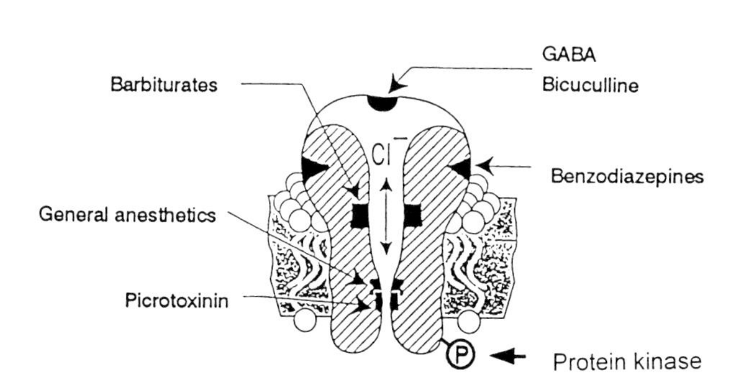

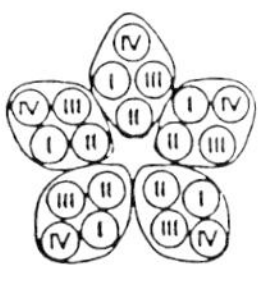

GABA$_B$

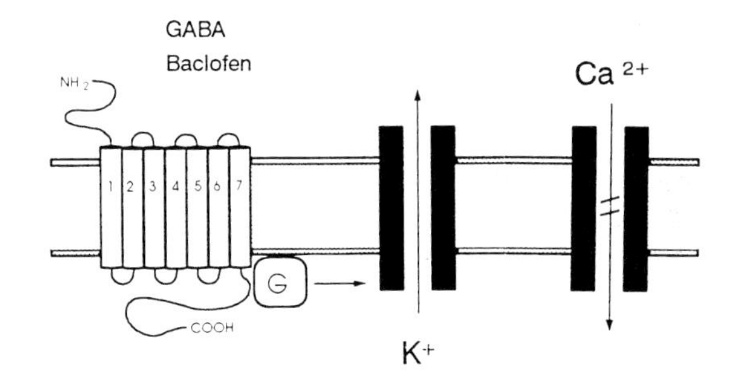

GABA$_C$

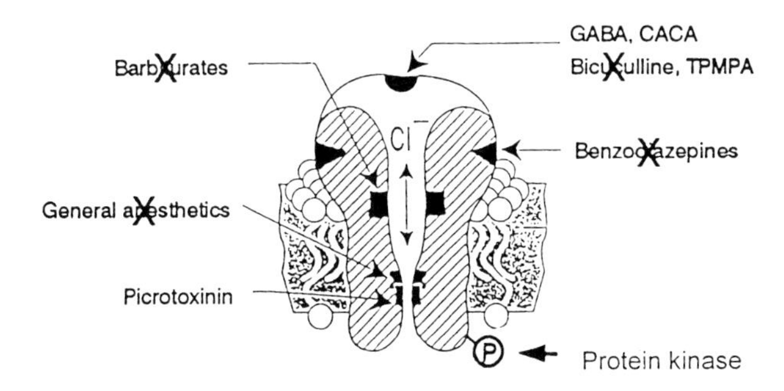

H. Terminology for $GABA_C$ Receptors

It now appears that GABA gates at least three types of GABA receptor that are distinct both pharmacologically and structurally (Fig. 7). Although the term '$GABA_C$' is widely accepted and represents a logical and convenient extension of the current GABA receptor nomenclature (JOHNSTON 1996), there is debate whether this receptor deserves separate classification or should be considered a subspecies of $GABA_A$ receptor Cl^- channels, as recommended by BARNARD et al. (1998). We are not in concert with this view, and favour the $GABA_C$ terminology, because:

1. *$GABA_C$ receptors are pharmacologically distinct.* Whilst $GABA_A$ and $GABA_B$ subtypes are defined by their respective sensitivities to bicuculline and baclofen (HILL and BOWERY 1981), $GABA_C$ receptors do not respond to those drugs. Also, they are selectively activated by *cis*-4-aminocrotonic acid (CACA) (JOHNSTON et al. 1975; FEIGENSPAN et al. 1993; QIAN and DOWLING 1993). TPMPA [(1,2,5,6-tetrahydropyridine-4-yl) methylphosphinic acid] has been identified as a potent and highly selective antagonist for $GABA_C$ receptors (MURATA et al. 1996; RAGOZZINO et al. 1996).
2. *$GABA_C$ receptors are structurally distinct.* Whilst fully functional $GABA_A$ receptors are composed of α-, β- and γ-subunits, $GABA_C$ receptors are assembled from ρ-subunits, known to mediate robust bicuculline-insensitive GABA responses in heterologous expression systems (CUTTING et al. 1991; WANG et al. 1994; OGURUSU and SHINGAI 1996). There is no evidence that the ρ-subunits coassemble with the $GABA_A$ receptor α-, β- and γ-subunits, or the glycine receptor β-subunit (SHIMADA et al. 1992; KUSAMA et al. 1993a).
3. *$GABA_C$ receptors are genetically distinct.* On human chromosomes, the genes for the $GABA_A$ subunits occur in clusters, each cluster containing genes for α, β and γ/ε (MCLEAN et al. 1995). The ρ-subunit genes are separated from these clusters (CUTTING et al. 1992).
4. *$GABA_C$ receptors are functionally distinct.* Electrophysiological responses from native or recombinant $GABA_C$ receptors differ markedly from $GABA_A$ receptors, most notably with respect to sensitivity, conductance, gating and desensitization (see BORMANN and FEIGENSPAN 1995 for review).
5. *$GABA_C$ receptors show distinct cellular localization.* Synaptic $GABA_C$ receptors consist of ρ-subunits and do not colocalize with $GABA_A$ or glycine receptor subunits (KOULEN et al. 1998). $GABA_C$ receptors are specifically linked to the cytoskeleton via microtubule-associated protein (MAP-1B) (HANLEY et al. 1999).

I. Conclusions

Recent developments in the understanding of GABA receptors support and extend the original observations of Johnston and colleagues of the existence of bicuculline- and baclofen-insensitive GABA$_C$ receptors. Pharmacological, molecular biological and physiological evidence are in favour of a new class of GABA receptor (Bormann 2000). The GABA$_C$ receptors are composed of ρ-subunits and are highly enriched in the vertebrate retina, but present also in many other CNS structures. GABA$_C$ receptors are integral membrane channels that stabilize the resting potential of the cell by increasing the membrane conductance to Cl^-. Inhibition mediated by GABA$_C$ receptors is expected to be very pronounced, occurring at very low GABA concentrations and to be longer lasting than GABA$_A$-receptor mediated inhibition. More efforts are needed to exploit the full range of physiological and pharmacological implications of GABA$_C$ receptors.

Acknowledgement. This work was supported by the Deutsche Forschungsgemeinschaft (SFB 50g 'Neurovision') and the Fonds der Chemischen Industrie.

References

Albrecht BE, Darlison MG (1995) Localization of the ρ1- and ρ2-subunit messenger RNAs in chick retina by in situ hybridization predicts the existence of γ-aminobutyric acid type C receptor subtypes. Neurosci Lett 189:155–158
Albrecht BE, Breitenbach U, Stühmer T, Harvey RJ, Darlison MG (1997) In situ hybridization and reverse transcription-polymerase chain reaction and studies on the expression of the GABA$_C$ receptor ρ1- and ρ2-subunit genes in avian and rat brain. Eur J Neurosci 9:2414–2422
Amin J, Weiss DS (1994) Homomeric ρ1 GABA channels: activation properties and domains. Receptor Channels 2:227–236
Assaf SY, Chung SH (1984) Release of endogenous Zn^{2+} from brain tissue during activity. Nature 308:734–736
Barnard EA, Skolnick P, Olsen RW, Möhler H, Sieghart W, Biggio G, Braestrup C, Bateson AN, Langer SZ (1998) International Union of Pharmacology. XV. Subtypes of γ-aminobutyric acid$_A$ receptor: Classification on the basis of subunit structure and receptor function. Pharmacol Rev 50:291–313
Bettler B, Kaupmann K, Bowery N (1998) GABA$_B$ receptors: drugs meet clones. Curr Opin Neurobiol 8:345–350
Bormann J (1988) Electrophysiology of GABA$_A$ and GABA$_B$ receptor subtypes. Trends Neurosci 11:112–116
Bormann J (2000) The 'ABC' of GABA receptors. Trends Pharmacol Sci 21:16–19
Bormann J, Feigenspan A (1995) GABA$_C$ receptors. Trends Neurosci 18:515–519
Bormann J, Hamill OP, Sakmann B (1987) Mechanism of anion permeation through channels gated by glycine and γ-aminobutyric acid in mouse cultured spinal neurones. J Physiol (London) 385:243–286
Boue-Grabot E, Roudbaraki M, Bascles L, Tramu R, Bloch B, Garret M (1998) Expression of GABA receptor ρ subunits in rat brain. J Neurochem 70:899–907
Bowery NG (1989) GABA$_B$ receptors and their significance in mammalian pharmacology. Trends Pharmacol Sci 10:401–407
Calvo DJ, Vazquez AE, Miledi R (1994) Cationic modulation of ρ_1 type γ-aminobutyrate receptors expressed in *Xenopus* oocytes. Proc Natl Acad Sci USA 91:12725–12729
Chang Y, Amin J, Weiss DS (1995) Zinc is a mixed antagonist at homomeric ρ1 γ-aminobutyric acid-activated channels. Mol Pharmacol 47:595–602
Chebib M, Vandenberg RJ, Johnston GAR (1997) Analogues of γ-aminobutyric acid (GABA) and *trans*-4-aminocrotonic acid (TACA) substituted in the 2-position as GABA$_C$ receptor antagonists. Br. J Pharmacol 122:1551–1560

Cherubini E, Strata F (1997) $GABA_C$ receptors: a novel receptor family with unusual pharmacology. News Physiol Sci 12:136–141
Chun M, Wässle H (1989) GABA-like immunoreactivity in the cat retina: electron microscopy. J Comp Neurol 279:55–67
Coombs JS, Eccles JC, Fatt P (1955) The specific ionic conductances and the ionic movements across the motoneural membrane that produce the inhibitory postsynaptic potential. J Physiol (London) 130:326–373
Cutting GR, Lu L, O'Hara BF, Kasch LM, Montrose-Rafizadeh C, Donovan DM, Shimada S, Antonarakis SE, Guggino WB, Uhl GR, Kazazian HH Jr. (1991) Cloning of the γ-aminobutyric acid (GABA) ρ_1 cDNA: A GABA receptor subunit highly expressed in the retina. Proc Natl Acad Sci USA 88:2673–2677
Cutting GR, Curristin S, Zoghbi H, O'Hara BF, Seldin MF, Uhl GR (1992) Identification of a putative γ-aminobutyric acid (GABA) receptor subunit rho_2 cDNA and colocalization of the genes encoding rho_2 (GABRR2) and rho_1 (GABRR1) to human chromosome 6q14-q21 and mouse chromosome 4. Genomics 12:801–806
de la Villa P, Kurahashi T, Kaneko A (1995) L-glutamate-induced responses and cGMP-activated channels in three subtypes of retinal bipolar cells dissociated from the cat. J Neurosci 15:3571–3582
DeVries SH, Schwartz EA (1999) Kainate receptors mediate synaptic transmission between cones and 'Off' bipolar cells in a mammalian retina. Nature 397:157–160
Dong CJ, Werblin FS (1994) Dopamine modulation of $GABA_C$ receptor function in an isolated retinal neuron. J Neurophysiol 71:1258–1260
Dong CJ, Werblin FS (1995) Zinc downmodulates the $GABA_C$ receptor current in cone horizontal cells acutely isolated from the catfish retina. J Neurophysiol 73:916–919
Dong CJ, Werblin FS (1996) Use-dependent and use-independent blocking actions of picrotoxin and zinc at the $GABA_C$ receptor in retinal horizontal cells. Vision Res 36:3997–4005
Dong C-J, Picaud SA, Werblin FS (1994) GABA transporters and $GABA_C$-like receptors on catfish cone- but not rod-driven horizontal cells. J Neurosci 14:2648–2658
Drew CA, Johnston GAR (1992) Bicuculline- and baclofen-insensitive γ-aminobutyric acid binding to rat cerebellar membranes. J Neurochem 58:1087–1092
Drew CA, Johnston GAR, Weatherby RP (1984) Bicuculline-insensitive GABA receptors: studies on the binding of (–)-baclofen to rat cerebellar membranes. Neurosci Lett 52:317–321
Enz R, Bormann J (1995) A single point mutation decreases picrotoxinin sensitivity of the human GABA receptor $\rho 1$ subunit. Neuroreport 6:1569–1572
Enz R, Cutting GR (1998) Molecular composition of $GABA_C$ receptors. Vis Res 38:1431–1441
Enz R, Cutting GR (1999) $GABA_C$ receptor ρ subunits are heterogeneously expressed in the human CNS and form homo- and heterooligomers with distinct properties. Eur J Neurosci 11:41–50
Enz R, Brandstätter JH, Hartveit E, Wässle H, Bormann J (1995) Expression of GABA receptor $\rho 1$ and $\rho 2$ subunits in the retina and brain of the rat. Eur J Neurosci 7:1495–1501
Enz R, Brandstätter JH, Wässle H, Bormann J (1996) Immunocytochemical localization of the $GABA_C$ receptor ρ subunits in the mammalian retina. J Neurosci 16:4479–4490
Euler T, Wässle H (1998) Different contributions of $GABA_A$ and $GABA_C$ receptors to rod and cone bipolar cells in a rat retinal slice preparation. J Neurophysiol 79:1384–1395
Feigenspan A, Wässle H, Bormann J (1993) Pharmacology of GABA receptor Cl^- channels in rat retinal bipolar cells. Nature 361:159–163
Feigenspan A, Bormann J (1994a) Differential pharmacology of $GABA_A$ and $GABA_C$ receptors on rat retinal bipolar cells. Eur J Pharmacol Mol Pharmacol Sect 288:97–104

Feigenspan A, Bormann J (1994b) Facilitation of GABAergic signaling in the retina by receptors stimulating adenylate cyclase. Proc Natl Acad Sci USA 91: 10893–10897

Feigenspan A, Bormann J (1994c) Modulation of $GABA_C$ receptors in rat retinal bipolar cells by protein kinase C. J Physiol (London) 481:325–330

Feigenspan A, Bormann J (1998) GABA-gated Cl^- channels in the rat retina. Prog Ret Eye Res 17:99–126

Frederickson CJ (1989) Neurobiology of zinc and zinc-containing neurons. Int Rev Neurobiol. 31:145–238

Greferath U, Grünert U, Wässle H (1990) Rod bipolar cells in the mammalian retina show protein kinase C-like immunoreactivity. J Comp Neurol 301:433–442

Greka A, Koolen JA, Lipton SA, Zhang D (1998) Cloning and characterization of mouse GABA(C) receptor subunits. Neuroreport 9:229–232

Hanley JG, Koulen P, Bedford F, Gordon-Weeks PR, Moss SJ (1999) The protein MAP-1B links $GABA_C$ receptors to the cytoskeleton at retinal synapses. Nature 397:66–69

Harrison NL, Gibbons SJ (1994) Zn^{2+}: an endogenous modulator of ligand- and voltage-gated ion channels. Neuropharmacology 33:935–952

Haug FM (1967) Electron microscopical localization of the zinc in hippocampal mossy fibre synapses by a modified sulfide silver procedure. Histochemistry 8:355–368

Hill DR, Bowery NG (1981) ^{3}H-baclofen and ^{3}H-GABA bind to bicuculline-insensitive $GABA_B$ sites in rat brain. Nature 290:149–152

Johnston GAR (1996) $GABA_C$ receptors: relatively simple transmitter-gated ion channels? Trends Pharmacol Sci 17:319–323

Johnston GAR, Curtis DR, Beart PM, Game CJA, McCulloch RM, Twitchin B (1975) *cis*- and *trans*-4-Aminocrotonic acid as GABA agonist of restricted conformation. J Neurochem 24:157–160

Kaneda M, Mochizuki M, Aoki K, Kaneko A (1997) Modulation of $GABA_C$ response by Ca^{2+} and other divalent cations in horizontal cells of the catfish retina. J Gen Physiol 110:741–747

Kano M, Konnerth A (1992) Potentiation of GABA-mediated currents by cAMP-dependent protein kinase. Neuroreport 3:563–566

Karschin, Wässle H (1990) Voltage- and transmitter-gated currents in isolated rod bipolar cells of the rat retina. J Neurophysiol 63:860–876

Koulen P, Brandstätter JH, Kröger S, Enz R, Bormann J, Wässle H (1997) Immunocytochemical localization of the $GABA_C$ receptor ρ subunits in the cat, goldfish, and chicken retina. J Comp Neurol 380:520–532

Koulen P, Brandstätter JH, Enz R, Bormann J, Wässle H (1998) Synaptic clustering of $GABA_C$ receptor ρ-subunits in the rat retina. Eur J Neurosci 10:115–127

Kusama T, Spivak CE, Whiting P, Dawson VL, Schaeffer JC, Uhl GR (1993a) Pharmacology of GABA ρ_1 and the GABA α/β receptors expressed in *Xenopus* oocytes and COS cells. Br J Pharmacol 109:200–206

Kusama T, Wang T-L, Guggino WB, Cutting GR, Uhl GR (1993b) GABA ρ_2 receptor pharmacological profile: GABA recognition site similarities to ρ_1. Eur J Pharmacol Mol Pharmacol Sect 245:83–84

Kusama T, Wang JB, Spivak CE, Uhl GR (1994) Mutagenesis of the GABA $\rho 1$ receptor alters agonist affinity and channel gating. Neuroreport 5:1209–1212

Kusama T, Sakurai M, Kizawa Y, Uhl GR, Murakami H (1995) GABA ρ_1 receptor: inhibition by protein kinase C activators. Eur J Pharmacol Mol Pharmacol Sect 291:431–434

Kusama T, Hatama K, Sakurai M, Kizawa Y, Uhl GR, Murakami H (1998) Consensus phosphorylation sites of human GABA(c)/GABAρ receptors are not critical for inhibition by protein kinase C activation. Neurosci Lett 255:17–20

Legendre P, Westbrook GL (1990) Noncompetitive inhibition of γ-aminobutyric $acid_A$ channels by Zn. Mol Pharmacol 39:267–274

Lu L, Huang Y (1998) Separate domains for desensitization of GABA $\rho 1$ and $\rho 2$ subunits expressed in *Xenopus* oocytes. J Membr Biol 164:115–124
Lukasiewicz PD (1996) $GABA_C$ receptors in the vertebrate retina. Mol Neurobiol 12:181–194
Lukasiewicz PD, Werblin FS (1994) A novel GABA receptor modulates synaptic transmission from bipolar to ganglion and amacrine cells in the tiger salamander retina. J Neurosci 14:1213–1223
Lukasiewicz PD, Wong RO (1997) $GABA_C$ receptors on ferret retinal bipolar cells: a diversity of subtypes in mammals? Vis Neurosci 14:989–994
Lukasiewicz PD, Shields CR (1998) Different combinations of $GABA_A$ and $GABA_C$ receptors confer distinct temporal properties to retinal synaptic responses. J Neurophysiol 79:3157–3167
Lukasiewicz PD, Maple BR, Werblin FS (1994) A novel GABA receptor on bipolar cell terminals in the tiger salamander retina. J Neurosci 14:1202–1212
Macdonald RL, Olsen RW (1994) $GABA_A$ receptor channels. Annu Rev Neurosci 17:569–602
Marc RE, Stell WK, Bok D, Lam DMK (1978) GABAergic pathways in the goldfish retina. J Comp Neurol 182:221–246
Martina M, Strata F, Cherubini E (1995) Whole cell and single channel properties of a new GABA receptor transiently expressed in the hippocampus. J Neurophysiol 73:902–906
Massey SC, Linn DM, Kittila CA, Mirza W (1997) Contributions of $GABA_A$ receptors and $GABA_C$ receptors to acetylcholine release and directional selectivity in the rabbit retina. Vis Neurosci 14:939–948
Matthews G, Ayoub GS, Heidelberger R (1994) Presynaptic inhibition by GABA is mediated via two distinct GABA receptors with novel pharmacology. J Neurosci 14:1079–1090
McLean PJ, Farb DH, Russek SJ (1995) Mapping of the α_4 subunit gene (GABRA4) to human chromosome 4 defines an α_2-α_4-β_1-γ_1 gene cluster: Further evidence that modern $GABA_A$ receptor gene clusters are derived from an ancestral cluster. Genomics 26:580–586
Mienville JM, Vicini S (1987) A pyridazinyl derivative of γ-aminobutyric acid (GABA), SR 95531, is a potent antagonist of Cl^- channel opening regulated by $GABA_A$ receptors. Neuropharmacology 26:779–783
Murata Y, Woodward RM, Miledi R, Overman LE (1996) The first selective antagonist for a $GABA_C$ receptor. Bioorg Med Chem Lett 6:2073–2076
Nawy S, Jahr CE (1990) Suppression by glutamate of cGMP-activated conductance in retinal bipolar cells. Nature 346:269–271
Nawy S, Jahr CE (1991) cGMP-gated conductance in retinal bipolar cells is suppressed by the photoreceptor transmitter. Neuron 7:677–683
Nayeem N, Green TP, Martin IL, Barnard EA (1994) Quaternary structure of the native $GABA_A$ receptor determined by electron microscopic image analysis. J Neurochem 62:815–818
Nistri A, Sivilotti L (1985) An unusual effect of γ-aminobutyric acid on synaptic transmission of frog tectal neurons in vitro. Br J Pharmacol 85:917–922
Nomura A, Shigemoto R, Nakamura Y, Okamoto N, Mizuno N, Nakanishi S (1994) Developmentally regulated postsynaptic localization of a metabotropic glutamate receptor in rat rod bipolar cells. Cell 77:361–369
Ogurusu T, Shingai R (1996) Cloning of a putative γ-aminobutyric acid receptor (GABA) subunit rho 3 cDNA. Biochim Biophys Acta 1305:15–18
Ogurusu T, Taira H, Shingai R (1995) Identification of $GABA_A$ receptor subunits in the rat retina: cloning of the $GABA_A$ receptor subunit ρ_2-subunit cDNA. J Neurochem 65:964–968
Pan ZH, Lipton SA (1995) Multiple GABA receptor subtypes mediate inhibition of calcium influx at rat retinal bipolar cell terminals. J Neurosci 15:2668–2679

Picaud S, Pattnaik B, Hicks D, Forster V, Fontaine V, Sahel J, Dreyfus H (1998) GABA$_A$ and GABA$_C$ receptors in adult porcine cones: evidence from a photoreceptor-glia co-culture model. J Physiol (London) 513:33–42
Polenzani L, Woodward RM, Miledi R (1991) Expression of mammalian γ-aminobutyric acid receptors with distinct pharmacology in *Xenopus* oocytes. Proc Natl Acad Sci USA 88:4318–4322
Pourcho RG, Owczarzak MT (1989) Distribution of GABA immunoreactivity in the cat retina. Vis Neurosci 2:425–435
Qian H, Dowling JE (1993) Novel GABA responses from rod-driven retinal horizontal cells. Nature 361:162–164
Qian H, Dowling JE (1994) Pharmacology of novel GABA receptors found on rod horizontal cells of the white perch retina. J Neurosci 14:4299–4307
Qian H, Dowling JE (1995) GABA$_A$ and GABA$_C$ receptors on hybrid bass retinal bipolar cells. J Neurophysiol 74:1920–1928
Qian H, Hyatt G, Schanzer A, Hazra R, Hackham AS, Cutting GR, Dowling JE (1997) A comparison of GABA$_C$ and ρ subunit receptors from the white perch retina. Vis Neurosci 14:843–851
Qian H, Dowling JE, Ripps H (1998) Molecular and pharmacological properties of GABA-ρ subunits from white perch retina. J Neurobiol 37:305–320
Ragozzino D, Woodward RM, Murata Y, Eusebi F, Overman LE, Miledi R (1996) Design and in vitro pharmacology of a selective γ-aminobutyric acid$_C$ receptor antagonist. Mol Pharmacol 50:1024–1030
Schoepp DD, Conn PJ (1993) Metabotropic glutamate receptors in brain function and pathology. Trends Pharmacol Sci 14:361–369
Shimada S, Cutting GR, Uhl GR (1992) γ-Aminobutyric acid A or C receptor? γ-Aminobutyric acid ρ_1 receptor RNA induces bicuculline-, barbiturate-, and benzodiazepine-insensitive γ-aminobutyric acid responses in *Xenopus* oocytes. Mol Pharmacol 41:683–687
Shingai R, Yanagi K, Fukushima T, Sakata K, Ogurusu T (1996) Functional expression of GABA $\rho 3$ receptors in *Xenopus* oocytes. Neurosci Res 263:87–90
Shirasaki T, Klee MR, Nakaye T, Akaike N (1991) Differential blockade of bicuculline and strychnine on GABA- and glycine-induced responses in dissociated rat hippocampal pyramidal cells. Brain Res 561:77–83
Sigel E, Baur R, Trube G, Mohler H, Malherbe P (1990) The effect of subunit composition of rat brain GABA$_A$ receptors on channel function. Neuron 5:703–711
Sivilotti L, Nistri A (1989) Pharmacology of a novel effect of γ-aminobutyric acid on the frog optic tectum in vitro. Eur J Pharmacol 164:205–212
Sivilotti L, Nistri A (1991) GABA receptor mechanisms in the central nervous system. Prog Neuobiol 36:35–92
Strata F, Cherubini E (1994) Transient expression of a novel type of GABA response in rat CA3 hippocampal neurones during development. J Physiol (London) 480:493–503
Tachibana M, Okada T, Arimura T, Kobayashi K, Piccolino M (1993) Dihydropyridine-sensitive calcium current mediates neurotransmitter release from bipolar cells of the goldfish retina. J Neurosci 13:2898–2909
Unwin N (1995) Acetylcholine receptor channel imaged in the open state. Nature 373:37–43
Veruki ML, Yeh HH (1992) Vasoactive intestinal polypeptide modulates GABA$_A$ receptor function in bipolar cells and ganglion cells of the rat retina. J Neurophysiol 67:791–797
Wässle H, Koulen P, Brandstätter JH, Fletscher EL, Becker C-M (1998) Glycine and GABA receptors in the mammalian retina. Vis Res 38:1411–1430
Wang TL, Guggino WB, Cutting GR (1994) A novel γ-aminobutyric acid receptor subunit (ρ_2) cloned from human retina forms bicuculline-insensitive homooligomeric receptors in *Xenopus* oocytes. J Neurosci 14:6524–6531

Wang TL, Hackam AS, Guggino WB, Cutting GR (1995a) A single amino acid in γ-aminobutyric acid $\rho1$ receptors affects competitive and noncompetitive components of picrotoxin inhibition. Proc Natl Acad Sci USA 92:11751–11755
Wang TL, Hackam AS, Guggino WB, Cutting GR (1995b) A single histidine residue is essential for zinc inhibition of GABA $\rho1$ receptors. J Neurosci 15:7684–7691
Wegelius K, Reeben M, Rivera C, Kaila K, Saarma M, Pasternack M (1996) The $\rho1$ GABA receptor cloned from rat retina is down-modulated by protons. Neuroreport 7:2005–2009
Wegelius K, Pasternack M, Hiltunen JO, Rivera C, Saarma M, Reeben M (1998) Distribution of GABA receptor ρ subunit transcripts in the rat brain. Eur J Neurosci 10:350–357
Wellis DP, Werblin FS (1995) Dopamine modulates $GABA_C$ receptors mediating inhibition of calcium entry into and transmitter release from bipolar cell terminals in tiger salamanders. J Neurosci 15:4748–4761
Westbrook GL, Mayer ML (1987) Micromolar concentrations of Zn^{2+} antagonize NMDA and GABA responses of hippocampal neurons. Nature 328:640–643
Woodward RM, Polenzani L, Miledi R (1992a) Effects of hexachlorocyclohexanes on γ-aminobutyric acid receptors expressed in *Xenopus* oocytes by RNA from mammalian brain and retina. Mol Pharmacol 41:1107–1115
Woodward RM, Polenzani L, Miledi R (1992b) Characterization of bicuculline/baclofen-insensitive γ-aminobutyric acid receptors expressed in *Xenopus* oocytes I. Effects of Cl^- channel inhibitors. Mol Pharmacol 42:165–173
Woodward RM, Polenzani L, Miledi R (1993) Characterization of bicuculline/baclofen-insensitive (ρ-like) γ-aminobutyric acid receptors expressed in *Xenopus* oocytes. II. Pharmacology of γ-aminobutyric $acid_A$ and γ-aminobutyric $acid_B$ receptor agonists and antagonists. Mol Pharmacol 43:609–625
Wu SM, Qiao X, Noebels JL, Yang XL (1993) Localization and modulatory actions of zinc in vertebrate retina. Vision Res 33:2611–1616
Xie X, Smart TG (1991) A physiological role for endogenous zinc in rat hippocampal synaptic neurotransmission. Nature 349:521–524
Yazulla S, Studholme KM, Wu J-Y (1987) GABAergic input to the synaptic terminals of mb1 bipolar cells in the goldfish retina. Brain Res 411:400–405
Yeh HH, Grigorenko EV, Veruki ML (1996) Correlation between a bicuculline-resistant response to GABA and $GABA_A$ receptor $\rho1$ subunit expression in rat retinal bipolar cells. Vis Neurosci 13:283–292
Zhang J, Slaughter MM (1995) Preferential suppression of the ON pathway by $GABA_C$ receptors in the amphibian retina. J Neurophysiol 74:1583–1592
Zhang D, Pan Z-H, Zhang X, Brideau AD, Lipton SA (1995) Cloning of a γ-aminobutyric acid type C receptor subunit in rat retina with a methionine residue critical for picrotoxinin channel block. Proc Natl Acad Sci USA 92:11756–11760

$GABA_B$ Receptors

CHAPTER 11

Structure of $GABA_B$ Receptors

B. BETTLER and K. KAUPMANN

A. Physiological Evidence for $GABA_B$ Receptor Subtypes

It is close to 20 years since the term $GABA_B$ was first introduced to define a metabotropic GABA receptor with a pharmacological profile distinct from that of the ionotropic $GABA_A$ and $GABA_C$ receptors (HILL and BOWERY 1981). It was subsequently shown that binding of agonists to $GABA_B$ receptors is sensitive to guanyl nucleotides, indicating that $GABA_B$ receptors are coupled to G-proteins. Many of the physiological roles of $GABA_B$ receptors can be attributed to the regulation of G-protein gated Ca^{2+} and K^+ channels (LÜSCHER et al. 1997; PONCER et al. 1997; SLESINGER et al. 1997; WU and SAGGAU 1997). Accordingly presynaptic $GABA_B$ receptor influence neurotransmission by suppression of neurotransmitter and neuropeptide release, presumably by diminution of a Ca^{2+} conductance. A Ca^{2+} independent interaction of $GABA_B$ receptors with the presynaptic secretion machinery was also proposed (CAPOGNA et al. 1996). Postsynaptic $GABA_B$ receptors hyperpolarize neurons by activating an outward K^+ current that underlies the late inhibitory postsynaptic potentials (IPSPs). Characteristically the late IPSP is slower in onset and has a prolonged duration as compared to the fast IPSP, which derives from the Cl^--permeable $GABA_A$ receptors. Recent studies indicate that inwardly rectifying K^+ channels of the Kir3 type (formerly GIRK) are prominent effectors of postsynaptic $GABA_B$ receptors. For example, the late IPSP evoked by L-baclofen, a selective $GABA_B$ receptor agonist, is largely absent in Kir3.2 knockout mice (LÜSCHER et al. 1997). Similarly in *weaver* mutant mice, who carry a point mutation in the pore-forming region of the Kir3.2 subunit, the amplitude of the $GABA_B$ receptor-activated K^+ current is significantly attenuated (SLESINGER et al. 1997). The rapid time course of $GABA_B$ receptor-mediated K^+ channel (KAUPMANN et al. 1998b) and Ca^{2+} channel (MINTZ and BEAN 1993) regulation indicates a membrane-delimited pathway through the $\beta\gamma$-subunits of the G-protein, similar to other G-protein coupled receptors. In addition to ion channel modulation, $GABA_B$ receptors were shown to negatively couple to adenylyl cyclase and to inhibit forskolin-stimulated cAMP levels (WOJCIK and NEFF 1984). No direct coupling to phospholipase C and the release of Ca^{2+} from internal stores has yet been demonstrated.

Considering differences in drug efficacies it was proposed that the presynaptic $GABA_B$ receptors are heterogeneous and distinct from the postsynaptic receptors (CUNNINGHAM and ENNA 1996; BONANNO et al. 1997; DEISZ et al. 1997; ZHANG et al. 1997). Furthermore there is data to support the idea that pre- and postsynaptic $GABA_B$ receptors differ in their coupling preferences. For example, the action of postsynaptic $GABA_B$ receptors can be blocked by treatment with pertussis toxin (PTX), indicating a coupling to G_i/G_o-type G-proteins, while PTX is unable to uncouple presynaptic $GABA_B$ receptors from their effectors (DUTAR and NICOLL 1988).

The diverse modulatory actions of $GABA_B$ receptors, their localization at pre- and postsynaptic sites, at both inhibitory and excitatory synapses classify them as potential therapeutic targets in diseases such as, e.g. pain, epilepsy, spasticity and psychiatric illness. To date baclofen (Lioresal) is the only $GABA_B$ drug marketed and is used as a muscle relaxant for treatment of spasticity in spinal injury and multiple sclerosis. However the cloning efforts have revived commercial interest into $GABA_B$ receptors because this facilitates the search for novel therapeutic indications and more selective drugs.

B. Pharmacology, Structure and Distribution of Cloned $GABA_B$ Receptors

I. Cloned $GABA_B$ Receptors

The isolation of a $GABA_B$ receptor protein proved difficult due to the lack of radioligands that bind the receptor irreversibly or with high affinity. Moreover, a scarce coupling of $GABA_B$ receptors to effectors in *Xenopus* oocytes rendered expression cloning strategies such as those commonly used for the isolation of neurotransmitter receptors impracticable. It was not until 1997 that the development of the high-affinity antagonist [^{125}I]CGP64213 allowed the isolation of $GABA_BR1a$ (BR1a) using an expression cloning approach (KAUPMANN et al. 1997). Subsequently the $GABA_BR1b$ (BR1b) cDNA was isolated using homology screening. BR1a and BR1b derive from the same gene by alternative splicing (PETERS et al. 1998), and are larger than most G-protein coupled receptors and comparable in size to the metabotropic glutamate receptors (mGluRs). The mature BR1b protein differs from BR1a in that 18 different residues replace the N-terminal 147 ones. The BR1a specific region contains two copies of short consensus repeats (SCRs) about 60 amino acid residues each, also known as complement control protein (CCP) or sushi repeats (Fig. 1) (BETTLER et al. 1998; HAWROT et al. 1998). These repeats exist in a wide variety of complement and adhesion proteins, principally the selectins. The sushi domains may direct protein-protein interactions and, e.g. serve as an extracellular targeting signal for BR1a. Additional splice variants, designated $GABA_BR1c$ (BR1c) and $GABA_BR1d$ (BR1d), generate isoforms with sequence differences in presumed extracelluar and intracellular domains

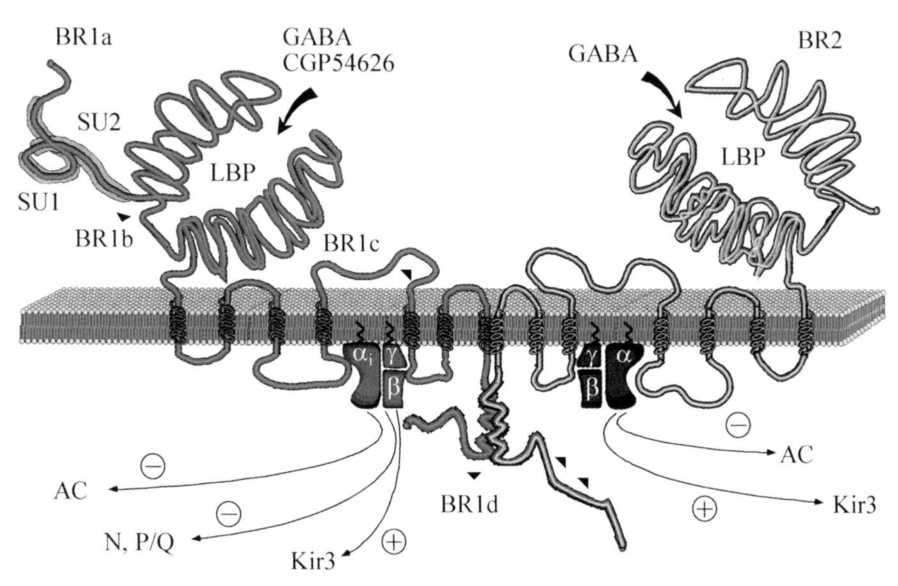

Fig. 1. Structural model and major effector systems of heteromeric BR1+BR2 receptors. BR1 and BR2 appear to form a tightly associated complex via an interaction of coiled-coil domains in the C-terminal tails. The BR1a specific region contains two copies of consensus repeats known as sushi repeats (SU1, SU2). Common to the N-terminal extracellular domains of BR1a/-b and BR2 is a region with homology to bacterial periplasmic proteins (LBP) that constitutes the ligand binding domain. Based on the crystal structure of an LBP and LIVBP, one predicts that this domain forms two globular lobes (see Fig. 2 for homology modelling). The model predicts that upon ligand binding the two lobes bend towards one another, thereby producing a conformational change that promotes G-protein activation. The conformational change may be directly transmitted through the transmembrane domains or, alternatively, the activated binding domain may interact with other extracellular domains like a tethered ligand. BR1 and BR2 both contain a functional GABA binding sites but only BR1 receptors are inhibited by known $GABA_B$ antagonists, such as, e.g, CGP54626. Seven membrane-spanning regions follow the domain implicated in GABA binding. Four splice variants (*arrowheads*) are described for rat BR1 (named BR1a–d) and two C-terminal isoforms are known for human BR2. Activation of native $GABA_B$ receptors can cause a decrease in Ca^{2+} conductance through N, P/Q type Ca^{2+} channels, an increase in K^+ conductance through Kir3 channels and changes of cAMP levels by negative coupling to adenylyl cyclase (AC)

(Isomoto et al. 1998; Pfaff et al. 1999). Database searches with the BR1 sequence information led to the discovery of $GABA_BR2$ (BR2) (Jones et al. 1998; Kaupmann et al. 1998a; White et al. 1998; Kuner et al. 1999; Marshall et al. 1999) which exhibits 35% amino acid sequence similarity to BR1. Two C-terminal splice variants were reported for the human BR2 receptors. Hydrophobicity analysis of the cloned $GABA_B$ receptors revealed a topological organization typical of G-protein coupled receptors, with seven transmembrane domains, an extracellular N-terminal domain and a C-terminal cytoplasmic domain. $GABA_B$ receptors share extended sequence similarity with the mGluRs, the Ca^{2+}-sensing receptor, a family of vomeronasal receptors and periplasmic bacterial amino acid binding proteins, such as the leucine isoleucine valine binding protein (LIVBP) and the leucine binding protein (LBP) (BETTLER et al. 1998; GALVEZ et al. 1999).

The genomic localization, tissue expression and function of the human *$GABA_BR1$* gene identifies it as a positional candidate for neurobehavioral disorders with a genetic locus on 6p21.3 (mouse chromosome 17B3), such as schizophrenia, juvenile myoclonic epilepsy, multiple sclerosis and dyslexia (GOEI et al. 1998; GRIFA et al. 1998; KAUPMANN et al. 1998a). So far association analysis of exonic variants of the *$GABA_BR1$* gene and families with idiopathic generalized epilepsy did not unravel any amino acid substitutions that are causal in disease (SANDER et al. 1999). The *$GABA_BR2$* gene maps to human and mouse chromosome 9q22.2–22.3 and 4B, respectively. This chromosomal localization does not point at neurologic disorders with a likely involvement of $GABA_B$ receptors.

II. Binding Pharmacology

The pharmacology of the cloned $GABA_B$ receptors was first studied using radioligand binding. The two prominent BR1 variants, BR1a and BR1b, demonstrate high affinities to all known $GABA_B$ antagonists and sensitivity to the agonists GABA, APPA and baclofen. BR1a and BR1b are unlikely to represent pharmacological subtypes, as their agonist and antagonist binding affinities match closely. Remarkably, while the antagonist pharmacology of the BR1a/-b and native $GABA_B$ receptors are similar, the agonist affinity at the recombinant receptors is reduced by a factor of ≥ 100. The rank order of agonist binding affinities at BR1a/-b and native receptors is identical. This possibly indicates a shortage of the specific G-protein to promote the high-affinity conformation of the recombinant receptor. This could be explained by the demonstrated lack of BR1a/-b cell surface expression in heterologous cells (COUVE et al. 1998; WHITE et al. 1998). Agonists display a reduced efficacy at the BR1c receptor (PFAFF et al., 1999), but as with BR1d, no thorough pharmacological analysis is available yet. BR2 does not bind any [^{3}H]-agonists and, by itself, does not provide an explanation for the native high-affinity agonist sites. Moreover BR2 protein does not bind any of the available $GABA_B$ antagonist radioligands with measurable affinity either. It was therefore impossible to demonstrate that BR2 represents a $GABA_B$ receptor using radioligand binding. However this became feasible with the development of functional assay systems for heterologous $GABA_B$ receptors (see below).

III. Molecular Determinants of Ligand Binding

Sequence comparison reveals that, like the mGluRs, the extracellular domain of $GABA_B$ receptors shares structural similarity with bacterial amino acid binding proteins (KAUPMANN et al. 1997; GALVEZ et al. 1999). These bacterial proteins bind ions and nutrients in the periplasm and deliver them via transporter proteins across the plasma membrane. The crystal structure of the bacterial proteins indicates that two globular lobes that are connected through a hinge region form the amino acid binding pocket. Several lines of evidence support that the ligand binding site of $GABA_B$ receptors has evolved from these ancestral bacterial amino acid binding proteins. For example a soluble protein encompassing the extracellular N-terminal domain of BR1b closely reproduces the binding pharmacology of wild-type $GABA_B$ receptors (MALITSCHEK et al. 1999). This demonstrates that the N-terminal extracellular domain can correctly fold when dissociated from the transmembrane domains and contains all the structural information that is necessary and sufficient for agonist and antagonist binding. A three-dimensional model of the ligand-binding site of $GABA_B$ receptors was constructed based on the known structure of LBP and LIVBP (Fig. 2) (GALVEZ et al. 1999). The validity of this model was subjected to experimental verification. Mutagenesis of amino acid residues in the vicinity of the presumed ligand-binding pocket has highlighted several residues that appear crucial for binding. Serine 246, a residue homologous to Serine 79 in LBP that forms a hydrogen bond with the ligand, is critical for antagonist binding. Similarly the mutation of Serine 269 was found to differentially affect the affinity of various GABA analogs. Finally, the mutation of Serine 247 and Glutamine 312 were found to increase the affinity of agonists and to decrease the affinity of antagonists, respectively. The effects of these point mutations clearly support an evolutionary relationship between the ligand binding sites of the LBP/LIVBP and $GABA_B$ receptors.

C. Functional Studies with Recombinant $GABA_B$ Receptors

I. Individually Expressed BR1 and BR2 Receptors

Although the cloned $GABA_B$ receptors showed many of the expected properties in terms of structure and pharmacology, they only reluctantly reproduced the signalling properties of native receptors in transfected mammalian cells. Biochemical studies indicated that activation of BR1a receptors in HEK293 cells inhibits adenylyl cyclase activity (KAUPMANN et al. 1997). Although the inhibition of forskolin stimulated cAMP production was weak (30%) it was clearly inhibited by $GABA_B$ antagonists. BR2 couples to adenylyl cyclase slightly more efficiently (approximately 60% inhibition), demonstrating that BR2 is a bona fide $GABA_B$ receptor (KUNER et al. 1999). The coupling of the cloned receptors to presumed effector ion channels proved even more difficult. Like BR1 (KAUPMANN et al. 1998b), BR2 fails to activate Kir3 channels in oocytes (JONES et al. 1998; KAUPMANN et al. 1998a; WHITE et

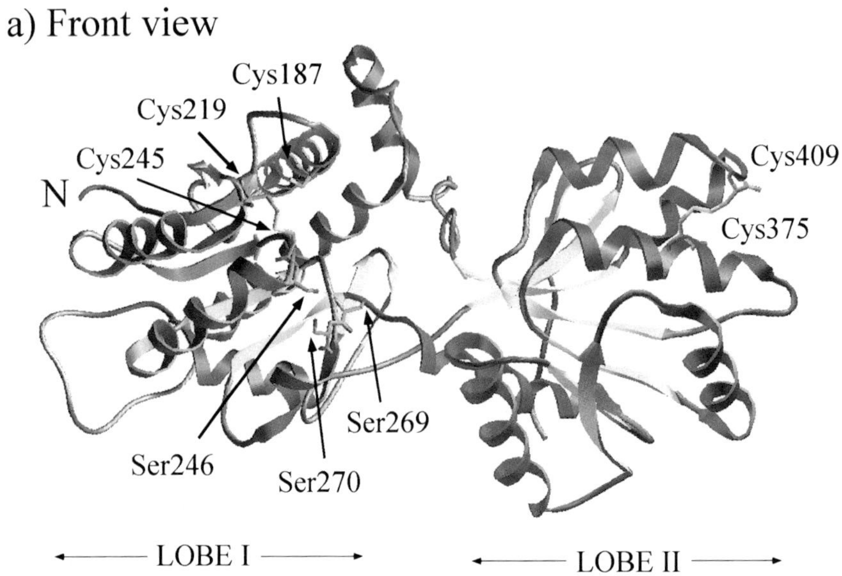

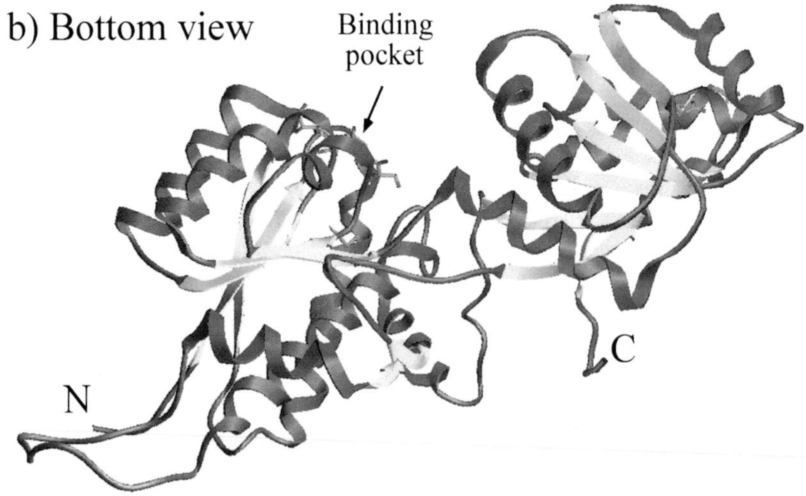

Fig. 2a,b. Ribbon view of a three-dimensional model of the BR1 binding domain. Alpha helices are *dark*, beta strands are *light*. The model was constructed by homology modelling using the co-ordinates of the bacterial periplasmic proteins that bind leucine (LBP, pdb accession number 2LBP) or leucine, isoleucine or valine (LIVBP, pdb accession number 2LIV), both of which have been crystallized in an open state. According to the proposed model, the $GABA_B$ binding domain constitutes two globular lobes (lobe-I and lobe-II) that are connected by three linkers. This structure is stabilised by two disulphide bridges [Cys219-Cys245, and Cys375-Cys409, numbers are according to the BR1a sequence (KAUPMANN et al. 1997); the initiation methionine is residue 1]. GABA is proposed to bind to lobe-I, where Ser246 is forming a hydrogen bond with the ligand. Mutation of Ser269 and Ser270 interferes with ligand binding. Most likely, these two residues do not directly contact the ligand but are important for a correct folding of the binding site: **a** front view; **b** bottom view. *N* and *C* indicate the N- and C-terminal residues (Courtesy of Dr J.P. Pin)

al. 1998), and it does so only inefficiently in HEK293 cells (JONES et al. 1998; KAUPMANN et al. 1998a). The failure of BR1a/-b receptors to couple to signalling pathways may again be explained by poor cell surface expression (COUVE et al. 1998; WHITE et al. 1998). However BR2 efficiently translocates to the cell membrane (WHITE et al. 1998) and therefore the low rate of Kir3 coupling was unexpected, given that this assay represents a sensitive read-out for many cloned G-protein coupled receptors. The lack of robust coupling therefore suggested the involvement of auxiliary factors that are limiting or missing in non-neuronal expression systems.

II. Heteromeric BR1 + BR2 Receptors

The strong overlap of the in situ hybridization patterns (see below) indicated that BR1 and BR2 are co-expressed in many neuronal populations and that a co-expression was possibly needed for robust functional activity. Analysis of hybridization signals on adjacent brain sections provided direct evidence for a co-expression of BR1 and BR2 transcripts within individual neurons, e.g. in Purkinje cells (JONES et al. 1998; KAUPMANN et al. 1998a). It was therefore explored whether an interaction between BR1 and BR2 could explain why efforts to express functionally the cloned receptors in isolation met largely with failure (JONES et al. 1998; KAUPMANN et al. 1998a; WHITE et al. 1998; KUNER et al. 1999; MARSHALL et al. 1999). Indeed while neither BR1a/b nor BR2 alone efficiently activated Kir3 channels, their co-expression in HEK293 cells and *Xenopus* oocytes yielded robust GABA evoked currents. Co-expression of BR1 and BR2 in heterologous cells also allowed for robust stimulation of GTPγ[^{35}S] binding (WHITE et al. 1998). All these functional responses exhibited pharmacological properties reminiscent of those reported for abundant native $GABA_B$ receptors. Independent evidence that BR1 and BR2 interact with each other derived from the search for putative BR1 trafficking factors (WHITE et al. 1998; KUNER et al. 1999). Using the C-terminal domain of BR1 as bait in a yeast two-hybrid screen, the BR2 protein was isolated. The BR1 and BR2 receptors tightly interact via coiled-coil structures in their C-terminal tails, a dimerization signal that is also used by leucine zipper transcription factors. Additional experiments in yeast indicated that the BR1 and BR2 interaction is specific and that neither receptor forms homodimers.

When BR1a or BR1b are expressed together with BR2 an up to tenfold increase in agonist and partial agonist binding potency is observed in the inhibition of [^{125}I]CGP64213 antagonist binding. When expressed together BR2 allows BR1a/-b to translocate to the cell surface (WHITE et al. 1998). Therefore the observed increase in agonist binding potency could arise from a more efficient coupling of the heteromeric receptor to G-proteins. The remaining tenfold discrepancy in apparent agonist binding potency between heteromeric recombinant and native receptors (see above) may be explained by receptor modification (e.g. phosphorylation) or differences in the relative expression levels of G-proteins and receptors (KENAKIN 1997). Immunoprecipitation experiments with native receptors revealed that BR1a and BR1b can assemble with BR2 individually (KAUPMANN et al. 1998a). Immunoelectron microscopy using specific antibodies provided further evidence for heteromeric $GABA_B$ receptors in vivo by showing an extensive co-localization of BR1 and BR2 proteins at Purkinje cell dendritic spines. This supports that the heteromeric receptor represents the predominant native $GABA_B$ receptor but does not rule out the occurrence of homomeric receptors.

D. Temporal and Spatial Distribution of Cloned $GABA_B$ Receptors

Since heteromerization is a prerequisite for robust functional coupling, at least in heterologous cells, it is important to find that the distribution of BR1 and BR2 transcripts in the brain, as studied by in situ hybridization, is largely overlapping. The in situ hybridization pattern qualitatively parallel those of $GABA_B$ agonist (e.g. WILKIN et al. 1981; GEHLERT et al. 1985; CHU et al. 1990; TURGEON and ALBIN 1993) and antagonist binding sites (TOWERS et al. 1997; KAUPMANN et al. 1998a; BISCHOFF et al. 1999), suggesting that BR1 and BR2 constitute the majority of native $GABA_B$ binding sites. The distribution of individual splice variants can differ quite drastically. In the cerebellum transcripts of BR1a are confined to the granule cell layer that comprises the cell bodies of the parallel fibers, which are excitatory to the Purkinje cell dendrites in the molecular layer. By comparison BR1b transcripts are mostly expressed in Purkinje cells, the dendrites of which possess $GABA_B$ receptors that would be postsynaptic to GABAergic Basket and Stellate cells or glutamatergic parallel fibers. Similarly in dorsal root ganglia the density of BR1a, but not BR1b, transcripts is high and confined to the neuronal cell bodies. This supports the association of BR1a with presynaptic receptors in the primary afferent terminals.

Some studies started to address the temporal and subcellular distribution of the BR1 and BR2 proteins using immunohistochemistry. The BR1a/b and BR2 protein levels appear to be differentially regulated during postnatal development and the relative ratios vary between tissues over time (MALITSCHEK et al. 1998; FRITSCHY et al. 1999). At GABAergic synapses in the rat retina, BR1 is localized at pre-, post and extrasynaptic sites, demonstrating that these receptors do not represent exclusive pre- or postsynaptic subtypes (KOULEN et al. 1998). In the cerebellum BR1b and BR2 protein expression is mostly restricted to the Purkinje cell dendrites and spines (KAUPMANN et al. 1998a; FRITSCHY et al. 1999). Surprisingly in Purkinje cells the BR1b and BR2 proteins are localized in the vicinity of excitatory synapses and the BR1 proteins is largely absent at GABAergic inputs. Altogether, current data suggest that the cloned receptors are present at a variety of synaptic sites, at both inhibitory and excitatory synapses.

E. Concluding Remarks and Future Directions

It is apparent that the extent of genetic diversity in the $GABA_B$ receptor gene family is less than that of the mGluR family. Heterologous coupling of $GABA_B$ receptors to Kir3 and adenylyl cyclase, together with the demonstration that BR1a/-b containing receptors inhibit high voltage-activated Ca^{2+} channels (MORRIS et al. 1998), indicate that all major actions of native $GABA_B$ receptors could relate to the cloned receptors. Possibly the targeting of receptor splice variants to distinct subcellular sites dictates effector preferences and compensates for the lack of extensive genetic diversity. Future knockout experiments will discriminate the effects of individual receptor variants and shed light on the degree of functional redundancy. As several diseases have been linked to the $GABA_BR1$ gene it is conceivable that loss-or gain-of-function mutations could produce disease phenotypes.

The cloning of BR1 and BR2 has not led to an immediate understanding of the pharmacological heterogeneity of native $GABA_B$ receptors. Several BR1 and BR2 splice variants have been identified that could potentially assemble into a number of heteromers with different pharmacological properties. Furthermore the implications for dimerization in $GABA_B$ receptor function are unclear. The possibility to bind two G-protein provides the opportunity to integrate various signals in diverse cellular contexts. For example the synergistic activation of two G-proteins may allow the integration of signals normally insufficient to affect metabolic events. It is also conceivable that distinct G-proteins bind to the heteromeric receptor, further increasing the ways in which regulatory inputs could initiate different sets of signaling events. A greater understanding of these inter- and intramolecular signal transduction events will certainly derive from efforts to crystallize functional domains of $GABA_B$ receptors.

References

Bettler B, Kaupmann K, Bowery NG (1998) $GABA_B$ receptors: drugs meet clones. Curr Opin Neurobiol 8:345–350

Bischoff S, Leonhard S, Reymann N, Schuler V, Felner A, Bittiger H, Shigemoto R, Kaupmann K, Bettler B (1999) Spatial distribution of $GABA_BR1$ Receptor mRNA and binding sites in the rat brain. J Comp Neurol 412:1–16

Bonanno G, Fassio A, Schmid G, Severi P, Sala R, Raiteri M (1997) Pharmacologically distinct $GABA_B$ receptors that mediate inhibition of GABA and glutamate release in human neocortex. Br J Pharmacol 120:60–64

Capogna M, Gahwiler BH, Thompson SM (1996) Presynaptic inhibition of calcium-dependent and -independent release elicited with ionomycin, gadolinium, and alpha-latrotoxin in the hippocampus. J Neurophysiol 75:2017–2028

Chu DC, Albin RL, Young AB, Penney JB (1990) Distribution and kinetics of $GABA_B$ binding sites in rat central nervous system: a quantitative autoradiographic study. Neuroscience 34:341–357

Couve A, Filippov AK, Connolly CN, Bettler B, Brown DA, Moss SJ (1998) Intracellular retention of recombinant $GABA_B$ receptors. J Biol Chem 273:26361–26367

Cunningham MD, Enna SJ (1996) Evidence for pharmacologically distinct $GABA_B$ receptors associated with cAMP production in rat brain. Brain Res 720:220–224

Deisz RA, Billard JM, Zieglgansberger W (1997) Presynaptic and postsynaptic $GABA_B$ receptors of neocortical neurons of the rat in vitro: differences in pharmacology and ionic mechanisms. Synapse 25:62–72

Dutar P, Nicoll RA (1988) Pre- and postsynaptic $GABA_B$ receptors in the hippocampus have different pharmacological properties. Neuron 1:585–591

Fritschy JM, Meskenaite V, Weinmann O, Honer M, Benke D, Möhler H (1999) $GABA_B$-receptor splice variants GB1a and GB1b in rat brain: developmental regulation, cellular distribution and extrasynaptic localization. Eur J Neurosci 11:761–768

Galvez T, Joly C, Parmentier ML, Malitschek B, Kaupmann K, Kuhn R, Bittiger H, Froestl W, Bettler B, Pin J-P (1999) Mutagenesis and modeling of the $GABA_B$ receptor extracellular domain support a venus flytrap mechanism for ligand binding. J Biol Chem 274:13362-13369

Gehlert DR, Yamamura HI, Wamsley JK (1985) Gamma-aminobutyric acid B receptors in the rat brain: quantitative autoradiographic localization using [^{3}H]-baclofen. Neurosci Lett 56:183–188

Goei VL, Choi J, Ahn J, Bowlus CL, Raha-Chowdhury R, Gruen JR (1998) Human gamma-aminobutyric acid B receptor gene: Complementary DNA cloning, expression, chromosomal location, and genomic organization. Biol Psychiatry 44:659–666

Grifa A, Totaro A, Rommens JM, Carella M, Roetto A, Borgato L, Zelante L, Gasparini P (1998) GABA (gamma-amino-butyric-acid) neurotransmission: Identification and fine mapping of the human $GABA_B$ receptor gene. Biochem Biophys Res Commun 250:240–245

Hawrot E, Yuanyuan X, Shi Q-L, Norman D, Kirkitadze M, Barlow PN (1998) Demonstration of a tandem pair of complement protein modules in $GABA_B$ receptor 1a. FEBS Lett 432:103–108

Hill DR, Bowery NG (1981) ^{3}H-Baclofen and ^{3}H-GABA bind to bicuculline-insensitive $GABA_B$ sites in rat brain. Nature 290:149–152

Isomoto S, Kaibara M, Sakurai-Yamashita Y, Nagayama Y, Uezono Y, Yano K, Taniyama K (1998) Cloning and tissue distribution of novel splice variants of the $GABA_B$ receptor. Biochem Biophys Res Commun 253:10–15

Jones KA, Borowsky B, Tamm JA, Craig DA, Durkin MM, Dai M, Yao W-J, Johnson M, Gunwaldsen C, Huang L-Y, Tang C, Shen O, Salon JA, Morse K, Laz T, Smith KE, Nagarathnam D, Noble SA, Branchek TA, Gerald C (1998) $GABA_B$ receptors function as a heteromeric assembly of the subunits $GABA_BR1$ and $GABA_BR2$. Nature 396:674–679

Kaupmann K, Huggel K, Heid J, Flor PJ, Bischoff S, Mickel SJ, McMaster G, Angst C, Bittiger H, Froestl W, Bettler B (1997) Expression cloning of $GABA_B$ receptors uncovers similarity to metabotropic glutamate receptors. Nature 386:239–246

Kaupmann K, Malitschek B, Schuler V, Heid J, Froestl W, Beck P, Mosbacher J, Bischoff S, Kulik A, Shigemoto R, Karschin A, Bettler B (1998a) $GABA_B$-receptor subtypes assemble into functional heteromeric complexes. Nature 396:683–687

Kaupmann K, Schuler V, Mosbacher J, Bischoff S, Bittiger H, Heid J, Fröstl W, Leonhardt T, Pfaff T, Karschin A, Bettler B (1998b) Human $GABA_B$ receptors are differentially expressed and regulate inwardly rectifying K^+ channels. Proc Natl Acad Sci USA 95:14991–14996

Kenakin T (1997) Differences between natural and recombinant G-protein coupled receptor systems with varying receptor/G-protein stoichiometry. Trends Pharmacol Sci 18:456–464

Koulen P, Malitschek B, Kuhn R, Bettler B, Wässle H, Brandstätter JH (1998) Presynaptic and postsynaptic localization of $GABA_B$ receptors in neurons of the rat retina. Eur J Neurosci 10:1446–1456

Kuner R, Kohr G, Grunewald S, Eisenhardt G, Bach A, Kornau HC (1999) Role of heteromer formation in $GABA_B$ receptor function. Science 283:74–77

Lüscher C, Jan LY, Stoffel M, Malenka RC, Nicoll RA (1997) G protein-coupled inwardly rectifying K^+ channels (GIRKs) mediate postsynaptic but not presynaptic transmitter actions in hippocampal neurons. Neuron 19:687–695

Malitschek B, Rüegg D, Heid J, Kaupmann K, Bittiger H, Fröstl W, Bettler B, Kuhn R (1998) Developmental changes in agonist affinity at $GABA_BR1$ receptor variants in rat brain. Mol & Cell Neurosci 12:56–64

Malitschek B, Schweizer C, Keir M, Heid J, Froestl W, Kuhn R, Henley J, Pin J-P, Kaupmann K, Bettler B (1999) The N-terminal domain of $GABA_B$ receptors is sufficient to specify agonist and antagonist binding. Mol Pharmacol 56:448–454

Marshall FH, Jones KA, Kaupmann K, Bettler B (1999) $GABA_B$ receptors– the first 7TM heterodimers. Trends Pharmacol Sci 20:396–399

Mintz IM, Bean BP (1993) $GABA_B$ receptor inhibition of P-type Ca^{2+} channels in central neurons. Neuron 10:889–898

Morris SJ, Beatty DM, Chronwall BM (1998) $GABA_BR1a/R1b$-type receptor antisense deoxynucleotide treatment of melanotropes blocks chronic $GABA_B$ receptor inhibition of high-voltage-activated Ca^{2+} channels. J Neurochem 71:1329–1332

Peters HC, Kämmer G, Volz A, Kaupmann K, Ziegler A, Bettler B, Epplen JT, Sander T, Riess O (1998) Mapping, genomic structure, and polymorphisms of the human *$GABA_BR1$* receptor gene: evaluation of its involvement in idiopathic generalized epilepsy. Neurogenetics 2:47–54

Pfaff T, Malitschek B, Kaupmann K, Prézeau L, Pin JP, Bettler B, Karschin A (1999) Alternative splicing generates a novel isoform of the rat metabotropic $GABA_BR1$ receptor. Eur J Neurosci 11:2874-2882

Poncer JC, McKinney RA, Gahwiler BH, Thompson SM (1997) Either N- or P-type calcium channels mediate GABA release at distinct hippocampal inhibitory synapses. Neuron 18:463–472

Sander T, Peters C, Kämmer G, Samochowiec J, Zirra M, Mischke A, Ziegler A, Kaupmann K, Bettler B, Epplen JT, Riess O (1999) Association analysis of exonic variants of the gene encoding the $GABA_B$ receptor and idiopathic generalized epilepsy. Am J Med Genet 88:305–310

Slesinger PA, Stoffel M, Jan YN, Jan LY (1997) Defective g-amino butyric acid type B receptor-activated inwardly rectifying K^+ currents in cerebellar granule cells isolated from weaver and GIRK2 null mutant mice. Proc Natl Acad Sci USA 94:12210–12217

Towers S, Meoni P, Billinton A, Kaupmann K, Bettler B, Urban L, Bowery NG, Spruce A (1997) $GABA_B$ receptor expression in spinal cord and dorsal root ganglia of neuropathic rats. Society for Neuroscience, Abstract 23:(1) 955

Turgeon SM, Albin RL (1993) Pharmacology, distribution, cellular localization, and development of $GABA_B$ binding in rodent cerebellum. Neuroscience 55:311–323

White JH, Wise A, Main MJ, Green A, Fraser NJ, Disney GH, Barnes AA, Emson P, Foord SM, Marshall FH (1998) Heteromerization is required for the formation of a functional $GABA_B$ receptor. Nature 396:679–682

Wilkin GP, Hudson AL, Hill DR, Bowery NG (1981) Autoradiographic localization of $GABA_B$ receptors in the cerebellum. Nature 294:584–587

Wojcik WJ, Neff NH (1984) γ-Aminobutyric acid B receptors are negatively coupled to adenylate cyclase in brain and in the cerebellum these receptors may be associated with granule cells. Mol Pharmacol 25:24–28

Wu LG, Saggau P (1997) Presynaptic inhibition of elicited neurotransmitter release. Trends Neurosci 20:204–212

Zhang J, Shen W, Slaughter MM (1997) Two metabotropic gamma-aminobutyric acid receptors differentially modulate calcium currents in retinal ganglion cells. J Gen Physiol 110:45–58

CHAPTER 12
Pharmacology of $GABA_B$ Receptors

N.G. BOWERY

A. Introduction

It is now about 20 years since we first examined the Cl^--dependent action of GABA on peripheral neurones and, in particular, on nerve terminals of sympathetic fibres. At the time we were trying to mimic the established depolarizing action on presynaptic terminals in the spinal cord (CURTIS 1978) which results in suppression of transmitter release. The outcome of our studies was to show that GABA could indeed inhibit the evoked release of ^3H-noradrenaline from sympathetic nerve terminals in isolated atria of the rat. This was particularly evident in the presence of a presynaptic α_2-adrenoceptor antagonist like yohimbine (BOWERY and HUDSON 1979; BOWERY et al. 1981). However, the GABA receptor responsible for this effect appeared to be distinct from that already described with a pharmacological profile which was strikingly different from that of the receptor mediating the fast action of GABA on spinal neurones or elsewhere in the brain. The effect could not be blocked by bicuculline, was not mimicked by isoguvacine and was only activated by high concentrations of the normally potent agonist, muscimol. Most striking of all was that the clinically used GABA analogue, baclofen (β-chlorophenyl GABA), was stereospecifically active in suppressing the release of the sympathetic amine. No evidence for any GABA-like activity had previously been shown with this compound and certainly there was no reason to believe that it could mimic the Cl^--dependent action of GABA even at very high concentrations.

Baclofen was introduced into therapeutics in the 1970s as an antispastic agent. This originated from a search to find a compound which would cross the blood brain barrier and mimic the inhibitory action of GABA (KEBERLE and FAIGLE 1972; BEIN 1972). Fortunately the primary screen used to detect the activity of baclofen was a functional in vivo assay to obtain central muscle relaxant activity which, it was assumed, would result from mimicking the action of GABA. This predated the original GABA receptor binding assay which would have failed to show baclofen as a positive "hit".

The action of GABA that we had observed in the atrial preparation did not involve any neuronal depolarization as was originally predicted but instead was dependent on the presence of external [Ca^{++}].

Subsequent experiments in other isolated tissues, including mammalian brain slices (BOWERY et al. 1980), soon made us realise that we had a novel receptor for GABA. This was confirmed when we were able to use membrane binding studies to show its presence on neuronal membranes and brain slices (HILL and BOWERY 1981; WILKIN et al. 1981) and it was then that we designated the term "$GABA_B$" for this receptor to contrast with the bicuculline-sensitive receptor which we designated "$GABA_A$" (HILL and BOWERY 1981).

The $GABA_B$ receptor ($GABA_{B1}$) was ultimately cloned some 16 years later by Bettler and colleagues (KAUPMANN et al. 1997) but subsequent studies, reported by four independent groups in December 1998 and January 1999, showed that the $GABA_B$ receptor exists as a heterodimer with a second "receptor" apparently linked to $GABA_{B1}$ through coiled coil domains at the C-terminal (JONES et al. 1998; WHITE et al. 1998; KAUPMANN et al. 1998a; KUNER et al. 1999) in a stoichiometric 1:1 ratio. This new receptor subunit has been designated $GABA_{B2}$ and has many of the structural features of $GABA_{B1}$ including a large molecular weight (110 KDa), seven transmembrane domains, a long extracellular chain at the N-terminus and 35% homology with 54% similarity (see BETTLER and KAUPMANN, Chap. 11, this volume). At present it is not clear what part(s) of the heterodimer determine the pharmacological profile of the $GABA_B$ receptor but since there appear to be at least three splice variants of each of the units the various combinations may have different characteristics. Whether this aligns to any of the proposed receptor subtyping that has been suggested (see later) remains to be seen.

B. Physiological Role

The contribution of $GABA_B$ receptors to inhibitory synaptic events in the mammalian brain is manifest throughout the cerebral axis. Both presynaptic and postsynaptic sites have been implicated and whilst the latter derives from GABAergic innervation of neuronal $GABA_B$ sites the former probably stems from the action of GABA released from an adjacent synapse at least at heteroreceptors (ISAACSON et al. 1993).

The result of $GABA_B$ receptor stimulation is normally a long-lasting neuronal hyperpolarization, mediated by an increase in membrane conductance to K^+, and a reduction in the excitatory postsynaptic potential (EPSP) produced by a decrease in the release of excitatory transmitter. This decrease is presumed to result from a reduction in presynaptic Ca^{++} conductance as a consequence of $GABA_B$ site activation although other mechanisms may contribute in part. No evidence for terminal innervation exists in higher centres and thus "wash-over" from adjacent synapses has been implicated. This seems not unreasonable as the estimated synaptic concentration of GABA is in the millimolar range whilst the affinity of GABA for $GABA_B$ sites is the submicromolar range. However, innervation of presynaptic $GABA_B$ sites does appear to occur in the spinal cord where primary afferent output is modulated by GABAergic interneurones which synapse on to the afferent fibre terminals (BARBER et al. 1978). Thus, $GABA_B$ agonists suppress the evoked release of substances such as substance P (SP) and glutamate which are believed to be sensory transmitters and are colocalised in primary afferent terminals of the dorsal horn (see later). Whilst the primary role of $GABA_B$ systems appears to reside in the CNS some of the actions of GABA outside the brain also have a physiological basis. The enteric nervous system of the intestine may be a particularly important focus. GABA neurones as well as an abundance of $GABA_B$ receptors are present and the action of $GABA_B$ agonists has been well documented in this system (see ONG and KERR 1990). Other effects on peripheral organs are probably of more pharmacological significance although central $GABA_B$ mechanisms do appear to influence peripheral cardiovascular and respiratory function as well as hormone release (see BOWERY 1993; FERREIRA et al. 1996; REY-ROLDAN et al. 1996).

C. $GABA_B$ Receptor Distribution and Localization in CNS

Receptor autoradiography of native $GABA_B$ receptors and immunohistochemistry of $GABA_{B1}$ and $GABA_{B2}$ protein indicate comparable distributions in the mammalian brain (BOWERY et al. 1987; CHU et al. 1990; FRITSCHY et al. 1999; MARGETA-MITROVIC et al. 1999; PRINCIVALLE et al. 1999; SLOVITER et al. 1999). Results so far indicate that the mRNA for $GABA_{B1}$ and $GABA_{B2}$ are also similarly distributed although in some brain regions, such as the caudate putamen, $GABA_{B1}$ mRNA is present whereas $GABA_{B2}$ mRNA appears to be absent (CLARK et al. 1998). In addition it has been noted that there is a low level of $GABA_{B2}$ mRNA relative to $GABA_{B1}$ mRNA in the hypothalamus (JONES et al. 1998). This may mean that another subunit, so far unidentified, dimerizes with $GABA_{B1}$ or possibly the level of mRNA for $GABA_{B2}$ is very low and this determines the level of expression of $GABA_{B1}$ in its role as a trafficking protein. In contrast to $GABA_{B1}$ an additional protein may not be required to dimerize with $GABA_{B2}$ where $GABA_{B1}$ levels are low, which would assume that $GABA_{B2}$ can act as a receptor and not just as a trafficking protein.

The highest densities of $GABA_B$ binding sites in mammalian brain occur in the thalamic nuclei, the molecular layer of the cerebellum, the cerebral cortex and interpeduncular nucleus (BOWERY et al. 1987; CHU et al. 1990). This reflects binding to either pre- or post-synaptic sites. Similar high densities of $GABA_B$ binding have been described in laminae II and III of the spinal cord (PRICE et al. 1987) where the binding sites appear to be largely associated with small diameter primary afferent terminals. These sites probably receive synaptic inputs from GABAergic interneurones (see MALCANGIO and BOWERY 1996) supporting the idea that $GABA_B$ receptors modulate afferent transmitter release.

Determination of the distribution of mRNAs for $GABA_{B1a}$ and $GABA_{B1b}$ using in situ hybridisation techniques has revealed that $GABA_{B1a}$ is probably more related to the generation of presynaptic $GABA_B$ receptors than $GABA_{B1b}$ which may be more relevant to the production of postsynaptic $GABA_B$ receptor in certain brain regions. Thus, in neurones of the rat dorsal root ganglion (DRG), which are the cell bodies of the primary afferent fibres which project to the dorsal horn of the spinal cord, >90% of the total $GABA_B$ mRNA is of the $GABA_{B1a}$ type. $GABA_{B1b}$ levels in the DRG are very low providing less than 10% of the total $GABA_{B1}$ mRNA (Towers et al. 1999). A similar pattern has emerged in the rat and human cerebellum. $GABA_{B1a}$ mRNA was detected over the granule cell bodies the axons of which form the parallel fibres. These innervate the dendrites of the Purkinje cells in the molecular layer. Receptors on the nerve terminals of the excitatory granule cells would be formed from this mRNA in the granule cell body (Kaupmann et al. 1998b; Billinton et al. 1999). Conversely in these same studies $GABA_{B1b}$ mRNA was found to be located over the soma of the Purkinje cells which express $GABA_B$ receptors on their dendrites in the molecular layer. These sites are probably postsynaptic to GABAergic stellate cells.

D. $GABA_B$ Receptor Coupling to Adenylate Cyclase

By definition, metabotropic receptors are coupled indirectly to their effector mechanism(s) and $GABA_B$ receptors are no exception as they are coupled via G-proteins to adenylate cyclase (Karbon et al. 1984; Hill et al. 1984; Hill 1985; Xu and Wojcik 1986) as well as to neuronal membrane K^+ and Ca^{++} channels (see Inoue et al. 1985; Andrade et al. 1986; Dolphin et al. 1990; Bindokas and Ishida 1991; Gage 1992) (see later). $GABA_B$ receptor activation has a dual action on adenylate cyclase. Inhibition of forskolin-activated and basal neuronal adenylate cyclase activity is well established (e.g. Xu and Wojcik 1986) and enhancement of cAMP formation, produced by G_s coupled receptor agonists such as isoprenaline, is also a well documented response to $GABA_B$ receptor activation in brain slice preparations (Karbon et al. 1984). This dual action of $GABA_B$ receptor agonists is also manifest in vivo. Using a microdialysis technique in freely moving rats, Hashimoto and Kuriyama (1997) were able to show that baclofen could reduce the increase in cAMP generated by infusion of forskolin in the cerebral cortex. This effect was mimicked by GABA and blocked by the $GABA_B$ antagonist CGP 54626. As in slice preparations baclofen also potentiated the generation of cAMP by isoprenaline.

The physiological significance of these two effects, particularly enhancement of cAMP generation, has yet to be fully established but they occur independently of any channel events and are presumed to be mediated via separate G-protein subunits. Enzyme inhibition derives from the α subunit whilst the enhancement stems from generation of the $\beta\gamma$ subunits (Lefkowitz 1992).

Whatever the significance of the $GABA_B$ receptor on adenylate cyclase the effects in isolated systems has provided a useful pharmacological assay to characterise the receptor. Moreover, the dual effect on cAMP generation might even involve distinct $GABA_B$ receptor subtypes which has provided a basis for suggesting receptor heterogeneity (CUNNINGHAM and ENNA 1996).

At present the nature of adenylate cyclase/G-protein coupling to the $GABA_B$ receptor heterodimer is not known. Presumably both $GABA_{B1}$ and $GABA_{B2}$ are G-protein coupled. $GABA_{B1}$, expressed in a CHO cell line, was originally demonstrated to behave like the native receptor in controlling cAMP levels even though the receptor could not be expressed on the cell membrane (KAUPMANN et al. 1997). By comparison it appears that $GABA_{B2}$ can be expressed on the cell membrane without the need for $GABA_{B1}$ (KAUPMANN et al. 1998a). Electrophysiological recordings have indicated that functional receptors are present in $GABA_{B2}$ expressed in the absence of $GABA_{B1}$ (KAUPMANN et al. 1998a; KUNER et al. 1999). This would suggest that G-protein coupling is occurring in both $GABA_{B1}$ and $GABA_{B2}$ but whether the G-proteins are the same has yet to be determined.

E. Ca^{++} and K^+ Channel Coupling to $GABA_B$ Sites

Receptor activation increases K^+ conductance but decreases Ca^{++} conductance with the former primarily associated with postsynaptic sites (e.g. LUSCHER et al. 1997) and the latter with presynaptic sites (e.g. CHEN and VAN DEN POL 1998; TAKAHASHI et al. 1998) associated with P/Q and N type channels (e.g. SANTOS et al. 1995; LAMBERT and WILSON 1996). In the feline lumbar spinal cord (–)-baclofen reduces excitatory neurotransmitter release from 1a afferent fibres and decreases the duration of orthodromic action potentials of the same fibres (CURTIS et al. 1997). Both of these presynaptic effects, which are mediated via $GABA_B$ receptors, are consistent with a reduction in the influx of Ca^{++} in the terminals of the 1a afferents (CURTIS et al. 1997). This mechanism is probably the most frequently associated with presynaptic $GABA_B$ sites (DOZE et al. 1995; WU and SAGGAU 1995; ISAACSON 1998; TAKAHASHI et al. 1998), but a mechanism independent of Ca^{++} inhibition has been described in rodent CA1 hippocampal pyramidal cells. At this site $GABA_B$ receptor activation can inhibit tetrodotoxin-resistant GABA release independent of any effect on Ca^{++} or K^+ channels (JAROLIMEK and MISGELD 1997). The authors suggest that activation of protein kinase C (PKC) may be responsible. In fact GABA, acting via $GABA_B$ receptors, has been shown to induce a rapid increase in PKC activity in rat hippocampal slices but this was only apparent in the early postnatal period (P1–P14). After P21 $GABA_B$ receptor activation had the opposite effect and reduced the PKC activity (TREMBLAY et al. 1995).

Presynaptic $GABA_B$ receptors have been suggested to act as regulators of transmitter release enabling sustained transmission to occur at high stimulus frequencies (BRENOWITZ et al. 1998). Normally, synapses with a high probability of transmitter release are subject to depression (OTIS et al. 1996) but the presence or absence of $GABA_B$ receptors may determine how they operate enabling discrimination between types of transmission.

$GABA_B$ receptor activation at postsynaptic sites is associated with more than one type of K^+ channel (WAGNER and DEKIN 1993) and even Ca^{++} channel events appear to be involved in some postsynaptic responses (HARAYAMA et al. 1998). Conversely a K^+(A) current has been suggested to be coupled to $GABA_B$ receptors on presynaptic terminals in hippocampal cultures (SAINT et al. 1990). However, the majority view supports changes in membrane K^+ flux as the primary mechanism mediating the postsynaptic action of $GABA_B$ receptor agonists. Even direct measurement of extracellular K^+ concentrations support this point. OBROCEA and MORRIS (1998) have demonstrated that $GABA_B$ receptor activation in guinea-pig hippocampal slices produces a significant increase in extracellular $[K^+]$ consistent with the rise in K^+ conductance attributed to postsynaptic $GABA_B$ site stimulation.

High intracellular concentrations of Cl^- have been shown to depress $GABA_B$-mediated increases in neuronal K^+ conductance (LENZ et al. 1997) which could provide a basis for a cellular interaction between $GABA_B$ and $GABA_A$ receptors. This influence of Cl^- may be at the level of the G-protein or directly on the K^+ channel (LENZ et al. 1997).

Low threshold Ca^{++} T-currents, which are inactivated at normal resting membrane potentials, may also be involved in the response to $GABA_B$ receptor activation at least within the thalamus (SCOTT et al. 1990). $GABA_B$ receptor activation produces a postsynaptic hyperpolarisation of long duration which initiates Ca^{++} spiking activity in thalamocortical cells. This action has been implicated in the production of the spike and wave activity detected on the surface of the cortex and associated with the generation of absence epilepsy (CRUNELLI and LERESCHE 1991).

F. Pharmacological Effects – $GABA_B$ Receptor Agonists

A variety of effects have been attributed to the action of $GABA_B$ receptor agonists and $GABA_B$-mediated synaptic events. These are listed in Table 1. Not least of these is the centrally mediated muscle relaxant or antispastic action for which baclofen has been used clinically for over 25 years. The basis of this action of the $GABA_B$ agonist appears to derive from its ability to reduce the release of excitatory neurotransmitter on to motoneurones in the ventral horn of the spinal cord. Its effectiveness has made it the drug of choice in treating spasticity irrespective of the cause. However, it is not without significant side effects in certain patients, making it poorly tolerated. This has been overcome to a large extent by intrathecal infusion of very low amounts of the drug and there are now numerous clinical centres employing this technique for the treatment of spasticity associated with tardive dystonia, brain and spinal cord injury, cerebral palsy, tetanus, multiple sclerosis and stiff-man syndrome (e.g.

Table 1. Consequences of $GABA_B$ receptor activation

Decreased release of hormones:	Integrative actions:
Corticotrophin-releasing hormone	Antinociception
Melanocyte-stimulating hormone	Memory retention and consolidation decreased
Gastric acid	Epileptogenesis
Prolactin-releasing factor	Panic attacks decreased
Luteinizing hormone	Antitussive
Decreased release of neurotransmitters:	Hiccup suppression
Catecholamines	Muscle relaxation
5HT	Brown fat thermogenesis
GABA	Cocaine craving reduced
Glutamate	Ethanol and diazepam withdrawal symptoms reduced
Acetylcholine	Heroine intake reduced
Somatostatin	Intestinal peristalsis reduced
Substance P	Induced gastric cancers reduced
CGRP	Oviduct and uterine contraction
Cellular effects:	Food intake increased
Synaptic slow IPSPS	Bronchiolar relaxation
Increase in neuronal K^+ conductance	Hypotension
Decrease in neuronal Ca^{++} conductance	Yawning
Inhibition of adenylate cyclase	5HT-induced head twitch reduced exacerbation of absence seizures
Enhancement of hormone-induced cAMP levels	
Modulation of the generation of long term potentiation	

Penn and Mangieri 1993; Ochs et al. 1989; Dressler et al. 1997; Meythaler et al. 1997; Armstrong et al. 1997; Francois et al. 1997; Becker et al. 1995; Albright et al. 1996; Paret et al. 1996; Dressnandt and Conrad 1996; Azouvi et al. 1996; Seitz et al. 1995).

β-[4-Chorophenyl]GABA (baclofen) was the selective agonist which was first shown not only to have efficacy at the $GABA_B$ receptor but also that it was stereospecifically active (Bowery et al. 1980, 1981). Unfortunately, relatively few compounds have subsequently emerged with selective activity for $GABA_B$ sites and even fewer with greater efficacy or affinity for the receptor than baclofen.

3-Aminopropyl phosphinic acid (2APPA) and its methyl homologue (AMPPA, SKF 97541) were reported to be 3–7 times more potent at $GABA_B$ receptors than (–) baclofen, the active isomer. A variety of phosphinic based agonist ligands have been produced (Froestl et al. 1995a) which have varying potencies but which have not really provided unequivocal support for the possible separation of distinct receptor subtypes.

The paucity of potent and selective agonists has limited their application as potentially effective therapeutic agents although other clinical effects have been reported with baclofen. For example, it has been shown to be very effective in the treatment of otherwise intractable hiccups (e.g. Guelaud et al. 1995; Marino 1998; Nickerson et al. 1997; Kumar and Dromerick 1998) and this effect is believed to stem from an inhibition of the hiccup reflex arc. This possibly involves GABAergic inputs from the nucleus raphe magnus as indicated by studies performed in the feline medulla (OSHIMA et al. 1998).

Another interesting effect elicited by baclofen in man is an antitussive action in low oral doses (DICPINIGAITIS and DOBKIN 1997) which confirms earlier reports of an antitussive action in the guinea-pig (BOLSER et al. 1994).

A recent clinical observation has been the demonstration that baclofen can reduce pain due to stroke or spinal cord injury and musculoskeletal pain. In both painful conditions baclofen was administered by intrathecal infusion (TAIRA et al. 1995; LOUBSER and AKMAN 1996). Although pain relief has also been noted in trigeminal neuralgia in man (FROMM 1994) as well as in a rodent model (IDÄNPÄÄN HEIKKILÄ and GUILBAUD 1999), its usefulness as an analgesic has always been questioned (see HANSSON and KINNMAN 1996).

Nevertheless, in animal acute pain models it has long been known to have an antinociceptive action. These include the tail flick, acetic acid writhing, formalin and hot plate tests in rodents (e.g. CUTTING and JORDAN 1975; LEVY and PROUDFIT 1979; SERRANO et al. 1992; PRZESMYCKI et al. 1998). Even in chronic neuropathic pain models in rats, baclofen clearly exhibits an antinociceptive or anti-allodynic response (SMITH et al. 1994; WIESENFELD HALLIN et al. 1997; CUI et al. 1998). The locus of this action is probably, in part, within higher centres of the brain (LIEBMAN and PASTOR 1980; THOMAS et al. 1995) but there is no doubt that a contribution from an action within the spinal cord is also important (SAWYNOK and DICKSON 1985; HAMMOND and WASHINGTON 1993; DIRIG and YAKSH 1995; THOMAS et al. 1996). The majority of $GABA_B$ receptors in the rat dorsal horn of the spinal cord appear to be located on small diameter afferent fibre terminals (PRICE et al. 1987) where their activation decreases the evoked release of sensory transmitters such as substance P and glutamate (KANGRA et al. 1991; MALCANGIO and BOWERY 1993, 1994; TEOH et al. 1996). This suppression of transmitter release would contribute to the antinociceptive action of baclofen after systemic or intrathecal administration. Whilst this could well explain the antinociceptive effect of $GABA_B$ receptor agonists in acute pain models it is not obvious why baclofen should be much less effective in chronic pain in man (e.g. HANSSON and KINNMAN 1996). It might be that the $GABA_B$ receptor is rapidly down-regulated following systemic administration of the necessarily high doses. Alternatively the receptor may be uncoupled from its associated G-proteins preventing functional activation. This might explain why baclofen is more effective when administered intrathecally in man as only very low amounts are required.

Recently it has been reported that baclofen may be very effective in the treatment of cocaine addiction, reducing the craving for the drug. In rats, baclofen, administered at doses of 1–5 mg/kg, suppressed the self-administration of cocaine without affecting responding for food reinforcement (ROBERTS and ANDREWS 1997; SHOAIB et al. 1998). This is an important observation which could have major consequences in the future therapy for drug addiction (LING et al. 1998).

The potential benefits associated with $GABA_B$ agonist administration are not confined to the CNS but may derive from actions on peripheral organs as well. For example, in asthma it has been suggested that there is a dysfunction of presynaptic $GABA_B$ systems which might normally attenuate cholinergic contraction of airway smooth muscle (TOHDA et al. 1998).

G. Pharmacological Effects – $GABA_B$ Receptor Antagonists

The actions of $GABA_B$ receptor antagonists in man have yet to be assessed as none have, thus far, been tested as therapeutic agents. However a number of predictions based on animal models can be made (BOWERY 1993).

$GABA_B$ antagonists improve cognitive performance in a variety of animal paradigms (MONDADORI et al. 1993; CARLETTI et al. 1993; GETOVA et al. 1997; YU et al. 1997; NAKAGAWA and TAKASHIMA 1997); but see BRUCATO et al. (1996). By contrast, $GABA_B$ agonists clearly impair learning behaviour in animal models (TONG and HASSELMO 1996; AROLFO et al. 1998; MCNAMARA and SKELTON 1996; NAKAGAWA et al. 1995) and this induced amnesia appears to be mediated via G-protein linked receptors as the impairment produced by baclofen in mice can be blocked by pertussis toxin administered intracerebroventricularly (GALEOTTI et al. 1998).

Another potentially important effect of the antagonists is in the suppression of absence seizures. Marescaux and colleagues (MARESCAUX et al. 1992) have shown that $GABA_B$ antagonists administered systemically or directly into the thalamus prevent the spike and wave discharges manifest in the EEG of genetic absence rats (GAERS). Similar observations have been made in the lethargic mouse (HOSFORD et al. 1992) and also in rats injected with gamma-hydroxybutyric acid which produces seizure activity reminiscent of absence epilepsy (SNEAD 1992). In all cases $GABA_B$ antagonists dose-dependently reduced the seizure activity. These and other data have prompted the suggestion that $GABA_B$ mechanisms may be involved in the generation of the Absence syndrome. Deinactivation of Ca^{++} T currents in thalamocortical neurones by prolonged membrane hyperpolarization has been suggested to be the underlying mechanism (CRUNELLI and LERESCHE 1991).

At much higher doses $GABA_B$ antagonists can, conversely, produce convulsant seizures in rats (VERGNES et al. 1997) but how and if this relates to blockade of possible subtypes of $GABA_B$ receptors is unknown. Moreover, not every antagonist appears to produce the same effect. For example, we have failed to observe any convulsant activity with SCH 50911 at doses 10- to 100-fold higher than the dose which completely blocks absence seizures in the genetic absence rat (RICHARDS and BOWERY 1996).

The production of absence-like seizures by γ-hydroxybutyric acid in rats appears to be due to a weak $GABA_B$ receptor agonist action (BERNASCONI et al. 1992, 1999). This property also appears to explain its ability to reduce the firing rate of dopamine neurones in the substantia nigra (Erhardt et al. 1998) and may also be responsible for mediating its abuse potential (Bernasconi et al. 1999).

Three other potential areas for $GABA_B$ antagonist intervention are anxiety, depression and neurodegeneration but the evidence for these indications is currently very limited. The possible significance in depression has previously been reviewed in 1993 and 1995 (Bowery 1993; Kerr and Ong 1995). Although the evidence from animal models was equivocal the potential still remains as indicated in both of these reviews. Unfortunately little has changed since then to support or refute the idea.

$GABA_B$ antagonists and agonists could both have the potential to produce neuroprotection. Lal et al. (1995) suggest that the baclofen, and not an antagonist could be cytoprotective in a cerebral ischaemia model in gerbils. However very large doses, well in excess of that producing muscle relaxation, were required and these were administered 5 min before as well as 24 h and 48 h after the insult. Extensive studies with antagonists remain to be performed.

Whilst the therapeutic indications for $GABA_B$ receptor antagonists appear to be limited, this may well change once the compounds are approved for medical use. However this still depends on the design of suitable agents. The design of selective $GABA_B$ receptor antagonists with increasing receptor affinity and improved pharmacokinetic profile has been an important process, so far, in establishing the significance and structure of $GABA_B$ sites rather than as potential therapeutic agents. Kerr and colleagues in Australia and Froestl and colleagues in Switzerland have primarily been responsible for this major contribution to the $GABA_B$ story. The former group produced the original selective antagonists, phaclofen and 2-hydroxy saclofen (Kerr et al. 1987, 1988) whilst Froestl and Mickle's group subsequently made all the major high affinity compounds in the search for effective antagonists. They provided the first antagonist to cross the blood brain barrier after intraperitoneal injection, CGP 35348 (Olpe et al. 1990) and this was quickly followed by CGP 36742 which was shown to be centrally-active after oral administration in rats (Olpe et al. 1993). However, both of these compounds and others in the same series have low potency even though they are selective for the $GABA_B$ receptor. The most crucial breakthrough in the discovery of antagonists came with the production of compounds with affinities about 10,000 times higher than any previous antagonist. This major advance stemmed from the substitution of a dichlorobenzene moiety into the existing molecules. This produced a profusion of compounds with affinities in the nanomolar or even subnanomolar range (Froestl et al. 1995b). Perhaps the most notable compounds among these are CGP 55845, CGP 54626 and CGP 62349 although many more were produced. This series eventually led to the development of the iodinated high affinity antagonist ^{125}I-CGP 64213 which was used in the elucidation of the structure of $GABA_{B1}$, the first half of the $GABA_B$ receptor dimer to be discovered (Kaupmann et al. 1997). The only other compound exhibiting significant CNS activity after peripheral administration is SCH 50911 (Bolser et al. 1995).

H. Subtypes of Receptor

$GABA_B$ receptors are unlikely to be homogeneous but at present it is unclear what are functionally distinct receptor subtypes. The recent data obtained from elucidation of the structure of the receptor has not provided any clear basis for receptor heterogeneity. However, many electrophysiological studies in mammalian brain suggest that there are subtle distinctions between pre- and post-synaptic receptors (DUTAR and NICOLL 1988; HARRISON et al. 1990; COLMERS and WILLIAMS 1988; THOMPSON and GAHWILER 1992; DEISZ et al. 1997; CHAN et al. 1998). Also, evidence from transmitter release studies suggests differences between receptors on different nerve terminals and between heteroreceptors and autoreceptors (GEMIGNANI et al. 1994; ONG et al. 1998; BONANNO et al. 1998) as does neurochemical evidence from the dual action of $GABA_B$ agonists on adenylate cyclase in brain slices (CUNNINGHAM and ENNA 1996). Nevertheless, it remains to be seen how these apparent functional distinctions can be equated with the lack of diversity in receptor structure. A major problem in defining and establishing any differences in pharmacological characteristics is the lack of ligands with specificity for the proposed receptor subtypes. Although certain antagonists select for the four subtypes described by GEMIGNANI et al. (1994) on synaptosomes, these same compounds have not been reported to produce the same separation in other neuronal systems. Equally the suggested distinctions in other systems such as cAMP generation in brain slices (CUNNINGHAM and ENNA 1996) are not necessarily supported by, e.g. electrophysiological recording studies in brain slices. Thus, whilst subtypes have been described the effects of pharmacological agents do not seem robust enough to make unequivocal decisions about the status of multiple $GABA_B$ receptors in the brain.

References

Albright AL, Barry MJ, Fasick P, Barron W, Shultz B (1996) Continuous intrathecal baclofen infusion for symptomatic generalized dystonia. Neurosurgery 38:934–938

Andrade R, Malenka RC, Nicoll RA (1986) A G protein couples serotonin and GABAB receptors to the same channels in hippocampus. Science. 234:1261–1265

Armstrong RW, Steinbrok P, Cochrane DD, Kube SD, Fife SE (1997) Intrathecally administered baclofen for treatment of children with spasticity of cerebral origin. Journal of Neurosurgery. 87:409–414

Arolfo MP, Zanudio MA, Ramirez OA (1998) Baclofen infused in rat hippocampal formation impairs spatial learning. Hippocampus 8:109–113

Azouvi P, Mane M, Thiebaut JB, Denys P, Remyneris O, Bussel B (1996) Intrathecal baclofen administration for control of severe spinal spasticity: Functional improvement and long-term follow-up. Archives of Physical Medicine and Rehabilitation. 77:35–38

Barber RP, Vaughn JE, Saito K, McLaughlin BJ, Roberts E (1978) GABAergic terminals are presynaptic to primary afferent terminals; in the substantia gelatinosa of the rat spinal cord. Brain Research 141:35–55

Becker WJ, Harris CJ, Long ML, Ablett DP, Klein GM, DeForge DA (1995) Long-term intrathecal baclofen therapy in patients with intractable spasticity. Canadian Journal of Neurological Sciences 22:208–217

Bein HJ (1972) Pharmacological differentiations of muscle relaxants. In: Birkmayer W. Spasticity: A topical survey. Hans Huber, Vienna

Bernasconi R, Lauber J, Marescaux C, Vergnes M, Martin P, Rubio V, Leonhardt T, Reymann N, Bittiger H (1992) Experimental absence seizures: potential role of gamma-hydroxybutyric acid and $GABA_B$ receptors. J Neural Transm [Suppl] 35:155–177

Bernasconi R, Mathivet P, Bischoff S, Marescaux C (1999) Gamma-hydroxybutyric acid: an endogenous neuromodulator with abuse potential? Trends in Pharmacological Sciences 20:135–141

Billinton A, Upton N, Bowery NG (1999) $GABA_B$ receptor isoforms GBR1a and GBR1b, appear to be associated with pre- and post-synaptic elements respectively in rat and human cerebellum. British Journal of Pharmacology. 126:1387–1392

Bindokas VP, Ishida AT (1991) (–)-Baclofen and gamma-aminobutyric acid inhibit calcium currents in isolated retinal ganglion cells. Proc Natl Acad Sci USA 88:10759–10763

Bolser DC, DeGennaro FC, O'Reilly S, Chapman RW, Kreutner W, Egan RW, Hey JA (1994) Peripheral and central sites of action of GABA-B agonists to inhibit the cough reflex in the cat and guinea pig. British Journal of Pharmacology. 113: 1344–1348

Bolser DC, Blythin DJ, Chapman RW, Egan RW, Hey JA, Rizzo C, Kuo SC, Kreutner W (1995) The pharmacology of SCH 50911: a novel, orally-active GABA-beta receptor antagonist. J Pharmacol Exp Ther 274:1393–1398

Bonanno G, Fassio A, Sala R, Schmid G, Raiteri M (1998) $GABA_B$ receptors as potential targets for drugs able to prevent excessive excitatory amino acid transmission in the spinal cord. European Journal of Pharmacology. 362:143–148

Bowery NG, Hill DR, Hudson AL, Doble A, Middlemiss DN, Shaw JS, Turnbull MJ (1980) (–)Baclofen decreases neurotransmitter release in the mammalian CNS by an action at a novel GABA receptor. Nature 283:92–94

Bowery NG, Doble A, Hill DR, Hudson AL, Shaw JS, Turnbull MJ, Warrington R (1981) Bicuculline-insensitive GABA receptors on peripheral autonomic nerve terminals. Eur. J Pharmacol 71:53–70

Bowery NG, Hudson AL, Price GW (1987) $GABA_A$ and $GABA_B$ receptor site distribution in the rat central nervous system. Neuroscience 20:365–383

Bowery NG (1993) $GABA_B$ receptor pharmacology. Annu. Rev. Pharmacol Toxicol. 33:109–147

Bowery NG, Hudson AL (1979) Gamma-aminobutyric acid reduces the evoked release of ^{3}H-noradrenaline from sympathetic nerve terminals. British Journal of Pharmacology. 66:108P

Brenowitz S, David J, Trussell L (1998) Enhancement of synaptic efficacy by presynaptic $GABA_B$ receptors. Neuron. 20:135–141

Brucato FH, Levin ED, Mott DD, Lewis DV, Wilson WA, Swartzwelder HS (1996) Hippocampal long-term potentiation and spatial learning in the rat: Effects of $GABA_B$ receptor blockade. Neuroscience 74:331–339

Carletti R, Libri V, Bowery NG (1993) The $GABA_B$ antagonist CGP 36742 enhances spatial learning performance and antagonises baclofen-induced amnesia in mice. British Journal of Pharmacology. 109:74P

Chan PKY, Leung CKS, Yung WH (1998) Differential expression of pre- and postsynaptic $GABA_B$ receptors in rat substantia nigra pars reticulata neurones. European Journal of Pharmacology. 349:187–197

Chen G, Van den Pol AN (1998) Presynaptic $GABA_B$ autoreceptor modulation of P/Q-type calcium channels and GABA release in rat suprachiasmatic nucleus neurons. Journal of Neuroscience 18:1913–1922

Chu DCM, Albin RL, Young AB, Penney JB (1990) Distribution and kinetics of $GABA_B$ binding sites in rat central nervous system: a quantitative autoradiographic study. Neuroscience 34:341–357

Clark JA, Mezey E, Lam AS, Bonner TI (1998) Functional expression and distribution of GB2 a second $GABA_B$ receptor. Soc. Neurosci. Abstr. 24:795.8

Colmers WF, Williams JT (1988) Pertussis toxin pretreatment discriminates between pre- and postsynaptic actions of baclofen in rat dorsal raphe nucleus in vitro. Neuroscience Letters. 93:300–306

Crunelli V, Leresche N (1991) A role for the $GABA_B$ receptors in excitation and inhibition of thalamocortical cells. Trends in Neurosciences. 14:16–21

Cui JG, Meyerson BA, Sollevi A, Linderoth B (1998) Effect of spinal cord stimulation on tactile hypersensitivity in mononeuropathic rats is potentiated by simultaneous $GABA_B$ and adenosine receptor activation. Neuroscience Letters. 247:183–186

Cunningham MD, Enna SJ (1996) Evidence for pharmacologically distinct $GABA_B$ receptors associated with cAMP production in rat brain. Brain Research 720:220–224

Curtis DR (1978) Pre-and non-synaptic activities of GABA and related amino acids in the mammalian nervous system. In: Fonnum F (ed) Amino acids as chemical transmitters. Plenum, New York, pp 747

Curtis DR, Gynther BD, Lacey G, Beattie DT (1997) Baclofen: reduction of presynaptic calcium influx in the cat spinal cord in vivo. Experimental Brain Research 113:520–533

Cutting DA, Jordan CC (1975) Alternative approaches to analgesia: baclofen as a model compound. British Journal of Pharmacology 54:171–179

Deisz RA, Billard JM, Zieglgänsberger W (1997) Presynaptic and postsynaptic $GABA_B$ receptors of neocortical neurons of the rat in vitro: Differences in pharmacology and ionic mechanisms. Synapse. 25:62–72

Dicpinigaitis PV, Dobkin JB (1997) Antitussive effect of the GABA-agonist baclofen. Chest. 111:996–999

Dirig DM, Yaksh TL (1995) Intrathecal baclofen and muscimol, but not midazolam, are antinociceptive using the rat-formalin model. J Pharmacol Exp Ther 275:219–227

Dolphin AC, Huston E, Scott RH (1990) $GABA_B$-mediated inhibition of calcium currents: a possible role in presynaptic inhibition. In: Bowery NG, Bittiger H, Olpe HR (eds) $GABA_B$ receptors in mammalian function. J Wiley, Chichester, pp 259–271

Doze VA, Cohen GA, Madison DV (1995) Calcium channel involvement in $GABA_B$ receptor-mediated inhibition of GABA release in area CA1 of the rat hippocampus. Journal of Neurophysiology. 74:43–53

Dressler D, Oeljeschlager RO, Ruther E (1997) Severe tardive dystonia: treatment with continuous intrathecal baclofen administration. Movement Disorders. 12:585–587

Dressnandt J, Conrad B (1996) Lasting reduction of severe spasticity after ending chronic treatment with intrathecal baclofen. Journal of Neurology Neurosurgery and Psychiatry. 2:168–173

Dutar P, Nicoll RA (1988) Pre- and postsynaptic $GABA_B$ receptors in the hippocampus have different pharmacological properties. Neuron. 1:585–591

Erhardt S, Andersson B, Nissbrandt H, Engberg G (1998) Inhibition of firing rate and changes in the firing pattern of nigral dopamine neurons by gamma-hydroxybutyric acid (GHBA) are specifically induced by activation of $GABA_B$ receptors. Naunyn-Schmiedebergs Archives of Pharmacology. 357:611–619

Ferreira SA, Scott CJ, Kuehl DE, Jackson GL (1996) Differential regulation of luteinizing hormone release by gamma-aminobutyric acid receptor subtypes in the arcuate-ventromedial region of the castrated ram. Endocrinology. 137:3453–3460

Francois B, Clavel M, Desachy A, Vignon P, Salle JY, Gastinne H (1997) Continuous intrathecal baclofen in tetanus – An alternative management. Presse Medicale. 26:1045–1047

Fritschy JM, Meskenaite V, Weinmann O, Honer M, Benke D, Mohler H (1999) $GABA_B$-receptor splice variants GB1a and GB1b in rat brain: developmental regulation, cellular distribution and extrasynaptic localization. European Journal of Neuroscience 11:761–768

Froestl W, Mickel SJ, Hall RG, Von Sprecher G, Strub D, Baumann PA, Brugger F, Gentsch C, Jaekel J, Olpe HR, Rihs G, Vassout A, Waldmeier PC, Bittiger H (1995a) Phosphinic acid analogues of GABA. 1. New potent and selective $GABA_B$ agonists. J Med. Chem. 38:3297–3312
Froestl W, Mickel SJ, Von Sprecher G et al.(1995b) Phosphinic acid analogues of GABA. 2. Selective, orally active $GABA_B$ antagonists. J Med. Chem. 38:3313–3331
Fromm GH (1994) Baclofen as an adjuvant analgesic. Journal of Pain Symptom Management. 9:500–509
Gage PW (1992) Activation and modulation of neuronal K^+ channels by GABA. Trends Neurosci. 15:46–51
Galeotti N, Ghelardini C, Bartolini A (1998) Effect of pertussis toxin on baclofen- and diphenhydramine-induced amnesia. Psychopharmacology. 136:328–334
Gemignani A, Paudice P, Bonanno G, Raiteri M (1994) Pharmacological discrimination between gamma-aminobutyric acid type B receptors regulating cholecystokinin and somatostatin release from rat neocortex synaptosomes. Mol. Pharmacol 46:558–562
Getova D, Bowery NG, Spassov V (1997) Effects of $GABA_B$ receptor antagonists on learning and memory retention in a rat model of absence epilepsy. European Journal of Pharmacology. 320:9–13
Guelaud C, Similowski T, Bizec JL, Cabane J, Whitelaw WA, Derenne JP (1995) Baclofen therapy for chronic hiccup. European Respiratory Journal. 8:235–237
Hammond DL, Washington JD (1993) Antagonism of L-baclofen-induced antinociception by CGP 35348 in the spinal cord of the rat. Eur. J Pharmacol 234:255–262
Hansson P, Kinnman E (1996) Unmasking mechanisms of peripheral neuropathic pain in a clinical perspective. Pain Review. 3:272–292
Harayama N, Shibuya I, Tanaka K, Kabashima N, Ueta Y, Yamashita H (1998) Inhibition of N- and P/Q-type calcium channels by postsynaptic $GABA_B$ receptor activation in rat supraoptic neurones. Journal of Physiology. 509:371–383
Harrison NL, Lambert NA, Lovinger DM (1990) Presynaptic $GABA_B$ receptors on rat hippocampal neurons. Bowery NG, Bittiger H, Olpe H-R. $GABA_B$ receptors in mammalian function. J Wiley, Chichester, pp 208–221
Hashimoto T, Kuriyama K (1997) In vivo evidence that $GABA_B$ receptors are negatively coupled to adenylate cyclase in rat striatum. J Neurochem 69:365–370
Hill DR, Bowery NG, Hudson AL (1984) Inhibition of $GABA_B$ receptor binding by guanyl nucleotides. J Neurochem 42:652–657
Hill DR (1985) $GABA_B$ receptor modulation of adenylate cyclase activity in rat brain slices. British Journal of Pharmacology. 84:249–257
Hill DR, Bowery NG (1981) ^{3}H-baclofen and ^{3}H-GABA bind to bicuculline-insensitive $GABA_B$ sites in rat brain. Nature 290:149–152
Hosford DA, Clark S, Cao Z, Wilson WA, Jr., Lin F, Morrisett RA, Huin A (1992) The role of $GABA_B$ receptor activation in absence seizures of lethargic (*lh/lh*) mice. Science. 257:398–401
Idänpään Heikkilä JJ, Guilbaud G (1999) Pharmacological studies on a rat model of trigeminal neuropathic pain: baclofen, but not carbamazepine, morphine or tricyclic antidepressants, attenuate the allodynia-like behaviours. Pain. 79:281–290
Inoue M, Matsuo T, Ogata N (1985) Possible involvement of K^+ conductance in the action of gamma-aminobutyric acid in the guinea-pig hippocampus. British Journal of Pharmacology. 86:515–524
Isaacson JS, Solís JM, Nicoll RA (1993) Local and diffuse synaptic actions of GABA in the hippocampus. Neuron. 10:165–175
Isaacson JS (1998) $GABA_B$ receptor-mediated modulation of presynaptic currents and excitatory transmission at a fast central synapse. Journal of Neurophysiology. 80:1571–1576
Jarolimek W, Misgeld U (1997) $GABA_B$ receptor-mediated inhibition of tetrodotoxin-resistant GABA release in rodent hippocampal CA1 pyramidal cells. Journal of Neuroscience 17:1025–1032

Jones KA, Borowsky B, Tamm JA, Craig DA, Durkin MM, Dai M, Yao WJ, Johnson M, Gunwaldsen C, Huang LY, Tang C, Shen QR, Salon JA, Morse K, Laz T, Smith KE, Nagarathnam D, Noble SA, Branchek TA, Gerald C (1998) $GABA_B$ receptors function as a heteromeric assembly of the subunits $GABA_BR1$ and $GABA_BR2$. Nature 396:674–679
Kangra I, Jiang MC, Randic M (1991) Actions of (–)-baclofen on rat dorsal horn neurons. Brain Research 562:265–275
Karbon EW, Duman RS, Enna SJ (1984) $GABA_B$ receptors and norepinephrine-stimulated cAMP production in rat brain cortex. Brain Research 306:327–332
Kaupmann K, Huggel K, Heid J, Flor PJ, Bischoff S, Mickel SJ, McMaster G, Angst C, Bittiger H, Froestl W, Bettler B (1997) Expression cloning of $GABA_B$ receptors uncovers similarity to metabotropic glutamate receptors. Nature 386:239–246
Kaupmann K, Malitschek B, Schuler V, Heid J, Froest W, Beck P, Mosbacher J, Bischoff S, Kulik A, Shigemoto R, Karschin A, Bettler B (1998a) $GABA_B$-receptor subtypes assemble into functional heteromeric complexes. Nature 396:683–687
Kaupmann K, Schuler V, Mosbacher J, Bischoff S, Bittiger H, Heid J, Froestl W, Leonhard S, Pfaff T, Karschin A, Bettler B (1998b) Human gamma-aminobutyric acid type B receptors are differentially expressed and regulate inwardly rectifying K^+ channels. Proc Natl Acad Sci USA 95:14991–14996
Keberle H, Faigle JW (1972) Synthesis and structure-activity relationship of the gamma-aminobutyric acid derivatives. In: Birkmayer W (ed) Spasticity: A topical survey. Hans Huber, Vienna
Kerr DI, Ong J (1995) $GABA_B$ receptors. Pharmacology & Therapeutics. 67:187–246
Kerr DIB, Ong J, Prager RH, Gynther BD, Curtis DR (1987) Phaclofen: a peripheral and central baclofen antagonist. Brain Research 405:150–154
Kerr DIB, Ong J, Johnston GAR, Abbenante J, Prager RH (1988) 2-Hydroxy-saclofen: an improved antagonist at central and peripheral $GABA_B$ receptors. Neurosci Lett 92:92–96
Kumar A, Dromerick AW (1998) Intractable hiccups during stroke rehabilitation. Archives of Physical Medicine and Rehabilitation. 79:697–699
Kuner R, Köhr G, Grünewald S, Eisenhardt G, Bach A, Kornau HC (1999) Role of heteromer formation in $GABA_B$ receptor function. Science. 283:74–77
Lal S, Shuaib A, Ijaz S (1995) Baclofen is cytoprotective to cerebral ischemia in gerbils. Neurochem. Res. 20:115–119
Lambert NA, Wilson WA (1996) High-threshold Ca^{2+} currents in rat hippocampal interneurones and their selective inhibition by activation of $GABA_B$ receptors. Journal of Physiology. 492:115–127
Lefkowitz RJ (1992) G proteins. The subunit story thickens. Nature 358:372
Lenz RA, Pitler TA, Alger BE (1997) High intracellular Cl^- concentrations depress G-protein-modulated ionic conductances. Journal of Neuroscience 17:6133–6141
Levy RA, Proudfit HK (1979) Analgesia produced by microinjection of baclofen and morphine at brain stem sites. European Journal of Pharmacology. 57:43–55
Liebman JM, Pastor G (1980) Antinociceptive effects of baclofen and muscimol upon intraventricular administration. European Journal of Pharmacology. 61:225–230
Ling W, Shoptaw S, Majewska D (1998) Baclofen as a cocaine anti-craving medication: a preliminary clinical study. Neuropsychopharmacology. 18:403–404
Loubser PG, Akman NM (1996) Effects of intrathecal baclofen on chronic spinal cord injury pain. Journal of Pain & Symptom Management. 12:241–247
Luscher C, Jan LY, Stoffel M, Malenka RC, Nicoll RA (1997) G protein-coupled inwardly rectifying K^+ channels (GIRKs) mediate postsynaptic but not presynaptic transmitter actions in hippocampal neurons. Neuron. 19:687–695
Malcangio M, Bowery NG (1993) Gamma-aminobutyric $acid_B$, but not gamma-aminobutyric $acid_A$ receptor activation, inhibits electrically evoked substance P-like immunoreactivity release from the rat spinal cord in vitro. Journal of Pharmacology and. Experimental Therapeutics. 266:1490–1496

Malcangio M, Bowery NG (1994) Spinal cord SP release and hyperalgesia in monoarthritic rats: Involvement of the GABAB receptor system. British. Journal. of. Pharmacology. 113:1561–1566

Malcangio M, Bowery NG (1996) GABA and its receptors in the spinal cord. Trends in Pharmacological Sciences. 17:457–462

Marescaux C, Vergnes M, Bernasconi R (1992) $GABA_B$ receptor antagonists: potential new anti-absence drugs. J Neural Transm. 35:179–188

Margeta-Mitrovic M, Mitrovic I, Riley RC, Jan LY, Basbaum AI (1999) Immunohistochemical localization of $GABA_B$ receptors in the rat central nervous system. Journal of Comparative Neurology. 405:299–321

Marino RA (1998) Baclofen therapy for intractable hiccups in pancreatic carcinoma. American Journal of Gastroenterology. 93:2000

McNamara RK, Skelton RW (1996) Baclofen, a selective $GABA_B$ receptor agonist, dose-dependently impairs spatial learning in rats. Pharmacology, Biochemistry and Behavior. 53:303–308

Meythaler JM, McCary A, Hadley MN (1997) Prospective assessment of continuous intrathecal infusion of baclofen for spasticity caused by acquired brain injury: a preliminary report. Journal of Neurosurgery. 87:415–419

Mondadori C, Jaekel J, Preiswerk G (1993) CGP 36742: The first orally active $GABA_B$ blocker improves the cognitive performance of mice, rats, and rhesus monkeys. Behavioral. and. Neural Biology. 60:62–68

Nakagawa Y, Ishibashi Y, Yoshii T, Tagashira E (1995) Involvement of cholinergic systems in the deficit of place learning in Morris water maze task induced by baclofen in rats. Brain Research 683:209–214

Nakagawa Y, Takashima T (1997) The $GABA_B$ receptor antagonist CGP36742 attenuates the baclofen-and scopolamine-induced deficit in Morris water maze task in rats. Brain Research 766:101–106

Nickerson RB, Atchison JW, Van Hoose JD, Hayes D (1997) Hiccups associated with lateral medullary syndrome. A case report. American Journal of Physical Medicine and Rehabilitation. 76:144–146

Obrocea GV, Morris ME (1998) Changes in $[K^+]_o$ evoked by baclofen in guinea pig hippocampus. Canadian Journal of Physiology and Pharmacology. 76:148–154

Ochs G, Struppler A, Meyerson BA, Linderoth G, Gybels J (1989) Intrathecal baclofen for long term treatment of spasticity: an multicentre study. Journal of Neurology, Neurosurgery & Psychiatry. 52:933–939

Olpe H-R, Steinmann MW, Ferrat T, Pozza MF, Greiner K, Brugger F, Froestl W, Mickel SJ, Bittiger H (1993) The actions of orally active $GABA_B$ receptor antagonists on GABAergic transmission in vivo and in vitro. Eur. J Pharmacol 233:179–186

Olpe HR, Karlsson G, Pozza MF, Brugger F, Steinmann M, Van Riezen H, Fagg G, Hall RG, Froestl W, Bittiger H (1990) CGP 35348: a centrally active blocker of $GABA_B$ receptors. European Journal of Pharmacology. 187:27–38

Ong J, Marino V, Parker DAS, Kerr DIB (1998) Differential effects of phosphonic analogues of GABA on $GABA_B$ autoreceptors in rat neocortical slices. Naunyn-Schmiedebergs Archives of Pharmacology. 357:408–412

Ong J, Kerr DI (1990) GABA-receptors in peripheral tissues. Life Sci. 46:1489–1501

Oshima T, Sakamoto M, Tatsuta H, Arita H (1998) GABAergic inhibition of hiccup-like reflex induced by electrical stimulation in medulla of cats. Neuroscience Research. 30:287–293

Otis TS, Zhang S, Trussell LO (1996) Direct measurement of AMPA receptor desensitization induced by glutamatergic synaptic transmission. Journal of Neuroscience 16:7496–7504

Paret G, Tirosh R, Benzeev B, Vardi A, Brandt N, Barzilay Z (1996) Intrathecal baclofen for severe torsion dystonia in a child. Acta Paediatrica. 85:635–637

Penn RD, Mangieri EA (1993) Stiff-man syndrome treated with intrathecal baclofen. Neurology. 43:2412

Price GW, Kelly JS, Bowery NG (1987) The location of $GABA_B$ receptor binding sites in mammalian spinal cord. Synapse. 1:530–538
Princivalle A, Regondi MC, Frassoni C, Bowery NG, Spreafico R (1999) Distribution of $GABA_B$ receptor protein in cortex and thalamus of adult rats and during postnatal development. Neuropharmacol
Przesmycki K, Dzieciuch JA, Czuczwar SJ, Kleinrok Z (1998) An isobolographic analysis of drug interaction between intrathecal clonidine and baclofen in the formalin test in rats. Neuropharmacol. 37:207–214
Rey-Roldan EB, Lux-Lantos AR, Gonzalez-Iglesias AE, Becu-Villalobos D, Libertun C (1996) Baclofen, a gamma-aminobutyric acid B agonist, modifies hormonal secretion in pituitary cells from infantile female rats. Life Sci. 58:1059–1065
Richards DA, Bowery NG (1996) Anti-seizure effects of the $GABA_B$ antagonist, SCH-50911, in the genetic absence epilepsy rat from Strasbourg (GAERS). Pharmacol Comm. 8:227–230
Roberts DCS, Andrews MM (1997) Baclofen suppression of cocaine self-administration: Demonstration using a discrete trials procedure. Psychopharmacology. 131:271–277
Saint DA, Thomas T, Gage PW (1990) $GABA_B$ agonists modulate a transient potassium current in cultured mammalian hippocampal neurons. Neuroscience Letters. 118:9–13
Santos AE, Carvalho CM, Macedo TA, Carvalho AP (1995) Regulation of intracellular [Ca^{2+}] and GABA release by presynaptic $GABA_B$ receptors in rat cerebrocortical synaptosomes. Neurochem. Int. 27:397–406
Sawynok J, Dickson C (1985) D-Baclofen is an antagonist at baclofen receptors mediating antinociception in the spinal cord. Pharmacology. 31:248–259
Scott RH, Wootton JF, Dolphin AC (1990) Modulation of neuronal T-type calcium channel currents by photoactivation of intracellular guanosine 5'-O(3-thio) triphosphate. Neuroscience 38:285–294
Seitz RJ, Blank B, Kiwit JCW, Benecke R (1995) Stiff-person syndrome with anti-glutamic acid decarboxylase autoantibodies – complete remission of symptoms after intrathecal baclofen administration. Journal of Neurology . 242:618–622
Serrano I, Ruiz RM, Serrano JS, Fernandez A (1992) GABAergic and cholinergic mediation in the antinociceptive action of homotaurine. General Pharmacology. 23:421–426
Shoaib M, Swanner LS, Beyer CE, Goldberg SR, Schindler CW (1998) The $GABA_B$ agonist baclofen modifies cocaine self-administration in rats. Behavioural Pharmacology. 9:195–206
Sloviter RS, Ali-Akbarian L, Elliot RC, Bowery BJ, Bowery NG (1999) Localization of $GABA_B$ (R1) receptors in the rat hippocampus by immunocytochemistry and high resolution autoradiography, with specific reference to its localization in identified hippocampal interneuron subpopulations. Neuropharmacol 38:1707–1721
Smith GD, Harrison SM, Birch PJ, Elliott PJ, Malcangio M, Bowery NG (1994) Increased sensitivity to the antinociceptive activity of (+/–)-baclofen in an animal model of chronic neuropathic, but not chronic inflammatory hyperalgesia. Neuropharmacol. 33:1103–1108
Snead OC III (1992) Evidence for $GABA_B$-mediated mechanisms in experimental generalized absence seizures. Eur. J Pharmacol 213:343–349
Taira T, Kawamura H, Tanikawa T, Iseki H, Kawabatake H, Takakura K (1995) A new approach to control central deafferentation pain: Spinal intrathecal baclofen. Stereotact. Funct. Neurosurg. 65:101–105
Takahashi T, Kajikawa Y, Tsujimoto T (1998) G-protein-coupled modulation of presynaptic calcium currents and transmitter release by a $GABA_B$ receptor. Journal of Neuroscience 18:3138–3146
Teoh H, Malcangio M, Bowery NG (1996) GABA, glutamate and substance P-like immunoreactivity release: Effects of novel $GABA_B$ antagonists. British Journal of Pharmacology. 118:1153–1160

Thomas DA, McGowan MK, Hammond DL (1995) Microinjection of baclofen in the ventromedial medulla of rats: antinociception at low doses and hyperalgesia at high doses. J Pharmacol Exp Ther 275:274–284

Thomas DA, Navarrete IM, Graham BA, McGowan MK, Hammond DL (1996) Antinociception produced by systemic R(+)-baclofen hydrochloride is attenuated by CGP 35348 administered to the spinal cord or ventromedial medulla of rats. Brain Research 718:129–137

Thompson SM, Gahwiler BH (1992) Comparison of the actions of baclofen at pre- and postsynaptic receptors in the rat hippocampus in vitro.. J Physiol (Lond). 451:329–345

Tohda Y, Ohkawa K, Kubo H, Muraki M, Fukuoka M, Nakajima S (1998) Role of GABA receptors in the bronchial response: studies in sensitized guinea-pigs. Clinical and Experimental Allergy. 28:772–777

Tong AC, Hasselmo ME (1996) Effects of long term baclofen treatment on recognition memory and novelty detection. Behavioural Brain Research 74:145–152

Towers S, Princivalle A, Billinton A, Edmunds M, Bettler B, Urban L, Castro-Lopes JM, Bowery NG (2000) $GABA_B$ receptor protein and mRNA distribution in rat spinal cord and dorsal root ganglia. European Journal of Neuroscience (in press)

Tremblay E, Ben-Ari Y, Roisin MP (1995) Different $GABA_B$-mediated effects on protein kinase C activity and immunoreactivity in neonatal and adult rat hippocampal slices. J Neurochem 65:863–870

Vergnes M, Boehrer A, Simler S, Bernasconi R, Marescaux C (1997) Opposite effects of $GABA_B$ receptor antagonists on absences and convulsive seizures. European Journal of Pharmacology 332:245–255

Wagner PG, Dekin MS (1993) $GABA_B$ receptors are coupled to a barium-insensitive outward rectifying potassium conductance in premotor respiratory neurons. J Neurophysiol 69:286–289

White JH, Wise A, Main MJ, Green A, Fraser NJ, Disney GH, Barnes AA, Emson P, Foord SM, Marshall FH (1998) Heterodimerization is required for the formation of a functional $GABA_B$ receptor. Nature 396:679–682

Wiesenfeld Hallin Z, Aldskogius H, Grant G, Hao JX, Hokfelt T, Xu XJ (1997) Central inhibitory dysfunctions: Mechanisms and clinical implications. Behavioral and Brain Sciences 20:420–430

Wilkin GP, Hudson AL, Hill DR, Bowery NG (1981) Autoradiographic localization of $GABA_B$ receptors in rat cerebellum. Nature 294:584–587

Wu LG, Saggau P (1995) $GABA_B$ receptor-mediated presynaptic inhibition in guinea-pig hippocampus is caused by reduction of presynaptic Ca2+ influx. Journal of Physiology 485:649–657

Xu J, Wojcik WJ (1986) Gamma aminobutyric acid B receptor-mediated inhibition of adenylate cyclase in cultured cerebellar granule cells: blockade by islet-activating protein. J Pharmacol Exp Ther 239:568–573

Yu ZF, Cheng GJ, Hu BR (1997) Mechanism of colchicine impairment on learning and memory, and protective effect of CGP36742 in mice. Brain Research 750:53–58

CHAPTER 13

GABA$_B$ Receptor Signaling Pathways

S.J. ENNA

A. Introduction

The GABA$_B$ receptor was first identified and characterized on the basis of its sensitivity to baclofen and insensitivity to bicuculline, benzodiazepines, and other agents known to interact with the GABA$_A$ site (BOWERY et al. 1980). Earlier and subsequent electrophysiological studies with baclofen revealed that it causes a neuronal hyperpolarization and an increase in membrane conductance (CURTIS et al. 1974; NEWBERRY and NICOLL 1985). Unlike the GABA$_A$ receptor, which is a Cl^- ionophore, the electrophysiological responses to baclofen are due to changes in K^+ and Ca^{++} conductances (NEWBERRY and NICOLL 1985). Moreover, GABA$_B$ receptor activation inhibits the evoked release of a number of transmitters from brain tissue, including glutamate, serotonin, dopamine and GABA itself (BOWERY et al. 1980; GRAY and GREEN 1987; HUSTON et al. 1990; PENDE et al. 1993). Taken together, these data provided compelling evidence that the GABA$_A$ and GABA$_B$ receptors represent pharmacologically, physiologically and molecularly distinct entities. The subsequent cloning of these sites provided unequivocal confirmation of this hypothesis (BARNARD 1995; MOHLER 1995; KAUPMANN et al. 1997).

Following the initial discovery that GABA, acting through a baclofen-sensitive receptor, influences neuronal activity in a manner distinct from GABA$_A$ receptor agonists, experiments were undertaken to define the GABA$_B$ effector system. Up to that time amino acid neurotransmitters, such as glutamate, glycine, aspartate and GABA, all appeared to activate inotropic receptors. However, it soon became apparent that GABA$_B$ sites are coupled to G proteins, suggesting they are metabotropic (HILL et al. 1984; KARBON et al. 1984; ANDRADE et al. 1986; WOJCIK and NEFF 1984). Indeed, characterization of GABA$_B$ receptor-mediated second messenger responses yielded new insights into intracellular signaling pathways which have subsequently been found of relevance to a number of systems, including those for other amino acid transmitters. Moreover, the recent discovery that GABA$_B$ receptors function as heteromers loosely linked at the carboxyl-terminal cytoplasmic tail provides the first in vivo evidence of such coupling for metabotropic receptors (KAUPMANN and BETTLER 1998; JONES et al. 1998; KAUPMANN et al. 1998; WHITE et al. 1998; KUNER et al. 1999). The aim of this chapter is to provide an overview of the intracellular effects of $GABA_B$ receptor stimulation as they relate to the physiological responses to this substance. The underlying theme is that the intracellular responses to $GABA_B$ agonists may all be mediated by subunits of G_o and G_i which are liberated upon receptor activation. Those desiring more detailed information on individual aspects of this topic are urged to consult other sources (ENNA and BOWERY 1997; BETTLER et al. 1998; MALCANGIO and BOWERY 1995).

B. Second Messenger Production

I. Overview

Once it was appreciated that $GABA_B$ receptors are metabotropic, attempts were made to determine which second messenger system(s) are regulated by this receptor. Early work suggested that baclofen reduces cGMP levels in cerebellum, although it remains unclear whether this is a direct or indirect effect of the drug (MAILMAN et al. 1978). While baclofen has been found to enhance phosphoinositide (PI) metabolism in dorsal root ganglia, by itself it does not appear to influence PI levels in neuronal tissue (DOLPHIN et al. 1989; GODFREY et al. 1988). Rather, $GABA_B$ receptor activation in rat brain cerebral cortex or hippocampus inhibits histamine-induced increases in PI turnover and enhances PI production stimulated by norepinephrine (GODFREY et al. 1988; CRAWFORD and YOUNG 1988; CORRADOTTI et al. 1987). Indeed, studies have shown that long-term potentiation in the cerebral cortex requires co-activation of $GABA_B$ receptors with those positively coupled to IP_3 formation (i.e., norepinephrine and serotonin) (KOMATSU 1996). The mechanism linking $GABA_B$ receptor activation with regulation of neurotransmitter-stimulated PI turnover in brain has yet to be defined. Further, it is unclear to what extent these effects contribute to the physiological effects of $GABA_B$ receptor agonists under normal circumstances. The data seem to support a coincident signaling between $GABA_B$ and other receptor systems with regard to PI turnover in some cells. Further work is necessary, however, to define more precisely the possible relationship between $GABA_B$ receptors, cGMP, and PI turnover.

As opposed to these second messengers, there is little question that $GABA_B$ receptor activation influences cAMP production in brain tissue (CUNNINGHAM and ENNA 1997). Indeed, virtually all intracellular responses initiated by $GABA_B$ receptor stimulation appear to be related to activation of G proteins known to influence production of this second messenger.

II. cAMP

Some of the earliest findings suggesting that $GABA_B$ receptors are coupled to G proteins were the discovery that agonist binding to this site is inhibited by guanyl nucleotides and that baclofen inhibits basal adenylyl cyclase activity in

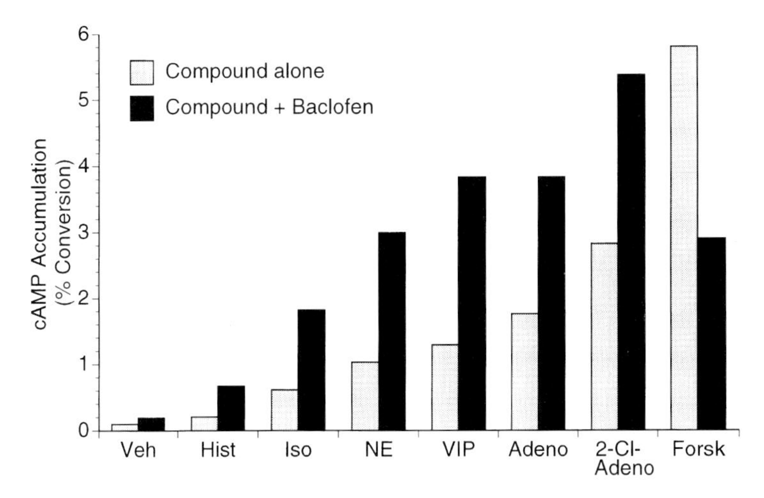

Fig. 1. Effect of baclofen on neurotransmitter and drug-induced cAMP formation in rat brain cerebral cortical slices. *Open bars* represent cAMP formation in the presence of a saturating concentration of the identified neurotransmitter or drug alone, whereas *solid bars* represent cAMP accumulation when the tissue was exposed to the saturating concentration of the neurotransmitter or drug in the presence of a saturating concentration of baclofen. Hist, histamine; iso, isoproterenol; NE, norepinephrine; VIP, vasoactive intestinal peptide; aden, adenosine; forsk, forskolin. Adapted from KARBON and ENNA (1985)

rat brain membranes (HILL et al. 1984; WOJCIK and NEFF 1984). While the latter discovery suggests the GABA$_B$ receptor is negatively coupled to adenylyl cyclase, it was found by others that stimulation of these sites enhances the increase in cAMP accumulation that occurs when brain slices are simultaneously exposed to agents known to stimulate receptors positively coupled to this enzyme (Fig. 1) (KARBON et al. 1984; KARBON and ENNA 1985; HILL 1985; DUMAN et al. 1986). Thus, while a saturating concentration of baclofen has little effect on basal cAMP levels in rat brain cerebral cortical slices, it increases, twofold or more, the amount of cAMP produced in the presence of a saturating concentration of histamine, isoproterenol, norepinephrine, VIP, adenosine, or 2-Cl-adenosine, all of which are known to stimulate receptors which activate adenylyl cyclase (Fig. 1). In contrast, forskolin-stimulated cAMP accumulation is inhibited by baclofen (Fig. 1).

These effects of baclofen are stereoselective, with only the physiologically active isomer influencing cAMP production. Furthermore, only GABA$_B$, but not GABA$_A$, agonists and antagonists are effective in these cAMP assays (KARBON and ENNA 1985). Although it has been reported that the GABA$_B$ receptors mediating the augmentation of cAMP production may be pharmacologically distinct from those responsible for inhibiting the response to forskolin, others have been unable to demonstrate pharmacological differences in this regard, leaving open the question as to whether these actions are mediated by different $GABA_B$ sites (SCHERER et al. 1988; CUNNINGHAM and ENNA 1996; KNIGHT and BOWERY 1996). Nonetheless, these data suggest that GABA, through an interaction with $GABA_B$ receptors, may either enhance or inhibit cAMP production, depending upon circumstances.

To understand the relationship between $GABA_B$ receptors and cAMP production, studies were undertaken to determine the G protein(s) affiliated with this site. The results indicate that $GABA_B$ receptors are associated with G_o and G_{i1}, but not G_{i2} or G_s (MORISHITA et al. 1990; XU and WOJCIK 1986; WOJCIK et al. 1989). While this is consistent with the finding that baclofen inhibits adenylyl cyclase in brain membrane, and forskolin-stimulated cAMP accumulation in brain slices, coupling with these G proteins fails, in itself, to explain the ability of $GABA_B$ agonists to augment cAMP production in response to other neurotransmitters.

A model to explain the differential effects of $GABA_B$ receptor stimulation on cAMP production evolved from discoveries made following the cloning and characterization of various isoforms of adenylyl cyclase (TANG and GILMAN 1991, 1992; TANG et al. 1992; TAUSSIG et al. 1993; YOSHIMURA and COOPER 1993). Of particular relevance was the finding that different types of adenylyl cyclase are differentially regulated by G protein subunits. For example, types II and IV adenylyl cyclase, both of which are found in brain, are only partially activated by $G_{s\alpha}$, requiring $G_{\beta\gamma}$ to be fully stimulated. Since G_o represents 1%–2% of membrane protein, it is a rich source of $G_{\beta\gamma}$ (TANG and GILMAN 1991). Inasmuch as $GABA_B$ receptors are coupled to both G_i and G_o, activation of these sites results in the liberation of $G_{\alpha i}$, $G_{\alpha o}$, and significant quantities of $G_{\beta\gamma}$, which have variable influences on cAMP production, depending upon the state of the adenylyl cyclases when $GABA_B$ receptors are stimulated (Fig. 2).

Thus, if $GABA_B$ receptors are activated simultaneously with a receptor system (e.g., β-adrenergic) that liberates $G_{s\alpha}$ in the same cell, the $G_{i\beta\gamma}$ and $G_{o\beta\gamma}$ resulting from $GABA_B$ receptor stimulation, together with the $G_{s\alpha}$ released by β-adrenoceptor stimulation, fully activates these adenylyl cyclases, yielding a greater production of cAMP than would be possible with the $G_{s\alpha}$-releasing agent alone (coincident signaling). The interaction between $G_{s\alpha}$ and $G_{\beta\gamma}$ in activating adenylyl cyclases overwhelms any inhibitory effect on these enzymes resulting from the $GABA_B$ receptor-mediated release of $G_{i\alpha}$, yielding a net increase in second messenger accumulation.

On the other hand, when $GABA_B$ receptors are stimulated in the absence of $G_{s\alpha}$, the inhibitory effect of $G_{i\alpha}$ on adenylyl cyclase predominates. This explains the results with forskolin, a diterpine that directly activates various forms of adenylyl cyclase. Since forskolin stimulates these enzymes in the absence of $G_{s\alpha}$, there is no synergy between the $G_{\beta\gamma}$ liberated by $GABA_B$ receptor activation and forskolin-stimulated adenylyl cyclase. Rather, what is observed experimentally is inhibition of forskolin-stimulated cAMP accumulation by $GABA_B$ agonists, reflecting the receptor-mediated release of $G_{i\alpha}$ (Figs. 1 and 2).

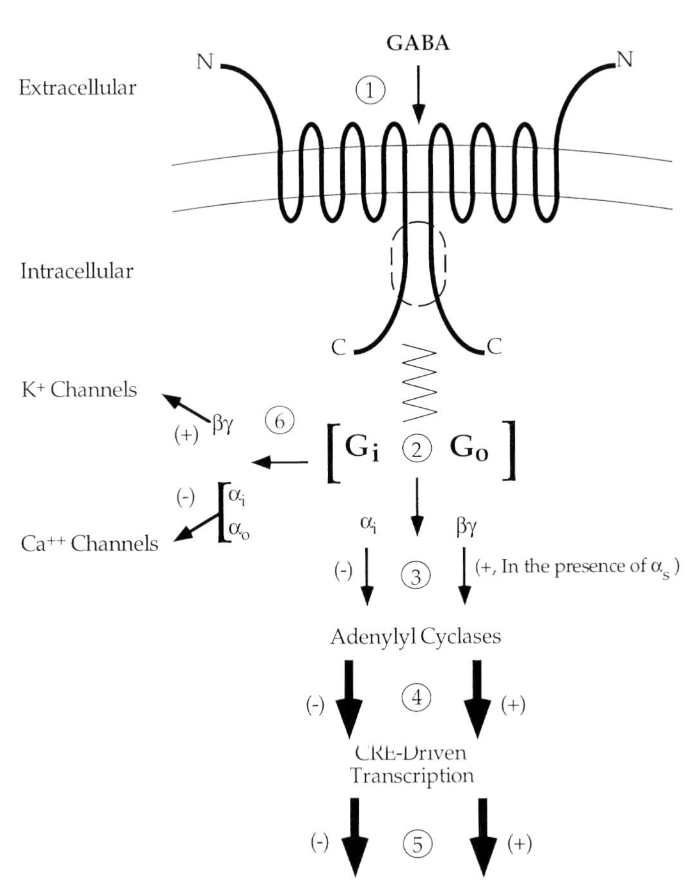

Fig. 2. Schematic representation of same intracellular responses to $GABA_B$ receptor activation. *1* GABA attachment to the $GABA_B$ receptor recognition site. *2* Activation of G_o and/or G_i protein, with consequent liberation of $G_{\alpha o}$, $G_{\alpha i}$ and $G_{\beta\gamma}$. *3* α_i directly inhibits adenylyl cyclase, whereas $G_{\beta\gamma}$, in the present of $G_{\alpha s}$, stimulates some isoforms of adenylyl cyclase. *4* Increase or decrease in cAMP formation leads to an increase or decrease, respectively, in cAMP-responsive element (CRE)-driven gene transcription. *5* Changes in CRE-driven gene transcription increases or decreases the production of a number of neurotransmitter-related peptides and proteins. *6* $G_{\alpha i}$ and $G_{\alpha o}$ inhibit Ca^{++} channels, whereas $G_{\beta\gamma}$, activates K^+ channels

It is noteworthy that at saturating concentrations of baclofen the cAMP response to forskolin is inhibited only 50%–60% (Fig. 1) (Cunningham and Enna 1996). This contrasts with the effects of LY354740, a group II metabotropic glutamate receptor agonist which inhibits forskolin-stimulated cAMP production by 90% or more (Schoepp et al. 1998). Thus, it is possible that baclofen is only a partial agonist at the $GABA_B$ receptor which liberates $G_{i\alpha}$. Alternatively, forskolin may be activating isoforms of adenylyl cyclase that are not inhibited by the $G_{i\alpha}$ liberated by $GABA_B$ agonists, but which are influenced by those G protein subunits associated with group II metabotropic glutamate receptors. If, however, baclofen is only a partial agonist at this site but a full agonist at receptors responsible for liberating $G_{o\beta\gamma}$, this would lend further support to the notion that these two $GABA_B$ receptors may be pharmacologically distinct (Cunningham and Enna 1996).

This model explaining the dual effect of $GABA_B$ receptors on cAMP accumulation has been substantiated in *Xenopus* oocytes expressing poly $(A)^+$ RNA taken from rat brain cerebral cortex (Uezono et al. 1997). While baclofen is inactive in this system when applied alone, it significantly enhances the cAMP response to isoproterenol or VIP. Moreover, the augmenting response to baclofen is enhanced further when type II adenylyl cyclase is coexpressed in these oocytes, whereas the response to the $GABA_B$ agonist is abolished in the presence of pertussis toxin, demonstrating the involvement of G_i or G_o.

The physiological relevance of the effects of $GABA_B$ agonists on cAMP production is suggested by in vivo studies (Hashimoto and Kuriyama 1997). Thus, as measured by microdialysis, baclofen inhibits forskolin-induced increases in cAMP efflux from rat brain corpus striatum and enhances the amount of second messenger released when the brain region is perfused with isoproterenol. Inasmuch as these findings are identical to those obtained using rat brain slices in vitro, they demonstrate that GABA, through an interaction with $GABA_B$ receptors, inhibits or enhances in vivo cAMP accumulation in brain tissue, while having little effect by itself on the production of this second messenger. Thus, with regard to cAMP production, GABA serves more as a neuromodulator than as a neurotransmitter when activating $GABA_B$ sites.

Taken together, these data indicate a complex series of intracellular signaling events resulting from $GABA_B$ receptor stimulation. These yield different biochemical and physiological responses depending upon the type of adenylyl cyclase present in the cell and the presence or absence of $G_{s\alpha}$ generated from other sources.

III. Gene Transcription

While $GABA_B$ receptors are normally studied in the context of short-term effects on intracellular signaling or changes in ion conductance, it is conceivable that $GABA_B$ receptor-mediated modifications in cAMP production could ultimately influence gene expression. Indeed, it has been shown that forskolin-stimulated gene transcription in primary cultures of cerebellar granule neurons is inhibited by baclofen (BARTHEL et al. 1995). It was further demonstrated that this effect is secondary to baclofen-induced inhibition of forskolin-stimulated cAMP production, leading to the conclusion that GABA$_B$ receptor activity regulates cAMP-responsive element (CRE) binding protein-mediated gene transcription in brain. Given this, and the data indicating that activation of GABA$_B$ receptors may lead to either inhibition or enhancement of cAMP formation, it is conceivable that GABA$_B$ receptor agonists could either increase or decrease gene transcription (Fig. 2). Since CRE-driven transcription results in the production of a number of proteins important for neurotransmission, such as tyrosine hydroxylase, GABA$_B$ receptor activation could contribute to maintaining function in a number of neurotransmitter systems.

Studies aimed at examining the antinociceptive effects of GABA$_B$ agonists have revealed that chronic pain, or administration of GABA$_B$ agonists and antagonists, modifies neurokinin-1 and GABA$_B$ R1 and R2 receptor mRNA expression in the rat spinal cord (MCCARSON and ENNA 1996, 1999; ENNA et al. 1998) (Fig. 3). For example, R2 GABA$_B$ receptor mRNA is significantly elevated in both the ipsilateral and contralateral lumbar spinal cord dorsal horns 24h after a formalin injection into the right hindpaw of the rat (Fig. 3).

While an increase in neurokinin-1 mRNA may be due to GABA$_B$ receptor-mediated inhibition of substance P release rather than to GABA$_B$ agonist-induced enhancement in cAMP formation and the resultant CRE-driven gene transcription, it is likely the latter mechanism is responsible for the increase

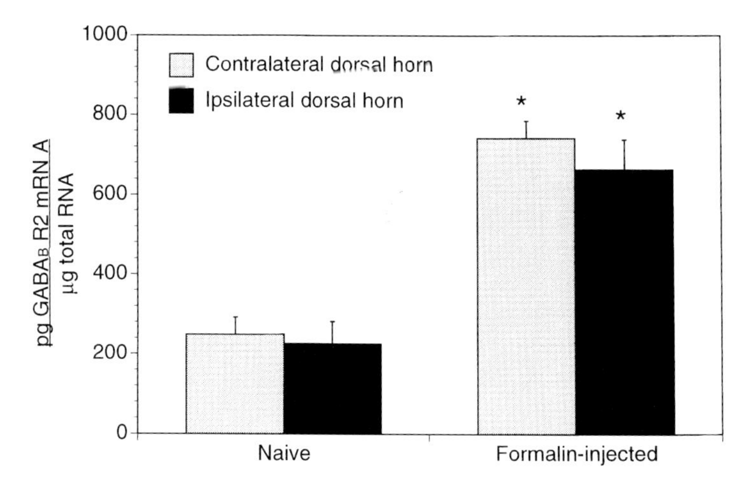

Fig. 3. Expression of GABA$_B$ R2 mRNA in rat lumbar spinal cord 24h following formalin injection into the right hindpaw. *Open bars* represent RNA levels in the contralateral dorsal horn, whereas *solid bars* represent the ipsilateral dorsal horn. Levels in the formalin-treated animals are significantly higher than in the untreated (naive) subjects. Adapted from MCCARSON and ENNA 1999

in $GABA_B$ R1 and R2 receptor mRNA in spinal sensory systems during chronic pain. Assuming that these changes in mRNA are indicative of an increase in the production of receptor proteins, such alterations probably play an important role in regulating the mediation, and perception, of chronic pain. Thus, intracellular signaling pathways activated or inhibited by $GABA_B$ receptor stimulation are capable of inducing long-term changes in synaptic activity, as well as short-term alterations in neuronal excitability.

C. Calcium Channels

Immediate responses to $GABA_B$ receptor activation include neuronal hyperpolarization and a reduction in excitatory postsynaptic potentials. Whereas the former appears due to a receptor-mediated increase in K^+ conductance, the latter may be secondary to an inhibition of the release of excitatory neurotransmitters by modification of presynaptic Ca^{++} currents (Newberry and Nicoll 1985; Dunlap 1981). While the effects on these two ions may be independent of one another, there are data suggesting that, in some cases, the modification in calcium action potentials may be secondary to baclofen-induced changes in K^+ conductance (Desarmenien et al. 1984). In any event, there is no question that activation of $GABA_B$ receptors modifies calcium currents in a variety of systems.

The $GABA_B$ receptor-mediated effect on Ca^{++} appears due to a direct effect of $G_{o\alpha}$ or $G_{i\alpha}$ on the Ca^{++} channels, and is independent of the production of cAMP (Fig. 2) (Hescheler et al. 1987; Surprenant et al. 1990). In contrast, $G_{\beta\gamma}$ subunits have little effect on these channels. In particular, it appears that $G_{o\alpha}$ is primarily responsible for mediating the inhibitory effect of $GABA_B$ receptor activation on neuronal calcium channels (Campbell et al. 1993; Menon-Johansson et al. 1993).

A variety of Ca^{++} channels are affected by $GABA_B$ receptor activation, depending upon the cell type and the system examined (Deisz and Lux 1985; Heidelberger and Matthews 1991; Menon-Johnsson et al. 1993; Mintz and Bean 1993). This includes T-, L-, N-, and P/Q-type Ca^{++} channels. For example, it has been demonstrated that baclofen-induced inhibition of GABA release from the suprachiasmatic nucleus in vitro is due to modulation of P/Q-type Ca^{++} channels in the axon terminal, whereas postsynaptic $GABA_B$ receptors inhibit N- and P/Q-type voltage-dependent Ca^{++} channels in rat supraoptic nucleus (Chen and van den Pol 1998; Harayama et al. 1998). In both cases, the presynaptic response to baclofen is blocked entirely, but the postsynaptic response only partially, by pretreatment of the tissue with pertussis toxin. These data suggest that regulation of pre- and postsynaptic Ca^{++} currents by $GABA_B$ receptors requires activation of a G protein, most likely G_o, although the postsynaptic effect may involve other mediators as well. It has also been proposed that prolonged $GABA_B$ receptor-mediated hyperpolarization de-inactivates T-type Ca^{++} channels in the thalamus, which may explain $GABA_B$

agonist-induced absence seizures and the antiepileptic effects of $GABA_B$ receptor antagonists (CRUNELLI and LERESCHE 1991; MARESCAUX et al. 1992).

While there is substantial evidence suggesting that $GABA_B$ receptor-mediated inhibition of neurotransmitter release is due primarily to blockade of presynaptic Ca^{++} channels by $G_{o\alpha}$, some data indicate this effect on presynaptic Ca^{++} channels may not, in all cases, fully explain regulation of neurotransmitter release (TAKAHASHI et al. 1998; DITTMAN and REGEHR 1996; HUSTON et al. 1995; SCANZIANI et al. 1992).

The role of $G_{o\alpha}$, and possibly $G_{i\alpha}$, in mediating the effects of $GABA_B$ receptor agonists on Ca^{++} channels underscores the importance of G_i and G_o in $GABA_B$ receptor signaling pathways. While the type of Ca^{++} channel affected may vary depending on the cell system, and the physiological responses to inhibition or deinactivation of channel activity may differ depending on the brain region and synaptic location, the common property shared by all of these ion channels is their regulation by G protein subunits liberated as a result of $GABA_B$ receptor stimulation.

D. Potassium Channels

One of the earliest observations regarding the action of baclofen is its ability to induce a late inhibitory postsynaptic potential (IPSP) in rat hippocampal cells in vitro (ALGER and NICOLL 1982; NEWBERRY and NICOLL 1985). This characteristic distinguishes $GABA_B$ from $GABA_A$ receptors since the latter induces a fast inhibitory postsynaptic potential/current. Subsequent work revealed the $GABA_B$ receptor-mediated late IPSP is due to a conductance increase in K^+ ions (HOWE et al. 1987). This, in turn, was ultimately attributed to a $GABA_B$ receptor-mediated activation of inwardly rectifying K^+ channels (GIRKs). Studies with oocytes transfected with poly $(A)^+$ rat cerebellar RNA and cRNAs for GIRKs revealed that baclofen elicits inwardly rectifying K^+ currents only if both GIRK1 and GIRK2 are coexpressed in the same cell, but not with either alone (UEZONO et al. 1998). Likewise, a point mutation, or complete knockout, of GIRK2 (Kir3.2) results in a decrease in $GABA_B$ receptor function in mouse hippocampal slices (JAROLIMEK et al. 1998; LUSCHER et al. 1997). The change in the $GABA_B$ receptor response in this study was limited to postsynaptic sites, with the presynaptic action of baclofen being unaffected in the GIRK2 knockout mouse hippocampus. This suggests the interaction with these inwardly rectifying K^+ channels by $GABA_B$ receptors accounts only for the postsynaptic effects of GABA. Besides $GABA_B$ sites, a number of other receptors have similar effects on postsynaptic K^+ channels, including $serotonin_{1A}$ and adenosine A1 receptors (LUSCHER et al. 1997).

There is ample evidence suggesting that $GABA_B$ receptors regulate inwardly rectifying K^+ channels through activation of G proteins, but is independent of cAMP formation (ANDRADE et al. 1986; THOMPSON and GAHWILER 1992; O'CALLAGHAN et al. 1996). Work with GIRKs expressed in oocytes suggests these channels are activated by $G_{\beta\gamma}$, most likely $G_{\beta1\gamma1}$, but not G_α or $G_{\beta1\gamma2}$ (Fig. 2) (REUVENY et al. 1994; TAKAO et al. 1994). Deletion experiments revealed the C-terminus of the GIRK is the regulatory region for the $G_{\beta\gamma}$ subunit (TAKAO et al. 1994). The importance of K^+ channel activation in the pharmacological response to $GABA_B$ agonists is demonstrated by the finding that the antinociceptive response to baclofen is completely abolished in mice following administration of a GIRK antisense oligodeoxyribonucleotide (anti-mKv 1.1) (GALEOTTI et al. 1997).

Thus it appears that the $GABA_B$ receptor-induced late IPSP is due to a $G_{\beta\gamma}$-mediated activation of inwardly rectifying K^+ channels. In contrast, the presynaptic effects of $GABA_B$ receptors is due to G_α-mediated inhibition of Ca^{++} channels.

E. Conclusion

Studies on the $GABA_B$ receptor/effector system have yielded new insights into transmitter-mediated signaling processes. Chief among these is the discovery that activation of $GABA_B$ receptors results in either inhibition or enhancement of cAMP formation. Subsequently it was found that other receptors coupled to G_i and G_o, such as metabotropic glutamate receptors, share this property.

$GABA_B$ receptor studies have also provided evidence that G protein coupled receptors function as heteromers. This opens new possibilities for regulation of receptor expression and pharmacological selectivity.

Another major finding emanating from work on $GABA_B$ receptor signaling is that G proteins appear to be primarily responsible for mediating the intracellular response to GABA. Thus, liberated G_α and $G_{\beta\gamma}$ subunits both participate in regulating cAMP formation which, ultimately, influences gene transcription. Likewise, both G_α and $G_{\beta\gamma}$ are directly responsible for the immediate effects of $GABA_B$ agonists on cellular activity, with the former inhibiting presynaptic and postsynaptic Ca^{++} channels which, in turn, influence neurotransmitter release, and the latter activating postsynaptic K^+ channels to induce neuronal hyperpolarization. The widespread distribution of $GABA_B$ receptors in brain and spinal cord, and the multiplicity of both short- and long-term effects resulting from $GABA_B$ receptor activation, reinforce the importance of this inhibitory neurotransmitter in maintaining central nervous system function. As more is learned about the differences among $GABA_B$ receptor subtypes in terms of structure, function, and, perhaps, intracellular signaling pathways, it will be possible to design new agents to pharmacologically manipulate this receptor system for therapeutic gain.

Acknowledgments. I thank Dr. Kenneth E. McCarson, Dr. Beth Levant and Ms. Maxine Floyd for their assistance in the preparation of this manuscript.

References

Alger BE, Nicoll R (1982) Pharmacological evidence for two kinds of GABA receptor on rat hippocampal pyramidal cells studied in vitro. J Physiol (Lond) 328: 125–141

Andrade R, Malenka RC, Nicoll RR (1986) A G protein couples serotonin and GABAB receptors to the same channels in hippocampus. Science 234:1261–1265

Barnard EA (1995) The molecular biology of GABA$_A$ receptors and their structural determinants. In: Biggio G, Sanna E, Serra M, Costa E (eds) GABA$_A$ receptors and anxiety. From neurobiology to treatment advances in biochemical psychopharmacology, Raven Press, New York, pp 1–16

Barthel F, Campard PK, Demeneix BA, Feltz P, Loeffler J (1995) GABA$_B$ receptors negatively regulate transcription in cerebellar granular neurons through cyclic AMP responsive element binding protein dependent mechanisms. Neuroscience 70:417–427

Bettler B, Kaupman K, Bowery N (1998) GABA$_B$ receptors: drugs meet clones. Curr Opin Neurobiol 8:345–350

Bowery NG, Hill DR, Hudson AL, Doble A, Middlemiss DN, Shaw J, Turnbull M (1980) (–)Baclofen decreases neurotransmitter release in mammalian CNS by an action at a novel GABA receptor. Nature 283:92–94

Campbell V, Berrow N, Dolphin AC (1993) GABAB receptor modulation of Ca2+ currents in rat sensory neurones by the G protein G(o): antisense oligonucleotide studies. J Physiol 470: 1–11

Chen G, van den Pol AN (1998) Presynaptic GABA$_B$ autoreceptor modulation of P/Q-type calcium channels and GABA release in rat suprachiasmatic nucleus neurons. J Neurosci 18: 1913–1922

Corradotti R, Ruggiero M, Chiarugi VP, Pepeu G (1987) GABA-receptor stimulation enhances norepinephrine-induced polyphosphoinositide metabolism in rat hippocampal slices. Brain Res 411:196–199

Crawford ML, Young JM (1988) GABAB receptor-mediated inhibition of histamine H1-receptor-induced inositol phosphate formation in slices of rat cerebral cortex. J Neurochem 51:1441–1447

Crunelli V, Leresche N (1991) A role for GABAB receptors in excitation and inhibition of thalamocortical cells. Trends Neursci 14:16–21

Cunningham MD, Enna SJ (1996) Evidence for pharmacologically distinct GABA$_B$ receptors associated with cAMP production in rat brain. Brain Res 720:220–224

Cunningham M, Enna SJ (1997) Cellular and biochemical responses to GABA$_B$ receptor activation. In: Enna SJ, Bowery NG (eds) The GABA receptors (2nd edn), Humana Press, Totowa, New Jersey, pp 237–258

Curtis DR, Game CJA, Johnston GAR, McCulloch RM (1974) Central effects of β(*p*-chlorophenyl)-γ-aminobutyric acid. Brain Res 70:493–499

Deisz DA, Lux HD (1985) γ-Aminobutyric acid induced depression of Ca^{2+} currents of chick sensory neurons. Neurosci Lett 56:205–210

Desarmenien M, Feltz P, Occhipinti G, Santangelo F, Schlichter R (1984) Coexistence of GABA$_A$ and GABA$_B$ receptors on Aδ and C primary afferents. Br J Pharmacol 81:327–333

Dittman JS, Regehr WG (1996) Contributions of calcium-dependent and calcium-independent mechanisms to presynaptic inhibition at a cerebellar synapse. J Neurosci 16:1623–1633

Dolphin AC, McGuirk SM, Scott RH (1989) An investigation into the mechanisms of inhibition of calcium channel currents in cultured sensory neurones of the rat by guanine nucleotide analogues and (–)–baclofen. Br J Pharmacol 97:263–273

Duman RS, Karbon EW, Harrington C, Enna SJ (1986) An examination of the involvement of phospholipases A2 and C in the alpha-adrenergic and gamma-aminobutyric acid receptor modulation of cyclic AMP accumulation in rat brain slices. J Neurochem 47:800–810

Dunlap K (1981) Two types of γ-aminobutyric acid receptor on embryonic sensory neurones. Br J Pharmacol 74:579–585

Enna SJ, Bowery NG (eds) (1997) The GABA receptors (2nd edn), Humana Press, Totowa, New Jersey

Enna SJ, Harstad EB, McCarson KE (1998) Regulation of neurokinin-1 receptor expression by $GABA_B$ receptor agonists. Life Sci 62:1525–1530

Galoetti N, Ghelardini C, Papucci L, Capaccioli S, Quattrone A, Bartolini A (1997) An antisense oligonucleotide on the mouse *Shaker*-like potassium channel Kv 1.1 gene prevents antinociception induced by morphine and baclofen. J Pharmacol Exp Ther 281:941–949

Godfrey PP, Grahame-Smith DG, Gray JA (1988) GABAB receptor activation inhibits 5-hydroxytryptamine-stimulated inositol phospholipid turnover in mouse cerebral cortex. Eur J Pharmacol 152:185–188

Gray, JA, Green AR (1987) GABAB-receptor mediated inhibition of potassium-evoked release of endogenous 5-hydroxytryptamine from mouse frontal cortex. Br J Pharmacol 91:517–522

Harayama N, Shibuya I, Tanaka K, Kabashima N, Ueta Y, Yamashita H (1998) Inhibition of N- and P/Q-type calcium channels by postsynaptic $GABA_B$ receptor activation in rat supraoptic neurones. J Physiol 509:371–383

Hashimoto T, Kuriyama K (1997) In vivo evidence that $GABA_B$ receptors are negatively coupled to adenylate cyclase in rat striatum. J Neurochem 69:365–370

Heidelberger R, Matthews G (1991) Inhibition of calcium influx and calcium current by γ-aminobutyric acid in single synaptic terminals. Proc Natl Acad Sci USA 88:7135–7139

Hescheler J, Rosenthal W, Trautwein W, Schultz G (1987) The GTP-binding protein, G_o, regulates neuronal calcium channels. Nature 325:445–447

Hill DR (1985) GABAB receptor modulation of adenylate cyclase activity in rat brain slices. Br J Pharmacol 84:249–257

Hill DR, Bowery NG, Hudson AL (1984) Inhibition of GABAB receptor binding by guanyl nucleotides. J Neurochem 42:652–657

Howe JR, Sutor B, Zieglgansberger W (1987) Characteristics of long-duration inhibitory postsynaptic potentials in rat neocortical neurons in vitro. Cell Mol Neurobiol 7:1–18

Huston E, Cullen GP, Burley JR, Dolphin AC (1995) The involvement of multiple calcium channel sub-types in glutamate release from cerebellar granule cells and its modulation by GABAB receptor activation. Neuroscience 68:465–478

Huston E, Scott RH, Dolphin AC (1990) A comparison of the effect of calcium channel ligands and $GABA_B$ agonists and antagonists on transmitter release and somatic calcium currents in cultured neurons. Neuroscience 38:721–729

Jarolimek W, Baurle J, Misgeld U (1998) Pore mutation in a G-protein-gated inwardly rectifying K^+-dependent inhibition in *Weaver* hippocampus. J Neurosci 18: 4001–4007

Jones KA, Borowsky B, Tamm JA, Craig DA, Durkin MM, Dai M, Yao WJ, Johnson M, Gunwaldsen C, Huang LY, Tang C, Shen Q, Salon JA, Morse K, Laz T, Smith KE, Nagaraathnam DD, Noble TA, Branchek TA, Gerald C (1998) Functional $GABA_B$ receptors require co-expression of $GABA_BR1$ and $GABA_BR2$. Nature 396:674–678

Karbon EW, Duman RS, Enna SJ (1984) $GABA_B$ receptors and norepinephrine-stimulated cAMP production in rat brain cortex. Brain Res 306:327–332

Karbon EW, Enna SJ (1985) Characterization of the relationship between γ-aminobutyric acid B agonists and transmitter-coupled cyclic nucleotide-generating systems in rat brain. Mol Pharm 27:53–59

Kaupmann K, Bettler B (1998) Heteromerization of $GABA_B$ receptors: a new principle for G protein-coupled receptors. CNS Drug Rev 4:376–379

Kaupmann K, Huggel K, Heid J, Flor PJ, Bischoff S, Mickel SJ, McMaster G, Angst C, Bittiger H, Froestl W, Bettler B (1997) Expression cloning of $GABA_B$ receptors uncovers similarity to metabotrophic glutamate receptors. Nature 386:239–246

Kaupmann K, Malitschek B, Schuler V, Heid J, Froestl W, Beck P, Mosbacher J, Bischoff S, Kulik A, Shigemoto R, Kaischin A, Bettler B (1998) GABA$_B$ receptor subtypes assemble into functional heteromeric complexes. Nature 396:683–687
Knight AR, Bowery NG (1996) The pharmacology of adenylyl cyclase modulation by GABA$_B$ receptors in rat brain slices. Neuropharmacology 35:703–712
Komastsu Y (1996) GABA$_B$ receptors, monoamine receptors, and postsynaptic inositol triphosphate-induced Ca^{2+} release are involved in the induction of long-term potentiation and visual cortical inhibitory synapses. J Neurosci 16:6342–6352
Kuner R, Kohr G, Grunewald S, Eisenhardt G, Bach A, Kornau HC (1999) Role of heteromer formation in GABA$_B$ receptor function. Science 283:74–77
Luscher C, Jan LY, Stoffel M, Malenka R, Nicoll R (1997) G protein-coupled inwardly rectifying K^+ channels (GIRKs) mediate postsynaptic but not presynaptic transmitter actions in hippocampal neurons. Neuron 19:687–695
Mailman RB, Mueller RA, Breese GR (1978) The effects of drugs which alter GABA-ergic function on cerebellar guanosine-3',5'-monophosphate content. Life Sci 623–627
Malcangio M, Bowery NG (1995) Possible therapeutic application of GABAB receptor agonists and antagonists. Clin Neuropharmacol 18:285–305
Marescaux C, Vergenes M, Bernasconi R (1992) GABAB receptor antagonists: potential new anti-absence drugs. J Neural Transm Suppl 35:179–188
McCarson KE, Enna SJ (1996) Relationship between GABA$_B$ receptor activation and neurokinin receptor expression in spinal cord. Pharmacol Rev Comm 8:191–194
McCarson KE, Enna SJ (1999) Nociceptive regulation of GABA$_B$ receptor gene expression in rat spinal cord. Neuropharmacology 38:1767–1773
Menon-Johansson AS, Berrow N, Dolphin AC (1993) G_0 transduces GABA$_B$-receptor modulation of N-type calcium channels in cultured dorsal root ganglion neurons. Pflugers Arch 425:335–343
Mintz IM, Bean BP (1993) GABA$_B$ receptor inhibition of P-type Ca^{2+} channels in central neurons. Neuron 10:889–898.
Mohler H, Knoflach F, Payson J, Motejlek K, Benke D, Lurscher B, Fritschy JM (1995) Heterogeneity of GABA$_A$-receptors: cell-specific expression, pharmacology, and regulation. Neurochem Res 20:631–636
Morishita R, Kato K, Asano T (1990) GABAB receptors couple to G proteins G_o, G_o* and G_{i1}, but not G_{i2}. FEBS Lett 271:231–235
Newberry NR, Nicoll RA (1985) Comparison of the actions of baclofen with gamma-aminobutyric acid on rat hippocampal pyramidal cells in vitro. J Physiol 360:161–185
O'Callaghan JFX, Jarolimek W, Lewen A, Misgeld U (1996) (–)-Baclofen-induced and constitutively active inward rectifying potassium conductances in cultured rat midbrain neurons. Pflugers Arch-Eur J Physiol 433:49–57
Pende M, Lanza M, Bonnano G, Raiteri M (1993) Release of endogenous glutamic and aspartic acids from cerebrocortex synaptosomes and its modulation through activation of gamma-aminobutyric acid B (GABAB) receptor subtype. Brain Res 604:325–330.
Reuveny E, Slesinger PA, Inglese J, Morales JM, Iniguez-Lluhi JA, Lefkowitz RJ, Bourne HR (1994) Activation of cloned muscarinic potassium channel by G protein beta gamma subunits. Nature 370:143–146
Scanziani M, Capogna M, Gahwiler BH, Thompson SM (1992) Presynaptic inhibition of miniature excitatory synaptic currents by baclofen and adenosine in the hippocampus. Neuron 9:919–927
Scherer RW, Ferkany JW, Enna SJ (1988) Evidence for pharmacologically distinct subsets of GABA$_B$ receptors. Brain Res Bull 21:439–443
Schoepp DD, Johnson BG, Wright RA, Salhoff CR, Mann JA (1998) Potent, stereoselective, and brain region selective modulation of second messengers in rat brain by (+)LY354740, a novel group II metabotropic glutamate receptor agonist. Naunyn Schmiedebergs Arch Pharmacol 358:175–180

Surprenant A, Shen K-Z, North RA, Tatsumi H (1990) Inhibition of calcium currents by noradrenaline, somatostatin and opioids in guinea pig submucosal neurones. J Physiol 431:585–608
Takao K, Yoshi M, Kanda A, Kokubun S, Nukada T (1994) A region of the muscarinic-gated atrial K^+ channel critical for activation by G protein beta gamma subunits. Neuron 13:747–755
Takahashi T, Kajikawa Y, Tsujimoto T (1998) G-protein-coupled modulation of presynaptic calcium currents and transmitter release by a GABA$_B$ receptor. J Neurosci 18:3138–3146
Tang W, Gilman AG (1991) Type-specific regulation of adenylyl cyclase by G protein beta-gamma subunits. Science 254:102–108
Tang WJ, Gilman AG (1992) Adenylyl cyclases. Cell 70:869–872
Tang WJ, Iniguez-Lluhi JA, Mumby S, Gilman AG (1992) Regulation of mammalian adenylyl cylase by G-protein alpha and beta gamma subunits. Cold Spring Harb Symp Quant Biol 57:135–144
Taussig R, Quarmby LM, Gilman AG (1993) Regulation of purified type I and type II adenylyl cyclases by G protein beta gamma subunits. J Biol Chem 268:9–12
Thompson SM, Gahwiler BH (1992) Comparison of the actions of baclofen at pre- and postsynaptic receptors in rat hippocampus in vitro. J Physiol 451:329–345
Uezono Y, Akihara M, Kaibara M, Kawano C, Shibuya I, Ueda Y, Yanagihara N, Toyohira Y, Yamashita H, Taniyama K, Izumi F (1998) Activation of inwardly rectifying K^+ channels by GABA-B receptors expressed in *Xenopus* oocytes. Neuro Report 9:583–587
Uezono Y, Ueda Y, Ueno S, Shibuya I, Yanagihara N, Toyohira Y, Yamashita H, Izumi F (1997) Enhancement by baclofen of the G_s-coupled receptor-mediated cAMP production in *Xenopus* oocytes expressing rat brain cortex poly $(A)^+$ RNA: a role of G-protein $\beta\gamma$ subunits. Biochem Biophys Res Comm 241:476–480
White JH, Wise A, Main MJ, Green A, Fraser NJ, Disney GH, Barnes AA, Emson P, Foord SM, Marshall FH (1998) Heterodimerization is required for the formation of a functional GABA$_B$ receptor. Nature 396:679–682
Wojcik WJ, Neff NH (1984) Gamma-aminobutyric acid B receptors are negatively coupled to adenylate cyclase in brain, and in cerebellum these receptors may be associated with granule cells. Mol Pharmacol 25:24–28
Wojcik WJ, Ulivi M, Paez X, Costar E (1989) Islet-activating protein inhibits the beta-adrenergic receptor facilitation elicited by gamma-aminobutyric acid B receptors. J Neurochem 53:753–758
Xu J, Wojcik WJ (1986) Gamma-aminobutyric acid B receptor mediated inhibition of adenylate cyclase in cultured cerebellar granule cells: blockade by islet-activating protein. J Pharmacol Exp Ther 239:568–573
Yoshimura M, Cooper DM (1993) Type-specific stimulation of adenyl cyclase by protein kinase C. J Biol Chem 268:4604–4607

GABA Transporters

CHAPTER 14

Structure and Function of GABA Transporters

B.I. KANNER

A. Introduction

Neurotransmitters are transported across two types of membranes:

1. Plasma membranes of nerve endings (presynaptic), dendrites (postsynaptic) and glial cells (see KANNER and SCHULDINER 1987; PALACIN et al. 1998 for reviews)
2. Membranes of intracellular storage organelles (see SCHULDINER et al. 1995 for a review)

Transport into storage organelles is powered by the electrochemical proton gradient and does not require sodium. Its major function is to concentrate the neurotransmitter from the cytoplasm into the storage organelles in preparation for exocytotic release. In addition to the family of vesicular transporters for biogenic amines and acetylcholine (SCHULDINER et al. 1995), recently the first member of a new family of vesicular transporters – carrying GABA and glycine – has been cloned (MCINTIRE et al. 1997; SAGNE et al. 1997).

Sodium-coupled transporters of neurotransmitters, located in neuronal and glial membranes surrounding the synapse, are thought to play a major role in maintaining low synaptic levels of the transmitter (for a review see KANNER and SCHULDINER 1987). Recently, this has been shown directly for the dopamine transporter using homozygous mice in which the transporter was disrupted (GIROS et al. 1996). Transporters of many neurotransmitters, including GABA, norepinephrine, serotonin, dopamine and glycine, belong to a large superfamily of sodium- and chloride-dependent neurotransmitter transporters (see UHL 1992 for a review). The noted exceptions are the transporters for glutamate which, together with small neutral amino acid transporters as well as prokaryotic glutamate and dicarboxylic acid transporters, form a separate family (KANNER 1993). The sodium-coupled neurotransmitter transporters are of considerable medical interest. Since they function to regulate activity of neurotransmitters by removing them from the synaptic cleft, specific transporter inhibitors can potentially be used as novel drugs for treatment of neurological diseases. For instance, attenuation of GABA removal will prolong the effect of this inhibitory transporter, thereby potentiating its action. Consequently, inhibitors of GABA transport could represent a novel class of anti-epileptic drugs. Well-known inhibitors that interfere with the functioning of biogenic amine transporters include antidepressants such as fluoxetine (Prozac) and citralopram, and stimulants such as amphetamines and cocaine.

In this chapter we shall review our knowledge on the structure and function of a prototype of the sodium- and chloride-coupled neurotransmitter transporters, the GABA transporter GAT-1.

B. Stoichiometry

GABA is accumulated by electrogenic co-transport with sodium and chloride. The electrogenicity of the process has been shown directly (KAVANAUGH et al. 1992; MAGER et al. 1993). We have been able to demonstrate directly that both sodium as well as chloride ions are cotransported with GABA by the transporter. This has been accomplished using a partly purified transporter preparation which was reconstituted into liposomes and the use of Dowex columns to terminate the reactions. These proteoliposomes catalyzed GABA- and chloride-dependent $^{22}[Na^+]$ transport, as well as GABA- and sodium-dependent $^{36}[Cl^-]$ translocation (KEYNAN and KANNER 1988). Using this system the stoichiometry has also been determined kinetically, i.e. by comparing the initial rate of the fluxes of $[^3H]$-GABA, $^{22}[Na^+]$ and $^{36}[Cl^-]$. The results are similar to those found using the thermodynamic method, yielding an apparent stoichiometry of 2.5 Na^+: 1 Cl^-: 1 GABA (RADIAN and KANNER 1983; KEYNAN and KANNER 1988). This is in harmony with the predicted restrictions; if GABA is translocated in the zwitterionic form – the predominant one at physiological pH – an electrogenic cotransport of the three species requires a stoichiometry of nNa^+:mCl^-: GABA with $n>m$. Many other neurotransmitter transporters, including those for norepinephrine, dopamine, serotonin, choline and glycine, require chloride in addition to sodium for optimal activity (KUHAR and ZARBIN 1978).

C. Reconstitution and Purification

Using the reconstitution methodology which enables one to reconstitute many samples simultaneously and rapidly, and employing sodium and chloride dependent GABA transport as an assay, one of the subtypes of the GABA transporter has been purified to an apparent homogeneity (RADIAN and KANNER 1985; RADIAN et al. 1986). It is a glycoprotein and has an apparent molecular weight of 70–80 kDa. This $GABA_A$ transporter retains all the properties observed in membrane vesicles, and represents the first cloned neurotransmitter transporter GAT-1 (see also Sect. E).

D. Biochemical Characterisation of the GABA Transporter

The effect of proteolysis on the function of the transporter was examined. It was purified using all steps except for the lectin chromatography (RADIAN et al. 1986). After papain treatment and lectin chromatography, GABA transport activity was eluted with *N*-acetyl glucosamine. The characteristics of transport were the same as that of the pure transporter (KANNER et al. 1989).

In order to define which regions of the transporter were cleaved, antibodies were raised against synthetic peptides corresponding to several regions of the rat brain GABA transporter. Both amino and carboxyl termini are predicted to be located in the cytoplasm. The antibodies recognized the intact transporter on Western blots. The papainized transporter runs on sodium dodecyl sulfate-polyacrylamide gels as a broad band with an apparent molecular mass between about 58 kDa and 68 kDa as compared to 80 kDa for the untreated transporter. The transporter fragment was recognized by all the antibodies, except for that raised against the amino terminus. Pronase cleaves the transporter to a relatively sharp 60 kDa band, which reacts with the antibodies against the internal loops but not with either the amino- or the carboxyl-termini. This pronase-treated transporter, upon isolation by lectin chromatography, was reconstituted. It exhibits full GABA transport activity. This activity exhibits the same features as the intact system including an absolute dependence on sodium and chloride as well as electrogenicity. Thus the amino- and carboxyl-terminal parts of the transporter are not required for functionality (MABJEESH and KANNER 1992).

Fragments of the (Na^++Cl^-)-coupled $GABA_A$ transporter, now known as GAT-1, were produced by proteolysis of membrane vesicles and reconstituted preparations from rat brain (MABJEESH and KANNER 1993). The former were digested with pronase, the latter with trypsin. Fragments with different apparent molecular masses were recognized by sequence directed antibodies raised against this transporter. When GABA was present in the digestion medium the generation of these fragments was almost entirely blocked (MABJEESH and KANNER 1993). At the same time, the neurotransmitter largely prevented the loss of activity caused by the protease. The effect was specific for GABA; protection was not afforded by other neurotransmitters. It was only observed when the two cosubstrates, sodium and chloride, were present on the same side of the membrane as GABA (MABJEESH and KANNER 1993). The results indicate that the transporter may exist in two conformations. In the absence of one or more of the substrates, multiple sites located throughout the transporter are accessible to the proteases. In the presence of all three substrates – conditions favouring the formation of the translocation complex – the conformation is changed such that these sites become inaccessible to protease action. Further evidence on the ability of GAT-1 to undergo conformational changes upon substrate binding will be discussed in Sect. G.

E. A New Superfamily of Na-Dependent Neurotransmitter Transporters

Partial sequencing of the purified $GABA_A$ transporter allowed the cloning of the first member of the new family of Na-dependent neurotransmitter transporters (GUASTELLA et al. 1990). After expression cloning of the noradrenaline transporter, it became clear that it had significant homology with the $GABA_A$ transporter (PACHOLCZYK et al. 1991). The use of functional c-DNA expression assays and amplification of related sequences by polymerase chain reaction (PCR) resulted in the cloning of additional transporters which belong to this family, such as the dopamine and serotonin transporters, additional GABA transporters, transporters of glycine, proline, taurine, betaine, creatine and orphan transporters, whose substrates are still unknown (BLAKELY et al. 1991; HOFFMAN et al. 1991; KILTY et al. 1991; SHIMADA et al. 1991; USDIN et al. 1991; BORDEN et al. 1992; CLARK et al. 1992; FREMEAU et al. 1992; GUASTELLA et al. 1992; LIU et al. 1992a,b, 1993a,b; LOPEZ-CORCUERA et al. 1992; SMITH et al. 1992; UCHIDA et al. 1992; UHL et al. 1992; YAMAUCHI et al. 1992; GUIMBAL and KILIMANN 1993). Another glycine transporter cDNA encoding for a 799 amino acid protein has been isolated. This is significantly longer than most members of the superfamily. It appears to encode for the 100kDa glycine transporter which was purified and reconstituted (LOPEZ-CORCUERA et al. 1991).

F. Topology

When GAT-1 was cloned (GUASTELLA et al. 1990), its protein sequence was analysed using hydropathy plotting to identify transmembrane α-helices. According to this analysis the transporter is composed of twelve putative transmembrane α-helices. The lack of a signal peptide suggests that both amino- and carboxyl-termini face the cytoplasm. These regions contain consensus phosphorylation sites that may be involved in the regulation of the transport process. The second extracellular loop between helices 3 and 4 is the largest, and it contains three consensus N-linked glycosylation sites. All three sites are in fact used (BENNETT and KANNER 1997). The same topology was predicted for the other transporters of the family because of the high similarity of their hydropathy plots.

The proposed topology was examined experimentally by N-glycosylation insertion scanning mutagenesis (BENNETT and KANNER 1997). The three endogenous glycosylation sites were removed by site-directed mutagenesis. The deglycosylated transporter, which ran faster on SDS-polyacrylamide gels due to its reduced mass, had almost the same transport activity as the wild type. This construct was used to insert N-linked glycosylation sites at various positions and, using the mobility assay, the glycosylation status for the various expressed constructs was determined. If the expressed construct is glycosylated and active in GABA transport, the site inserted is extracellular. This approach enabled us to confirm the predicted topology from transmembrane domain 4 till the carboxyl terminus. An unexpected result was found in the amino terminal part, where the predicted first intracellular loop was found to be glycosylated. The interpretation of this result is ambiguous because the construct was devoid of transport activity. If this loop is external this would lead to a different topology in the amino terminal part. The predicted transmembrane domain 1 becomes a reentrant loop (for the inside), transmembrane domain 2 becomes domain 1 and in addition to transmembrane domain 3 another has to be postulated in order for the large loop, containing the endogenous glycosylation sites, to be on the outside (BENNETT and KANNER 1997). A similar model was proposed (OLIVARES et al. 1997) based on experiments with the glycine transporter GlyT1. Subsequently a modified model for GAT-1 was proposed (YU et al. 1998) with the predicted transmembrane domain 2 serving as a reentrant loop. A detailed study on the related serotonin transporter (CHEN et al. 1998) provides quite convincing evidence that the originally predicted topology (GUASTELLA et al. 1990) is correct. Therefore, it appears that this will also be the case for all the members of the sodium- and chloride-dependent neurotransmitter transporter family.

G. Structure-Function Relationships

It has been shown previously that parts of amino- and carboxyl-termini of the $GABA_A$ transporter are not required for function (MABJEESH and KANNER 1992). In order to define these domains, a series of deletion mutants was studied in the GABA transporter (BENDAHAN and KANNER 1993). Transporters truncated at either end until just a few amino acids distance from the beginning of helix 1 and the end of helix 12, retained their ability to catalyse sodium and chloride-dependent GABA transport. These deleted segments did not contain any residues conserved among the different members of the superfamily. Once the truncated segment included part of these conserved residues, the transporter's activity was severely reduced. However, the functional damage was not due to impaired turnover or impaired targeting of the truncated proteins (BENDAHAN and KANNER 1993).

The substrate translocation performed by the various members of the superfamily is sodium- and usually chloride-dependent. In addition, some of the substrates contain charged groups as well. Therefore, charged amino acids in the membrane domain of the transporters may be essential for their normal function. This was tested using the GABA transporter (PANTANOWITZ et al. 1993). Out of five charged amino acids within its membrane domain, only one, $arginine_{69}$ in helix 1, is absolutely essential for activity. It is not merely the positive charge that is important, since even its substitution to other positively charged amino acids does not restore activity. The functional damage is not caused by impaired turnover or impaired targeting of the mutated protein. The three other positively charged amino acids and the only negatively charged one are not critical (PANTANOWITZ et al. 1993).

The transporters of biogenic amines contain an additional negatively charged residue in helix 1. Replacement of aspartate-79 in the dopamine transporter with alanine, glycine or glutamate significantly reduced the uptake of dopamine and MPP^+ (parkinsonism-inducing neurotoxin), and binding of CFT (cocaine analog) without affecting B_{max} (KITAYAMA et al. 1992). Further support for the idea that aspartate-79 in helix 1 interacts with dopamine's amino group during the transport process has been obtained recently (BARKER et al. 1999). In all the amino acid transporters of the family, including GAT-1, the equivalent position of aspartate-79 of the dopamine transporter is occupied by glycine. In GAT-1 mutation of this glycine to aspartate or alanine leads to inactive transporters (E.R. Bennett and B.I. Kanner, unpublished observations).

Studies of other proteins indicate that, in addition to charged amino acids, aromatic amino acids containing π-electrons are also involved in maintaining the structure and function of these proteins (SUSSMAN and SILMAN 1992). Therefore, tryptophan residues in the membrane domain of the GABA transporter were mutated into serine as well as leucine (KLEINBERGER-DORON and KANNER 1994). Mutations at the 68 and 222 positions (in helix 1 and helix 4, respectively) led to a decrease of over 90% of the GABA uptake. Mutation at position 68 led to increased sodium affinity (MAGER et al. 1996).

We have identified a single tyrosine residue that is critical for GABA recognition and transport. It is completely conserved throughout the superfamily, and even substitution to the other aromatic amino acids, phenylalanine (Y140F) and tryptophan (Y140W), results in completely inactive transporters. Electrophysiological characterisation reveals that both mutant transporters exhibit the sodium-dependent transient currents associated with sodium binding, as well as the chloride-dependent lithium leak currents characteristic of GAT-1. On the other hand, in both mutants GABA is neither able to induce a steady-state transport current nor to block their transient currents. The non-transportable analogue SKF 100330A potently inhibits the sodium-dependent transient in the wild type GAT-1 but not in the Y140W transporter. It partly blocks the transient of Y140F. Thus, although sodium and chloride binding are unimpaired in the tyrosine mutants, they have a specific defect in the binding of GABA. The total conservation of the residue throughout the family suggests that tyrosine 140 may be involved in the liganding of the amino group, the moiety common to all the neurotransmitters (BISMUTH et al. 1997).

We have explored the role of the hydrophilic loops connecting the putative transmembrane domains. Deletions of randomly picked non-conserved single amino acids in the loops connecting helices 7 and 8 or 8 and 9 result in inactive transport upon expression in HeLa cells. However, transporters where these amino acids are replaced with glycine retain significant activity. The expression levels of the inactive mutant transporters were similar to those of the wild-type, but one of these, ΔVal-348, appears to be defectively targeted to the plasma membrane. Our data are compatible with the idea that a minimal length of the loops is required, presumably to enable the transmembrane domains to interact optimally with each other (KANNER et al. 1994). Furthermore, it is possible that parts of some of the loops may line the translocation pathway of the transporter. Consistent with this is the critical role of residue glutamate 101 located in the first intracellular loop of GAT-1 in GABA transport. Its replacement to aspartate leaves only 1% of the transport activity (KESHET al. 1995). The fifth extracellular loop of the GABA transporters plays a role in substrate selectivity. GAT-1 is inhibited by ACHC, but not by β-alanine (KEYNAN et al. 1992). Replacement of the residues of this external loop by those from GAT-3, which is sensitive to β-alanine, leads to an increased sensitivity of GAT-1 to this analog (TAMURA et al. 1995).

Transport by GAT-1 is sensitive to the polar sulfhydryl-reagent (2-aminoethyl) methanethiosulfonate. Following replacement of endogenous cysteines to other residues by site-directed mutagenesis, we have identified cysteine-399 as the major determinant of the sensitivity of the transporter to sulfhydryl modification. Cysteine-399 is located in the intracellular loop connecting putative transmembrane domains 8 and 9. Binding of both sodium and chloride leads to a reduced sensitivity to sulfhydryl reagents, whereas subsequent binding of GABA increases it. Strikingly binding of the non-transportable GABA analogue SKF100330A gives rise to a marked protection against sulfhydryl modification. These effects were not observed in C399S transporters. Under standard conditions GAT-1 is almost insensitive toward the impermeant [2-(trimethylamonium)ethyl] methanethiosulfonate. However, in a chloride-free medium addition of SKF100330A renders wild type GAT-1, but not C399S, very sensitive to this impermeant reagent. These observations indicate that the accessibility of cysteine-399 is highly dependent on the conformation of GAT-1 (GOLOVANEVSKY and KANNER 1999).

H. Conclusions

A series of breakthroughs, including the purification of some of the sodium-coupled neurotransmitter transporters, followed by the cloning of their cDNAs, have considerably improved our understanding of the structure of these transporters. Studies using site-directed mutagenesis revealed the importance of specific residues in the function of these transporters. Additional mutations and further functional characterisation of all the mutated transporters should help to understand the functional contribution of different segments of these proteins to the overall transport process. Applying independent structural approaches will complement and extend our knowledge of the structure and function of these transporters.

Acknowledgment. The work cited from the author's laboratory was supported by grants from the Israel Science Foundation, NINDS National Institutes of Health, The US-Israel Binational Science Foundation, BMBF, Germany, the EC-TMR program 1994–1998 and the Bernard Katz Minerva Center for Cellular Biophysics.

References

Barker EL, Moore KR, Rakhshan F, Blakely RD (1999) Transmembrane domain I contributes to the permeation pathway for serotonin and ions in the serotonin transporter. J Neurosci 19:4705–4717

Bendahan A, Kanner BI (1993) Identification of domains of a cloned rat brain GABA transporter which are not required for its functional expression. FEBS Lett 318:41–44

Bennett ER, Kanner BI (1997) The membrane topology of GAT-1, a (Na^+-Cl^-)-coupled γ-aminobutyric acid transporter from brain. J Biol Chem 272:1203–1210

Bismuth Y, Kavanaugh MP, Kanner BI (1997) Tyrosine 140 of the γ-aminobutyric acid transporter GAT-1 plays a critical role in neurotransmitter recognition. J Biol Chem 272:16046–16102

Blakely RD, Benson HE, Fremeau RT Jr, Caron MG, Peek MM, Prince HK, Bradley CC (1991) Cloning and expression of a functional serotonin transporter from rat brain. Nature 353:66–70

Borden LA, Smith KE, Hartig PR, Branchek TA, Weinshank RL (1992) Molecular heterogeneity of the GABA transport system. J Biol Chem 267:21098–21104

Chen JG, Chen SL, Rudnick G (1998) Determination of external loop topology in the serotonin transporter by site-directed chemical labeling. J Biol Chem 273:12675–12681

Clark JA, Deutch AY, Gallipoli PZ, Amara SG (1992) Functional expression and CNS distribution of a β-alanine sensitive neuronal GABA transporter. Neuron 9:337–348

Fremeau RT Jr, Caron MG, Blakely RD (1992) Molecular cloning and expression of a high affinity l-proline transporter expressed in putative glutamatergic pathways of rat brain. Neuron 8:915–926

Giros B, Jaber M, Jones SR, Wightman RM and Caron MG (1996) Hyperlocomotion and indifference to cocaine and amphetamine in mice lacking the dopamine transporter. Nature 379:606–612

Golovanevsky V, Kanner BI (1999) The reactivity of the γ-aminobutyric acid transporter GAT-1 is conformationally sensitive. Identification of a major target residue. J Biol Chem 274:23020–23026

Guastella J, Brecha N, Weigmann C, Lester HA (1992) Cloning, expression and localization of a rat brain high affinity glycine transporter. Proc Natl Acad Sci, USA 89:7189–7193

Guastella J, Nelson N, Nelson H, Czyzyk L, Keynan S, Miedel MC, Davidson NC, Lester HA, Kanner BI (1990) Cloning and expression of a rat brain GABA transporter. Science 249:1303–1306

Guimbal C, Kilimann MW (1993) A Na^+ dependent creatine transporter in rabbit brain, muscle, heart and kidney. cDNA cloning and functional expression. J Biol Chem 268:8418–8421

Hoffman BJ, Mezey E, Brownstein MJ (1991) Cloning of a serotonin transporter affected by antidepressants. Science 254:579–580

Kanner BI (1993) Glutamate transporters from brain: a novel neurotransmitter transporter family. FEBS Lett 325:95–99

Kanner BI, Bendahan A, Pantanowitz S, Su H (1994) The number of amino acid residues in hydrophilic loops connecting transmembrane domains of the GABA transporter GAT-1 is critical for its function. FEBS Lett 356:191–194

Kanner BI, Keynan S, Radian R (1989) Structural and functional studies on the sodium- and chloride-coupled γ-aminobutyric acid transporter. Deglycosylation and limited proteolysis. Biochemistry 28:3722–3727

Kanner BI, Schuldiner S (1987) Mechanism of transport and storage of neurotransmitters. CRC Crit Rev Biochem 22:1–39

Kavanaugh MP, Arriza JL, North RA, Amara SG (1992) Electrogenic uptake of γ-aminobutyric acid by a cloned transporter expressed in oocytes. J Biol Chem 267:22007–22009

Keshet GI, Bendahan A, Su H, Mager S, Lester HA, Kanner BI (1995) Glutamate-101 is critical for the function of the sodium and chloride-coupled GABA transporter GAT-1. FEBS Lett 371:39–42
Keynan S, Kanner BI (1988) γ-Aminobutyric acid transport in reconstituted preparations from rat brain: coupled sodium and chloride fluxes. Biochemistry 27:12–17
Keynan S, Suh Y-J, Kanner BI, Rudnick G (1992) Expression of a cloned γ-aminobutyric acid transporter in mammalian cells. Biochemistry 31:1974–1979
Kilty JE, Lorang D, Amara SG (1991) Cloning and expression of a cocaine-sensitive rat dopamine transporter. Science 254:578–579
Kitayama S, Shimada S, Xu H, Markham L, Donovan DM, Uhl GR (1992) Dopamine transporter site-directed mutations differentially alter substrate transport and cocaine binding. Proc Natl Acad Sci USA 89:7782–7785
Kleinberger-Doron N, Kanner BI (1994) Identification of tryptophan residues critical for the function and targeting of the γ-aminobutyric acid transporter (subtype A). J Biol Chem 269:3063–3067
Kuhar MJ, Zarbin MA (1978) Synaptosomal transport: a chloride dependence for choline, GABA, glycine and several other compounds. J Neurochem 31:251–256
Liu QR, Lopez-Corcuera B, Mandiyan S, Nelson H, Nelson N (1993a) Molecular characterization of four pharmacologically distinct γ-aminobutyric acid transporters in mouse brain. J Biol Chem 268:2104–2112
Liu QR, Lopez-Corcuera B, Nelson H, Mandiyan S, Nelson N (1992a) Cloning and expression of a cDNA encoding the transporter of taurine and β-alanine in mouse brain. Proc Natl Acad Sci USA 89:12145–12149
Liu QR, Mandiyan S, Lopez-Corcuera B, Nelson H, Nelson N (1993b) A rat brain cDNA encoding the neurotransmitter transporter with an unusual structure. FEBS Lett 315:114–118
Liu QR, Nelson H, Mandiyan S, Lopez-Corcuera B, Nelson N (1992b) Cloning and expression of a glycine transporter from mouse brain. FEBS Lett 305:110–114
Lopez-Corcuera B, Liu QR, Mandiyan S, Nelson H, Nelson N (1992) Expression of a mouse brain cDNA encoding novel γ-aminobutyric acid transporter. J Biol Chem 267:17491–17493
Lopez-Corcuera B, Vazquez J, Aragon C (1991) Purification of the sodium- and chloride-coupled glycine transporter from central nervous system. J Biol Chem 266:24809–24814
Mabjeesh NJ, Kanner, BI (1992) Neither amino nor carboxyl termini are required for function of the sodium- and chloride-coupled γ-aminobutyric acid transporter from rat brain. J Biol Chem 267:2563–2568
Mabjeesh NJ, Kanner BI (1993) The substrates of a sodium- and chloride-coupled γ-aminobutyric acid transporter protect multiple sites throughout the protein against proteolytic cleavage. Biochemistry 32:8540–8546
Mager S, Kleinberger-Doron N, Keshet, GI, Davidson N, Kanner BI, Lester HA (1996) Ion binding and permeation at the GABA transporter GAT-1. J Neurosci 16:5405–5414
Mager S, Naeve J, Quick M, Labarca C, Davidson N, Lester HA (1993) Steady states, charge movements and rates for a cloned GABA transporter expressed in *Xenopus* oocytes. Neuron 10:177–188
McIntire SL, Reimer RJ, Schuski K, Edwards RH, Jorgensen IM (1997) Identification and characterization of the vesicular GABA transporter. Nature 389:870–876
Olivares L, Aragon C, Gimenez C, Zafra F (1997) Analysis of the transmembrane topology of the glycine transporter GlyT1. J Biol Chem 272:1211–1217
Pacholczyk T, Blakely RD, Amara SG (1991) Expression cloning of a cocaine and antidepressant-sensitive human noradrenaline transporter. Nature 350:350–354
Palacin M, Estevez R, Bertan J, Zorzano A (1998) Molecular biology of mammalian plasma membrane amino acid transporters. Physiol Rev 78:969–1054

Pantanowitz S, Bendahan A, Kanner BI (1993) Only one of the charged amino acids located in the transmembrane a helices of the γ-aminobutyric acid transporter (subtype A) is essential for its activity. J Biol Chem 268:3222–3225
Radian R, Bendahan A, Kanner BI (1986) Purification and identification of the functional sodium- and chloride-coupled γ-aminobutyric acid transport glycoprotein from rat brain. J Biol Chem 261:15437–15441
Radian R, Kanner BI (1983) Stoichiometry of sodium- and chloride-coupled γ-aminobutyric acid transport by synaptic plasma membrane vesicles isolated from rat brain. Biochemistry 22:1236–1241
Radian R, Kanner, BI (1985) Reconstitution and purification of the sodium- and chloride-coupled γ-aminobutyric acid transporter from rat brain. J Biol Chem 260:11859–11865
Sagné C, El-Mestikawy S, Isembert MF, Hamon M, Henry JP, Giros B and Gasnier B (1997) Cloning of a functional vesicular GABA and glycine transporter by screening of genomic databases. FEBS Lett 412:177–183
Schuldiner S, Shirvan A, Linial M (1995) Vesicular neurotransmitters: from bacteria to humans. Physiol Rev 75:369–392
Shimada S, Kitayama S, Lin CL, Patel A, Nanthakumar E, Gregor P, Kuhar M, Uhl G (1991) Cloning and expression of a cocaine-sensitive dopamine transporter complementary DNA. Science 254:576–578
Smith KE, Borden LA, Hartig PA, Branchek T, Weinshank RL (1992) Cloning and expression of a glycine transporter reveal colocalization with NMDA receptors. Neuron 8:927–935
Sussman JL, Silman I (1992) Acetylcholinesterase: structure and use as a model for specific cation–protein interactions. Curr Opin Struc Biol 2:721–729
Tamura S, Nelson H, Tamura A, Nelson N (1995) Short external loops as potential substrate binding site of γ-aminobutyric acid transporters. J Biol Chem 270: 28712–28715
Uchida S, Kwon HM, Yamauchi A, Preston AS, Marumo F, Handler JS (1992) Molecular cloning of the cDNA for an MDCK cell Na^+- and Cl^--dependent taurine transporter that is regulated by hypertonicity. Proc Natl Acad Sci, USA 89:8230–8234
Uhl GR (1992) Neurotransmitters (plus): a promising new gene family. Trends Neurosci 15:265–285
Uhl GR, Kitayama S, Gregor P, Nanthakumer E, Persico A, Shimada S (1992) Neurotransmitter transporter family cDNAs, in a rat midbrain library: 'orphan transporters' suggest sizable structural variations. Molec Brain Res 16:353–359
Usdin TB, Mezey E, Chen C, Brownstein MJ, Hoffman BJ (1991) Cloning of the cocaine sensitive bovine dopamine transporter. Proc Natl Acad Sci USA 88:11168–11171
Yamauchi A, Uchida S, Kwon HM, Preston AS, Robey RB, Garcia-Perez A, Burg MB, Handler JS (1992) Cloning of a Na^+ and Cl^- dependent betaine transporter that is regulated by hypertonicity. J Biol Chem 267:649–652
Yu N, Cao Y, Mager S, Lester HA (1998) Topological localization of cysteine 74 in the GABA transporter, GAT-1, and its importance in ion binding and permeation. FEBS Lett 426:174–178

CHAPTER 15

Pharmacology of GABA Transporters

J.E. CLARK and W.A. CLARK

A. Introduction

It is widely accepted that γ-aminobutyric acid (GABA) is the major inhibitory neurotransmitter in the mammalian central nervous system (CNS). Disruptions in GABAergic neurotransmission are implicated in a number of neurological and psychiatric disorders including epilepsy, schizophrenia, and affective disorders (BRAESTRUP and NIELSEN 1982; PERRY et al. 1973; SPOKES 1980; LOSCHER and SCHWARTZ-PORSCHE 1986; REYNOLDS et al. 1990; HAMBERGER et al. 1991; BENES et al. 1992; SIMPSON et al. 1989, 1992; DURING et al. 1995; RIBAK et al. 1979; MELDRUM 1975; LLOYD et al. 1977; ENNA et al. 1976). There are then several clinical situations in which GABAmimmetic agents may prove therapeutically useful. In particular, because potentiation of GABAergic function is recognized as a means of producing anticonvulsant activity (MELDRUM 1995), it is reasonable to expect that low extracellular GABA concentrations be associated with poor seizure control (PETROFF et al. 1996). Development of specific compounds which block reuptake or metabolism of GABA or stimulate particular GABA receptor subtypes will likely be useful in the treatment of conditions where a deficit in GABAergic tone is implicated.

GABA is produced from glutamic acid by glutamic acid decarboxylase (GAD) and sequestered in a vesicular compartment (Fig. 1). Upon release from a presynaptic terminal, GABA may bind to two classes of receptors: $GABA_A$ ligand gated ion channels and $GABA_B$ G protein coupled receptors. GABAergic neurotransmission is terminated primarily by a specific high-affinity transport mechanism (IVERSEN and NEAL 1968), the discovery of which aided in establishing the neurotransmitter status of this amino acid. Following transport into glia or reuptake into the presynaptic terminal, GABA is converted to succinic acid semi-aldehyde (SSA) by GABA transaminase. Figure 1 illustrates a number of possible sites for pharmacological intervention in the GABAergic synapse. Whereas specific agents have been developed to interact with many of these target proteins (for review see KROGSGAARD-LARSEN and BUNDGAARD 1991), the remainder of this review will focus on compounds targeting GABA transport, the application of these agents in basic research

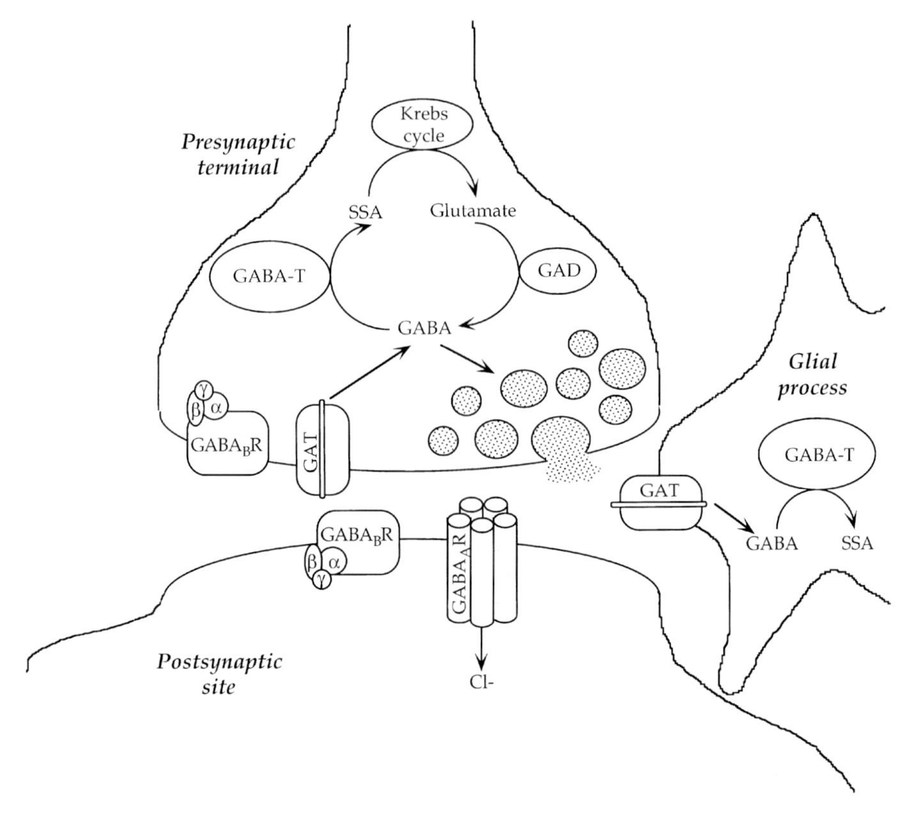

Fig. 1. Schematic model of a GABAergic synapse illustrating sites potentially susceptible to pharmacological manipulation. GABA, γ-aminobutyric acid; GABA-T, GABA aminotransferase; $GABA_AR$, $GABA_A$ receptor channel; $GABA_BR$, $GABA_B$ G-protein coupled receptor; GAD, glutamic acid decarboxylase; GAT, GABA transporter; SSA, succinic acid semialdehyde; α, β, γ, trimeric guanine nucleotide binding protein subunits

for determining physiological roles of GABA transport, and their therapeutic potential in treating disease states where an increase in GABAergic input is indicated.

B. Physiological Relevance of GABA Transporters

A number of investigations have elucidated specific physiological roles for GABA transporters. Inhibition of transport in rat hippocampus prolonged the decay phase of both $GABA_A$- and $GABA_B$-mediated postsynaptic potentials and increased the magnitude of $GABA_B$-mediated responses (ISAACSON et al. 1993; DINGLEDINE and KORN 1985; SOLIS and NICOLL 1992). In toad and catfish retinal horizontal cells, calcium-independent GABA efflux through reversal of the transporter appeared to be the major mode of GABA release (SCHWARTZ 1987). Reversal of GABA transport in rat hippocampus by depolarization and/or reversal of the sodium gradient resulted in activation of GABA receptors (GASPARY et al. 1998). In the clinical sphere, decreased transporter-mediated GABA efflux was detected in the affected hippocampus of human subjects having temporal lobe epilepsy. This reduction in calcium-independent GABA efflux stemmed from a decreased number of GABA transporters and appeared to contribute to reduced inhibitory tone (DURING et al. 1995). Finally, BERNSTEIN and QUICK (1999) recently demonstrated that extracellular GABA modulated GABA transporter function. Specifically, exogenous GABA caused a dose-dependent increase in transporter number apparently by slowing transporter turnover.

Despite the multitude of studies which confirm that GABAergic function can be altered as a direct result of inhibition of GABA transport, we must note that small increases in extracellular GABA concentrations are not readily detectable *in vivo* when transport is blocked. Whereas dose-dependent increases in extracellular GABA were detected with i.p. administration of high doses of tiagabine, NNC-711, or SK&F 89976A, no such changes were detectable at lower doses that are known to have anticonvulsant effects (WALDMEIER et al. 1992; RICHARDS and BOWERY 1996). Therefore, the physiological alterations that are discerned with administration of GABA transport inhibitors *in vivo* and *in vivo* are not consistently accompanied by easily detectable elevations in extracellular GABA. However, this does not exclude the possibility that small changes in extracellular GABA do occur at doses of uptake inhibitors known to have anticonvulsant effects. In particular, interpretation of results from microdialysis experiments must take into account the limitations of this method. Much of dialysate GABA is derived from metabolic, rather than synaptic, pools (SAYIN et al. 1995). This confound is exacerbated by the spatial limitations of the microdialysis approach and long sampling intervals necessary to measure extracellular GABA. Yet even in the absence of conclusive data, a large body of experimental evidence strongly suggests that blockade of transporters at low doses of transport inhibitors has physiologically relevant effects on GABAergic transmission. Thus, GABA transporters are reasonable targets for the development of compounds for the treatment of diseases where enhanced inhibition is required.

C. 'Neuronal'- and 'Glial'-Specific GABA Transport Inhibitors

Early studies of native GABA transporters were aided by the use of such GABA analogs as β-alanine, *cis*-3-aminocyclohexanecarboxylic acid (ACHC), and L-2,4-diaminobutyric acid (L-DABA) (Fig. 2), that are specific competitive inhibitors of GABA transport (SCHON and KELLY 1974, 1975; IVERSEN and KELLY 1975; BOWERY et al. 1976). These investigations yielded two principal

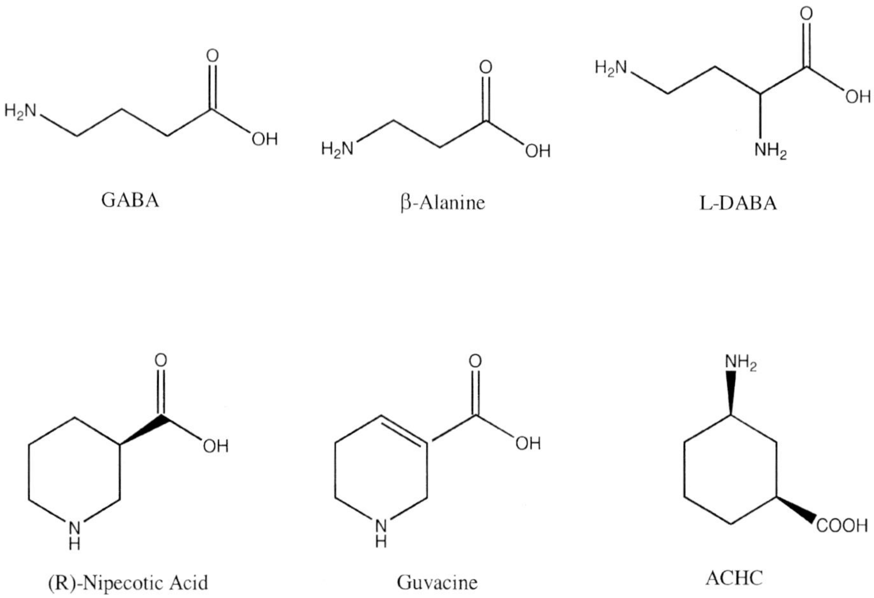

GABA β-Alanine L-DABA

(R)-Nipecotic Acid Guvacine ACHC

Fig. 2. GABA and classical competitive inhibitors of GABA transport

findings. First, reuptake of GABA was determined to be the primary means of maintaining low extracellular GABA concentrations. Second, transport mechanisms appeared heterogeneous and were broadly classified based on pharmacological sensitivities to these GABA uptake inhibitors. Transport into neurons was effectively inhibited by ACHC (Bowery et al. 1976) or by L-DABA (Iversen and Kelly 1975; Larsson et al. 1983) whereas transporters in central and peripheral glia transported β-alanine and were inhibited by β-alanine (Schon and Kelly 1974, 1975; Gavrilovic et al. 1984). GABA transporters were hence defined as 'neuronal' or 'glial' based upon these criteria. However, a number of discrepancies arose in the literature indicating that transport processes exhibited greater complexity than this pharmacological classification allowed. Studies of primary cultures of rat retinal Müller cells (Iversen and Kelly 1975), cerebellar stellate astrocytes (Cummins et al. 1982; Levi et al. 1983), and oligodendrocytes (Reynolds and Herschowitz 1986) revealed that not all glial GABA transport was sensitive to β-alanine, nor was β-alanine a substrate for all glial GABA transport systems. In fact, transport in rat retinal Müller cells and cerebellar stellate astrocytes was sensitive to the putatively neuronal transport-selective agents ACHC and L-DABA (Iversen and Kelly 1975; Levi et al. 1983). These complexities of GABA transport processes underscored the need for cloning and expression of the transporter species to resolve the observed inconsistencies.

D. GABA Transporter Heterogeneity

Recent molecular cloning studies identified a family of high affinity GABA transporters having unique primary sequences and pharmacological profiles. Five transporters were identified (GAT-1 (GUASTELLA et al. 1990), GAT-B or GAT-3 (CLARK et al. 1992; BORDEN et al. 1992), GAT-2 (BORDEN et al. 1992), BGT-1 (YAMAUCHI et al. 1992), and TAUT (SMITH et al. 1992)), which transported GABA with varying affinities (Table 1). GAT-1, originally isolated from rat brain, transported GABA with high affinity and was sensitive to the GABA transport inhibitors ACHC and L-DABA (K_M = 7.0 μmol/l) (GUASTELLA et al. 1990). GAT-3 was isolated from rat midbrain, transported both GABA and β-alanine with relatively high affinity (K_M = 2.3 μmol/l and 6.7 μmol/l, respectively) (CLARK et al. 1992), and was inhibited by β-alanine. GAT-2, also isolated from rat brain, transported GABA with relatively high affinity (K_M = 8 μmol/l) (BORDEN et al. 1992), and was sensitive to β-alanine (IC_{50} = 19 μmol/l) (BORDEN et al. 1994a). BGT-1, cloned first from Madin-Darby canine kidney (MDCK) cells (YAMAUCHI et al. 1992) and later from neonatal mouse brain (LIU et al. 1993), transported both GABA and the osmolyte betaine. The relative affinity of BGT-1 for GABA was several fold higher than that for betaine (K_M = 93 μmol/l and 398 μmol/l, respectively) (YAMAUCHI et al. 1992). TAUT, isolated both from MDCK cells and rat brain, exhibited high affinity transport for taurine (K_M = 10 μmol/l), low affinity for GABA (K_M = 1 mmol/l), and was inhibited by β-alanine (IC_{50} = 100 μmol/l) (SMITH et al. 1992). The identification of a subfamily of transporters that transported β-alanine with high affinity suggested that the glial GABA transporter had been identified. However, data from *in situ* hybridization histochemistry and immunocytochemistry for the cloned transporters could not be reconciled with the pharmacological characterization of native transporters as strictly 'neuronal' or 'glial.' The most abundant GABA transporter message in the rat brain, GAT-1, was found principally in neurons (DURKIN et al. 1995; RATTRAY and PRIESTLEY 1993). Yet GAT-1 mRNA was also identified in certain specialized glial cells including Müller cells of the retina (BRECHA and WEIGMANN 1994) and Bergmann glia in the cerebellum (RATTRAY and PRIESTLEY 1993), providing an explanation for the earlier finding that some glia were capable of transporting L-DABA and ACHC (IVERSEN and KELLY 1975; LEVI et al. 1983). Similarly, GAT-3 mRNA was identified in both neuronal and glial cell populations throughout the rat brain (DURKIN et al. 1995; CLARK et al. 1992), indicating that select putative 'neuronal' GABA transporters also transport β-alanine. In summary, molecular studies have identified a class of Na^+- and Cl^--dependent transporters that transport GABA with varying affinities and that cannot be characterized strictly as neuronal or glial based on pharmacological sensitivities. These molecular details of GABA transport are fairly recent developments and significantly add to our fundamental knowledge of the actions of transport inhibitors from studies predating the cloning achievements.

Table 1. IC_{50} values of GABA uptake inhibitors at native and cloned GABA transporters

	IC_{50} (µmol/l)												
	Crude synaptosomes	Neurons in culture	Astrocytes in culture	*Rat GAT-1*	*Human GAT-1*	*Murine GAT-1*	Rat GAT-2	Murine GAT-3	*Rat GAT-3*	*Human GAT-3*	*Murine GAT-4*	Human BGT-1	Murine GAT-2
Nipecotic acid	6.0[a]	16.8[c]	60[a]	24[f]	8.0[f,i]	70*[j]	38[f,i]	124*[j]	159[f]	106[f,i]	201*[j]	2370[f,i]	2785*[j]
	3.9[b]	–	33[c]	–	–	–	–	–	–	–	–	–	–
	3.79[c]	–	–	–	–	–	–	–	–	–	–	–	–
	2.79[e]	–	–	–	–	–	–	–	–	–	–	–	–
Guvacine	5.0[a]	–	25[a]	39[f]	14[f,i]	–	58[f,i]	–	378[f]	119[f]	–	1870[f,i]	–
	4.3[b]	–	–	–	–	–	–	–	–	–	–	–	–
SK&F 100330A	0.21[b]	1.772[c]	0.559[c]	–	–	–	–	–	191[h]	–	–	–	–
	0.331[c]	–	–	–	–	–	–	–	–	–	–	–	–
SK&F 89976A	0.2[b]	–	–	0.64[f]	0.13[f,i]	–	550[f,i]	–	203[h]	944[f]	–	7210[f,i]	–
	–	–	–	–	–	–	–	–	4390[f]	1990[i]	–	–	–
CI-966	0.104[d]	4.64[d]	0.304[d]	1.2[f]	0.26[f,i]	–	297[f]	–	1140[f]	333[f,i]	–	300[f,i]	–
	–	–	–	–	–	–	1280[i]	–	–	–	–	–	–
NNC-711	0.047[e]	1.238[e]	0.636[e]	0.38[f]	0.04[f]	–	171[f]	–	349[f]	1700[f]	–	622[f]	–
Tiagabine	0.067[c]	0.446[c]	0.182[c]	0.64[f]	0.07[f,i]	0.11*[j]	1410[f,i]	>100*[j]	362[h]	917[f,i]	>100*[j]	1670[f,i]	>100*[j]
	–	–	–	–	–	–	–	–	2040[f]	–	–	–	–
EGYT-3886	–	–	–	–	26[g,i]	–	30[g,i]	–	–	46[g,i]	–	39[g,i]	–
SNAP 5114	–	–	–	–	388[g,i]	>30*[j]	21[g,i]	20 *[j]	–	5.0[g,i]	6.6*[j]	140[g,i]	22*[j]
NNC 05-2045	–	–	–	–	–	27*[j]	–	14*[j]	–	–	6.1*[j]	–	1.6*[j]
NNC 05-2090	–	–	–	–	–	19*[j]	–	41*[j]	–	–	15*[j]	–	1.4*[j]

* K_i values (µmol/l).
[a] KROGSGAARD-LARSEN (1980).
[b] YUNGER et al. (1984).
[c] BRAESTRUP et al. (1990).
[d] TAYLOR and SEDMAN (1991).
[e] SUZDAK et al. (1992).
[f] BORDEN et al. (1994a).
[g] BORDEN et al. (1994b).
[h] CLARK et al. (1994).
[i] DHAR et al. (1994).
[j] THOMSEN et al. (1997).

E. Lipophilic GABA Transport Inhibitors

I. THPO

Compounds with increased potency and specificity are persistently sought to aid in examining the physiological relevance of GABA transport and its contributions to GABAergic neurotransmission. 4,5,6,7-Tetrahydroisoxazolo[4,5-*c*]pyridin-3-ol (THPO), an analogue of the potent $GABA_A$ receptor agonist muscimol (KROGSGAARD-LARSEN and JOHNSTON 1975), specifically inhibited GABA transport and displayed no measurable affinity for $GABA_A$ receptors (KROGSGAARD-LARSEN 1980). THPO was one of the first specific GABA uptake inhibitors found to display anticonvulsant properties in rodent models of epilepsy (CROUCHER et al. 1983), and its actions were initially attributed to specific inhibition of glial GABA transport (KROGSGAARD-LARSEN 1980). This work confirmed previous data with less lipophilic glial and neuronal GABA uptake inhibitors exhibiting anticonvulsant properties in the rodent (FREY et al. 1979; HORTON et al. 1979). Taken together, data from studies with THPO and related compounds delineated the GABA transporter as a potential target for the development of anticonvulsants and highlighted the need for more potent, specific, lipophilic compounds.

II. Prodrugs of Nipecotic Acid, Hydroxynipecotic Acid, and Isoguvacine

Nipecotic acid and guvacine (Fig. 2) are specific GABA transport inhibitors and substrates for all of the GABA transporters identified to date. However, these compounds are neither selective nor potent, nor do they penetrate the blood-brain barrier well. In an attempt to generate more lipophilic GABA transport inhibitors that would more readily cross the blood-brain barrier, prodrugs of the transporter specific agents nipecotic acid, hydroxynipecotic acid, and isoguvacine were developed (FREY et al. 1979; LOSCHER 1982; FALCH and KROGSGAARD-LARSEN 1981; CROUCHER et al. 1983; BONINA et al. 1999). Systemic administration of (±)-nipecotic acid or (±)-*cis*-4-hydroxynipecotic acid provided no protection against audiogenic seizures in genetically susceptible mice (DBA/2 mice) (CROUCHER et al. 1983; BONINA et al. 1999). In contrast, i.p. administration of (±)-nipecotic acid pivaloyloxymethyl ester or (±)-*cis*-4-hydroxynipecotic acid methyl ester protected against both chemically- and sound-induced seizures (CROUCHER et al. 1983). The authors noted, however, that following systemic administration of these esters, a range of cholinergic side effects and an apparent 'GABA toxicity' were observed. Recently, another prodrug of nipecotic acid, nipecotic acid tyrosine ester, protected against audiogenic seizures in a dose-dependent manner with no apparent cholinergic side effects when administered i.p. in DBA/2 mice (BONINA et al. 1999). Nipecotic acid tyrosine ester was a more potent anticonvulsant (tonic seizure, ED_{50} = 0.13 mmol/kg and clonic seizure, ED_{50} = 0.173 mmol/kg)

(BONINA et al. 1999) than nipecotic acid pivaloyloxymethyl ester and *cis*-4-hydroxynipecotic acid methyl ester (tonic and clonic seizures: ED_{50} = 1.7 mmol/kg and approximately 3.0 mmol/kg, respectively) (CROUCHER et al. 1983). However, nipecotic acid tyrosine ester was less potent than the GAT-1-selective uptake inhibitor tiagabine (tonic seizure, ED_{50} = 1 μmol/kg and clonic seizure, ED_{50} = 5 μmol/kg) (BONINA et al. 1999). While prodrugs have largely been more effective in penetrating the blood-brain barrier than the parent compounds, data generated with these compounds raise different concerns for drug development. Some undesirable characteristics of prodrugs include low potency, ability to act as false transmitters at GABAergic terminals, and serious side effects following generation of toxic products, all of which pose challenges in the current development of potent GABA uptake inhibitors for systemic administration.

III. Nipecotic Acid and Guvacine Derivatives

In further efforts to design lipophilic GABA uptake inhibitors having higher potency for *in vivo* study and potential therapeutic use, YUNGER et al. (1984) initiated synthesis of compounds structurally related to nipecotic acid and guvacine (Fig. 2). Armed with evidence that many neuroleptics reportedly block synaptosomal accumulation of GABA (FJALLAND 1978), Yunger and colleagues substituted lipophilic side chains, reminiscent of those found in the structures of some neuroleptics, onto the nitrogen atom of amino acids known to block GABA uptake. This work resulted in a novel series of selective and potent agents that readily penetrated the blood-brain barrier after peripheral administration. Attachment of a 4,4-diphenyl-3-butenyl group to the amines of nipecotic acid and guvacine resulted in the compounds *N*-(4,4-diphenyl-3-butenyl)-nipecotic acid (SK&F 89976A) (Fig. 3) and *N*-(4,4-diphenyl-3-butneyl)-guvacine (SK&F 100330A) (Fig. 4). Both compounds were potent competitive inhibitors of GABA uptake and were not substrates for the carrier(s). Each agent was ~20-fold more potent than the cognate parent compound in competing GABA interaction with carrier(s) in rat diencephalic membranes (Table 1) (YUNGER et al. 1984). In addition, each novel compound was found to have potent and relatively long-acting anticonvulsant activity in rats and mice following oral or i.p. administration (YUNGER et al. 1984). Electrophysiological studies determined that these compounds increased GABA-mediated inhibition *in vivo* in the rat CNS (ALBERTSON and JOY 1987). Further work confirmed that these compounds potentiated GABAergic tone by blocking GABA uptake resulting in increased seizure thresholds (SWINYARD et al. 1991).

Several other potent, selective, and relatively lipophilic derivatives of nipecotic acid and guvacine were developed using a strategy similar to that used for synthesis of the SK&F compounds. N-Alkylation of guvacine to yield [1-[2-bis[4-(trifluoromethyl)phenyl]-methoxy]ethyl]-1,2,5,6-tetrahydro-3-pyridine-carboxylic acid, or CI-966 (Fig. 4) (BJORGE et al. 1990), resulted in

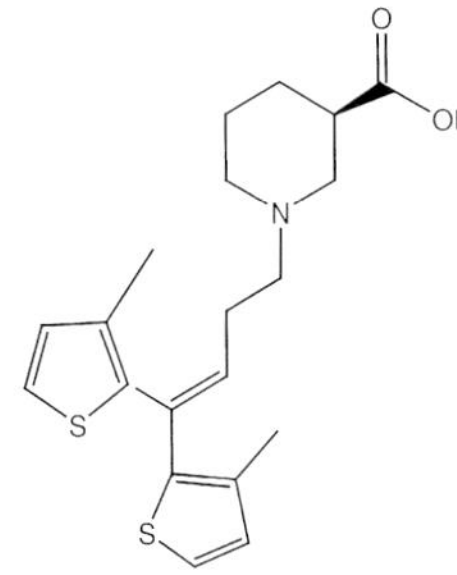

Tiagabine 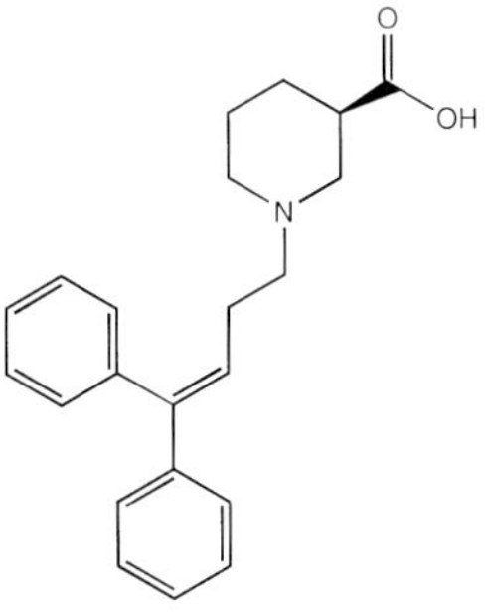 SK&F-89976A 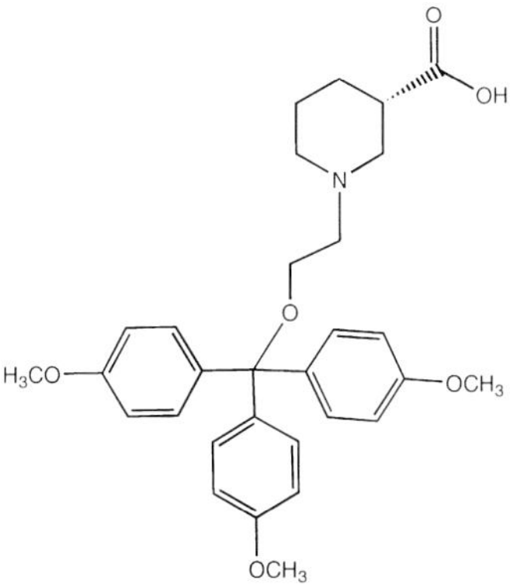 (S)-SNAP-5114

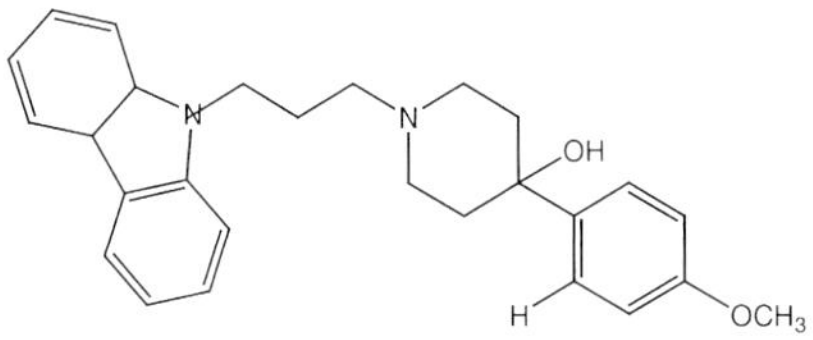

NNC 05-2045 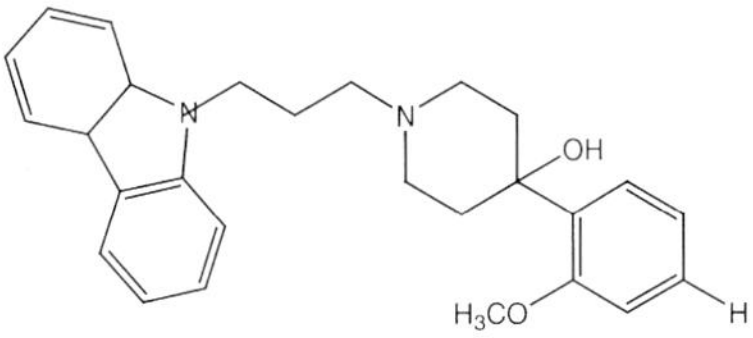 NNC 05-2090

Fig. 3. Lipophilic derivatives of nipecotic acid

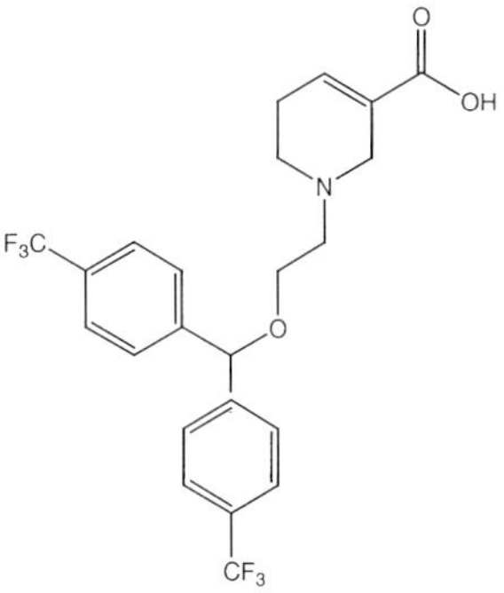

CI-966 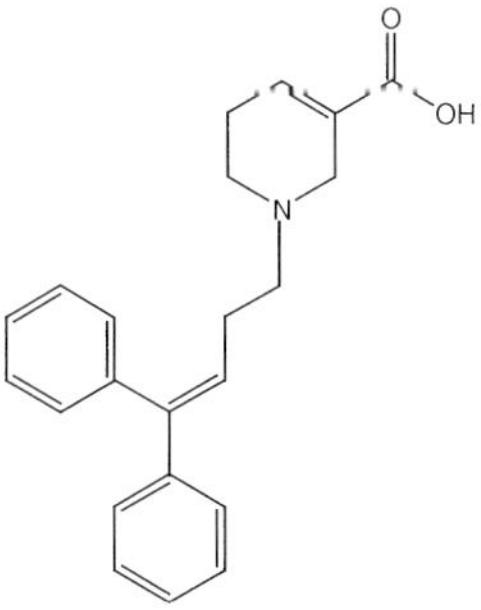 SK&F 100330A 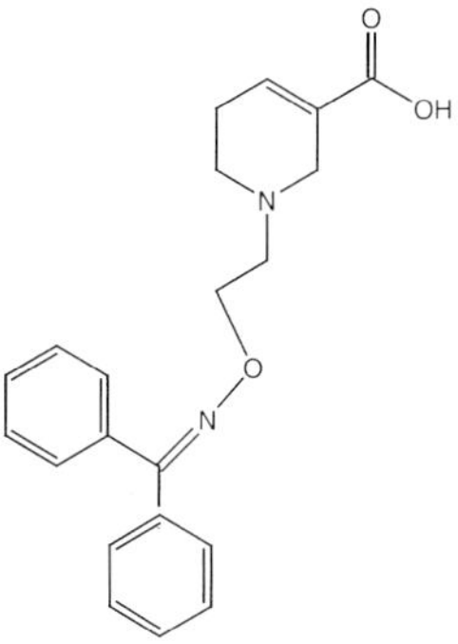 NNC-711

Fig. 4. Lipophilic derivatives of guvacine

an agent more potent than guvacine at inhibiting GABA uptake (Table 1) (for review see Taylor and Sedman 1991), and which did not serve as a transport substrate. CI-966 exhibited potent anticonvulsant activity in several rodent models of seizure following systemic administration. Inhibition of hippocampal population spikes elicited by microiontophoretic application of GABA in CA1 was significantly enhanced in rats given systemic injections of CI-966 (Ebert and Krnjevic 1990). These effects were attributed to a deficit in GABA clearance from the synapse due to GABA transport blockade by CI-966. Preliminary clinical studies in human subjects were discontinued, however, due to adverse neurological and psychological effects lasting for several days which arose at higher doses of CI-966 (for review see Taylor and Sedman 1991). The potent ability of CI-966 to block transport and the resulting action of excess synaptic GABA on GABA receptors may have elicited these adverse effects. Another guvacine derivative, 1-(2-(((diphenylmethylene)amino)oxy)ethyl)-1,2,5,6-tetrahydro-3-pyridinecarboxylic acid, or NNC-711 (Fig. 4), displayed potent and selective GABA uptake inhibition with anticonvulsant activity in rodent models of seizure (Suzdak et al. 1992). NNC-711 was ~85-fold more potent than the parent compound guvacine in inhibiting GABA uptake into a crude synaptosome preparation (Table 1) and remains the most potent GABA uptake inhibitor reported to date (Suzdak et al. 1992; Borden 1996). The nipecotic acid derivative (*R*)-*N*-[4,4-bis(3-methyl-2-thienyl)but-3-en-1-yl]nipecotic acid, also tiagabine or NO 328 (Fig. 3), was found to be a potent non-competitive inhibitor, but not a substrate, of GABA transport (Braestrup et al. 1987). Tiagabine was ~60-fold more potent than its parent compound nipecotic acid at inhibiting GABA uptake in crude rat brain synaptosomes (Table 1) (Braestrup et al. 1990). Furthermore, tiagabine exhibited a relatively broad anticonvulsant activity in several rodent models of seizure at doses that did not produce sedation or motor debilitation, i.e., side effects commonly observed with other derivatives of nipecotic acid and guvacine (Nielsen et al. 1991). At doses 10- to 14-fold those yielding anticonvulsant effects, tiagabine produced motor impairment. The ratio between anticonvulsant activity and motor disruption, or therapeutic index, was thus greater for tiagabine than for SK&F 100330A or any of the reference anti-epileptic drugs tested, suggesting that tiagabine may circumvent some neurological side-effects in humans (Nielsen et al. 1991). Indeed, tiagabine has been approved in the United States as add-on therapy for refractory epilepsy (for review see Leach and Brodie 1998) and is currently under investigation as a monotherapy for childhood and newly diagnosed epilepsy. Apparently the most specific antiepileptic drug in clinical use, tiagabine, is the only GABA transport inhibitor that has been studied extensively both *in vitro* and *in vivo* and found thus far to be both therapeutically useful and safe.

F. Specific GABA Transport Inhibitors

I. Compounds Selective for GAT-1

The recent identification of four unique transporters having moderate to high affinity for GABA facilitated the examination of the selectivity of these potent nipecotic acid and guvacine derivatives in the hope of determining their molecular site(s) of action. Transport studies with each of the cloned rat and human GABA transporters revealed that SK&F 89976A, SK&F 100330A, CI-966, NC-711, and tiagabine were highly selective for GAT-1 (Table 1) (CLARK and AMARA 1994; BORDEN et al. 1994a) and displayed relatively low affinities for GAT-2, GAT-3, and BGT-1. These data strongly suggest that the anticonvulsant effects of these agents are sequelae of GAT-1 transporter blockade and emphasize the critical importance of this protein as an exciting target for pharmacological manipulation. In addition, these results provoke questions concerning the potential physiological roles which the other GABA carriers may fulfill.

II. Compounds Selective for GAT-2, GAT-3, and BGT-1

The lack of selective agents for investigating the physiological relevance of GAT-2, GAT-3, and BGT-1 in GABAergic neurotransmission prompted at least two groups to search for novel compounds with specificity for these other carriers. DHAR et al. (1994) reported that the bicycloheptane EGYT-3886 ([(–)-2-phenyl-2-[(dimethylamino)ethoxy]-(1*R*)-1,7,7-trimethylbicyclo[2.2.1]heptan] was a nonselective inhibitor of all of the cloned transporters and shared many structural features with CI-966. With this in mind, a number of triarylnipecotic acid derivatives were synthesized mimicking the structural features of EGYT-3886 and CI-966. From this series of compounds, (*S*)-1-[2-[tris(4-methoxyphenyl)methoxy]ethyl]-3-piperidinecarboxylic acid ((S)-SNAP-5114) (Fig. 3) was identified as a novel ligand with selectivity for GAT-3, exhibiting 4-, 40-, and 28-fold selectivity for GAT-3 vs GAT-2, GAT-1, and BGT-1, respectively (Table 1). Similarly, two novel nipecotic acid derivatives, 1-(3-(9*H*-carbazol-9-yl)-1-propyl)-4-(4-methoxyphenyl)-4-piperidinol (NNC 05-2045) and 1-(3-(9*H*-carbazol-9-yl)-1-propyl)-4-(2-methoxyphenyl)-4-piperidinol (NNC 05-2090) (Fig. 3), displayed mild selectivity for mGAT-2 (BGT-1) and mGAT-4 (GAT-3) (Table 1) (THOMSEN et al. 1997). NNC 05–2090 was 14-, 30- and 11-fold more selective for BGT-1 vs GAT-1, GAT-2, and GAT-3, respectively, and proved to be the most selective BGT-1 transport inhibitor reported to date. In rodent models of seizure, both NNC compounds were found to have dose-dependent anticonvulsant effects that differ from those observed for inhibitors of GAT-1 (DALBY et al. 1997). Inhibition of GAT-3 was the likely mechanism of action in the observed anti-epileptic effects of these two nipecotic acid derivatives. It must be noted, however, that both of these compounds exhibited nmol/l to μmol/l affinities for α_1 adrenergic receptors and D_2 dopamine receptors, and NNC 05–2045 displayed nmol/l affinity for sigma receptors (DALBY et al. 1997). Although these authors ruled out dopaminergic and sigma receptor mechanisms as responsible for anticonvulsant effects of NNC-2045, given that adrenoreceptor agonism has been shown to reduce seizure severity (BROWNING 1987; LAIRD and JOBE 1987; MCNAMARA et al. 1987), an α_1 receptor mechanism could not be excluded. Therefore, it is possible that the observed anticonvulsant effects were partly mediated through adrenergic agonism. With the availability of (S)-SNAP-5114, NNC 05-2045, and NNC 05-2090, the individual contributions made by each GABA transporter to GABAergic neurotransmission in the CNS are now beginning to be differentiated. The promising anticonvulsant properties of these agents verify that a number of GABA transporter subtypes may prove to be extremely useful therapeutic targets. Certainly, the development of additional novel, potent, and selective non-GAT-1 transport inhibitors would greatly facilitate such studies.

G. GABA Uptake Inhibitors as Experimental Tools

I. GABA Transport Inhibition and Sleep

GABA uptake inhibitors have proven to be valuable tools in discerning the degree of involvement of GABAergic neurotransmission in various physiological states. For example, potentiation of GABA activity by benzodiazepines or sustained application of GABA into extracellular fluid was shown to promote sleep (SCHERSCHLICHT and PIERE 1988; JUHASZ et al. 1989). Localized and sustained perfusion of the GABA uptake inhibitor THPO into the thalamic relay nucleus in awake cats reduced wakefulness (JUHÁSZ et al. 1991). This latter study strongly supported a role for GABA in sleep by revealing that manipulation of endogenous GABA levels had sleep promoting effects similar to those observed upon application of exogenous GABA (JUHASZ et al. 1989). Separately, the GABA uptake inhibitor SK&F 89976A was used to examine the mechanism of action of modafinil (FERRARO et al. 1996), an agent used for the treatment of hypersomnia in narcoleptic patients (BASTUJI and JOUVET 1988). While the vigilance promoting effects of modafinil were attributed to an increase in dopamine release, this work tested the hypothesis that the effects on dopamine release were mediated through GABAergic mechanisms. Indeed, an increase in GABAergic tone secondary to administration of the uptake inhibitor SK&F 89976A, the $GABA_B$ receptor antagonist phaclofen, or the $GABA_A$ receptor agonist muscimol blocked the actions of modafinil on dopamine release (FERRARO et al. 1996). In contrast, blockade of GABAergic neurotransmission with the $GABA_A$ receptor antagonist bicuculline augmented the effect of modafinil on dopamine release. Thus, the vigilance promoting effects of modafinil stemmed from an inhibition of tonic GABA release that in turn disinhibited dopamine release.

More recently, tiagabine was used in an attempt to determine which actions of GABAergic compounds (benzodiazepines, agonist modulators of $GABA_A$ receptors, and $GABA_A$ receptor agonists) may govern specific sleep related changes (LANCEL et al. 1998). In contrast to benzodiazepines, tiagabine administration caused a marked dose-dependent enhancement of EEG power density in all frequency bands during non-rapid eye movement sleep (non-REMS), and had minimal effects on EEG activity during wakefulness and REMS. These data indicated that tiagabine promoted overall synchronization of EEG signals and was unlikely to enhance the amplitude and duration of inhibitory postsynaptic potentials. The effects of tiagabine were most similar to those of the $GABA_A$ receptor agonists muscimol and 4,5,6,7-tetrahydroisoxazolo(5,4-*c*)pyridin-3-ol (THIP), and any differences in their effects were attributed to the activation of $GABA_B$ receptors by excess synaptic GABA (Lancel et al. 1998). Because alterations in endogenous extracellular GABA concentrations result in physiologically relevant activations of both $GABA_A$ and $GABA_B$ receptors, GABA transport inhibitors have become essential tools in examining GABAergic components in sleep.

II. Depolarizing Effects of GABA and Inhibition of GABA Uptake

Although GABA typically elicits hyperpolarizations as a result of the inward flow of chloride ions through $GABA_A$ receptors, a number of groups have shown that GABA can have depolarizing effects in neurons *in vitro* (Hu and Davies 1997; Davies and Shakesby 1999; Phillips et al. 1998; Avoli and Perreault 1987; Cherubini et al. 1991; van den Pol et al. 1996). In particular, NNC-711 and tiagabine have been used to probe the excitatory effects of GABA in rodent brain. NNC-711 and tiagabine each elicited depolarizations in cortical slice preparations from DBA/2 and BALB/c mice (Hu and Davies 1997; Davies and Shakesby 1999) and NNC-711 potentiated GABA-induced depolarizations (Davies and Shakesby 1999). These depolarizations were calcium-dependent, blocked by tetrodotoxin, and inhibited by the $GABA_A$ receptor antagonist bicuculline. Since GABA, muscimol, and THIP each elicited depolarizations in the same cortical preparation, the depolarizations appeared to be mediated by $GABA_A$ receptors. Although the mechanism(s) underlying GABA-induced depolarizing events in these preparations remain unclear, evidence has accumulated to suggest that these responses are reliable in adult neuronal preparations. GABA uptake inhibitors are particularly useful in exploring these events as they permit the manipulation of endogenous GABA levels for examination of the physiological relevance of these phenomena.

H. Conclusion

GABA transporters are physiologically important proteins whose functions directly impact the inhibitory tone of the CNS. Disruption of GABAergic neurotransmission has been correlated with a number of neurological and psychiatric disorders. Therefore, it is reasonable to suggest that compounds which modulate GABAergic tone by altering GABA uptake may be useful in treating some of these clinical conditions. Indeed, tiagabine is a prime example of a selective and potent GABA transport inhibitor that is prescribed currently as an add-on therapy for refractory epilepsy. The availability of several potent and selective GAT-1 transport inhibitors has advanced our understanding of the contributions of GAT-1 to GABAergic transmission and its potential as a therapeutic target. However, due to the lack of diverse potent inhibitors selective for GAT-2, GAT-3, and BGT-1, our understanding of the contributions made by these transporters to GABAergic function lags well behind that of GAT-1. Development of compounds specific for these targets is essential for achieving a more complete understanding of GABA neurotransmission. These efforts may also uncover nuances within the GABAergic system which might be exploited in the treatment of epilepsy, schizophrenia, and affective disorders.

References

Albertson TE, Joy RM (1987) Increased inhibition in dentate gyrus granule cells following exposure to GABA-uptake blockers. Brain Res 435:283–292

Avoli M, Perreault P (1987) A GABAergic depolarizing potential in the hippocampus disclosed by the convulsant 4-aminopyridine. Brain Res 400:191–195

Bastuji H, Jouvet M (1988) Successful treatment of idiopathic hypersomnia and narcolepsy with modafinil. Prog Neuropsychopharmacol Biol Psychiatry 12:695–700

Benes FM, Vincent SL, Alsterberg G, Bird ED, SanGiovanni JP (1992) Increased $GABA_A$ receptor binding in superficial layers of cingulate cortex in schizophrenics. J Neurosci 12:924–929

Bernstein EM, Quick MW (1999) Regulation of γ-aminobutyric acid (GABA) transporters by extracellular GABA. J Biol Chem 274:889–895

Bjorge S, Black A, Bockbrader H, Chang T, Gregor VE, Lobbestael SJ, Nugiel D, Pavia MR, Radulovic L, Woolf T (1990) Synthesis and metabolic profile of CI-966: A potent, orally-active inhibitor of GABA uptake. Drug Develop Res 21:189–193

Bonina FP, Arenare L, Palagiano F, Saija A, Nava F, Trombetta D, de Caprariis P (1999) Synthesis, stability, and pharmacological evaluation of nipecotic acid prodrugs. J Pharm Sci 88:561–567

Borden LA (1996) GABA transporter heterogeneity: Pharmacology and cellular localization. Neurochem Int 29:335–356

Borden LA, Murali Dhar TG, Smith KE, Weinshank RL, Branchek TA, Gluchowski C (1994a) Tiagabine, SK&F 89976-A, CI-966, and NNC-711 are selective for the cloned GABA transporter GAT-1. Eur J Pharmacol 269:219–224

Borden LA, Dhar TG, Smith KE, Branchek TA, Gluchowski C, Weinshank RL (1994b) Cloning of the human homologue of the GABA transporter GAT-3 and identification of a novel inhibitor with selectivity for this site. Recep Chan 2:207–213

Borden LA, Smith KE, Hartig PR, Branchek TA, Weinshank RL (1992) Molecular heterogeneity of the γ-aminobutyric acid (GABA) transport system. Cloning of two novel high affinity GABA transporters from rat brain. J Biol Chem 267: 21098–21104

Bowery NG, Jones GP, Neal MJ (1976) Selective inhibition of neuronal GABA uptake by *cis*-3-aminocyclohexane carboxylic acid. Nature 264:281–284

Braestrup C, Nielsen EB, Sonnewald U, Knutsen LJS, Andersen KE, Jansen JA, Frederiksen K, Andersen PH, Mortensen A, Suzdak PD (1990) (R)-*N*-[4,4-bis(3-methly-2-thienyl)but-3-en-1-y]nipecotic acid binds with high affinity to the brain γ-aminobutyric acid uptake carrier. J Neurochem 54:639–647

Braestrup C, Nielsen EB, Wolffbrandt KH, Andersen KE, Knutsen LJS, Sonnewald U (1987) Modulation of GABA receptor interaction with GABA uptake inhibitors. In: Rand MJ, Raper C (eds) Pharmacology. Elsevier, New York, pp 125–128

Braestrup C, Nielsen M (1982) Neurotransmitters and CNS disease: Anxiety. Lancet 2:1030–1034

Brecha NC, Weigmann C (1994) Expression of GAT-1, a high-affinity gamma-aminobutyric acid plasma membrane transporter in the rat retina. J Comp Neurol 345:602–611
Browning,RA (1987) The role of neurotransmitters in electroshock seizure models. In: Jobe PC, Laird HE II (eds) Neurotransmitters and epilepsy. Humana Press, New Jersey, pp 115–118
Cherubini E, Gaiarsa JL, Ben-Ari Y (1991) GABA: an excitatory transmitter in early postnatal life. Trends Pharmacol Sci 14:515–519
Clark JA, Amara SG (1994) Stable expression of a neuronal gamma-aminobutyric acid transporter, GAT-3, in mammalian cells demonstrates unique pharmacological properties and ion dependence. Mol Pharmacol 46:550–557
Clark JA, Deutch AY, Gallipoli PZ, Amara SG (1992) Functional expression and CNS distribution of a β-alanine-sensitive neuronal GABA transporter. Neuron 9: 337–348
Croucher MJ, Meldrum, BS, Krogsgaard-Larsen P (1983) Anticonvulsant activity of GABA uptake inhibitors and their prodrugs following central or systemic administration. Eur J Pharmacol 89:217–228
Cummins CJ, Glover RA, Sellinger OZ (1982) Beta-alanine uptake is not a marker for brain astroglia in culture. Brain Res 239:299–302
Dalby NO, Thomsen C, Fink-Jensen A, Lundbeck J, SØkilde B, Man CM, SØrensen PO, Meldrum B (1997) Anticonvulsant properties of two GABA uptake inhibitors NNC 05-2045 and NNC 05-2090, not acting preferentially on GAT-1. Epilepsy Res 28:51–61
Davies JA, Shakesby A (1999) Blockade of GABA uptake potentiates GABA-induced depolarizations in adult mouse cortical slices. Neurosci Lett 266:201–204
Dhar TGM, Borden LA, Tyagarajan S, Smith KE, Branchek TA, Weinshank RL, Gluchowski C (1994) Design, synthesis and evaluation of substituted triarylnipecotic acid derivatives as GABA uptake inhibitors: Identification of a ligand with moderate affinity and selectivity for the cloned human GABA transporter GAT-3. J Med Chem 37:2334–2342
Dingledine R, Korn SJ (1985) γ-Aminobutyric acid uptake and the termination of inhibitory synaptic potentials in the rat hippocampal slice. J Physiol 366:387–409
During MJ, Ryder KM, Spencer DD (1995) Hippocampal GABA transporter function in temporal-lobe epilepsy. Nature 376:174–177
Durkin MM, Smith KE, Borden LA, Weinshank RL, Branchek TA, Gustafson EL (1995) Localization of messenger RNAs encoding three GABA transporters in rat brain: an in situ hybridization study. Mol Brain Res 33:7–21
Ebert U, Krnjevic K (1990) Systemic CI-966, a new γ-aminobutyric acid uptake blocker, enhances γ-aminobutyric acid action in CA1 pyramidal layer in situ. Can J Physiol 68:1194–1199
Enna SJ, Bennett Jr. JP, Bylund DB, Snyder SH, Bird ED, Iversen LL (1976) Alterations of brain neurotransmitter receptor binding in Huntington's chorea. Brain Res 116:531–537
Falch E, Krogsgaard-Larsen P (1981) Esters of isoguvacine as potential prodrugs. J Med Chem 24:285–289
Ferraro L, Tanganelli S, O'Connor WT, Antonelli T, Rambert F, Fuxe K (1996) The vigilance promoting drug modafinil increases dopamine release in the rat nucleus accumbens via the involvement of a local GABAergic mechanism. Eur J Pharmacol 306:33–39
Fjalland B (1978) Inhibition by neuroleptics of uptake of 3H-GABA into rat brain synaptosomes. Acta Pharmacol Toxicol (Copenh) 42:73–76
Frey H-H, Popp C, Löscher W (1979) Influence of inhibitors of the high affinity GABA uptake on seizure thresholds in mice. Neuropharmacol 18:581–590
Gaspary HL, Wang W, Richerson GB (1998) Carrier-mediated GABA release activates GABA receptors on hippocampal neurons. J Neurophys 80:270–281

Gavrilovic J, Raff M, Cohen J (1984) GABA uptake by purified rat Schwann cells in culture. Brain Res 303:183–185

Guastella J, Nelson N, Nelson H, Czyzyk L, Keynan S, Miedel MC, Davidson N, Lester H, Kanner B (1990) Cloning and expression of a rat brain GABA transporter. Science 249:1303–1306

Hamberger A, Nyström B, Larsson S, Silfvenius H, Nordborg C (1991) Amino acids in the neuronal microenvironment of focal human epileptic lesions. Epilepsy Res 9:32–43

Horton RW, Collins JF, Anlezark GM, Meldrum BS (1979) Convulsant and anticonvulsant actions in DBA/2 mice of compounds blocking the reuptake of GABA. Eur J Pharmacol 59:76–83

Hu RQ, Davies JA (1997) Tiagabine hydrochloride, an inhibitor of gamma-aminobutyric acid (GABA) uptake, induces cortical depolarizations in vitro. Brain Res 753:260–268

Isaacson JS, Solís JM, Nicoll RA (1993) Local and diffuse synaptic actions of GABA in the hippocampus. Neuron 10:165–175

Iversen LL, Kelly JS (1975) Uptake and metabolism of γ-aminobutyric acid by neurones and glial cells. Biochem Pharmacol 24:933–938

Iversen LL, Neal MJ (1968) The uptake of [^{3}H]GABA by slices of rat cerebral cortex. J Neurochem 15:1141–1149

Juhasz G, Emri Z, Kekesi K, Pungor K (1989) Local perfusion of the thalamus with GABA increases sleep and induces long-lasting inhibition of somatosensory event-related potentials in cats. Neurosci Lett 103:229–233

Juhász G, Kékesi KA, Emri Z, Ujszászi J, Krogsgaard-Larsen P, Schousboe A (1991) Sleep promoting effect of a putative glial γ-aminobutyric acid uptake blocker applied in the thalamus of cats. Eur J Pharmacol 209:131–133

Krogsgaard-Larsen P (1980) Inhibitors of the GABA uptake systems. Mol Cell Biochem 31:105–121

Krogsgaard-Larsen P, Bundgaard H (1991) Synaptic mechanisms as pharmacological targets. 11.3 The GABA neurotransmitter systems: Pathophysiological aspects. In: Krogsgaard-Larsen P, Bundgaard H (eds) A textbook of drug design and development. Harwood Academic Publishers, Copenhagen, pp 391–433

Krogsgaard-Larsen P, Johnston GA (1975) Inhibition of GABA uptake in rat brain slices by nipecotic acid, various isoxazoles and related compounds. J Neurochem 25:797–802

Laird HE II, Jobe PC (1987) The genetically epilepsy prone rat. In: Jobe PC, Laird HE II (eds) Neurotransmitters and epilepsy. Humana Press, New Jersey, pp 57–94

Lancel M, Faulhaber J, Deisz RA (1998) Effect of the GABA uptake inhibitor tiagabine on sleep and EEG power spectra in the rat. Br J Pharmacol 123:1471–1477

Larsson OM, Johnston GAR, Schousboe A (1983) Differences in uptake kinetics of *cis*-3-aminocyclohexane carboxylic acid into neurons and astrocytes in primary cultures. Brain Res 260:279–285

Leach JP, Brodie MJ (1998) Tiagabine. The Lancet 351:203–207

Levi G, Wilkin GP, Ciotti MT, Johnstone S (1983) Enrichment of differentiated, stellate astrocytes in cerebellar interneuron cultures as studied by GFAP immunofluorescence and autoradiographic uptake patterns with [^{3}H]D-aspartate and [^{3}H]GABA. Dev Brain Res 10:227–241

Liu Q-R, López-Corcuera B, Mandiyan S, Nelson H, Nelson N (1993) Molecular characterization of four pharmacologically distinct α-aminobutyric acid transporters in mouse brain. J Biol Chem 268:2106–2112

Lloyd KG, Drekaler S, Bird ED (1977) Alterations in ^{3}H-GABA binding in Huntington's Chorea. Life Sci 21:747–754

Loscher W (1982) Comparative assay of anticonvulsant and toxic potencies of sixteen GABAmimetic drugs. Neuropharmacol 21:803–810

Loscher W, Schwartz-Porsche D (1986) Low levels of γ-aminobutyric acid in cerebrospinal fluid of dogs with epilepsy. J Neurochem 46:1322–1325

McNamara JO, Bonhaus DW, Crain BJ, Gellman RL, Shin C (1987) Biochemical and pharmacological studies of neurotransmitters in the kindling model. In: Jobe PC, Laird HE II (eds) Neurotransmitters and epilepsy. Humana Press, New Jersey, pp 119–160

Meldrum BS (1975) Epilepsy and γ-aminobutyric acid-mediated inhibition. Int Rev Neurobiol 17:1–36

Meldrum BS (1995) Neurotransmission in epilepsy. Epilepsia 36:S30–S35

Nielsen EB, Suzdak PD, Andersen KE, Knutsen LJS, Sonnewald U, Braestrup C (1991) Characterization of tiagabine (NO-328), a new potent and selective GABA uptake inhibitor. Eur J Pharmacol 196:257–266

Perry TL, Hansen S, Kloster M (1973) Huntington's chorea: Deficiency of γ-aminobutyric acid in brain. New England J Med 288:337–342

Petroff OA, Rothman DL, Behar KL, Mattson RH (1996) Low brain GABA level is associated with poor seizure control. Ann Neurol 40:908–911

Phillips I, Martin KF, Thompson KSJ, Heal DJ (1998) GABA-evoked depolarisations in the rat cortical wedge: involvement of $GABA_A$ receptors and HCO_3^- ions. Brain Res 798:330–332

Rattray M, Priestley JV (1993) Differential expression of GABA transporter-1 messenger RNA in subpopulations of GABA neurones. Neurosci Lett 156: 163–166

Reynolds GP, Czudek C, Andrews HB (1990) Deficit and hemispheric asymmetry of GABA uptake sites in the hippocampus in Schizophrenia. Biol Psychiatry 27: 1038–1044

Reynolds R, Herschowitz N (1986) Selective uptake of neuroactive amino acids by both oligodendrocytes and astrocytes in primary dissociated cell culture: a possible role for oligodendrocytes in neurotransmitter metabolism. Brain Res 371:253–266

Ribak CE, Harris AB, Vaughn JE, Roberts E (1979) Inhibitory, GABAergic nerve terminals decrease at sites of focal epilepsy. Science 205:211–214

Richards DA, Bowery NG (1996) Comparative effects of the GABA uptake inhibitors, tiagabine and NNC-711, on extracellular GABA levels in the rat ventrolateral thalamus. Neurochem Res 21:135–140

Sayin U, Timmerman W, Westerink BH (1995) The significance of extracellular GABA in the substantia nigra of the rat during seizures and anticonvulsant treatments. Brain Res 669:67–72

Scherschlicht A, Piere L (1988) Pharmacology of midazolam and other short acting benzodiazepines. In: Koella WR, Obal F, Schultz H, Visser P (eds) Sleep 86. Fischer, Stuttgart, pp 101ff

Schon F, Kelly JS (1974) The characterisation of [^{3}H]GABA uptake into the satellite glial cells of rat sensory ganglia. Brain Res 66:289–300

Schon F, Kelly JS (1975) Selective uptake of [^{3}H]β-Alanine by glia: Association with the glial uptake system for GABA. Brain Res 86:243–257

Schwartz EA (1987) Depolarization without calcium can release γ-aminobutyric acid from a retinal neuron. Science 238:350–355

Simpson MDC, Slater P, Deakin JFW, Royston MC, Skan WJ (1989) Reduced GABA uptake sites in the temporal lobe in schizophrenia. Neurosci Lett 107:211–215

Simpson MDC, Slater P, Royston C, Deakin JFW (1992) Regionally selective deficits in uptake sites for glutamate and gamma-aminobutyric acid in the basal ganglia in schizophrenia. Psychiatry Res 42:273–282

Smith KE, Borden LA, Wang C-HD, Hartig PR, Branchek TA, Weinshank RL (1992) Cloning and expression of a high affinity taurine transporter from rat brain. Mol Pharmacol 42:563–569

Solis JM, Nicoll RA (1992) Postsynaptic action of endogenous GABA released by nipecotic acid in the hippocampus. Neurosci Lett 147:16–20

Spokes EG (1980) Neurochemical alterations in Huntington's chorea: a study of post-mortem brain tissue. Brain 103:179–210

Suzdak PD, Frederiksen K, Andersen KE, Sørensen PO, Knutsen LJS, Nielsen EB (1992) NNC-711, a novel potent and selective γ-aminobutyric acid uptake inhibitor: pharmacological characterization. Eur J Pharmacol 223:189–198

Swinyard EA, White HS, Wolf HH, Bondinell WE (1991) Anticonvulsant profiles of the potent and orally active GABA uptake inhibitors SK&F 89976-A and SK&F 100330-A and four prototype antiepileptic drugs in mice and rats. Epilepsia 32:569–577

Taylor CP, Sedman AJ (1991) Pharmacology of the gamma-aminobutyric acid-uptake inhibitor CI-966 and its metabolites: Preclinical and clinical studies. In: Barnard E, Costa E (eds) Transmitter amino acid receptors: structures, transduction and models for drug development. Thieme Med. Pub. Inc., New York, pp 251–270

Thomsen C, Sørensen PO, Egebjerg J (1997) 1-(3-(9H-Carbazol-9-yl)-1-propyl)-4-(2-methoxyphenyl)-4-piperidinol, a novel subtype selective inhibitor of the mouse type II GABA-transporter. Br J Pharmacol 120:983–985

van den Pol AN, Obrietan K, Chen G (1996) Excitatory actions of GABA after neuronal trauma. J Neurosci 16:4283–4292

Waldmeier PC, Stöcklin K, Feldtrauer J-J (1992) Weak effects of local and systemic administration of the GABA uptake inhibitor, SK&F89976, on extracellular GABA in the rat striatum. Naunyn-Schmiedeberg's Arch Pharmacol 345:544–547

Yamauchi A, Uchida S, Kwon HM, Preston AS, Robey RB, Garcia-Perez A, Burg AB, Handler JS (1992) Cloning of a Na^+- and Cl^--dependent betaine transporter that is regulated by hypertonicity. J Biol Chem 267:649–652

Yunger LM, Fowler PJ, Zarevics P, Setler PE (1984) Novel inhibitors of γ-aminobutyric acid (GABA) uptake: Anticonvulsant actions in rats and mice. J Pharmacol Exp Ther 228:109–115

Section III
Pharmacology of the Glycine System

CHAPTER 16

Structures, Diversity and Pharmacology of Glycine Receptors and Transporters

H. BETZ, R.J. HARVEY and P. SCHLOSS

A. Introduction

I. The Neurotransmitter Glycine

The amino acid glycine is highly concentrated in the ventral and dorsal horns of the spinal cord, in many brain stem nuclei, and in sensory relay stations such as the cochlear nucleus and the retina. Traditional physiological studies have shown that glycine is a major inhibitory neurotransmitter that performs a vital role in the control of both motor and sensory pathways (APRISON 1990). In presynaptic nerve terminals of glycinergic interneurons in spinal cord and brain stem, cytosolic glycine is concentrated in small clear synaptic vesicles by an H^+-dependent vesicular glycine transporter. Excitation of these interneurons leads to the calcium-triggered fusion of these synaptic vesicles with the presynaptic plasma membrane, thus liberating glycine into the synaptic cleft. Glycine then binds to postsynaptic glycine receptors (GlyRs), causing gating of an integral anion channel that increases the chloride ion conductance of the plasma membrane. This postsynaptic action of glycine is selectively antagonized by the plant alkaloid strychnine. In mature neurons, where the chloride equilibrium potential approximates the resting potential, GlyR activation results in chloride ion influx. This neutralizes depolarization by sodium ion influx, thereby inhibiting the propagation of action potentials. However, a different response is found in the developing nervous system, where immature neurons contain very high intracellular chloride concentrations (WANG et al. 1994). Here, glycine-induced increases in chloride conductance cause Cl^- efflux, resulting in depolarization of the postsynaptic cell (see REICHLING et al. 1994; BOEHM et al. 1997). The presynaptic neurotransmitter pool of glycine is replenished by (i) synthesis from metabolic precursors by the enzyme serine hydroxymethyl-transferase and (ii) re-uptake from the synaptic cleft by presynaptic sodium-dependent glycine transporters. Genetic disruption of glycinergic neurotransmission in hereditary neuromotor disorder results in complex neurological phenotypes characterized by hypertonia and an exaggerated startle reflex (SHIANG et al. 1993; BUCKWALTER et al. 1994; KINGSMORE et al. 1994; RYAN et al. 1994, SAUL et al. 1994; FENG et al. 1998).

In addition to its role as an inhibition neurotransmitter substance, glycine also serves as a co-agonist of glutamate at excitatory synapses which contain the NMDA-subtype of glutamate receptors (JOHNSON and ASCHER 1987). The properties of the receptors mediating this co-transmitter function of glycine are discussed in the review by HOLLMANN in the *Handbook of Experimental Pharmacology*, vol. 141 (HOLLMANN 1999).

B. Structure and Diversity of Glycine Transporters

I. Structure of Plasma Membrane Glycine Transporters

Rapid re-uptake into the presynaptic terminal or surrounding glial cells via specific Na^+- and Cl^--dependent neurotransmitter transporters is the principal means of terminating the action of most neurotransmitters. Once in the cytosol, transmitters can be reloaded into synaptic vesicles via vesicular transport systems that are different from the neurotransmitter transporters in the plasma membrane (see below). The cloning of the Na^+-dependent rat γ-aminobutyric acid (GABA) and human norepinephrine transporters revealed that these polypeptides display a remarkable amino acid identity (GUASTELLA et al. 1990; PACHOLCZYK et al. 1991). Subsequent homology screening led to the isolation of additional highly related cDNAs encoding transporters for other neurotransmitters, such as glycine, dopamine, and serotonin (for reviews see CLARK and AMARA 1993; SCHLOSS et al. 1994; WORRAL and WILLIAMS 1994). All members of the Na^+- and Cl^--dependent neurotransmitter transporter family share a common, putative twelve transmembrane domain (TMD) structure with extended cytoplasmic N- and C-terminal regions. The latter contain putative phosphorylation sites that could be used for the regulation of transport activity. In addition, a long putative extracellular loop containing N-glycosylation sites is conserved between the third and fourth membrane-spanning domains (TMD3 and TMD4). However, while there is a common model for all transporters distal from the third TMD, there is still controversy over the topographical arrangement of the first three TMDs. The original model (Fig. 1A) which (based on hydropathy analysis) had been proposed for GAT1 (GUASTELLA et al. 1990) was later adopted for all other members of the neurotransmitter transporter family. This topology was confirmed for the human norepinephrine transporter by immunofluorescence studies with anti-peptide antibodies (BRÜSS et al. 1995), and for the serotonin and GABA transporters by analyzing the accessibility of cysteine residues (CHEN et al. 1997) and epitope tagging of COOH-terminal truncations (CLARK 1997), respectively. However, when the topology of the glycine transporter GLYT1 was investigated by introducing N-glycosylation sites along the polypeptide sequence followed by examining their use in an *in vitro* transcription/translation assay, a new arrangement of the first third of this protein was suggested (OLIVARES et al. 1997). Accordingly (Fig. 1B), TMD1 does not span the membrane, and thus the loop connecting TMD2 and TMD3 which was formerly believed to be intracellular is located extracellularly. TMD3 is thought to be

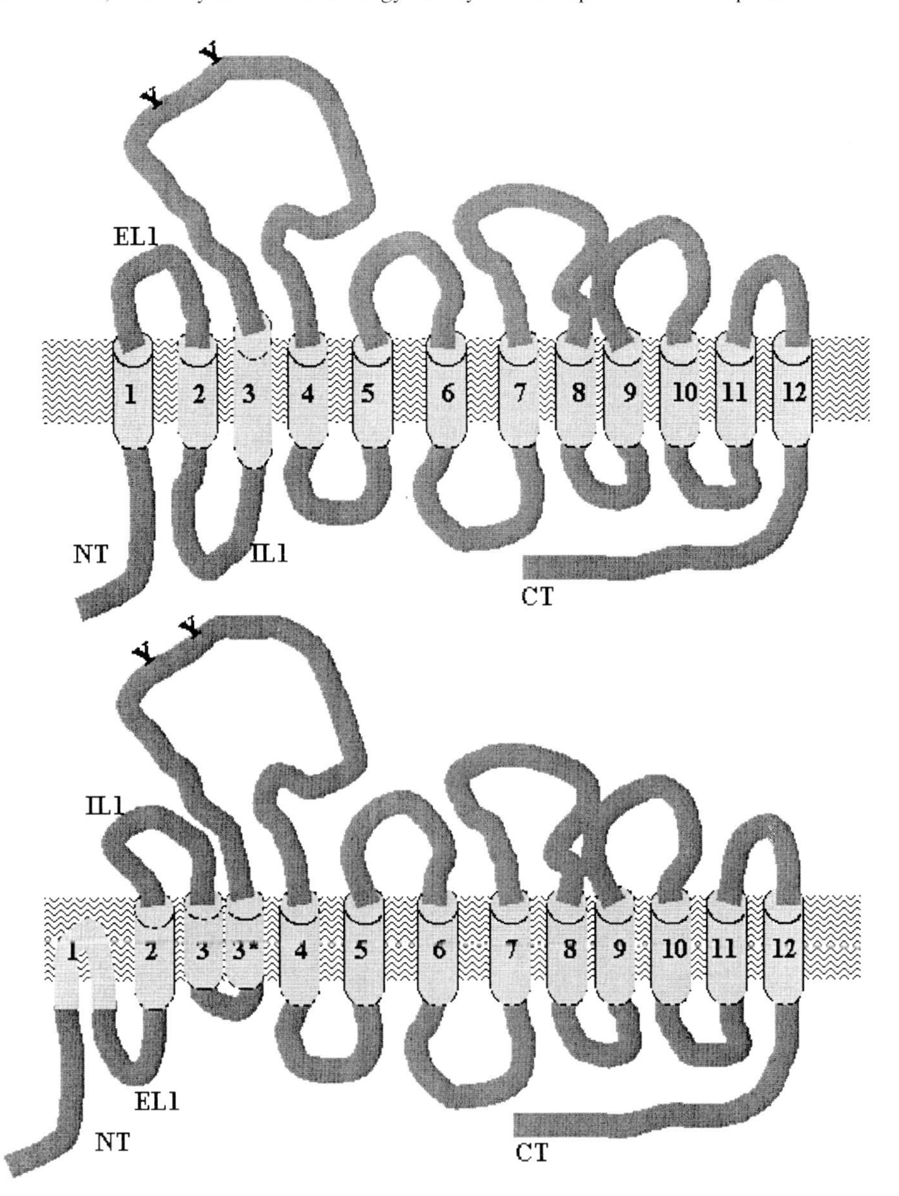

Fig. 1A,B. Possible topological arrangements of GLYT1. **A** In the originally proposed topology model based on hydropathy analysis, the N- and C-terminal (*NT* and *CT*) regions are localized in the cytoplasm, and the polypeptide traverses the membrane twelve times as indicated. **B** In the alternative model by OLIVARES et al. (1997) the first transmembrane domain does not fully cross the membrane, and thus the first extracellular loop (*EL1*) in model **A** is now located intracellularly, and the region connecting TMD2 and TMD3 (*IL1*) extracellularly. TMD 3 is thought to be large enough to span the membrane twice (*3*/*3**), which results in the correct extracellular location of the glycosylated (*Y*) loop between TMD3 and TMD4

large enough to span the membrane twice, which results in the correct extracellular location of the glycosylated loop between TMD3 and TMD4. Simultaneously, Bennett and Kanner (1997) predicted a similar model for GAT1. Additional experimental work is clearly necessary to establish firmly the transmembrane organization of these proteins.

II. Diversity and Regulation of Plasma Membrane Glycine Transporters

Whereas only one transporter type has been found for each of the biogenic amines norepinephrine, dopamine and serotonin, two different murine and human transporter genes (GLYT1 and GLYT2) have been identified for glycine (Guastella et al. 1992; Liu et al. 1992, 1993; Smith et al. 1992; Kim et al. 1994; Morrow et al. 1998). The GLYT1 gene encompasses 16 exons and generates three different glycine transporter isoforms (Borowsky et al. 1993; Kim et al. 1994; Adams et al. 1995). The GLYT1a and 1b/1c mRNAs are transcribed from two different promoters; the GLYT1c variant is produced by alternative splicing and encodes a protein with an extra N-terminal 54 amino acids relative to the isoform synthesized from the GLYT1b transcript (Adams et al. 1995). Thus, with the exception of their N-termini, the three isoforms are nearly identical, and heterologous expression of the corresponding cDNAs confers similar uptake properties. The GLYT2 gene also produces two mRNA variants by alternative use of exons 2a or 2b, respectively (Ponce et al. 1998). However, whereas the recombinant GLYT2a protein actively accumulates glycine into transfected cells, GLYT2b, which is only five amino acids longer than GLYT2a, seems to function only as a glycine exchanger in the heterologous expression system (Ponce et al. 1998).

Functional analysis of GLYT1b and GLYT2a stably expressed in HEK-293 cells was performed to compare the kinetics, pharmacological profiles, ion dependence, and electrical properties of the two isoforms (Lopez-Corcuera et al. 1998). GLYT1b exhibits a lower affinity for glycine (K_m: 447 μmol/l) than GLYT2b (K_m: 220 μmol/l), but both transporters translocate glycine with a stoichiometry of at least two sodium ions and one chloride ion per glycine molecule. Electrophysiological recordings using the whole cell patch-clamp technique revealed transport associated currents of ~16 pA and ~9 pA for GLYT1b and GLYT2a, respectively. Glycine transport by GLYT1b, but not GLYT2a, is sensitive to sarcosine (Fig. 2), which serves as a substrate for GLYT1b and evokes currents similar as glycine at this transporter isoform. Studies on the regulation of transport activity have, to date, only been performed for GLYT1. Treatment of GLYT1 in glioma cells or HEK-293 cells expressing GLYT1 with the protein kinase C (PKC) activator phorbol 12-myristate 13-acetate (PMA) decreased specific glycine accumulation (Gomeza et al. 1995; Sato et al. 1995b). This down-regulation resulted from a reduction of the maximal transport rate and was blocked by the PKC inhibitors 1-(5-isoquinolinsulfonyl)-2-methylpiperazine (H7) and staurosporine. Interest-

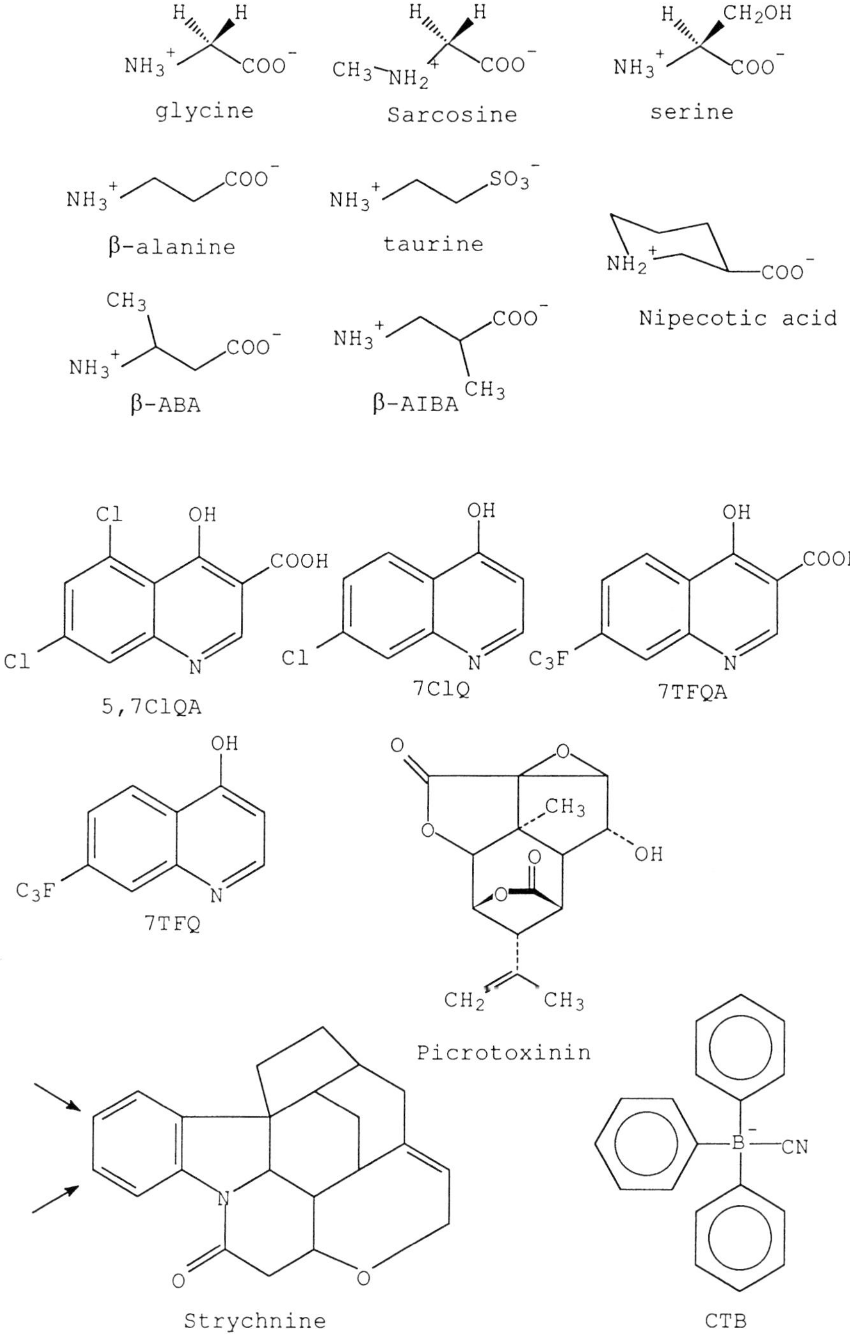

Fig. 2. Chemical structures of compounds that are useful pharmacological tools in the study of glycine receptors and transporters. *Upper panel*: the α-amino acids glycine, sarcosine, and serine and the β-amino acids β-alanine, taurine, β-aminobutyric acid (β-ABA), and β-aminoisobutyric acid (β-AIBA). *Middle panel*: the piperidine derivative nipecotic acid, and the quinolinic acid-based substances 5,7-dichloro-4-hydroxyquinoline-3-carboxylic acid (5,7ClQA), 7-chloro-4-hydroxyquinoline (7ClQ), 7-trifluoromethyl-4-hydroxyquinoline-3-carboxylic acid (7TFQA), and 7-trifluoromethyl-4-hydroxyquinoline (7TFQ). *Bottom panel*: the GlyR antagonists strychnine, picrotoxinin, and cyanotriphenylborate. Note that the aromatic ring positions indicated by *arrows* on the strychnine molecule can be replaced without affecting toxicity; this property has been utilized in affinity purification of the GlyR and the synthesis of fluorescent derivatives of strychnine

ingly, the inhibitory effect of PMA treatment was also observed after removing all predicted phosphorylation sites for PKC in GLYT1 by site-directed mutagenesis, suggesting that regulation via PKC may involve indirect phosphorylation mechanisms (SATO et al. 1995b). A PKC-induced down-regulation of transporter activity has also been reported for the GABA, dopamine and the serotonin transporters (SATO et al. 1995a; KITAYAMA et al. 1994; OSAWA et al. 1994; QUIAN et al. 1997; SAKAI et al. 1997). For the latter, it was shown that PMA-induced reduction of the maximal transport rate was due to a subcellular redistribution of transporter proteins from the plasma membrane to the trans-Golgi network (QUIAN et al. 1997). As for GLYT1, removal of all putative PKC phosphorylation sites did not abolish the phorbol ester mediated effect (SAKAI et al. 1997). It therefore appears that a general mechanism exists by which activation of presynaptic second messenger systems regulates the concentration of active neurotransmitters in the synaptic cleft by stimulating the internalization of cell surface neurotransmitter transporters.

III. Distribution of Plasma Membrane Glycine Transporters and Possible Physiological Function

In the last few years, several laboratories have analyzed the expression patterns of GLYT1 and GLYT2 in the CNS (GUASTELLA et al. 1992; SMITH et al. 1992; BOROWSKY et al. 1993; ADAMS et al. 1995; JURSKY and NELSON 1995; ZAFRA et al. 1995a,b). *In situ* hybridization and immunocytochemical techniques have shown that GLYT1 is widely expressed in the spinal cord, brainstem, and cerebellum, and to a lesser extent in the cortex and hippocampus. As revealed by high-resolution autoradiography and immunoelectron microscopy, mainly glial cells around both glycinergic and non-glycinergic neurons synthesize GLYT1. In contrast, the highest expression of GLYT2 is found in the spinal cord and brainstem, but exclusively on axons and terminal boutons. The cellular localization of GLYT2 correlates well with the distribution of GlyRs, i.e., areas devoid of GlyRs do not express GLYT2. This suggests that GLYT2 forms the presynaptic uptake system responsible for terminating glycinergic neurotransmission. Around such glycinergic synapses, GLYT1 might also contribute to the regulation of extracellular glycine produced by glial cells. Based on other *in situ* hybridization studies (SMITH et al. 1992) it has been suggested that GLYT1 is located at non-glycinergic synapses, and might regulate *N*-methyl-D-aspartate receptor (NMDAR) function by controlling glycine concentrations at the NMDAR modulatory glycine site. This hypothesis is supported by the recent finding that exogenous glycine as well as GLYT1 antagonists selectively enhanced the amplitude of the NMDA component of a glutamatergic excitatory postsynaptic current of hippocampal pyramidal neurons (BERGERON et al. 1998). Moreover, it has been shown that glycine uptake by GLYT1 dramatically reduces NMDAR currents evoked in *Xenopus* oocytes co-expressing the recombinant GLYT1 and NMDAR (SUPPLISSON and BERGMAN 1997). The results of these experiments show that GLYT1 can indeed reduce glycine near the membrane to concentrations below 1 μmol/l (the saturating concentration for activation of NMDARs) provided the glycine concentration in the bath solution is similar to that of the CSF (1–10 μmol/l). These findings make GLYT1 a feasible target for therapeutic agents directed toward diseases related to hypofunction of NMDARs.

IV. The Vesicular Glycine/GABA Transporter

As discussed above, glycinergic neurotransmission depends on regulated exocytosis of glycine, which in turn requires the packaging of this amino acid into small synaptic vesicles. This process is mediated by a vesicular transporter that is driven by the pH gradient (ΔpH) across the vesicular membrane. The cloning of genes encoding two different vesicular transporters for biogenic amines (VMAT1; VMAT2) and one for acetylcholine (VAChT) has revealed a new gene family of H^+-dependent transporter proteins (reviewed in VAROQUI and ERICKSON 1997). The uptake of GABA and glycine into synaptic vesicles has been shown to equally depend on the electrical gradient (ΔΨ) and ΔpH (FYKSE and FONNUM 1988, 1996). Recently, using a genetic and pharmacological approach in *Caenorhabditis elegans*, cDNAs encoding vesicular GABA transporters (VGATs) from worm and rat were isolated (MACINTIRE et al. 1997). Hydropathy plots for the predicted polypeptides suggest the existence of ten TMDs, with the N- and C-termini being located in the cytosol. Interestingly, the primary structures of the VGATs exhibit no significant homology to the vesicular transporters for monoamines or acetylcholine. Hetorologous expression of the *C. elegans* protein and its mammalian counterpart revealed vesicular GABA transport with adequate bioenergetic dependence on ΔΨ and ΔpH. GABA transport was only weakly inhibited by glycine (IC_{50}: 25 mmol/l), and no significant uptake of [^{3}H]glycine was detectable. In the same year, a mouse cDNA almost identical to that encoding VGAT was identified by screening genome databases (SAGNÉ et al. 1997). Because heterologous expression of this gene conferred [^{3}H]GABA as well as [^{3}H]glycine uptake, and *in situ* hybridization revealed co-distribution with both GABAergic and glycinergic neuronal markers, this transporter was termed "vesicular inhibitory amino acid transporter" (VIAAT). A detailed immunocytochemical analysis using specific antibodies against the N- and C-terminal epitopes of VGAT has shown that the protein is highly concentrated in the nerve endings of both GABAergic and glycinergic neurons in rat brain and spinal cord (CHAUDHRY et al. 1998). Taken together, these data corroborate the idea that both inhibitory neurotransmitters share a common vesicular uptake mechanism. As a consequence, glycine and GABA could be accumulated and released from the same nerve terminal. Indeed, co-release of glycine and GABA from the same terminal has been demonstrated recently in spinal cord by electrophysiological methods (JONAS et al. 1998). Co-transport of GABA and glycine also appears useful in systems that switch during development from GABAergic to mainly glycinergic neurotransmission, such as the lateral superior olive (KOTAK et al. 1998).

C. Structure and Diversity of Glycine Receptor Channels

I. GlyRs are Ligand-Gated Ion Channels of the nAChR Superfamily

Glycine receptors were originally isolated from rodent spinal cord using aminostrychnine-agarose columns (Pfeiffer et al. 1982). Purified adult GlyRs consist of two N-glycosylated integral membrane proteins of 48 kDa (α) and 58 kDa (β) as well as an associated peripheral membrane protein of 93 kDa, which was named gephyrin. The sequences of GlyR α and β subunits were determined using cDNA cloning strategies (Grenningloh et al. 1987, 1990a), and show considerable sequence similarity to subunits of the nicotinic acetylcholine receptor (nAChR), γ-aminobutyric acid type A ($GABA_A$) receptor, and serotonin type 3 ($5HT_3$) receptor (Betz 1990). Alignments of the members of this ligand-gated ion channel superfamily also revealed that certain structural motifs are well conserved, such as a dicysteine motif in the large N-terminal extracellular domain and four hydrophobic membrane-spanning domains (M1–M4) (see Fig. 2). Crosslinking experiments have shown that GlyRs are pentameric proteins (Langosch et al. 1988). In adult spinal cord, GlyRs contain three α (α1; see below) and two β subunits (Langosch et al. 1988; Kuhse et al. 1993), whereas embryonic receptors appear to be homo-oligomers containing exclusively α2 subunits (Hoch et al. 1989). The pentameric structure of the GlyRs resembles that of nAChR and $GABA_A$ receptor proteins, which are thought to represent pentamers of related subunits (see Betz 1990; Nayeem et al. 1994).

To date, several amino acid residues and protein subdomains have been identified that influence the assembly of recombinant GlyRs in heterologous expression systems (see below). Substitution of N38 (Fig. 3) with alanine, a putative glycosylation site in the GlyR α1 subunit, abolishes glycine activated currents (Akagi et al. 1991b), suggesting that N-linked glycosylation may influence receptor assembly. Mutation of cysteine residues in the conserved dicysteine motif (C138 and C152) (Fig. 3) shared with $GABA_A$, $GABA_C$, nAChR, and $5HT_3$ receptors (Akagi et al. 1991b) also eliminates

Fig. 3. A schematic representation of the transmembrane topology and location of functionally important residues in the human GlyR α1 subunit. *Structural motifs*: regions involved in subunit processing and receptor assembly are indicated by *filled green circles*; conserved cysteine residues that are believed to form disulphide bridges are denoted by *filled black circles*. *Natural GlyR point mutations* (*filled yellow circles*): A52S is found in the GlyR α1 subunit gene in *spasmodic* mice; mutations I244N, P250T, Q266H, R271Q/L, K276E, and Y279C are found in the human GlyR α1 subunit gene in different hyperekplexia families. *Agonist and antagonist binding site determinants* (*filled blue circles*): in the GlyR α1 subunit residues G160, K200 and Y202 are involved in strychnine binding, the efficacy of taurine is influenced by residues I111 and A212, while F159, Y161 and T204 are important determinants of agonist affinity and specificity. S267 is a target for alcohol and volatile anaesthetics. *Channel function*: G254 is a major determinant of main-state conductances and CTB block. *Intracellular modification* (*filled grey circle*): S391 in the α1 subunit is a part of a potential phosphorylation site for protein kinase C

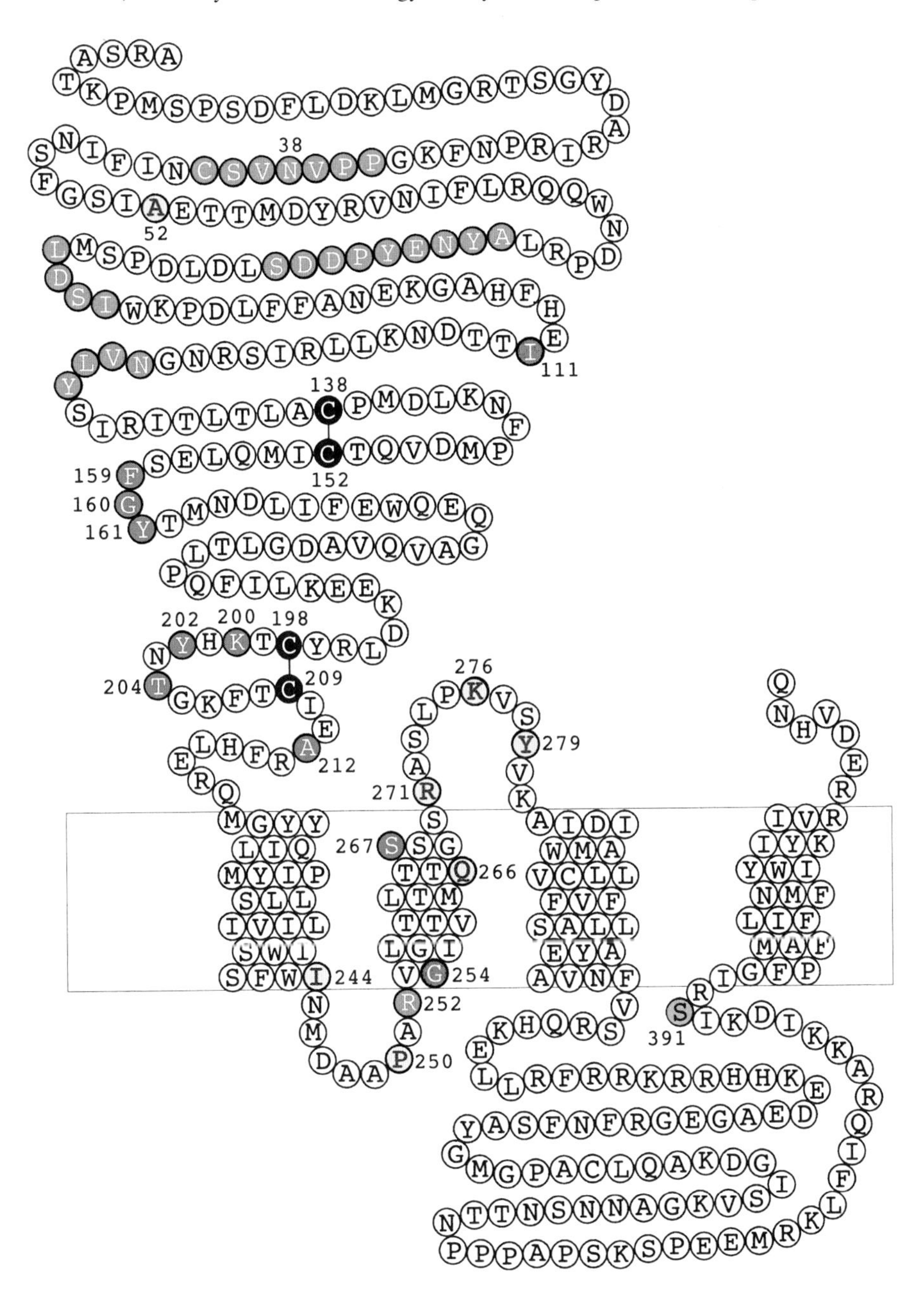
38
52
111
138
152
159
160
161
202
200
198
204
209
212
276
279
271
267
266
244
254
252
250
391

○391: serine

GlyR currents. Disruption of a second cysteine loop (C198–C209) (Fig. 3), that is only found in GlyRs and invertebrate glutamate gated chloride channels (CULLY et al. 1994), has similar effects (RAJENDRA et al. 1995b). These cysteine loops are generally considered to be crucial for stabilizing the assembly of ligand-gated ion channels. In the M2 domain of the human GlyR $\alpha 1$ subunit, substitution of R252 also prevents receptor assembly (LANGOSCH et al. 1993). Further information on the domains of GlyR α and β subunits that are important for subunit-subunit interactions (KUHSE et al. 1993) has been obtained from the analysis of chimeric subunits, which highlighted the importance of several discontinuous motifs in the large extracellular domain for the stoichiometric assembly of GlyR α and β subunits (Fig. 3). However, despite these studies, very little is known about the 3-D structure of GlyR subunits. Recently, parts of the human GlyR $\alpha 1$ subunit were modeled using a 1-D–3-D structure mapping approach based on a significant match with the biotin repressor (GREADY et al. 1997). This model predicts that the extracellular domain of the GlyR $\alpha 1$ subunit contains SH2-like (N-terminal) and SH3-like (membrane-proximal) domains, separated by a deep crevice which includes the dicysteine motif conserved in nAChR, $GABA_A$, $GABA_C$, and $5HT_3$ receptor subunits. This model, however, differs significantly from one generated for the agonist binding site of the homologous $\alpha 1$ subunit of the nAChR that is based on partial sequence similarities to copper binding proteins (TSIGELNY et al. 1997). Clearly, further structural information is required to allow for predictive modelling of the GlyR's agonist binding site.

II. Glycine Receptor Isoforms

The existence of multiple isoforms of the GlyR was first suggested by the disclosure of a neonatal rodent GlyR that exhibits a lower strychnine-binding affinity, an altered apparent molecular weight (49 kDa) of its α subunit and distinct immunological properties as compared to the adult receptor (BECKER et al. 1988). Molecular cloning studies confirmed that GlyRs are heterogeneous. At first, peptide sequences derived from purified adult spinal cord GlyRs enabled the isolation of cDNAs for the 48 kDa ($\alpha 1$) and 58 kDa (β) subunits (GRENNINGLOH et al. 1987, 1990a). Using homology screening approaches, further clones corresponding to two additional GlyR α subunits ($\alpha 2$ and $\alpha 3$) were isolated (GRENNINGLOH et al. 1990b; KUHSE et al. 1990b, 1991; AKAGI et al. 1991a). More recently, a fourth α subunit gene has been characterized (MATZENBACH et al. 1994). Localization of gene expression patterns using *in situ* hybridization techniques has revealed that the different α and β subunit genes exhibit spatially and temporally distinct patterns of expression in spinal cord, brain stem, higher brain regions and the retina (KUHSE et al. 1991; FUJITA et al. 1991; MALOSIO et al. 1991a,b; SATO et al. 1991, 1992; WATANABE and AKAGI 1995). For example, transcripts for the GlyR $\alpha 2$ subunit are abundant in the embryonic and neonatal brain and spinal cord while the $\alpha 1$ and $\alpha 3$ subunit genes appear to be expressed only postnatally. Surprisingly, the GlyR β subunit gene is very widely expressed, and β subunit transcripts are found in some adult brain regions (e.g., certain layers of the olfactory bulb and cerebellum) that are devoid of $\alpha1$, $\alpha2$ and $\alpha3$ subunit transcripts (FUJITA et al. 1991; MALOSIO et al. 1991a). While it is possible that additional GlyR α subunits remain to be identified, it is also conceivable that the β subunit forms part of another receptor complex.

Additional GlyR subunit diversity arises from alternative splicing of primary gene transcripts. Two forms of the rat (MALOSIO et al. 1991b), mouse (RYAN et al. 1994), and human (SHIANG et al. 1993) GlyR $\alpha1$ subunit ($\alpha1$ and $\alpha1_{ins}$) arise by the choice of one of two 3′ acceptor sites; the longer form contains an additional eight amino acids (SPMLNLFQ), including a serine residue that might serve as a target for phosphorylation by cAMP-dependent protein kinase (MALOSIO et al. 1991b). Alternative splicing of one of two homologous 68 bp exons in the rat, mouse and human $\alpha2$ subunit genes (KUHSE et al. 1991; MATZENBACH et al. 1994; CUMMINGS et al. 1998), which encodes part of the large extracellular domain, yields two further variants named α2A and α2B that differ in sequence by only two amino acids. The role of the α2A and α2B variants is presently unknown. A rat GlyR $\alpha2$ subunit variant ($\alpha2^*$) has also been described (KUHSE et al. 1990a) which contains a codon for glutamic acid (GAG) instead of the glycine codon (GGG) found at the equivalent position (residue 167 of the mature polypeptide) in other mouse (MATZENBACH et al. 1994), rat (AKAGI et al. 1991a; KUHSE et al. 1991) and human (GRENNINGLOH et al. 1990b; CUMMINGS et al. 1998) GlyR $\alpha2$ subunit cDNAs and/or genomic sequences. Recombinant GlyRs containing the $\alpha2^*$ subunit (see below) have a ~40-fold reduced glycine affinity and a ~560-fold reduced strychnine sensitivity (KUHSE et al. 1990a). Although it has been suggested (KUHSE et al. 1990a) that $\alpha2^*$-containing receptors represent the strychnine-insensitive neonatal GlyR isoform, sequencing of the mouse and human GlyR $\alpha2$ subunit genes (MATZENBACH et al. 1994; CUMMINGS et al. 1998) appears to contradict this proposal. In both species, only a single exon (exon 6) encoding this part of the protein exists, and this specifies a glycine codon (GGG) at position 167. In the case of the human subjects, DNAs from 40 individuals were sequenced. Hence, it is probable that the sequence exchange in the $\alpha2^*$ isoform is either (i) a rare allelic variant found only in the rat, or (ii) a result of a reverse transcription error during cDNA library construction. Alternative splicing of cassette exons also introduces extra peptide sequences into the large intracellular M3–M4 loops of the mouse GlyR β subunit (HECK et al. 1997) and the human GlyR $\alpha3$ subunit (NIKOLIC et al. 1998). The latter insertion (TEAFALEKFYRFSDM) is of particular interest as it influences the kinetics of GlyR desensitization.

III. Ligand-Binding Determinants

The first indications that the ligand-binding site resides on GlyR α subunits came from photoaffinity labeling experiments using the selective GlyR antagonist strychnine (GRAHAM et al. 1981, 1983). Protease mapping of

[^{3}H]strychnine-labeled rat GlyR preparations revealed that this antagonist was incorporated between amino acids 170 and 220 of the N-terminal domain of the GlyR α1 subunit (Graham et al. 1983; Ruiz-Gomez et al. 1990). Further information about ligand-binding site determinants has been collected from functional expression studies with recombinant GlyRs in both *Xenopus laevis* oocytes and mammalian cells (e.g., HEK-293). In these heterologous expression systems, GlyR α subunits assemble into homo-oligomeric chloride channels, which can be opened by micromolar concentrations of glycine, taurine, or β-alanine, and are blocked by nanomolar concentrations of strychnine (Schmieden et al. 1989; Sontheimer et al. 1989). In contrast, the GlyR β subunit is incapable of forming functional homomeric GlyRs (Pribilla et al. 1992; Bormann et al. 1993), but when included in heteromeric GlyRs alters crucial functional aspects of the ion channel. By comparing the properties of different GlyR α subunit variants (Kuhse et al. 1990a,b; Schmieden et al. 1989, 1992, 1993), it has become clear that several discontinuous domains of the α subunit extracellular domain are responsible for forming the ligand-binding pocket (Fig. 3). As stated above, residue G167 of the α2 polypeptide (equivalent to G160 in the α1 subunit) has been shown to define a crucial position for both agonist and antagonist (strychnine) binding (Kuhse et al. 1990a; Vandenberg et al. 1992). Subsequently, the two neighboring aromatic residues (F159 and Y161 in the α1 subunit) were found to account for agonist selectivity and antagonist efficacy (Schmieden et al. 1993). Two other residues in the GlyR α1 subunit, K200 and Y202, have also been identified as determinants of the strychnine binding site (Vandenberg et al. 1992), whereas residues I111 and A212 are important for the potency of the glycinergic agonists β-alanine and taurine (Schmieden et al. 1992).

Studies of spontaneous mutations in GlyR subunit genes (see also Chap. 12) have revealed additional residues involved in agonist binding (Fig. 3). For example, the mouse mutant *spasmodic* (Ryan et al. 1994; Saul et al. 1994) has a missense mutation in the GlyR α1 subunit gene. This results in an alanine to serine conversion at position 52, which in turn results in a ~6-fold reduction of glycine affinity, without affecting strychnine binding (Ryan et al. 1994; Saul et al. 1994). Point mutations of the GlyR α1 subunit gene found in the human neurological illness hyperekplexia (Shiang et al. 1993; Moorhouse et al. 1999; Saul et al. 1999) lie in exposed domains that may be responsible for linking agonist binding and channel gating. Homomeric GlyRs containing the substitutions R271L, R271Q, K276E, or Y279C (residues located in the M2–M3 loop), or Q266H (within M2) have a decreased sensitivity to glycine and a loss of β-alanine and taurine responses (Langosch et al. 1994; Rajendra et al. 1994; Laube et al. 1995b; Lynch et al. 1997; Moorhouse et al. 1999), but do not affect receptor expression as assessed by [^{3}H]strychnine binding. There is evidence that some of these mutations (R271L/Q, K276E and Q266H) reduce the single-channel conductance and/or open channel probability of the expressed GlyRs (Langosch et al. 1994; Rajendra et al. 1995a; Moorhouse et al. 1999), implying that the M2–M3 loop is vital for coupling signal transduction and ligand binding. Mutation P250T (Saul et al. 1999), in the M1–M2 loop, decreases single-channel conductances, but also affects the desensitization and resensitization properties of the expressed mutant GlyRs. The substitution I244N, within M1, also reduces channel gating (Lynch et al. 1997), but additionally impairs the efficiency of GlyR expression. These data have suggested a complex ligand-binding/signal transduction mechanism that involves both the large extracellular domain of GlyR subunits and the short intracellular and extracellular loops between M1 and M2, and M2 and M3, respectively.

IV. Ion Channel Function

Single-channel electrophysiological analysis has revealed that in addition to Cl^-, GlyR ion channels are permeable to other halides as well as nitrate, bicarbonate, and small organic ions. Ion substitution studies on native neuronal GlyRs have disclosed a permeability sequence of $SCN^- > I^- > NO_3^- > Br^- > Cl^- > HCO_3^- >$ acetate$> F^- >$ propionate (Bormann et al. 1987). The membrane spanning segments M1 to M3 are highly conserved between GlyR, $GABA_A$, and $GABA_C$ receptor subunits, strongly suggesting their importance in chloride channel function. In particular M2 has a high content of uncharged polar amino-acid residues such as serine and threonine, and is thought to constitute an α-helical hydrophilic chloride channel lining. Evidence in support of this theory was provided by studies using synthetic peptide corresponding to the M2 segment of the GlyR $\alpha 1$ subunit, which is capable of producing channel activity in liposomes and planar lipid bilayers (Langosch et al. 1991; Reddy et al. 1993). Additional studies have assigned determinants of resistance to channel blockade by the plant alkaloid picrotoxinin to the M2 segment of the β subunit (Pribilla et al. 1992). Residues within the C-terminal half of M2 in GlyR α and β subunits have also been shown to be responsible for the main-state conductances of homo- and hetero-oligomeric GlyRs. GlyR α subunit homomeric receptors show distinct main-state conductances of 86 ($\alpha 1$), 111 ($\alpha 2$), and 105 ($\alpha 3$) pS (Bormann et al. 1993). Mutation of glycine 254 in the GlyR $\alpha 1$ subunit (Fig. 3) to alanine (which is found in the equivalent position in $\alpha 2$ and $\alpha 3$ subunits) gave rise to a main-state conductance of 107 pS, showing that the identity of the amino acid at this position is a major determinant of channel conductance. The main-state conductances of heteromeric $\alpha 1\beta$, $\alpha 2\beta$, and $\alpha 3\beta$ GlyRs are significantly lower (44, 54, and 48 pS) than those of homomeric α subunit GlyRs (Bormann et al. 1993) and are consistent with values recorded from spinal neurons (Takahashi et al. 1992). These findings strongly suggest that most native GlyRs are heteromeric, and indicate that the M2 domains of both GlyR α and β subunits contain crucial determinants of ion channel function.

V. The Peripheral Membrane Protein Gephyrin

Most neurotransmitter receptors are found in dense clusters at postsynaptic specializations opposite nerve terminals releasing the appropriate neurotransmitter. For the GlyR, this specialized arrangement is believed to be achieved by gephyrin, a peripheral membrane protein of 93 kDa that co-purifies with GlyRs (Pfeiffer et al. 1982; Graham et al. 1985; Schmitt et al. 1987). Gephyrin co-distributes with GlyRs at inhibitory synapses (Zarbin et al. 1981; Triller et al. 1985, 1987; Altschuler et al. 1986) and serves as an anchor molecule, linking GlyRs to the subsynaptic cytoskeleton by binding polymerized tubulin with nanomolar affinity (Kirsch et al. 1991; Kirsch and Betz 1995). Molecular cloning studies have shown that several isoforms of gephyrin exist, which result from complex alternative splicing of four cassette exons (Prior et al. 1992).

Northern blot (Prior et al. 1992), *in situ* hybridization (Kirsch et al. 1993a), and immunocytochemical (Araki et al. 1988; Kirsch and Betz 1993) studies indicate that gephyrin is abundant not only in spinal cord, but also in brain; in addition, gephyrin transcripts are found in peripheral tissues, such as liver, kidney, heart, and lung. This suggests that gephyrin may play other roles in addition to GlyR clustering. Indeed, the primary sequence of gephyrin shows unexpected similarity to three *Escherichia coli* proteins (MoeA, Moab, and MogA), a *Drosophila melanogaster* protein (cinnamon), and an *Arabidopsis thaliana* protein (cnx1), all of which are involved in the synthesis of a molybdenum-containing co-factor (Moco) that is essential for the activity of molybdoenzymes (Prior et al. 1992; Betz 1998). Gene targeting of the mouse gephyrin gene (Feng et al. 1998) has recently shown that gephyrin is indeed required for the activity of the molybdoenzymes sulfite oxidase and xanthine dehydrogenase found in peripheral tissues. In addition, gephyrin binds molybdopterin with high affinity and can re-constitute Moco biosynthesis in Moco-deficient bacteria, a molybdenum-dependent mouse cell line, and in a Moco deficient plant mutant (Stallmeyer et al. 1999). Furthermore, gephyrin has been found at GABAergic synapses in hippocampus (Craig et al. 1996), spinal cord (Todd et al. 1996), and retina (Sassoe-Pognetto et al. 1995), suggesting that this multifunctional protein also plays a role in postsynaptic $GABA_A$ receptor clustering (see Essrich et al. 1998; Betz 1998).

Using cultured embryonic spinal neurons, it has been shown that the formation of membrane-associated gephyrin deposits precedes the postsynaptic localization of GlyRs (Kirsch et al. 1993b; Kirsch and Betz 1995, 1998). Elimination of gephyrin by treatment with antisense oligonucleotides (Kirsch et al. 1993b) or by targeted disruption of the gephyrin gene (Feng et al. 1998) prevents the correct synaptic clustering of GlyRs. Similarly, addition of strychnine or L-type Ca^{2+} channel blockers to spinal neuron cultures blocks gephyrin and GlyR cluster formation (Kirsch and Betz 1998), indicating that the activation of embryonic GlyRs, resulting in Ca^{2+} influx, is crucial for the formation of gephyrin and GlyR clusters at developing postsynaptic sites. In addition, compounds that disrupt the integrity of microtubules (e.g., demecolcine) and microfilaments (e.g., cytochalsin D), affect the packing density of gephyrin and GlyR specializations formed in these cultures (Kirsch and Betz 1995). These cytoskeletal structures appear to operate antagonistically: microtubules condense GlyR clusters, while microfilaments disperse them. Thus, a complex interaction between GlyRs, gephyrin and the cytoskeleton is responsible for the correct topology of inhibitory post-synaptic specializations in spinal cord. GlyRs are thought to bind to gephyrin via sequences located within the large intracellular loop of the GlyR β subunit. In overlay and transfection experiments (MEYER et al. 1995), this interaction was shown to involve a 33 amino-acid portion of the M3–M4 loop, with an 18 amino acid 'core sequence' (RSNDFSIVGSLPRDFELS) containing the minimal binding-site determinants. This core sequence is capable of conferring gephyrin-binding properties on $GABA_A$ (MEYER et al. 1995) and NMDA (KINS et al. 1999) receptor subunits, which would not normally link to gephyrin. Most recently, site-directed mutagenesis combined with a novel assay, using green fluorescent protein-gephyrin binding motif fusion proteins (KNEUSSEL et al. 1999), has indicated that the core motif may form an imperfect amphipathic helix, and that binding to gephyrin may be mediated by hydrophobic residues on one side of this structure. The gephyrin residues interacting with this hydrophobic domain are presently unknown.

VI. Pharmacology of GlyRs

1. Antagonism of GlyR Function by Strychnine

Alkaloids are basic heterocyclic compounds that are found in plants. One typical example is strychnine (Fig. 2), which is derived from the tree *Strychnos nux vomica* (poison nut) native to Sri Lanka, Australia, and India. Strychnine is a potent convulsant that acts by antagonizing glycinergic inhibition. Structurally related alkaloids, such as brucine, also act as competitive glycine antagonists at the inhibitory GlyR, and studies with a range of strychnine analogues have established detailed structure-function relationships (reviewed in BETZ 1985; BECKER 1992). Strychnine represents a unique tool in the investigation of postsynaptic GlyRs and is widely used to distinguish glycinergic from GABAergic inhibition. Glycine-displaceable [^{3}H]strychnine binding (YOUNG and SNYDER 1973) still constitutes the most reliable ligand binding assay for this receptor system. Further, strychnine provides a natural photoaffinity label for the GlyR as, upon UV illumination, [^{3}H]strychnine is incorporated into the ligand-binding α subunit (GRAHAM et al. 1981, 1983). Lastly, substitutions at the aromatic ring have little effect on the toxicity of strychnine (BETZ 1985) and have been exploited to generate affinity columns for GlyR purification (PFEIFFER et al. 1982) and fluorescent derivatives for visualizing the distribution of native neuronal GlyRs (ST JOHN and STEVENS 1993).

The physiological symptoms of strychnine poisoning (BECKER 1992) underline the importance of glycinergic inhibition in the control of both motor behavior and sensory processing. Onset of symptoms is usually within 15–20 min of ingestion. Consistent with a systemic disinhibition of motorneurons, sublethal strychnine poisoning leads to motor disturbances and increased muscle tone. These characteristic motor symptoms are accompanied by unusual sensory impressions; due to disinhibition of the respective afferent pathways, intoxicated patients report hyperacuity of vision and audition in addition to acute pain. Higher doses of strychnine (5–8 mg/kg body weight) cause convulsions and ultimately death by interference with pulmonary function, by depression of respiratory center activity, or both. Due to its high toxicity, strychnine was used as a rat poison for over 5 centuries. Strangely, the disinhibition of both motor and sensory pathways by strychnine has also been used therapeutically as a stimulant in humans. The *nux vomica* seeds are still used as a homeopathic medicine to treat stomach upsets, headache, nausea, and fever. Strychnine was even used as a performance-enhancing drug by Roman gladiators, who used the alkaloid in combination with wine to give them an edge in sporting combat.

2. Amino Acids and Piperidine Carboxylic Acid Compounds

In addition to glycine, β-alanine and taurine (Fig. 2) also display inhibitory activity when applied to neurons (e.g., BOEHM et al. 1997), and these endogeneous amino acids may well activate GlyRs *in vivo* (FLINT et al. 1998). The agonist and antagonist actions of several α- and β-amino acids have been studied in detail on homomeric $\alpha 1$ GlyRs expressed in *Xenopus* oocytes (SCHMIEDEN and BETZ 1995). The agonistic activity of α-amino acids (e.g., glycine, sarcosine, alanine, and serine) (Fig. 2) exhibits a marked stereoselectivity and is susceptible to substitutions at the C_α-atom. However, antagonism by α-amino acids is not enantiomer-dependent nor influenced by C_α-atom substitutions. In contrast, β-amino acids such as taurine, β-aminobutyric acid (β-ABA), and β-aminoisobutyric acid (β-AIBA), which are partial agonists at GlyRs (Fig. 2), show competitive inhibition at low concentrations whereas high concentrations elicit significant membrane currents. Hence, the partial agonist activity of a given β-amino acid on GlyRs may be determined by the relative amounts of the respective *cis/trans* isomers in the compound (SCHMIEDEN and BETZ 1995). Nipecotic acid (Fig. 2), and related compounds which contain a *trans*-β-amino acid configuration, behave as competitive GlyR antagonists.

3. Picrotoxinin, Cyanotriphenylborate, and Quinolinic Acid Derivatives

The plant alkaloid picrotoxin, derived from the plant *Anamirta cocculin*, is an equimolar mixture of picrotin and picrotoxinin (Fig. 2), and has been widely used to antagonize GABA responses. The action of picrotoxin at $GABA_A$ receptors shows a high degree of selectivity for picrotoxinin over picrotin, and is use-dependent and non-competitive. Picrotoxin therefore is considered a potent chloride channel blocker. Currently, picrotoxin is the best pharmacological tool to discriminate homo-oligomeric from heteromeric GlyRs (PRIBILLA et al. 1992; HANDFORD et al. 1996). Native GlyRs and heterologously expressed heteroligomeric $\alpha\beta$ subunit GlyRs are largely resistant to block by picrotoxin, whereas GlyR $\alpha 1$, $\alpha 2$, or $\alpha 3$ subunit homo-oligomers are sensitive to micromolar doses (IC_{50} ~25 μmol/l). PRIBILLA et al. (1992) have demonstrated that this resistance to picrotoxin blockade is due to multiple amino acid substitutions within the channel lining M2 segment of the β subunit. However, more recent studies suggest that, in contrast to its action at $GABA_A$ receptors, picrotoxin blockade of GlyR function (i) exhibits no selectivity between picrotoxinin and picrotin, and (ii) is competitive and not voltage-dependent (LYNCH et al. 1995). The latter findings would be consistent with an extracellular alkaloid binding site. These apparently contradictory results might be explained by studies involving the hyperekplexia mutations R271L and R271Q, which transform picrotoxin from an allosterically-acting competitive antagonist to an allosteric potentiator at low (0.01–3 μmol/l) concentrations and to a non-competitive antagonist at higher (~3 mmol/l) concentrations (LYNCH et al. 1995). This may be reconciled with picrotoxin binding both within the channel and on the extracellular domain.

In contrast to picrotoxinin, the organic anion cyanotriphenylborate (CTB; Fig. 2) is a purely non-competitive, use-dependent antagonist. Blockade is more pronounced at positive membrane potentials (RUNDSTRÖM et al. 1994) suggesting that it is an open-channel blocker. CTB can also be used to discriminate some GlyR subtypes. Homomeric GlyR α1 subunit receptors are readily blocked by CTB with an IC_{50} of 1.3 μmol/l whilst α2 subunit GlyRs are relatively insensitive (IC_{50}>>20 μmol/l) (RUNDSTRÖM et al. 1994). This difference has been traced to residue G254 in the M2 segment of the GlyR α1 subunit (Fig. 3).

Derivatives of quinolinic acid compounds based on 2-carboxy-4-hydroquinolines, which antagonize binding of the co-agonist glycine to the NMDAR, have also been tested as selective GlyR antagonists (SCHMIEDEN et al. 1996). In *Xenopus laevis* oocytes expressing GlyR α1 subunit homo-oligomers, the chloride-substituted derivatives 5,7-dichloro-4-hydroxyquinoline-3-carboxylic acid (5,7ClQA) and 7-chloro-4-hydroxyquinoline (7ClQ) inhibit glycine currents in a mixed high-affinity competitive and low-affinity non-competitive manner. The related compounds 7-trifluoromethyl-4-hydroxyquinoline-3-carboxylic acid (7TFQA) and 7-trifluoromethyl-4-hydroxyquinoline (7TFQ) exhibit purely competitive antagonism. These compounds (Fig. 2) not only represent novel tools to study native and recombinant GlyRs, but also suggest that the GlyR agonist/antagonist binding pocket may exhibit some structural similarity to that of the glycine binding site of the NMDAR.

4. Potentiation of GlyR Function by Anesthetics, Alcohol, and Zn^{2+}

Volatile anesthetics and ethanol enhance the effect of glycine at both native GlyRs (CELENTANO et al. 1988; AGUAYO and PANCETTI 1994) and recombinantly expressed homo-oligomeric α1 or α2 subunit GlyRs (MASCIA et al. 1996a,b). In contrast, such compounds inhibit heterologously expressed $GABA_C$ receptors formed from the ρ1 subunit (MIHIC and HARRIS 1996). This difference was exploited to identify the domain in the human GlyR α1 subunit that is responsible for enhancement of GlyR function by ethanol (Mihic et al. 1997). By constructing chimeric GlyR α1/GABA$_C$ receptor ρ1 subunits, and subsequently using site-directed mutagenesis, a single amino acid, S267, in the M2 region of the GlyR α1 subunit (Fig. 3) was shown to be sufficient to abolish enhancement of GlyR function by ethanol and the volatile anaesthetic enflurane (Mihic et al. 1997). The importance of S267 for potentiation by anaesthetics and ethanol has been underscored by a further study (Ye et al. 1998) which demonstrated that ethanol enhancement is inversely correlated with the molecular volume of the residue present at position 267.

The divalent cation Zn^{2+} is stored within synaptic vesicles in the mammalian central nervous system, and is released upon nerve terminal stimulation. Elevated concentrations of Zn^{2+} in the synaptic cleft result in multiple effects on neuronal excitability by inhibiting or potentiating current flow through ligand-gated and voltage-operated ion channels. Although the presence of Zn^{2+} had been demonstrated only in certain higher brain areas such as the hippocampus, recent findings show that both vesicular Zn^{2+} and metallothionein III, which is involved in regulating the availability of Zn^{2+}, are abundant in spinal cord (Velazquez et al. 1999). Zn^{2+} exhibits biphasic effects on both native GlyRs on rat spinal cord neurons and on recombinantly expressed homo-oligomeric and heteromeric GlyRs (Bloomenthal et al. 1994; Laube et al. 1995a). At low concentrations (0.5–10 μmol/l) Zn^{2+} potentiates glycine-activated chloride currents, while higher concentrations (>100 μmol/l) of Zn^{2+} inhibit the glycine response. These changes are accompanied by respective shifts in agonist dose-response curves, whereas inhibition by the competitive antagonist strychnine is not changed (Laube et al. 1995a). Analysis of glycine-gated single channel events indicates that Zn^{2+} alters the open probability of the GlyR without changing its unitary conductance (Laube et al. 2000). Using chimeric α/β GlyR subunit cDNA constructs, Laube et al. (1995a) found that the positive and negative modulatory effects of Zn^{2+} can be separated to different regions of the α subunits, and proposed that determinants of the potentiating Zn^{2+} binding site are localized between amino acids 74–86 of the GlyR α1 subunit.

More recent studies (Lynch et al. 1998; Laube et al. 2000) suggest that Zn^{2+} modulation of GlyR currents involves complex allosteric processes. In GlyRs incorporating mutations in the M1–M2 or M2–M3 loops (which are thought to transduce agonist binding to channel opening), Zn^{2+} potentiation was uncoupled from glycine-gated currents, while Zn^{2+} potentiation of taurine currents was preserved (Lynch et al. 1998). Interestingly, none of these mutations disrupted the ability of Zn^{2+} to inhibit glycine or taurine gated currents. Substitution of a critical aspartate residue, D80, in the α1 subunit abolished Zn^{2+} potentiation of glycine gated currents (Lynch et al. 1998; Laube et al. 2000); however, potentiation of taurine-gated currents by Zn^{2+} has been reported to remain intact in the mutant receptor (Lynch et al. 1998). This suggests that the potentiating site for Zn^{2+} on GlyRs may be complex and susceptible to numerous distinct mutations, making it difficult to locate by conventional methods. This would not appear to hold true for the inhibitory Zn^{2+} binding site on the GlyRs, which may involve histidine residues, like the recently reported inhibitory Zn^{2+} sites on $GABA_A$ and $GABA_C$ receptor subunits (WANG et al. 1995; WOOLTORTON et al. 1997; FISHER and MACDONALD 1998; HORENSTEIN and AKABAS 1998).

Some 5-HT_3 receptor ligands have also been reported to produce both potentiating and inhibitory effects on glycine currents recorded from cultured spinal neurons (CHESNOY-MARCHAIS 1996). Of these, the tropeines tropisetron and atropine displayed only competitive inhibition at micromolar concentrations when tested on recombinant GlyRs generated in *Xenopus* oocytes (MAKSAY et al. 1999). Notably, inhibition showed selectivity for the $\alpha 2$ subunit, suggesting that further exploration of these compounds might help to develop subtype-selective high-affinity antagonists.

D. Concluding Remarks

The presently available molecular and functional data demonstrate that our understanding of glycinergic inhibitory neurotransmission in the mammalian central nervous system has deepened enormously during the past decade. Despite a wealth of information on the expression and functional characteristics of different GlyR and glycine transporter isoforms, however, the pharmacology of glycinergic neurotransmission is still rather poor. Although a number of novel GlyR antagonists has become available recently in addition to the classical antagonist strychnine, compounds that selectively potentiate GlyR currents are still scarce. Such potentiators of GlyR function that mimic the actions of low concentrations of Zn^{2+} or high concentrations of ethanol, might serve as potent leads for the development of novel drugs that, in analogy to benzodiazepine at the $GABA_A$ receptor, boost inhibition of sensory afferents and/or motor outputs. Such compounds might have great promise for diverse medical indications including analgesia, muscle relaxation, and narcosis. Due to the sparse expression of GlyRs in higher brain areas, central side effects commonly observed with $GABA_A$ receptor modulators should be rare. Similarly, selective inhibition of the presynaptic glycine transporter GLYT2 also could provide for potentiation of glycinergic inhibitory pathways. In contrast, drugs targeting the glial GLYT1 variants should increase excitatory NMDAR activity. In conclusion, the GlyRs and glycine transporters described here constitute an novel yet poorly charted field for neuropharmacological development.

References

Adams RH, Sato K, Shimada S, Tohyama M, Püschel AW, Betz H (1995) Gene structure and glial expression of the glycine transporter GLYT1 in embryonic and adult rodents. J Neurosci 15:2524–2532

Aguayo LG, Pancetti FC (1994) Ethanol modulation of the γ-aminobutyric acid$_A$- and glycine-activated Cl^- current in cultured mouse neurons. J Pharmacol Exp Ther 270:61–69

Akagi H, Hirai K, Hishinuma F (1991a) Cloning of a glycine receptor subtype expressed in rat brain and spinal cord during a specific period of neuronal development. FEBS Lett 281:160–166

Akagi H, Hirai K, Hishinuma F (1991b) Functional properties of strychnine-sensitive glycine receptors expressed in *Xenopus* oocytes injected with a single mRNA. Neurosci Res 11:28–40

Altschuler RA, Betz H, Parakkal MH, Reeks KA, Wenthold KA (1986) Identification of glycinergic synapses in the cochlear nucleus through immunocytochemical localization of the postsynaptic receptor. Brain Res 369:316–320

Aprison MH (1990) The discovery of the neurotransmitter role of glycine. In: Ottersen OP, Storm-Mathiesen J (eds) Glycine neurotransmission. John Wiley, New York, pp 1–23

Araki T, Yamano M, Murakami T, Wanaka A, Betz H, Tohyama M (1988) Localization of glycine receptors in the rat central nervous system: an immunocytochemical analysis using monoclonal antibody. Neuroscience 25:613–624

Becker C-M (1992) Convulsants acting at the inhibitory glycine receptor. In: Herken H, Hucho F (eds) Handbook of experimental pharmacology, vol. 102, Springer, Berlin Heidelberg New York, pp 539–575

Becker CM, Hoch W, Betz H (1988) Glycine receptor heterogeneity in rat spinal cord during postnatal development. EMBO J 7:3717–3726

Bennett JP, Kanner BI (1997) The membrane typology of GAT-1, a ($Na^+ + Cl^-$)-coupled γ-aminobutyric acid transporter from rat brain. J Biol Chem 272:1203–1210

Bergeron R, Meyer TM, Coyle JT, Greene RW (1998) Modulation of *N*-methyl-D-aspartate receptor function by glycine transport. Proc Natl Acad Sci USA 95:15730–15734

Betz H (1985) The glycine receptor of rat spinal cord: exploring the site of action of the plant alkaloid strychnine. Angew Chem Int Ed 24:365–370

Betz H (1990) Ligand-gated ion channels in the brain: the amino acid receptor superfamily. Neuron 5:383–392

Betz H (1998) Gephyrin, a major player in GABAergic postsynaptic membrane assembly? Nat Neurosci 7: 541–543

Bloomenthal AB, Goldwater E, Pritchett DB, Harrison NL (1994) Biphasic modulation of the strychnine-sensitive glycine receptor by Zn^{2+}. Mol Pharmacol 46:1156–1159

Boehm S, Harvey RJ, von Holst A, Rohrer H, Betz H (1997) Glycine receptors in cultured chick sympathetic neurons are excitatory and trigger neurotransmitter release. J Physiol 504:683–694

Bormann J, Hamill OP, Sakmann B (1987) Mechanism of anion permeation through channels gated by glycine and γ-aminobutyric acid in mouse cultured spinal neurones. J Physiol 385:243–286

Bormann J, Rundström N, Betz H, Langosch D (1993) Residues within transmembrane segment M2 determine chloride conductance of glycine receptor homo- and hetero-oligomers. EMBO J 12:3729–3737

Borowsky B, Mezey E, Hoffman B (1993) Two glycine transporter variants with distinct localization in the CNS and peripheral tissues are encoded by a common gene. Neuron 10:851–863

Brüss M, Hammermann R, Brimijoins S, Bönisch H (1995) Antipeptide antibodies confirm the topology of the human norepinephrine transporter. J Biol Chem 270:9197–9201

Buckwalter MS, Cook SA, Davisson MT, White WF, Camper S (1994) A frameshift mutation in the mouse $\alpha 1$ glycine receptor gene (*Glra1*) results in progressive neurological symptoms and juvenile death. Hum Mol Genet 3:2025–2030

Celentano JJ, Gibbs TT, Farb DH (1988) Ethanol potentiates GABA- and glycine-induced chloride currents in chick spinal cord neurons. Brain Res 455:377–380

Chaudhry FA, Reimer RJ, Bellocchio EE, Danbolt NC, Osen KK, Edwards RH, Storm-Mathisen J (1998) The vesicular GABA transporter, VGAT, localizes to synaptic vesicles in sets of glycinergic as well as GABAergic neurons. J Neurosci 18:9733–9750

Chen JG, Liu-Chen S, Rudnick G (1997) External cysteine residues in the serotonin transporter. Biochemistry 36:1479–1486

Chesnoy-Marchais D (1996) Potentiation of chloride responses to glycine by three 5-HT_3 antagonists in rat spinal neurones. Br J Pharmacol 118:2115–2125

Clark JA (1997) Analysis of the transmembrane topology and membrane assembly of the GAT-1 gamma-aminobutyric acid transporter. J Biol Chem 272:14695–14704

Clark JA, Amara SG (1993) Amino acid neurotransmitter transporters: structure, function, and molecular diversity. Bioessays 15:323–332

Craig AM, Banker G, Chang W, McGrath ME, Serpinskaya AS (1996) Clustering of gephyrin at GABAergic but not glutamatergic synapses in cultured rat hippocampal neurons. J Neurosci 16:3166–3177

Cully DF, Vassilatis DK, Liu KK, Paress PS, Van der Ploeg LH, Schaeffer JM, Arena JP (1994) Cloning of an avermectin-sensitive glutamate-gated chloride channel from *Caenorhabditis elegans*. Nature 371:707–711

Cummings CJ, Dahle EJ, Zoghbi HY (1998) Analysis of the genomic structure of the human glycine receptor $\alpha 2$ subunit gene and exclusion of this gene as a candidate for Rett syndrome. Am J Med Genet 78:176–178

Essrich C, Lorez M, Benson J, Fritschy JM, Lüscher B (1998) Postsynaptic clustering of major $GABA_A$ receptor subtypes requires the $\gamma 2$ subunit and gephyrin. Nat Neurosci 1:563–571

Feng G, Tintrup H, Kirsch J, Nichol MC, Kuhse J, Betz H, Sanes JR (1998) Dual requirement for gephyrin in glycine receptor clustering and molybdoenzyme activity. Science 282:1321–1324.

Fisher JL, MacDonald RL (1998) The role of an α subtype M_2-M_3 His in regulating inhibition of $GABA_A$ receptor current by Zinc and other divalent cations. J Neurosci 18:2944–2953

Flint AC, Liu X, Kriegstein AR (1998) Nonsynaptic glycine receptor activation during early neocortical development. Neuron 20:43–53

Fujita M, Sato K, Sato M, Inoue T, Kozuka T, Tohyama M (1991) Regional distribution of the cells expressing glycine receptor β subunit mRNA in the rat brain. Brain Res 560:23–37

Fykse EM, Fonnum F (1988) Uptake of gamma-aminobutyric acid by a synaptic vesicle fraction isolated from rat brain. J Neurochem 50:1237–1242

Fykse EM, Fonnum F (1996) Amino acid neurotransmission: dynamics of vesicular uptake. Neurochem Res 21:1053–1060

Gomeza J, Zafra F, Olivares L, Gimenez C, Aragon C (1995) Regulation by phorbol esters of the glycine transporter (GLYT1) in glioblastoma cells. Biochim Biophys Acta 1233:41–46

Graham D, Pfeiffer F, Betz H (1981) UV light-induced cross-linking of strychnine to the glycine receptor of rat spinal cord membranes. Biochem Biophys Res Commun 102:1330–1335

Graham D, Pfeiffer F, Betz H (1983) Photoaffinity-labelling of the glycine receptor of rat spinal cord. Eur J Biochem 131:519–525

Graham D, Pfeiffer F, Simler R, Betz H (1985) Purification and characterization of the glycine receptor of rat spinal cord. Biochemistry 24:990–994

Gready JE, Ranganathan S, Schofield PR, Matsuo Y, Nishikawa K (1997) Predicted structure of the extracellular region of ligand-gated ion-channel receptors shows SH2-like and SH3-like domains forming the ligand-binding site. Protein Sci 6:983–998

Grenningloh G, Rienitz A, Schmitt B, Methfessel C, Zensen M, Beyreuther K, Gundelfinger ED, Betz H (1987) The strychnine-binding subunit of the glycine receptor shows homology with nicotinic acetylcholine receptors. Nature 328:215–220

Grenningloh G, Pribilla I, Prior P, Multhaup G, Beyreuther K, Taleb O, Betz H (1990a) Cloning and expression of the 58 kd β subunit of the inhibitory glycine receptor. Neuron 4:963–970

Grenningloh G, Schmieden V, Schofield PR, Seeburg PH, Siddique T, Mohandas TK, Becker C-M, Betz H (1990b) α Subunit variants of the human glycine receptor: primary structures, functional expression and chromosomal localisation of the corresponding genes. EMBO J 9:771–776

Guastella J, Nelson N, Nelson H, Czyzyk L, Keynan S, Miedel MC, Davidson N, Lester HA, Kanner BI (1990) Cloning and expression of a rat brain GABA transporter. Science 249:1303–1306

Guastella J, Brecha N, Weigman C, Lester HA, Davidson N (1992) Cloning, expression and localization of a rat brain high-affinity glycine transporter. Proc Natl Acad Sci USA 89:7189–7193

Handford CA, Lynch JW, Baker E, Webb GC, Ford JH, Sutherland GR, Schofield PR (1996) The human glycine receptor β subunit: primary structure, functional characterisation and chromosomal localisation of the human and murine genes. Mol Brain Res 25:211–219

Heck S, Enz R, Richter-Landsberg C, Blohm DH (1997) Expression and mRNA splicing of glycine receptor subunits and gephyrin during neuronal differentiation of P19 cells in vitro, studied by RT-PCR and immunocytochemistry. Dev Brain Res 98:211–220

Hoch W, Betz H, Becker C-M (1989) Primary cultures of mouse spinal cord express the neonatal isoform of the inhibitory glycine receptor. Neuron 3:339–348

Hollmann M (1999) Structure of ionotropic glutamate receptors. In: Jonas P, Monyer H (eds) Ionotropic glutamate receptors in the CNS (Handbook of experimental pharmacology, vol. 141). Springer, Berlin Heidelberg New York, pp 1–98

Horenstein J, Akabas MH (1998) Location of a high affinity Zn^{2+} binding site in the channel of $\alpha 1\beta 1$ γ-aminobutyric $acid_A$ receptors. Mol Pharmacol 53:870–877

Johnson JW, Ascher P (1987) Glycine potentiates the NMDA response in cultured mouse brain neurons. Nature 325:529–531

Jonas P, Bischofberger J, Sandkuhler J (1998) Corelease of two fast neurotransmitters at a central synapse. Science 281:419–424

Jursky F, Nelson N (1995) Localization of glycine neurotransmitter transporter (GLYT2) reveals correlation with the distribution of glycine receptor. J Neurochem 64:1026–1033

Kim KM, Kingsmore SF, Han H, Yang-Feng TL, Godinot N, Seldin MF, Caron MG, Giros B (1994) Cloning of the human glycine transporter type 1: molecular and pharmacological characterization of novel isoform variants and chromosomal localization of the gene in the human and mouse genomes. Mol Pharmacol 45:608–617

Kingsmore SF, Giros B, Suh D, Bieniarz M, Caron MG, Seldin MF (1994) Glycine receptor β-subunit gene mutation in *spastic* mice associated with LINE-1 element insertion. Nature Genet 7:136–142

Kins S, Kuhse J, Laube B, Betz H, Kirsch J (1999) Incorporation of a gephyrin-binding motif targets NMDA receptors to gephyrin-rich domains in HEK 293 cells. Eur J Neurosci 11:740–744

Kirsch J, Betz H (1993) Widespread expression of gephyrin, a putative glycine receptor-tubulin linker protein, in rat brain. Brain Res 621:301–310

Kirsch J, Betz H (1995) The postsynaptic localization of the glycine receptor-associated gephyrin is regulated by the cytoskeleton. J Neurosci 15:4148–4156

Kirsch J, Betz H (1998) Glycine-receptor activation is required for receptor clustering in spinal neurons. Nature 392:717–720

Kirsch J, Langosch D, Prior P, Littauer UZ, Schmitt B, Betz H (1991) The 93 kDa glycine receptor-associated protein binds to tubulin. J Biol Chem 266:22242–22245

Kirsch J, Malosio ML, Wolters I, Betz H (1993a) Distribution of gephyrin transcripts in the adult and developing rat brain. Eur J Neurosci 5:1109–1117
Kirsch J, Wolters I, Triller A, Betz H (1993b) Gephyrin antisense oligonucleotides prevent glycine receptor clustering in spinal neurons. Nature 366:745–748
Kitayama S, Dohi T, Uhl GR (1994) Phorbol esters alter functions of the expressed dopamine transporter. Eur J Pharmacol 268:115–119
Kneussel M, Hermann A, Kirsch J, Betz H (1999) Hydrophobic interactions mediate binding of the glycine receptor β-subunit to gephyrin. J Neurochem 72:1323–1326
Kotak VC, Korada S, Schwartz IR, Sanes DH (1998) A developmental shift from GABAergic to glycinergic transmission in the central auditory system. J Neurosci 18:4646–4655
Kuhse J, Schmieden V, Betz H (1990a) A single amino acid exchange alters the pharmacology of neonatal rat glycine receptor subunit. Neuron 5:867–873
Kuhse J, Schmieden V, Betz H (1990b) Identification and functional expression of a novel ligand binding subunit of the inhibitory glycine receptor. J Biol Chem 265:22317–22320
Kuhse J, Kuryatov A, Maulet Y, Malosio ML, Schmieden V, Betz H (1991) Alternative splicing generates two isoforms of the $\alpha 2$ subunit of the inhibitory glycine receptor. FEBS Lett 283:73–77
Kuhse J, Laube B, Magalei D, Betz H (1993) Assembly of the inhibitory glycine receptor: identification of amino acid sequence motifs governing subunit stoichiometry. Neuron 11:1049–1056
Langosch D, Thomas L, Betz H (1988) Conserved quaternary structure of ligand gated ion channels: the postsynaptic glycine receptor is a pentamer. Proc Natl Acad Sci USA 85:7394–7398
Langosch D, Hartung K, Grell E, Bamberg E, Betz H (1991) Ion channel formation by synthetic transmembrane segments of the inhibitory glycine receptor – a model study. Biochim Biophys Acta 1063:36–44
Langosh D, Herbold A, Schmieden V, Borman J, Kirsch J (1993) Importance of Arg-219 for correct biogenesis of $\alpha 1$ homooligomeric glycine receptors. FEBS Lett 336:540–544
Langosch D, Laube B, Rundström N, Schmieden V, Bormann J, Betz H (1994) Decreased agonist affinity and chloride conductance of mutant glycine receptors associated with human hereditary hyperekplexia. EMBO J 13:4223–4228
Laube B, Kuhse J, Rundström N, Kirsch J, Schmieden V, Betz H (1995a) Modulation by zinc ions of native rat and recombinant human inhibitory glycine receptors. J Physiol 483:613–619
Laube B, Langosch D, Betz H, Schmieden V (1995b) Hyperekplexia mutations of the glycine receptor unmask the inhibitory subsite for β-amino-acids. Neuroreport 6:897–900
Laube B, Kuhse J, Betz H (2000) Kinetic and mutational analysis of Zn^{2+} modulation of recombinant human inhibitory glycine receptors. J. Physiol. 522: 215–230
Liu QR, Nelson H, Mandiyan S, López-Corcuera B, Nelson N (1992) Cloning and expression of a glycine transporter from mouse brain. FEBS Lett 305:110–114
Liu QR, López-Corcuera B, Mandiyan S, Nelson H, Nelson N (1993) Cloning and expression of a spinal cord- and brain-specific glycine transporter with novel structure features. J Biol Chem 268:22802–22808
Lopez-Corcuera B, Martinzez-Maza R, Nunez E, Roux M, Supplisson S, Aragon C (1998) Differential properties of two stably expressed brain-specific glycine transporters. J Neurochem 71:2211–2219
Lynch JW, Rajendra S, Barry PH, Schofield PR (1995) Mutations affecting the glycine receptor agonist transduction mechanism convert the competitive antagonist, picrotoxin, into an allosteric potentiator. J Biol Chem 270:13799–13806
Lynch JW, Rajendra S, Pierce KD, Handford CA, Barry PH, Schofield PR (1997) Identification of intracellular and extracellular domains mediating signal transduction in the inhibitory glycine receptor. EMBO J 16:110–120

Lynch JW, Jacques P, Pierce KD, Schofield PR (1998) Zinc potentiation of the glycine receptor chloride channel is mediated by allosteric pathways. J Neurochem 71:2159–2168

Maksay G, Laube B, Betz H (1999) Selective blocking effects of tropisetron and atropine on recombinant glycine receptors. J Neurochem 73:802–806

Malosio ML, Marquèze-Pouey B, Kuhse J, Betz H (1991a) Widespread expression of glycine receptor subunit mRNAs in the adult and developing rat brain. EMBO J 10:2401–2409

Malosio ML, Grenningloh G, Kuhse J, Schmieden V, Schmitt B, Prior P, Betz H (1991b) Alternative splicing generates two variants of the α1 subunit of the inhibitory glycine receptor. J Biol Chem 266:2048–2053

Mascia MP, Machu TK, Harris RA (1996a) Enhancement of homomeric glycine receptor function by long-chain alcohols and anaesthetics. Br J Pharmacol 119: 1331–1336

Mascia MP, Mihic SJ, Valenzuela CF, Schofield PR, Harris RA (1996b) A single amino acid determines differences in ethanol actions on strychnine-sensitive glycine receptors. Mol Pharmacol 50:402–406

Matzenbach B, Maulet Y, Sefton L, Courtier B, Avner P, Guénet J-L, Betz H (1994) Structural analysis of mouse glycine receptor α subunit genes: identification and chromosomal localization of a novel variant, α4. J Biol Chem 269:2607–2612

McIntire SL, Reimer RJ, Schuske K, Edwards RH, Jorgensen EM (1997) Identification and characterization of the vesicular GABA transporter. Nature 389:870–876

Meyer G, Kirsch J, Betz H, Langosch D (1995) Identification of a gephyrin binding motif on the glycine receptor β subunit. Neuron 15, 563–572

Mihic SJ, Harris RA (1996) Inhibition of rho1 receptor GABAergic currents by alcohols and volatile anaesthetics. J Pharmacol Exp Ther 277:411–416

Mihic SJ, Ye Q, Wick MJ, Koltchine VV, Krasowski MD, Finn SE, Mascia MP, Valenzuela CF, Hanson KK, Greenblatt EP, Harris RA, Harrison NL (1997) Sites of alcohol and volatile anaesthetic action on $GABA_A$ and glycine receptors. Nature 389:385–389

Moorhouse AJ, Jacques P, Barry PH, Schofield PR (1999) The startle disease mutation Q266H, in the second transmembrane domain of the human glycine receptor, impairs channel gating. Mol Pharmacol 55:386–395

Morrow JA, Collie IT, Dunbar DR, Walker GB, Shahid M, Hill DR (1998) Molecular cloning and functional expression of the human glycine transporter GLYT2 and chromosomal localisation of the gene in the human genome. FEBS Lett 439:334–340

Nayeem N, Green TP, Martin IL, Barnard EA (1994) Quaternary structure of the native $GABA_A$ receptor determined by electron microscopic image analysis. J Neurochem 62:815–818

Nikolic Z, Laube B, Weber RG, Lichter P, Kioschis P, Poustka A, Mulhardt C, Becker C-M (1998) The human glycine receptor subunit α3. GLRA3 gene structure, chromosomal localization, and functional characterization of alternative transcripts. J Biol Chem 273:19708–19714

Olivares L, Aragon C, Gimenez C, Zafra F (1997) Analysis of the transmembrane topology of the glycine transporter GLYT1. J Biol Chem 272:1211–1217

Osawa I, Saito N, Koga T, Tanaka C (1994) Phorbol ester-induced inhibition of GABA uptake by synaptosomes and by *Xenopus* oocytes expressing GABA transporter (GAT1). Neurosci Res 19:287–293

Pacholczyk T, Blakely RD, Amara SG (1991) Expression cloning of a cocaine- and antidepressant-sensitive human noradrenaline transporter. Nature 350:350–354

Pfeiffer F, Graham D, Betz H (1982) Purification by affinity chromatography of the glycine receptor of rat spinal cord. J Biol Chem 257:9389–9393

Ponce J, Poyatos I, Aragon C, Gimenez C, Zafra F (1998) Characterization of the 5′ region of the rat brain glycine transporter GLYT2 gene: identification of a novel isoform. Neurosci Lett 242:25–28

Pribilla I, Takagi T, Langosch D, Bormann J, Betz H (1992) The atypical M2 segment of the β subunit confers picrotoxinin resistance to inhibitory glycine receptor channels. EMBO J 11:4305–4311
Prior P, Schmitt B, Grenningloh G, Pribilla I, Multhaup G, Beyreuther K, Maulet Y, Werner P, Langosch D, Kirsch J, Betz H (1992) Primary structure and alternative splice variants of gephyrin, a putative glycine receptor-tubulin linker protein. Neuron 8:1161–1170
Quian Y, Galli A, Ramamoorthy S, Risso S, DeFelice LJ, Blakely RD (1997) Protein kinase C activation regulates human serotonin transporters in HEK-293 cells via altered cell surface expression. J Neurosci 17:45–57
Rajendra S, Lynch JW, Pierce KD, French CR, Barry PH, Schofield PR (1994) Startle disease mutations reduce the agonist sensitivity of the human inhibitory glycine receptor. J Biol Chem 269:18739–18742
Rajendra S, Lynch JW, Pierce KD, French CR, Barry PH, Schofield PR (1995a) Mutation of an arginine residue in the human glycine receptor transforms β-alanine and taurine from agonists into competitive antagonists. Neuron 14:169–175
Rajendra S, Vandenberg RJ, Pierce KD, Cunningham AM, French PW, Barry PH, Schofield PR (1995b) The unique extracellular disulfide loop of the glycine receptor is a principal ligand binding element. EMBO J 14:2987–2998
Reddy GL, Iwamoto T, Tomich JM, Montal M (1993) Synthetic peptides and four-helix bundle proteins as model systems for the pore-forming structure of channel proteins. II. Transmembrane segment M2 of the brain glycine receptor is a plausible candidate for the pore-lining structure. J Biol Chem 268:14608–14615
Reichling DB, Kyrozis A, Wang J, MacDermott AB (1994) Mechanisms of GABA and glycine depolarization-induced calcium transients in rat dorsal horn neurons. J Physiol 476:411–421
Ruiz-Gomez A, Morato E, Garcia-Calvo M, Valdivieso F, Mayor F Jr (1990) Localization of the strychnine binding site on the 48-kilodalton subunit of the glycine receptor. Biochemistry 29:7033–7040
Ründstrom N, Schmieden V, Betz H, Bormann J, Langosch D (1994) Cyanotriphenylborate: subtype-specific blocker of glycine receptor chloride channels. Proc Natl Acad Sci USA 91:8950–8954
Ryan SG, Buckwalter MS, Lynch JW, Handford CA, Segura L, Shiang R, Wasmuth JJ, Camper SA, Schofield P, O'Connell P (1994) A missense mutation in the gene encoding the α1 subunit of the inhibitory glycine receptor in the *spasmodic* mouse. Nature Genet 7: 131–135
Sakai N, Sasaki K, Nakashita M, Honda S, Ikegaki N, Saito N (1997) Modulation of serotonin transporter activity by a protein kinase C activator and an inhibitor of type 1 and 2A serine/threonine phosphatases. J Neurochem 68:2618–2624
Sagné C, El Mestikawy S, Isambert MF, Hamon M, Henry JP, Giros B, Gasnier B (1997) Cloning of a functional vesicular GABA and glycine transporter by screening of genome databases. FEBS Lett 417:177–183
Sassoe-Pognetto M, Kirsch J, Grunert U, Greferath U, Fritschy JM, Mohler H, Betz H, Wassle H (1995) Colocalization of gephyrin and $GABA_A$-receptor subunits in the rat retina. J Comp Neurol 357:1–14
Sato K, Zhang JH, Saika T, Sato M, Tada K, Tohyama M (1991) Localization of glycine receptor α1 subunit mRNA-containing neurons in the rat brain: an analysis using in situ hybridization histochemistry. Neuroscience 43:381–395
Sato K, Kiyama H, Tohyama M (1992) Regional distribution of cells expressing glycine receptor α2 subunit mRNA in the rat brain. Brain Res 590:95–108
Sato K, Betz H, Schloss P (1995a) The recombinant GABA transporter GAT1 is down-regulated upon activation of protein kinase C. FEBS Lett 375:99–102
Sato K, Adams R, Betz H, Schloss P (1995b) Modulation of a recombinant glycine transporter (GLYT1b) by activation of protein kinase C. J Neurochem 65:1967–1973

Saul B, Schmieden V, Kling C, Mülhardt C, Gass P, Kuhse J, Becker C-M (1994) Point mutation of glycine receptor $\alpha 1$ subunit in the *spasmodic* mouse affects agonist responses. FEBS Lett 350:71–76

Saul B, Kuner T, Sobetzko D, Brune W, Hanefeld F, Meinck HM, Becker C-M (1999) Novel GLRA1 missense mutation (P250T) in dominant hyperekplexia defines an intracellular determinant of glycine receptor channel gating. J Neurosci 19:869–877

Schloss P, Püschel A, Betz H (1994) Neurotransmitter transporters: new members of known families. Curr Opinion Cell Biol 6:595–599

Schmieden V, Betz H (1995) Pharmacology of the inhibitory glycine receptor: agonist and antagonist actions of amino acids and piperidine carboxylic acid compounds. Mol Pharmacol 48:919–927

Schmieden V, Grenningloh G, Schofield PR, Betz H (1989) Functional expression in *Xenopus* oocytes of the strychnine binding 48 kd subunit of the glycine receptor. EMBO J 8:695–700

Schmieden V, Kuhse J, Betz H (1992) Agonist pharmacology of neonatal and adult glycine receptor α-subunits: identification of amino acid residues involved in taurine activation. EMBO J 11:2025–2032

Schmieden V, Kuhse J, Betz H (1993) Mutation of glycine receptor subunit creates β-alanine receptor responsive to GABA. Science 262:256–258

Schmieden V, Jezequel S, Betz H (1996) Novel antagonists of the inhibitory glycine receptor derived from quinolinic acid compounds. Mol Pharmacol 50:1200–1206

Schmitt B, Knaus P, Becker CM, Betz H (1987) The M_r 93,000 polypeptide of the postsynaptic glycine receptor complex is a peripheral membrane protein. Biochemistry 26:805–811

Shiang R, Ryan SG, Zhu YZ, Hahn AF, O'Connell P, Wasmuth JJ (1993) Mutations in the $\alpha 1$ subunit of the inhibitory glycine receptor cause the dominant neurologic disorder hyperekplexia. Nature Genet 5:351–358

Smith KE, Borden LA, Hartig PR, Branchek T, Weinshank RL (1992) Cloning and expression of a glycine transporter reveal colocalization with NMDA receptors. Neuron 8:927–935

Sontheimer H, Becker C-M, Pritchett DB, Schofield PR, Grenningloh G, Kettenmann H, Betz H, Seeburg PH (1989) Functional chloride channels by mammalian cell expression of rat glycine receptor subunit. Neuron 2:1491–1497

St John PA, Stephens SL (1993) Adult-type glycine receptors form clusters on embryonic rat spinal cord neurons developing in vitro. J Neurosci 13:2749–2757

Stallmeyer B, Schwarz G, Schulze J, Nerlich A, Reiss J, Kirsch J, Mendel RR (1999) The neurotransmitter receptor-anchoring protein gephyrin reconstitutes molybdenum cofactor biosynthesis in bacteria, plants, and mammalian cells. Proc Natl Acad Sci USA 96:1333–1338

Supplisson S, Bergman C (1997) Control of NMDA receptor activation by a glycine transporter co-expressed in *Xenopus* oocytes. J Neurosci 17:4580–4590

Takahashi T, Momiyama A, Hirai K, Hishinuma F, Akagi H (1992) Functional correlation of fetal and adult forms of glycine receptors with developmental changes in inhibitory synaptic receptor channels. Neuron 9:1155–1161

Todd AJ, Watt C, Spike RC, Sieghart W (1996) Colocalization of GABA, glycine, and their receptors at synapses in the rat spinal cord. J Neurosci 16:974–982

Triller A, Cluzeaud F, Pfeiffer F, Betz H, Korn H (1985) Distribution of glycine receptors at central synapses of the rat spinal cord. J Cell Biol 101:683–688

Triller A, Cluzeaud F, Korn H (1987) Gamma-aminobutyric acid-containing terminals can be apposed to glycine receptors at central synapses. J Cell Biol 104:947–956

Tsigelny I, Sugiyama N, Sine SM, Taylor P (1997) A model of the nicotinic receptor extracellular domain based on sequence identity and residue location. Biophys J 73:52–66

Vandenberg RJ, French CR, Barry PH, Shine J, Schofield PR (1992) Antagonism of ligand-gated ion channel receptors: two domains of the glycine receptor α subunit form the strychnine-binding site. Proc Natl Acad Sci USA 89:1765–1769

Varoqui H, Erickson JD (1997) Vesicular neurotransmitter transporters. Mol Neurobiol 15:165–192
Velazquez RA, Cai Y, Shi Q, Larson AA (1999) The distribution of Zinc selenite and expression of metallothionein-III mRNA in the spinal cord and dorsal root ganglia of the rat suggest a role for Zinc in sensory transmission. J Neurosci 19:2288–2300
Wang J, Reichling DB, Kyrozis A, MacDermott AB (1994) Developmental loss of GABA- and glycine-induced depolarization and Ca^{2+} transients in embryonic rat dorsal horn neurons in culture. Eur J Neurosci 6:1275–1280
Wang T-L, Hackam A, Guggino WB, Cutting GR (1995) A single histidine residue is essential for zinc inhibition of GABA $\rho 1$ receptors. J Neurosci 15:7684–7691
Watanabe E, Akagi H (1995) Distribution patterns of mRNAs encoding glycine receptor channels in the developing rat spinal cord. Neurosci Res 23:377–382
Wooltorton JRA, McDonald BJ, Moss SJ, Smart TG (1997) Identification of a Zn^{2+} binding site on the murine $GABA_A$ receptor complex: dependence on the second transmembrane domain of β subunits. J Physiol 505:633–640
Worrall DM, Williams DC (1994) Sodium ion-dependent transporters for neurotransmitters: a review of recent developments. Biochem J 297:425–436
Ye Q, Koltchine VV, Mihic SJ, Mascia MP, Wick MJ, Finn SE, Harrison NL, Harris RA (1998) Enhancement of glycine receptor function by ethanol is inversely correlated with molecular volume at position α267. J Biol Chem 273:3314–3319
Young AB, Snyder SH (1973) Strychnine binding associated with glycine receptors of the central nervous system. Proc Natl Acad Sci USA 70:2832–2836
Zafra F, Aragon C, Olivares L, Danbolt NC, Gimenez C, Storm-Mathisen J (1995a) Glycine transporters are differentially expressed among CNS cells. J Neurosci 15:3952–3969
Zafra F, Gomeza J, Olivares L, Aragon C, Gimenez C (1995b) Regional distribution and developmental variation of the glycine transporters GLYT1 and GLYT2 in the rat CNS. Eur J Neurosci 7:1342–1352
Zarbin MA, Wamsley JK, Kuhar MJ (1981) Glycine receptor: light microscopic autoradiographic localization with $[^3H]$ strychnine. J Neurosci 1:532–547

Subject Index

Printing (Computer to Film): Saladruck, Berlin
Binding: H. Stürtz AG, Würzburg